BIOCHEMISTRY

BIOCHEMISTRY

William M. Southerland, Ph.D.

Associate Professor
Department of Biochemistry
Howard University College of Medicine
Washington, D.C.

Churchill Livingstone
New York, Edinburgh, London, Melbourne

Library of Congress Cataloging-in-Publication Data

Southerland, William M.
   Biochemistry/William M. Southerland.
   p.   cm.—(Foundations of medicine)
   Includes bibliographical references.
   ISBN 0-443-08570-6
   1. Biochemistry   I. Title.   II. Series.
   [DNLM: 1. Biochemistry.   QU 4 S727b]
QP514.2.S59   1990
612'.015—dc20
DNLM/DLC   89-22049
for Library of Congress   CIP

© **Churchill Livingstone Inc.   1990**

All rights reserved. No part of this publication may be reproduced, stored in a retrieval system, or transmitted in any form or by any means, electronic, mechanical, photocopying, recording, or otherwise, without prior permission of the publisher (Churchill Livingstone Inc., 1560 Broadway, New York, NY 10036).

Distributed in the United Kingdom by Churchill Livingstone, Robert Stevenson House, 1–3 Baxter's Place, Leith Walk, Edinburgh EH1 3AF, and by associated companies, branches, and representatives throughout the world.

The Publishers have made every effort to trace the copyright holders for borrowed material. If they have inadvertently overlooked any, they will be pleased to make the necessary arrangements at the first opportunity.

Acquisitions Editor: *Beth Kaufman Barry*
Developmental Editor: *Margot Otway*
Production Designer: *Jill Little*
Production Supervisor: *Jocelyn Eckstein*

Printed in the United States of America

First published in 1990

*To my wife Brenda,
my son Kevin,
and my mother Geneva*

Preface

My intent in writing this text was to create a succinct textbook of medical biochemistry that would serve both as a core textbook for students during their course and as a helpful review prior to the National Boards. With the pressures most medical students now face, they have less time to study a comprehensive tome on biochemistry, and many already depend on class notes. It is my hope that this book will provide a core of essential information that can either supplement class notes or be used to tie together disparate lectures. In reviewing for board exams, many students depend only on questions and answers, and hence lose sight of the themes that tie together the details of biochemistry. By reviewing this text in addition to answering the questions, the student should obtain a better and longer-lasting understanding of the intricacies of biochemistry.

The text incorporates a variety of pedagogic aids intended to help the student absorb the material quickly and correlate it with related areas of biochemistry and physiology. Each chapter begins with a list of learning objectives that guide the student to the concepts that he or she should retain. Many text sections are preceded by a boxed Perspective that previews the material to be covered and relates it to the overall functioning of the organism. For example, sections on complex biochemical pathways are preceded by a Perspective that states the function and goal of the pathway and its importance in cell, tissue, and organismal homeostasis. At the end of each chapter is a selection of National Board-type self-assessment questions that will help the student to identify areas that need more study. Answers, explanations, and a reference to the text page on which the material is discussed are given for all questions at the end of the book.

Most chapters finish with one or more Clinical Correlations that describe the clinical effects that develop when particular biochemical transformations do not occur or are impaired. The biochemical lesion, its biochemical and physiologic consequences, and the clinical manifestations of the disturbance are described, and treatment strategies and their rationales are discussed. The grouping of the clinical correlations in a single section at the end of each chapter allows easy reference to information on metabolic disorders.

William M. Southerland, Ph.D.

Acknowledgments

The writing of this book was made more pleasant by the generous assistance of my colleagues and friends. Therefore, I wish to thank my colleagues in the Department of Biochemistry and the College of Medicine for thoughtfully reading the various chapters, for tactfully offering suggestions, and for providing ever-abiding encouragement. Their efforts helped me write a better book. In particular, I would like to thank Drs. Thomas E. Smith, Richard H. Pointer, Alan H. Mehler, Matthew George, Allen R. Rhoads, and Felix Grissom for helpful discussions along the way.

I am deeply grateful to Mrs. Mary L. Smith and Mrs. Geraldine Renfrow-Shearod who skillfully and painstakingly typed and corrected the original manuscript. They adopted this project as their own and their enthusiasm for it was delightful. I also thank Ms. Sylvia Hodge who graciously proofread many of the original manuscript chapters.

At Churchill Livingstone, I would like to thank Beth Kaufman Barry who believed in the original idea for the book, and Jill Little and Margot Otway whose superb design and editing skills guided the transition of the original manuscript into final book pages.

Finally, I must express my heartfelt gratitude to my wife, Brenda, who always challenged, encouraged, and kept the faith; and to my son Kevin, whose confidence in me was a continuous source of inspiration.

William M. Southerland, Ph.D.

Contents

1. **Amino Acids and Proteins**   1
   Amino Acid Structure/Amino Acid Chemistry/ Protein Structure/ Clinical Correlations

2. **Enzymes**   25
   Types of Enzyme-Catalyzed Reactions/How Enzymes Catalyze Reactions/Enzyme Kinetics/Enzyme Regulation/Clinical Correlations

3. **Energy Metabolism: Bioenergetics**   53
   Biochemical Thermodynamics/Cellular Energetics/Phosphate Bond Energy

4. **Carbohydrate Metabolism I**   61
   Monosaccharides, Oligosaccharides, and Polysaccharides/Glycogen Metabolism/Clinical Correlation

5. **Carbohydrate Metabolism II**   87
   Digestion, Absorption, and Cellular Uptake of Dietary Carbohydrates/Glycolysis/Gluconeogenesis/Metabolism of Pyruvate/ The Pentose Phosphate Pathway/Clinical Correlations

6. **Mitochondrial Oxidations**   121
   The Mitochondrion/The Tricarboxylic Acid Cycle/Electron Transport/ Oxidative Phosphorylation/Clinical Correlation

7. **Lipid Metabolism I**   149
   Fatty Acids and Triacylglycerols/Digestion of Dietary Lipids/ Lipoproteins/Fatty Acid Synthesis/Fatty Acid Oxidation/Clinical Correlations

8. **Lipid Metabolism II**   187
   Metabolism of Complex Lipids: Triacylglycerols/Phosphoglycerides/ Sphingolipids/Cholesterol/Prostaglandins/Leukotrienes/Clinical Correlation

9. Amino Acid Metabolism I — 227
Digestion of Proteins and Absorption and Tissue Distribution of Amino Acids/Ammonia Metabolism/The Urea Cycle/Metabolism of Amino Acid Carbon Chains/One-Carbon Chemistry/Clinical Correlations

10. Amino Acid Metabolism II — 261
Reactions of Amino Acids/Heme Metabolism/Clinical Correlations

11. Nucleotide Metabolism — 287
Structure and Function of Nucleotides/De Novo Synthesis/Deoxyribonucleotides/Nucleotide Degradation/Salvage Pathways/Coenzyme Synthesis/Clinical Correlation

12. Structure of Nucleic Acids — 321
Structure of DNA/Structure of RNA

13. DNA Replication, Mutation, and Recombination — 345
DNA Replication/Mutation and Repair/Recombination/Clinical Correlation

14. Gene Expression — 373
Transcription/Translation/Gene Expression in Mitochondria/Clinical Correlation

15. Biochemical Endocrinology — 403
Mechanisms of Hormone Action/Steroid Hormones/Hormones of Calcium Metabolism/Thyroid Hormones/Pancreatic Hormones/Catecholamines/Pituitary and Hypothalamic Hormones/Clinical Correlation

16. Blood Components — 439
Plasma Proteins/Hemoglobin/Blood Clotting/Fibrinolysis/Anticoagulants/Clinical Correlation

17. The Extracellular Fluid — 459
Regulation of Volume, Osmotic Pressure, and pH

18. Nutrition — 469
Energy-Yielding Nutrients/Water-Soluble Vitamins/Fat-Soluble Vitamins/Minerals/Clinical Correlation

19. Muscle Contraction — 491
Types of Muscle/The Contractile Unit/Mechanism of Contraction/Energy Sources for Contraction

20.	**Biochemistry of Vision**	**505**
	Structure and Metabolism of the Eye/Mechanism of Vision/Clinical Correlation	
21.	**Biologic Membranes**	**513**
	Membrane Composition/Interaction of Lipids with Water/Membrane Dynamics/Membrane Asymmetry/Membrane Transport	
22.	**Connective Tissue Macromolecules**	**529**
	Proteins/Proteoglycans/Clinical Correlations	
23.	**Immunoglobulins**	**545**
	Antibody Production/The Antigen-Antibody Reaction/Antibody Structure/The Mechanism of Antibody Diversity	
24.	**Biochemistry of Nerve Tissue**	**555**
	Structure of the Neuron/Metabolism of Nerve Tissue/The Action Potential/Synaptic Transmission	
25.	**Biochemistry of Viruses**	**567**
	Types of Viruses/Mechanisms of Viral Multiplication	

Appendix 1   **583**
    Water, Acids and Bases, and Buffers

Appendix 2   **587**
    Reduction Potentials

Answers   **591**

Index   **609**

1

AMINO ACIDS AND PROTEINS

Amino Acid Structure/Amino Acid Chemistry/
Protein Structure/Clinical Correlations

The student should be able to:

1. Draw the general structure of α-amino acids, and understand that the structural and chemical differences among the protein amino acids are due to the differences in their R groups.
2. Calculate the pI of a monoamino monocarboxylic amino acid.
3. Obtain the pI and pK values of an amino acid from the titration curve.
4. Summarize the principal features of each of the twenty protein amino acids that contribute to protein structure.
5. Describe a procedure for determining the amino-terminal residue of a polypeptide chain.
6. Define and distinguish primary, secondary, supersecondary, tertiary, and quaternary protein structure.
7. Describe the Edman degradation procedure.

Learning Objectives

Perspective: Amino Acids and Proteins

Proteins are the biomolecules that are most responsible for the diverse functions of the cell. They may be classified according to the type of function performed—for example, transport proteins, receptor proteins, structural proteins, and catalytic proteins (enzymes). Enzymes catalyze biochemical reactions; individual enzymes are generally quite specific as to the reactions they catalyze. Other types of proteins show similar functional specificity.

Proteins are built from twenty different amino acids, each with distinct chemical and structural properties. It is the sequence of amino acids that makes each protein a unique molecular species. The amino acid sequence largely determines the conformational, physical, and ultimately the functional properties of proteins. Therefore, some knowledge of the chemical and structural properties of the protein amino acids is essential to an understanding of the structure and function of proteins.

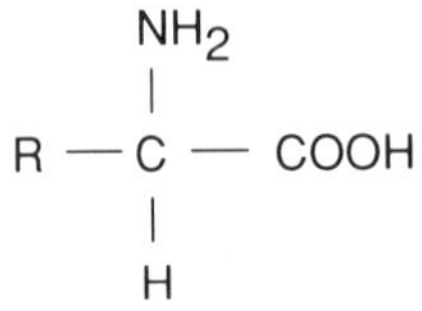

$$R - \underset{\underset{H}{|}}{\overset{\overset{NH_2}{|}}{C}} - COOH$$

Fig. 1-1. The general structure of α-amino acids. The R group represents the amino acid side chain.

AMINO ACIDS

Amino Acid Structure

By definition, amino acids have a carboxylic acid group and an amino group. Many substances in nature meet this requirement, but the most abundant amino acids are the twenty α-*amino acids* that are used to build proteins. In α-amino acids, the amine is attached to the carbon α to the carboxyl carbon.

Figure 1-1 shows the general structure of α-amino acids. Two features are immediately obvious:

1. Except in the case of the amino acid glycine, in which the R group is a hydrogen, α-amino acids have four different groups covalently linked to the α-carbon. Consequently, except in the case of glycine, the α-carbon is an asymmetric center, and the α-amino acids are optically active. The L-stereoisomer is used exclusively in protein synthesis.
2. The *R group* (also called the *side chain*) on the α-carbon is different for each amino acid, and is what gives them their different properties. The unique structure and chemistry of each protein depends largely on the steric, hydrophobic/hydrophilic, electrostatic, hydrogen-bonding, and acid-base properties of the amino acid side chains.

AMINO ACID CHEMISTRY

Acid-Base Properties

At physiologic pH, most amino acids are *zwitterions*; i.e., the α-amino and α-carboxyl groups are both ionized (Fig. 1-2). Therefore, unless there is an ionized side chain group, the net charge of the amino acid at physiologic pH is zero. If there is an ionized side chain, the net charge

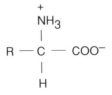

Fig. 1-2. The general structure of the zwitterionic form of an α-amino acid.

is +1 if the side chain is basic, and −1 if it is acidic. If the pH of the medium is varied, each ionizable group will vary between charged and neutral, and the amino acid will exist as different molecular species with different net charges. This phenomenon is illustrated by the titration curves of amino acids (Fig. 1-3).

The pK of an amino acid ionizable group (the pH at which it is half dissociated) is readily obtained from the titration curves. The isoelectric point pI—the pH at which the net charge on the amino acid is zero—can be calculated from the pK values. For a monoamino monocarboxylic amino acid the expression is

$$pI = \frac{pK_1 + pK_2}{2}$$

where pK_1 and pK_2 are the pKs of the carboxyl and amino groups, respectively.

The pKs of several ionizable side chain groups are important in the mechanism of action of many proteins. For example, transport proteins may recognize and/or bind the species to be transported by an electrostatic interaction involving one or more charged amino acid side chains. The activity of many enzymes also depends on the acid-base chemistry of key amino acid side chains located at the active site.

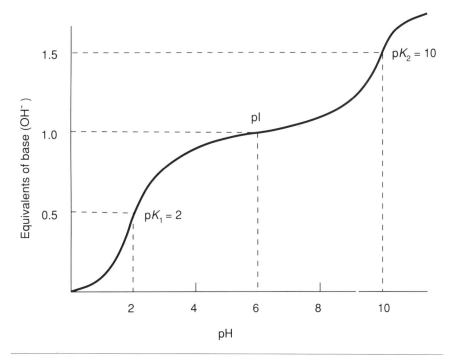

Fig. 1-3. A hypothetical titration curve for a monoamino monocarboxylic amino acid. Most amino acids have a pK_1 around 2.0 and a pK_2 between 8.0 and 10.0.

The Peptide Bond

In proteins and polypeptides, amino acids are linked by *peptide bonds* between the α-amino group of one amino acid and the α-carboxyl group of the next (Fig. 1-4). The peptide bond is an amide linkage, and is formed with the elimination of a molecule of water. A single polypeptide chain may contain up to several hundred amino acids joined linearly by peptide bonds. The carbonyl oxygens and the amide hydrogens of peptide bonds participate in hydrogen bonding interactions that are very important in generating and stabilizing protein structure. In addition, the carbonyl oxygen and the amide hydrogen are important in resonance stabilization of the peptide bond. As a result of resonance stabilization, the amide C–N bond has a certain amount of double-bond character; it is shorter than a normal C–N single bond, and there is no rotation about its axis. As a result, the peptide bond is planar. In naturally occurring polypeptides, most peptide bonds are in the *trans* configuration.

The Peptide Chain

Amino acids may be linked by peptide bonds to form oligopeptides (several amino acid residues) or polypeptides (many residues); proteins consist of from one to several polypeptides. The terminal residue bearing the free amino group is called the *amino-terminal* or *N-terminal* residue;

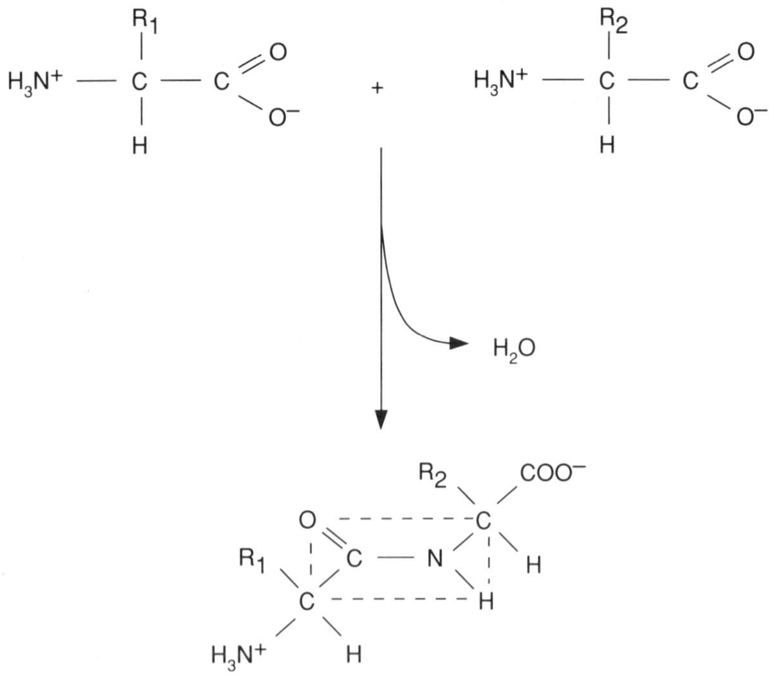

Fig. 1-4. Formation of the peptide bond. The box represents the plane of the bond.

the one bearing the free carboxyl group is called the *carboxyl-terminal* or *C-terminal* residue. The amino acid residues are numbered starting at the N-terminus; this corresponds to the direction in which amino acids are incorporated during protein synthesis.

Small peptides often break the rules that govern protein peptides. They may contain D-amino acids (for example, some antibiotics), and the amino acids may be linked by bonds other than the standard peptide bond (for example, the γ-glutamyl bond of glutathione).

THE PROTEIN AMINO ACIDS

Each of the twenty amino acids commonly found in proteins has certain properties that affect protein structure and function. These properties are determined primarily by the bulk and shape of the side chain, by its hydrophobicity or hydrophilicity, and by its ability to participate in electrostatic, acid-base, or hydrogen bonding interactions. Figure 1-5 shows the twenty amino acids commonly found in proteins.

Glycine

As mentioned above, the "side chain" of glycine is a hydrogen atom. This side chain does not sterically hinder or restrict the polypeptide; in fact, glycine increases chain flexibility and allows sharp bends. Glycine often occurs in locations where tight chain folding does not leave room for a larger side chain.

Aliphatic Amino Acids: Alanine, Valine, Isoleucine, Leucine

Alanine is a very abundant amino acid in proteins. Its small methyl side chain does not significantly hinder the flexibility of the polypeptide chain. Although it is aliphatic, the methyl side chain is not strongly

Glycine (Gly, G)

$$H-\underset{\underset{H}{|}}{\overset{\overset{NH_2}{|}}{C}}-COOH$$

Alanine (Ala, A)

$$CH_3-\underset{\underset{H}{|}}{\overset{\overset{NH_2}{|}}{C}}-COOH$$

Valine (Val, V)

$$\begin{matrix}CH_3\\ \\ CH_3\end{matrix}\!\!\diagup\!\!\!CH-\underset{\underset{H}{|}}{\overset{\overset{NH_2}{|}}{C}}-COOH$$

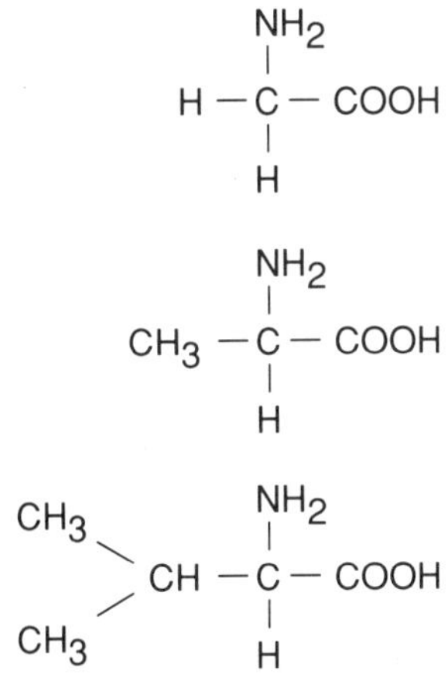

Fig. 1-5. The structures of the twenty protein amino acids, with their three-letter and one-letter abbreviations. (*Figure continues.*)

Leucine (Leu, L): (CH$_3$)$_2$CH—CH$_2$—C(NH$_2$)(H)—COOH

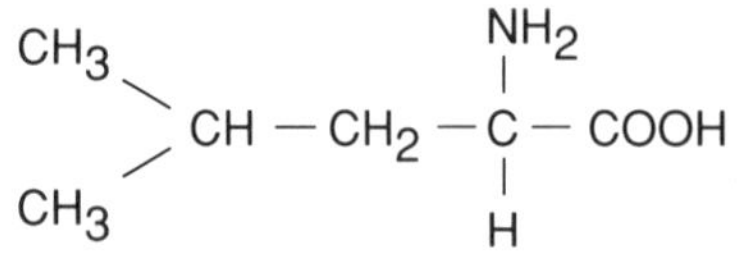

Isoleucine (Ile, I): CH$_3$—CH$_2$—CH(CH$_3$)—C(NH$_2$)(H)—COOH

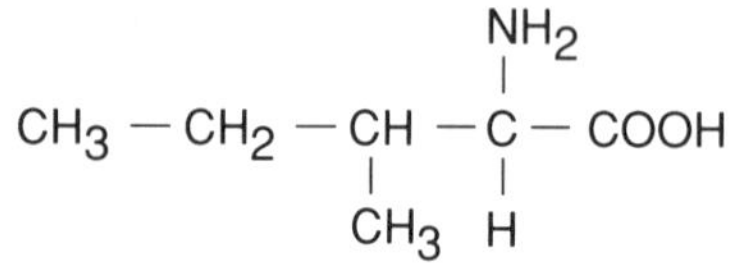

Phenylalanine (Phe, F): C$_6$H$_5$—CH$_2$—C(NH$_2$)(H)—COOH

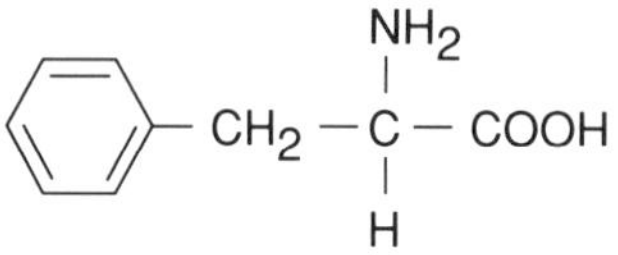

Tyrosine (Tyr, Y): HO—C$_6$H$_4$—CH$_2$—C(NH$_2$)(H)—COOH

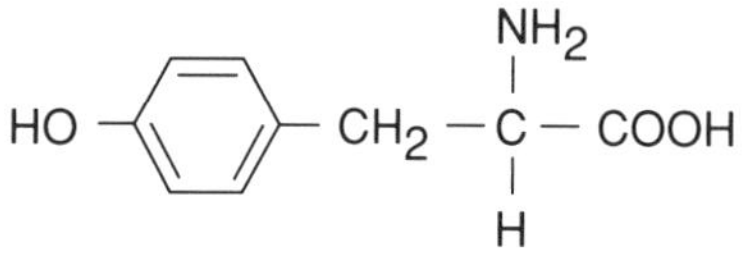

Tryptophan (Trp, W): (indole)—C=CH(NH)—CH$_2$—C(NH$_2$)(H)—COOH

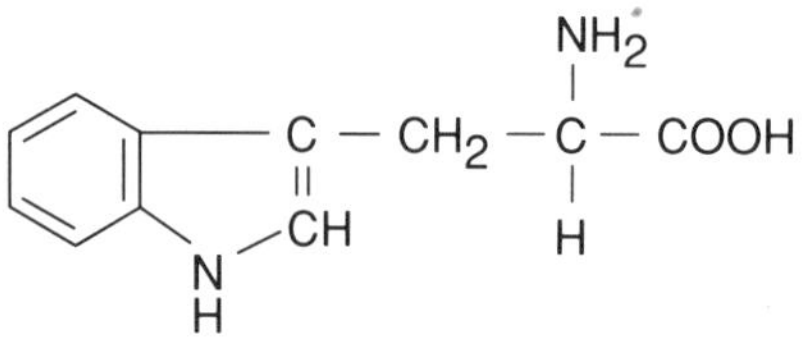

Proline (Pro, P):
$$\begin{array}{c} H_2C-NH \\ | \quad\quad \diagdown \\ \quad\quad\quad CH-COOH \\ | \quad\quad \diagup \\ H_2C-CH_2 \end{array}$$

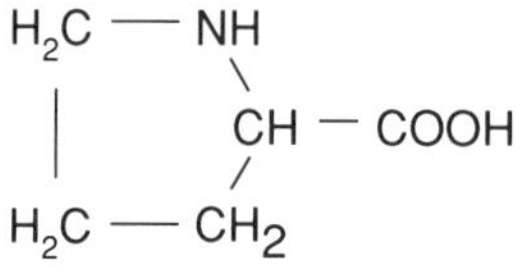

Cysteine (Cys, C): HS—CH$_2$—C(NH$_2$)(H)—COOH

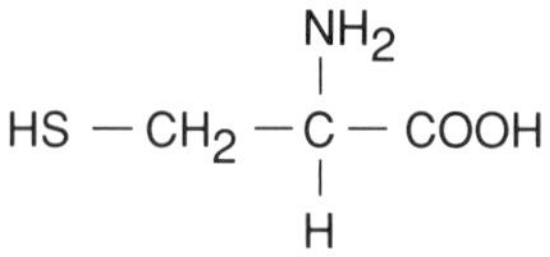

Methionine (Met, M): CH$_3$—S—CH$_2$—CH$_2$—C(NH$_2$)(H)—COOH

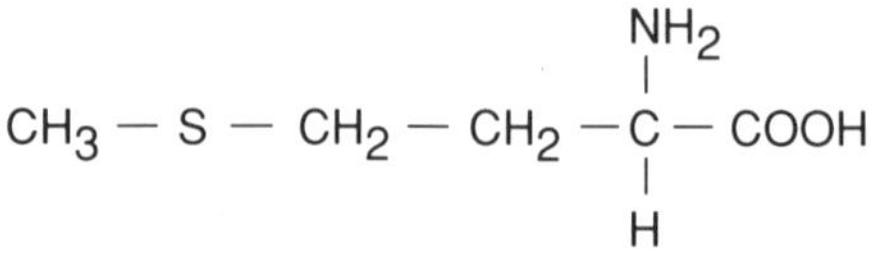

Serine (Ser, S): HO—CH$_2$—C(NH$_2$)(H)—COOH

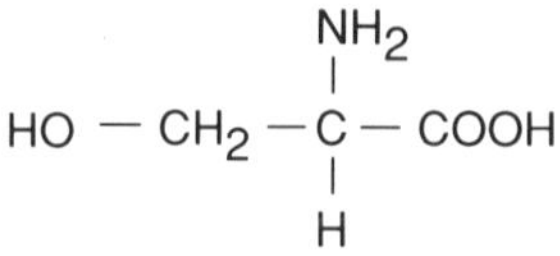

Fig. 1-5. (*Continued*)

Methionine (Met, M)	CH$_3$—S—CH$_2$—CH$_2$—C(NH$_2$)(H)—COOH
Serine (Ser, S)	HO—CH$_2$—C(NH$_2$)(H)—COOH
Threonine (Thr, T)	CH$_3$(HO)CH—C(NH$_2$)(H)—COOH
Asparagine (Asn, N)	H$_2$N—C(=O)—CH$_2$—C(NH$_2$)(H)—COOH
Glutamine (Gln, Q)	H$_2$N—C(=O)—CH$_2$—CH$_2$—C(NH$_2$)(H)—COOH
Aspartate (Asp, D)	$^-$O—C(=O)—CH$_2$—C(NH$_2$)(H)—COOH
Glutamate (Glu, E)	$^-$O—C(=O)—CH$_2$—CH$_2$—C(NH$_2$)(H)—COOH
Lysine (Lys, K)	H$_3$N$^+$—CH$_2$—CH$_2$—CH$_2$—CH$_2$—C(NH$_2$)(H)—COOH
Arginine (Arg, R)	(H$_2$N$^+$=)(H$_2$N—)C—NH—CH$_2$—CH$_2$—CH$_2$—C(NH$_2$)(H)—COOH
Histidine (His, H)	imidazole—CH$_2$—C(NH$_2$)(H)—COOH

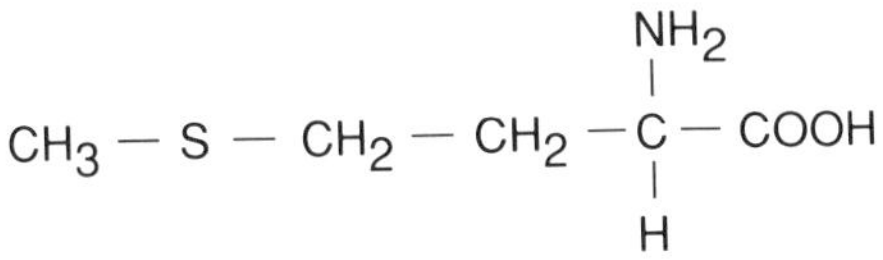

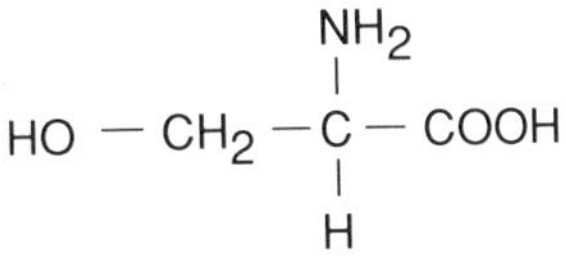

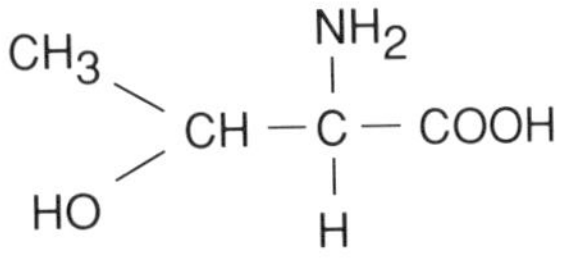

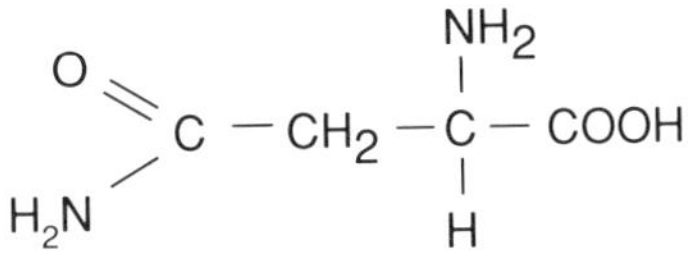

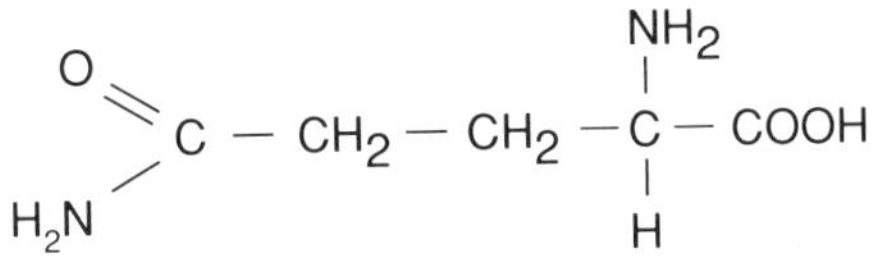

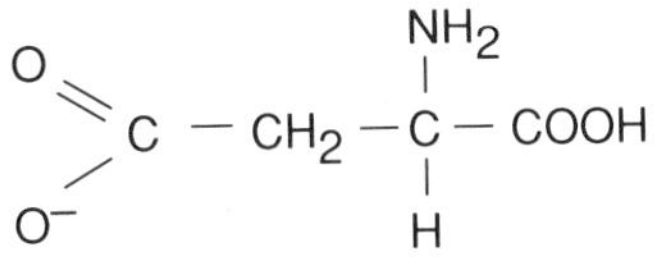

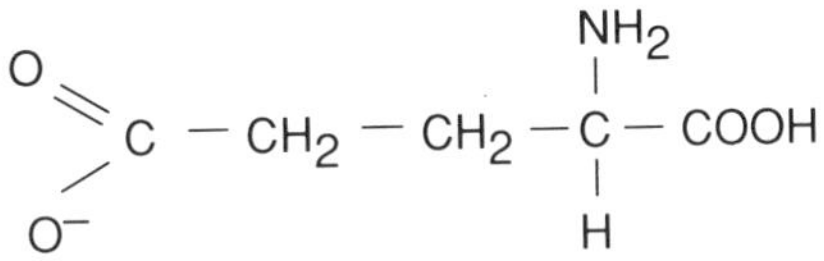

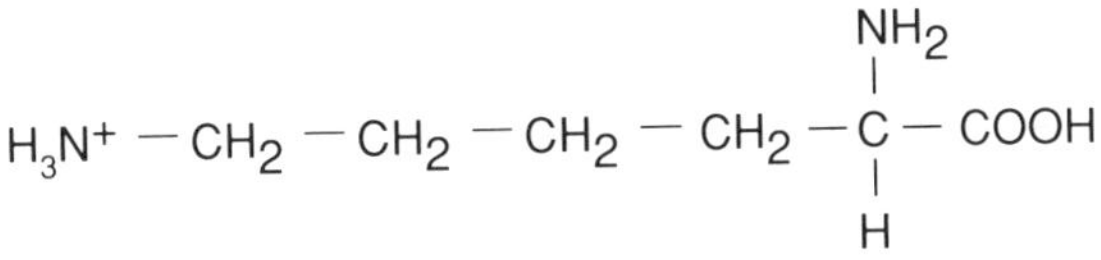

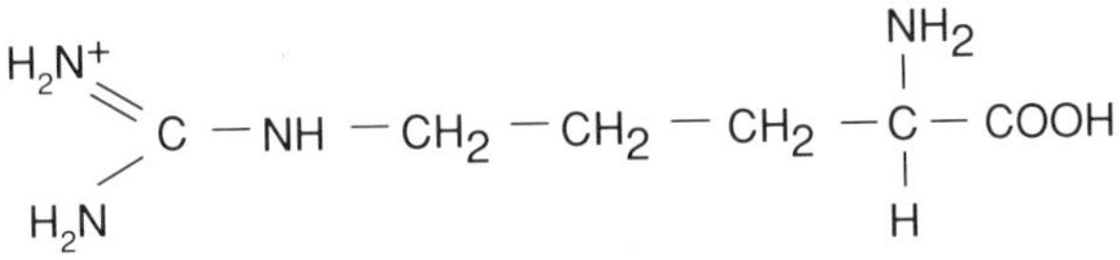

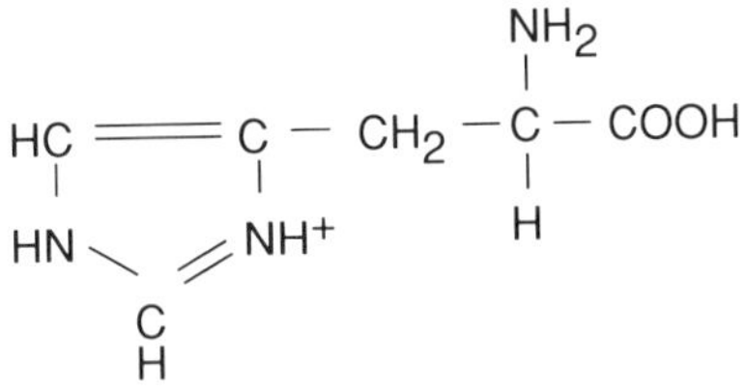

Fig. 1-5. (*Continued*)

hydrophobic, and alanine is found both on the surface of soluble (globular) proteins and in the hydrophobic interior. The larger aliphatic amino acids are quite hydrophobic and are rarely found on the surface of globular proteins.

Valine and isoleucine have bulky, branched side chains in which the branch points are close to the polypeptide backbone. Consequently, these residues restrict the movement and flexibility of the polypeptide chain. The branching also makes the side chains themselves more rigid. Rigid side chains are more easily fixed in a given position, which facilitates chain folding.

Leucine also has a branched side chain, but the branch point is farther from the α-carbon (and thus from the polypeptide chain). Therefore, the leucine side chain interacts less with neighboring side chains, and does not greatly hinder polypeptide chain movement.

Aromatic Amino Acids: Phenylalanine, Tyrosine, Tryptophan

The side chains of phenylalanine, tyrosine, and tryptophan are very rigid, and the proximity of the aromatic ring to the polypeptide chain decreases the flexibility of the chain in the vicinity of these amino acids. The aromatic amino acids also significantly contribute to the ultraviolet absorbance of proteins. The hydroxyl group of tyrosine can participate in hydrogen bonding. The side chains of phenylalanine and tryptophan are hydrophobic, whereas the phenolic group of tyrosine has significant hydrophilic properties.

Proline

The side chain of proline consists of a ring structure that joins the α-carbon and the α-amino group. Therefore, the amino group is secondary, and proline technically is an imino acid. Both the ring structure and the secondary α-amino group have important structural consequences in proteins. The ring structure inhibits rotation of the α-nitrogen of proline relative to its α-carbon (i.e., about the N–C^α bond). The resulting limited rotation forces peptide bonds involving proline to adopt a *cis* configuration rather than the common *trans* configuration. As a consequence, proline residues are important in generating sharp bends in the folded polypeptide chain. Since proline has a secondary amino group, peptide bonds involving proline have no peptide (amide) hydrogen. Therefore, these peptide bonds are not resonance stabilized, nor can they contribute via amide hydrogen bonding to the conformational stability of the folded polypeptide.

Sulfur-Containing Amino Acids: Cysteine and Methionine

The methionine sulfur atom is a strong nucleophile, but the aliphatic carbon chain is hydrophobic. The linear side chain of methionine does not constrain polypeptide chain flexibility.

Cysteine plays a unique and crucial role in the structure of many proteins. When protein folding brings two cysteine residues close together, they can react with each other to form a *disulfide bridge* (–S–S–), which stabilizes the folded structure. A pair of cysteines linked by a disulfide bond is called a *cystine residue*. Free cysteine side chains are polar, and can participate in hydrogen bonding.

Aliphatic Hydroxyl Amino Acids: Serine, Threonine

Serine and threonine have aliphatic side chains that contain hydroxyl groups. These side chains (like the phenolic side chain of tyrosine) are polar and can participate in hydrogen bonding. They occur both on the exterior and in the interior of soluble proteins. When they are present in the interior, they are usually involved in hydrogen bonding.

Amide Amino Acids: Glutamine, Asparagine

The amide-containing side chains of glutamine and asparagine are relatively polar and are also capable of participating in hydrogen bonds. Glutamine and asparagine are uncharged derivatives of the acidic amino acids glutamate and aspartate respectively.

Ionized Amino Acids: Glutamate, Aspartate, Lysine, Arginine

Glutamate, aspartate, lysine, and arginine have side chains that are almost entirely ionized at physiologic pH. Consequently, these amino acids are key participants in electrostatic or ionic interactions in proteins. They are usually found on the surface of globular proteins. Glutamate and aspartate are acidic and carry a negative charge, whereas lysine and arginine are basic and carry a positive charge.

Histidine

The histidine imidazole side chain contains a tertiary amine. Tertiary amines in general are potent nucleophilic catalysts. However, most tertiary amines have a high pK, and at neutral pH are protonated and catalytically inactive. The histidine tertiary amine is unusual. It is a strong nucleophile, but its pK is in the physiologic range. It can therefore be either charged or neutral under physiologic conditions, depending on the microenvironment, and is consequently very useful as a catalyst in biologic reactions. Histidine side chains are involved in the catalytic mechanisms of many enzymes.

Modified Amino Acids

Some proteins contain amino acids other than the common protein amino acids. These amino acids are formed by the post-translational modification of specific protein amino acids. The modified amino acids usually confer a specific structural or functional property on the mature protein (Chapter 14). For example, collagen contains substantial amounts of hydroxyproline and hydroxylysine residues (Chapter 22).

Amino Acid Side Chains and Protein Structure: General Considerations

The protein amino acids may be divided into three groups on the basis of the polarity of their side chains:

Nonpolar	Polar and Neutral	Polar and Ionized
Alanine	Tyrosine	Aspartate
Valine	Serine	Glutamate
Leucine	Threonine	Lysine
Isoleucine	Cysteine	Arginine
Phenylalanine	Methionine	Histidine
Tryptophan	Asparagine	
Proline	Glutamine	
Glycine		

Neutral and polar side chains are soluble in aqueous media, and are also soluble in nonpolar media provided the polar groups are hydrogen bonded to each other. Charged polar side chains are only soluble in aqueous media and nonpolar side chains are only soluble in nonpolar media. Consequently, the solubility of proteins in polar and nonpolar media is directly related to the proportion of polar and nonpolar side chains on the surface of the proteins.

Generally, soluble globular proteins are folded so that their surfaces consist predominately of polar amino acids, whereas nonpolar and hydrogen-bonded nonionized polar amino acids make up the interior. On the other hand, the segments of membrane proteins that are embedded in the lipid portion of the membrane have nonpolar surface amino acids.

Ionized side chains are very rare in the hydrophobic interior of proteins, and, when they occur, are usually involved in an electrostatic interaction (ion pair). Therefore, the net charge in the hydrophobic environment is almost always zero. Ion pairs in the interior of a protein usually play very important structural and/or functional roles.

PROTEINS

Perspective: Protein Structure

The term *protein structure* refers to the physical architecture of proteins. Several levels of structure are recognized: primary, secondary, tertiary, and quaternary. *Primary structure* is the linear sequence of amino acids along the polypeptide chain. *Secondary structure* refers to several patterns of localized folding of polypeptide chain segments that are common in proteins. *Tertiary structure* is the interaction pattern or arrangement of the areas of secondary structure and the areas of non-folded chain segments that gives

(continued)

each protein its unique three-dimensional conformation. In some proteins, individually folded polypeptide chains associate noncovalently to form a complex. Each polypeptide chain is referred to as a subunit, and the complex is called a multisubunit protein. Only multisubunit proteins have *quaternary structure*, which is the spatial arrangement of the individual subunits to produce the multisubunit protein.

Determining the N-terminal Residue

There are several chemical procedures for determining the amino terminal residue of a polypeptide chain. The most frequently used procedures involve 1-fluoro-2,4-dinitrobenzene or dansyl chloride. These substances react with the free α-amino group of the N-terminal residue to form a stable N-linked adduct. The entire protein is then subjected to complete acid digestion to release the free amino acids. The N-terminal amino acid will be labeled and the others will not. The labeled N-terminal residue can subsequently be identified by paper or thin-layer chromatography. Figure 1-6 illustrates this procedure.

Analysis of the N-terminal amino acid is important in ascertaining the identity of the different subunits of multisubunit proteins. Agents that react with N-terminal amino groups also react with the ε-amino group of lysine residues. However, lysine labeled at the ε-amino group can be separated from α-amino acid derivatives.

Primary Structure

The primary structure of proteins is determined by identifying the sequence of amino acids in the polypeptide chain. Before determining the amino acid sequence of a protein, as much information as possible about the protein is obtained, usually including the amino acid composition (the moles of each amino acid per mole of protein).

Amino acid composition determinations are based on the colorimetric or fluorometric quantitative analysis of free amino acids in solution. To perform the determination, the protein is first digested to free amino acids. The mixture of free amino acids is then applied to an amino acid analyzer column, where they are most often separated on the basis of charge (by ion exchange chromatography). The separated amino acids are then reacted with ninhydrin to produce a colored product (absorbance maximum at 570 nm), which is quantitated (Fig. 1-7). Proline reacts to give a colored substance with an absorbance maximum at 440 nm. Ninhydrin detection is also used for amino acids separated by paper and thin-layer chromatography. Amino acids can also be reacted with compounds that give a fluorescent derivative of the amino acid. The amino acids are then detected by observing the fluorescence emission. Fluorometric detection of amino acids has mostly been used with paper or thin-layer chromatography. More recently, fluorescence labeling has

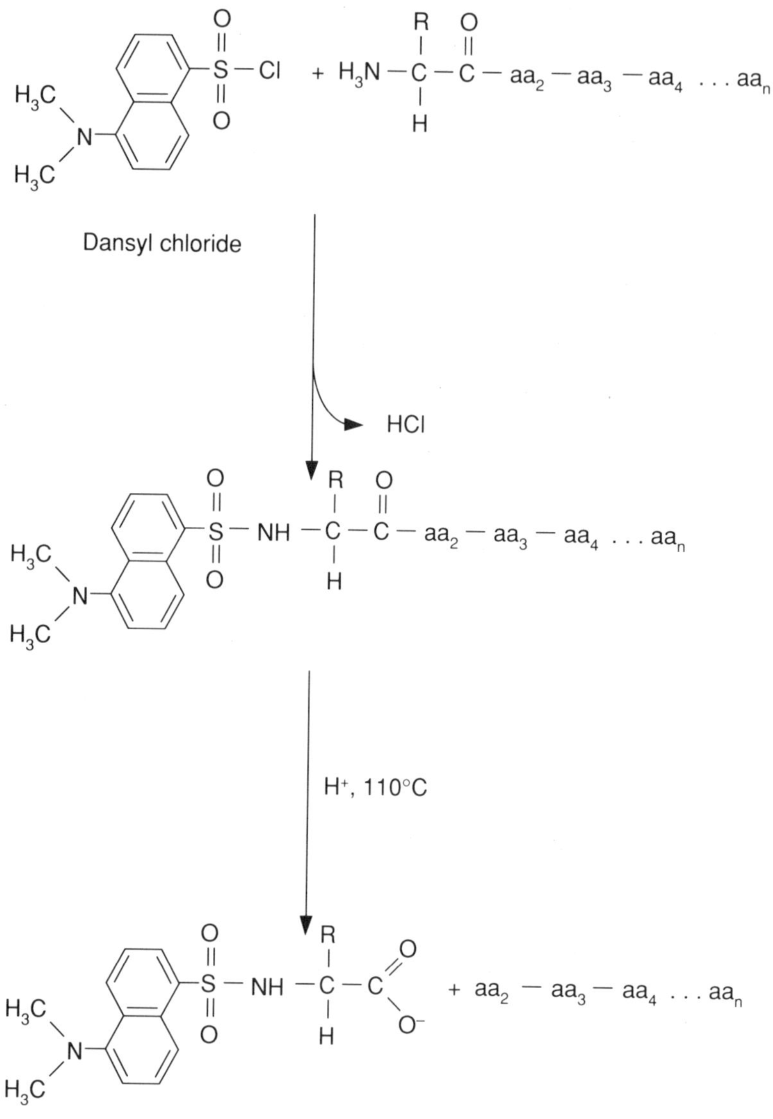

Fig. 1-6. N-terminal analysis using dansyl chloride. The N-terminal amino acid is dansylated, then the polypeptide chain is hydrolyzed to free amino acids with hot acid. The dansylated N-terminal amino acid is separated from the other amino acids in the hydrolytic mixture by polyamide paper chromatography (aa_2, aa_3, etc. are the amino acids).

been used in the determination of amino acids separated by high performance liquid chromatography (HPLC). HPLC procedures usually separate amino acids on the basis of their hydrophobic properties. The major difference between fluorescence detection procedures and colorimetric detection with ninhydrin is sensitivity. For example, amino acid detection using fluorescamine is approximately 10,000 times more sensitive than ninhydrin.

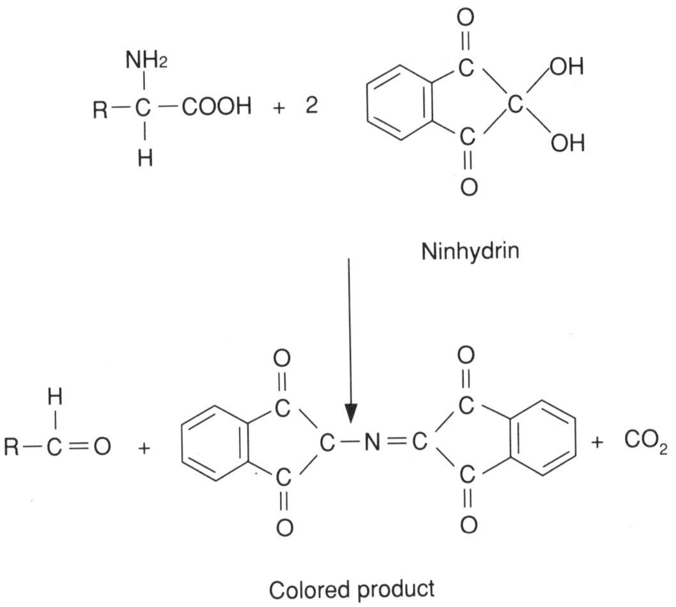

Fig. 1-7. The reaction of ninhydrin with amino acids. The colored product for all the amino acids except proline is quantitated by absorbance at 570 nm; the colored product obtained with proline is quantitated at 440 nm.

The amino acid sequence is determined from the amino terminus to the carboxyl terminus of polypeptide chains. This is achieved by releasing one amino acid at a time from the N-terminal end and identifying it. That is the basis of the *Edman degradation procedure*, which is used in determining amino acid sequences of proteins. First, the polypeptide is reacted with phenylisothiocyanate, which is specific for free amino groups. The product is the phenylthiocarbamoyl derivative of the polypeptide (the PTC-polypeptide). The presence of PTC destabilizes the peptide bond connecting the N-terminal residue to the rest of the peptide chain. That bond is then cleaved under mild conditions and the PTC-amino acid is released as a phenylthiohydantoin (PTH)-amino acid. The remaining peptide chain is left intact, and is ready for another reaction/cleavage cycle. Each released phenylthiohydantoin is identified, usually by gas or liquid chromatography. One cycle of the Edman degradation procedure is illustrated in Figure 1-8.

It is the Edman degradation procedure that is performed by protein sequencing instruments. Most protein sequencers are capable of completing about thirty cycles before identification of the released amino acid becomes too difficult due to the background interference of accumulated side reactions. Therefore, large proteins must first be cleaved into fragments small enough to be sequenced. To determine the correct order of these fragments, the protein is cleaved a second time by a different procedure into fragments that overlap with those from the first cleavage. The fragments from both cleavages are sequenced, and the

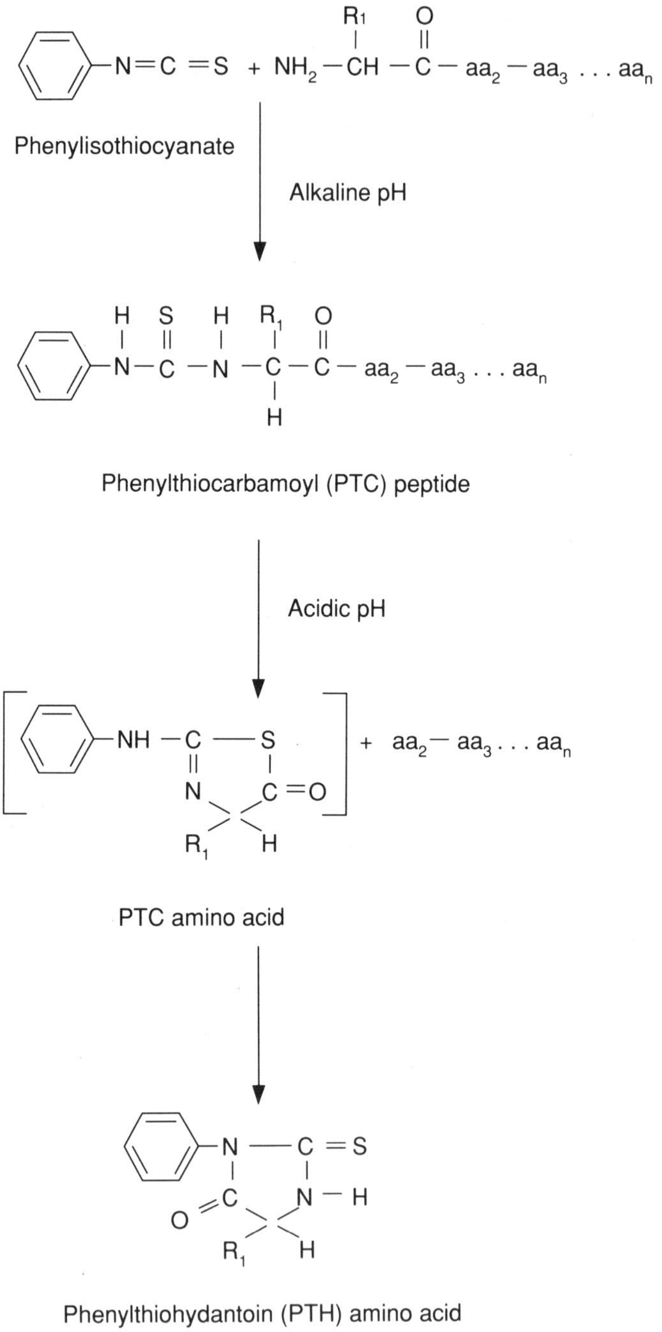

Fig. 1-8. The Edman degradation procedure for determining amino acid sequences of proteins. The PTC amino acid undergoes spontaneous rearrangement to the PTH derivative, which is subsequently identified using gas or liquid chromatography. The remaining peptide, consisting of one fewer amino acid, is now ready to start the procedure again.

areas of overlap between the two sets of fragments are compared to determine the sequence of the entire polypeptide chain. Three or even more sets of overlapping fragments are sometimes needed.

Secondary Structure

Most proteins contain areas where the polypeptide chain is folded into distinct patterns or conformations, called secondary structures. These peptide segments are usually small relative to the size of globular proteins. Therefore, they may be considered local conformations within proteins. Secondary structures are systematically stabilized by hydrogen bonding among the amide hydrogens and carbonyl oxygens of the peptide bonds. These properties characterize the secondary structure of proteins. The most common secondary structures in naturally occurring proteins are the α-*helix*, the β-*sheet*, and the *reverse turn*. It should be noted that some fibrous proteins consist almost entirely of secondary structure.

α-Helix and ^{3}10 Helix

The most abundant secondary structure in proteins is the α-helix (Fig. 1-9). In α-helices, the polypeptide chain appears to wind itself to the right around an imaginary cylinder; i.e., α-helices are right-handed. Each turn around the cylinder consists of 3.6 residues and causes a 5.6 Å rise of the helix. The average length of an α-helical segment in proteins is seven to ten residues.

The α-helix is stabilized by hydrogen bonds between each carbonyl oxygen and an amide hydrogen four residues away. The side chains of the helix residues all project out from the helical cylinder, and in most cases do not interfere with the helix. Proline residues, however, break or interrupt the helix, both because they do not have an amide hydrogen and therefore cannot hydrogen bond and because the proline side chain interferes sterically with the side chain of the preceding amino acid. Consequently, when proline is present in an area of helical structure, the proline peptide bond is restricted to a nonhelical conformation that is not susceptible to stabilization by the helix system of hydrogen bonds.

Another type of helix found in proteins is the ^{3}*10 helix*. These helices contain three residues per turn and usually are no more than two or three turns long. The ^{3}10 helix is less stable than the α-helix because of steric hindrance due to unfavorable stacking of side chains. It is in fact observed only rarely in proteins, usually at the ends of α-helices.

β-Sheets

The β-sheet (also called the *pleated sheet* or β-*pleated sheet*) is a secondary structure involving extended (i.e., not folded or coiled) polypeptide chain segments, which are aligned side by side to form a pleated sheet-like structure (Fig. 1-10). Adjacent chains are held together by hydrogen bonds between amide hydrogens and carbonyl oxygens, and the side chains project above and below the sheet. The polypeptide chains in β-

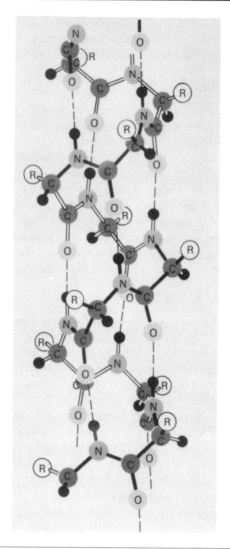

Fig. 1-9. The right-handed α-helix. (From Pauling L: The Nature of the Chemical Bond. 3rd Ed. Cornell University Press, Ithaca, NY, 1960, with permission.)

sheets may be either parallel or antiparallel (Fig. 1-10). The average length of the chain segments involved in β-sheets is about six residues. The number of strands participating in a β-sheet varies, but is usually less than seven. β-sheet regions in proteins are often somewhat twisted or curved. In some proteins, the sheet-like structure is rolled into a barrel-like shape, giving the β-*barrel* conformation.

Reverse Turns

The polypeptide chain of globular proteins is extensively folded. Much of the folding occurs at sharp turns or bends in the peptide chain. These sharp turns are called reverse turns (also *hairpin loops* or β-*turns*). A

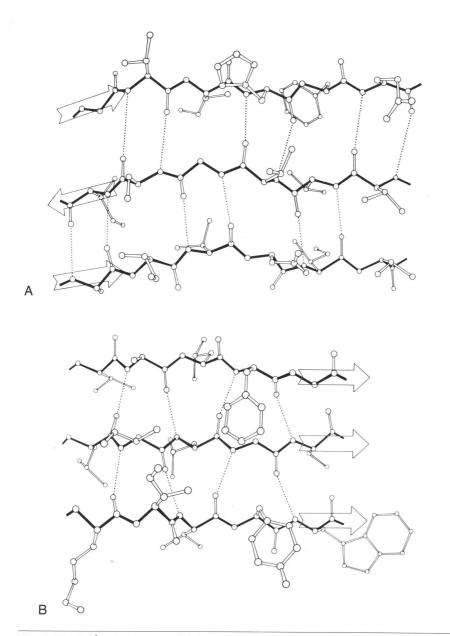

Fig. 1-10. β-Sheet structure. **(A)** An example of an antiparallel β-sheet from the enzyme Cu,Zn superoxide dismutase. **(B)** An example of a parallel β-sheet from the electron transport protein flavodoxin. Hydrogen bonds are indicated by dotted lines, and the polypeptide chain backbone by heavy solid lines. (From Richardson JS: The anatomy and taxonomy of protein structure. Adv Protein Chem 34:168, 1981, with permission.)

reverse turn involves four amino acids, and is stabilized by a hydrogen bond between the carbonyl oxygen of the first amino acid and the amide hydrogen of the fourth (Fig. 1-11). Most reverse turns occur at the surface of the protein and allow the polypeptide chain to abruptly change its direction.

Supersecondary Structures

Regions of aggregated secondary structure, called *supersecondary structures*, also occur in proteins. One type of supersecondary structure is the *coiled-coil α-helix*, in which two α-helices are wound around each other to form a left-handed superhelix (Fig. 1-12). This structure has been found primarily in fibrous proteins. Combinations of α-helix and β-sheet structures have also been observed. When parallel strands of a β-sheet are connected by an α-helical segment, the structure is referred to as a βαβ *unit* (Fig. 1-13). Still another structure observed in proteins is a combination of antiparallel β-sheet strands with reverse turns. This structure is referred to as a β-*meander* (Fig. 1-14).

Tertiary Structure

If it were possible to unravel the structure of a folded globular protein without disrupting its secondary structure, the result would be a linear

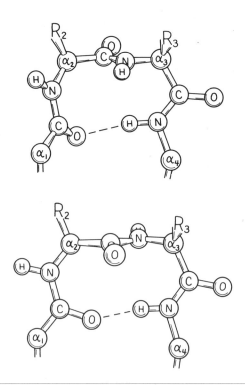

Fig. 1-11. The two types of reverse turn most frequently encountered in proteins. (From Richardson JS: The anatomy and taxonomy of protein structure. Adv Protein Chem 34:168, 1981, with permission.)

Amino Acids and Proteins   19

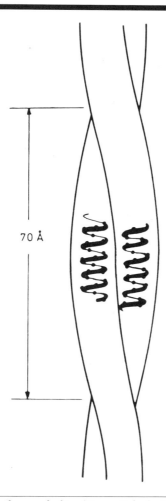

Fig. 1-12. The coiled-coil superhelix. (From Schulz GE, Schirmer RH: Principles of Protein Structure, Springer-Verlag, New York, 1979, with permission.)

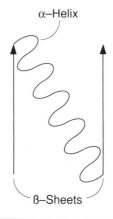

Fig. 1-13. The βαβ supersecondary structure. The helical segment usually lies above the plane of the β-strands.

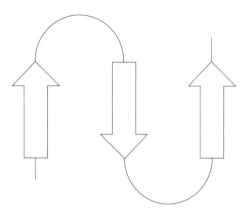

Fig. 1-14. The β-meander structure. Arrows indicate antiparallel β-sheet strands. β-meanders are seen most often in connecting strands of antiparallel β-sheets.

polypeptide chain interspersed with segments of secondary structure. The tertiary structure is obtained when this linear polypeptide of extended chain and secondary structure segments is folded to yield the globular structure of the native protein. It is the noncovalent forces of hydrogen bonds and electrostatic and hydrophobic interactions between different segments of the polypeptide chain that provide the driving force for generating the tertiary structure. The tertiary structure determines the overall shape and dimensions of proteins. Disulfide bonds linking different segments of the polypeptide chain are very important in stabilizing the tertiary structure of many proteins.

Very long polypeptide chains may fold into two or more regions of globular structure connected by strands of extended chain. Multiple globular structures that arise from the folding of a single polypeptide chain are called *structural domains*. Structural domains in a protein are usually only loosely associated with each other, and may function independently in the intact protein. The domain structure of the immunoglobulin G molecule is illustrated in Figure 1-15.

Quaternary Structure

Many proteins are complexes or aggregates of two or more polypeptide chains, known as *subunits*. The individual subunits may or may not be identical. The association of subunits to form the native protein constitutes quaternary structure.

For the subunits to aggregate correctly, their interacting surfaces must be complementary in both shape and side chain properties. Complex formation usually involves both hydrophobic and hydrophilic interactions. Identical subunits can interact in two different ways. If the contacting surfaces are identical, the interaction is *isologous*; if they are different, the interaction is *heterologous*. Subunits are usually arranged symmetrically (for example the tetrahedral arrangement of the four

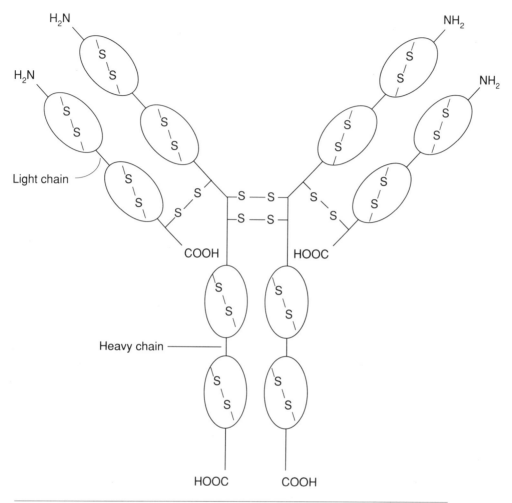

Fig. 1-15. Schematic representation of immunglobulin G, showing its domain structure. The immunoglobulin consists of two heavy chains and two light chains. Each light chain is folded into two domains, and each heavy chain into four. Each domain contains an intrachain disulfide bridge.

hemoglobin subunits), because symmetric arrangements are energetically favored. Individual subunits may undergo conformational changes in response to substrate or ligand binding, which convey information from one subunit to another (see the sections on Allosteric Regulation and on Cooperativity in Chapter 2).

CLINICAL CORRELATION: SICKLE CELL ANEMIA

Hemoglobin, the oxygen-transport protein of blood, is a tetramer composed of two α subunits and two β subunits (an $\alpha_2\beta_2$ tetramer). Sickle cell anemia is caused by a point mutation at residue 6 in the α-chain, which is changed from glutamic acid to valine. This substitution is a radical change because a charged polar amino acid is replaced by one that is hydrophobic. The resulting sickle cell hemoglobin (HbS) has a decreased affinity for oxygen and an increased tendency to aggregate when it is in the deoxygenated form. As a result, deoxygenated HbS aggregates to form long, filamentous fibers. Each fiber consists of six filaments coiled about a hollow central core. As the linear HbS fibers accumulate, the entire cell deforms into a sickle shape. Sickled red cells tend to clog in small blood vessels, causing local ischemia.

Individuals with sickle cell anemia suffer from a variety of chronic symptoms, and also from sporadic, severe "sickle cell crises." Chronic symptoms include reticulocytosis and sickling, which may begin as early as 10 to 12 weeks of age; enlargement of the spleen may occur as early as 5 to 6 months and persist for approximately 7 years. Some patients exhibit swelling of the hands and feet. Children with sickle cell anemia are also prone to infection.

Sickle cell crises supervene when some event that causes prolonged low oxygen tension triggers massive HbS aggregation and sickling. The symptoms of sickle cell crisis are variable; they may include sudden, severe pain in the long bones and joints or chest, sudden hemolysis (usually associated with an underlying infection), and sudden, massive enlargement of liver and spleen accompanied by an acute decrease in the hematocrit.

Currently, no effective treatment is available for sickle cell anemia. However, recommendations for patients with chronic hemolysis should also be followed by sickle cell anemia patients. These include good nutrition, early diagnosis and treatment of infections, and folic acid administration. The use of anti-sickling drugs appears promising, but has been generally hampered by unacceptably severe side effects.

It is quite clear that in the case of sickle cell anemia a single change in the primary structure of the hemoglobin β-chain causes a severe disease. The clinical symptoms associated with sickle cell anemia are manifested only in individuals homozygous for HbS.

SUGGESTED READING

1. Barrett GC (ed.): Chemistry and Biochemistry of the Amino Acids. Chapman and Hall, London, 1985
2. Creighton TE: Proteins Structure and Molecular Properties. Freeman, San Francisco, 1983
3. Schulz GE, Schirmer RH: Principles of Protein Structure, Springer-Verlag, New York, 1979

STUDY QUESTIONS

Directions: For each of the following multiple choice questions, choose the most appropriate answer.

1. Which of the following amino acids will most strongly increase the rigidity of a polypeptide chain?
   A. Alanine
   B. Valine
   C. Lysine
   D. Methionine

2. Alanine is placed in a solution that has a pH equal to the pK_2 of alanine. Alanine will be present in the form of the molecular species
   A. $^+NH_3CHCH_3COOH$ and $^+NH_3CHCH_3COO^-$
   B. $^+NH_3CHCH_3COO^-$ and NH_2CHCH_3COOH
   C. $^+NH_3CHCH_3COO^-$ and $NH_2CHCH_3COO^-$
   D. $^+NH_3CHCH_3COOH$ and $NH_2CHCH_3COO^-$

3. Which of the following is a type of supersecondary structure?
   A. β-Meander
   B. Domain
   C. α-Helix
   D. β-Sheet

4. Which of the following substances cannot be used in determining the N-terminal amino acid of a polypeptide chain?
   A. Dansyl chloride
   B. Ninhydrin
   C. 1-Fluoro-2,4-dinitrobenzene
   D. Phenylisothiocyanate

5. Most peptide bonds in naturally occurring proteins
   A. preferentially adopt a *trans* configuration
   B. preferentially adopt a *cis* configuration
   C. show no preference for a *cis* or *trans* configuration
   D. cannot assume a *cis* or *trans* configuration

Items 6 and 7. Consider the peptide:

Gly-Ser-Glu-Asp-Lys-Val-Pro

6. At neutral pH, the overall charge on this peptide will be
   A. -2
   B. -1
   C. 0
   D. $+1$

7. Under very basic conditions, the overall charge on the peptide will be
   A. −3
   B. −1
   C. 0
   D. +1

Directions: For questions 8 through 10, use the following key:
(A) if 1, 2, and 3 are correct
(B) if 1 and 3 are correct
(C) if 2 and 4 are correct
(D) if only 4 is correct
(E) if all four are correct

---

8. The peptide bond
   1. is resonance stabilized
   2. is an amide linkage
   3. is a covalent bond
   4. joins the α-amino group to the β-carbon

9. The individual subunits of multisubunit proteins are usually held together by
   1. covalent bonds
   2. hydrophobic interactions
   3. covalent bonds and hydrophilic interactions
   4. hydrophilic interactions

10. Elements that enhance the stability of α-helices are:
    1. The presence of three residues per turn
    2. Hydrogen bonds
    3. Proline residues
    4. The presence of 3.6 residues per turn

2

ENZYMES

Types of Enzyme-Catalyzed Reactions/
How Enzymes Catalyze Reactions/Enzyme Kinetics/
Enzyme Regulation/Clinical Correlations

The student should be able to:

Learning Objectives

1. Define energy of activation and explain how it is influenced during enzyme catalysis.
2. List three types of catalytic mechanism by which various enzymes can influence the energy of activation.
3. Define V_{max}, K_M, and the turnover number.
4. Distinguish competitive, uncompetitive, and noncompetitive inhibitors by their effects on V_{max} and K_M.
8. Diagram ping-pong, random sequential, and ordered sequential multisubstrate mechanisms.
5. Distinguish allosteric kinetic behavior from classic Michaelis-Menten kinetic behavior.
6. List four types of intracellular control of enzyme activity.
7. Define isozymes.

ENZYMES

Proteins are the workhorses of the cell. One of the cellular functions performed by proteins is catalysis. Proteins that catalyze cellular reactions are known as enzymes. Enzymes are highly specific catalysts. Moreover, their catalytic properties can be regulated, allowing the cell to control its metabolic pathways. Many enzymes catalyze only one reaction, although there are cases where an enzyme may catalyze analogous reactions with structurally similar but distinct reactants (substrates). Enzymes are divided into six categories on the basis of the types of reactions they catalyze (Table 2-1).

This chapter covers the following aspects of enzymes and enzyme catalysis:

How enzymes catalyze reactions
The determination of the rate of an enzyme-catalyzed reaction
The regulation of enzymatic catalysis

Table 2-1. The Six Classes of Enzyme

Oxidoreductases	Catalyze oxidation-reduction reactions
Transferases	Catalyze group transfer reactions of the type $A-G + Y \rightleftharpoons A + Y-G$
Hydrolases	Catalyze hydrolytic cleavage reactions
Lyases	Catalyze nonhydrolytic cleavage reactions
Isomerases	Catalyze the interconversion of isomers
Ligases	Catalyze the covalent linkage of two substrates

Perspective: How Enzymes Catalyze Reactions

1. Enzymes specifically bind their reactant or reactants (substrates). To catalyze a reaction, the enzyme must first recognize its correct substrate; thus, enzymes exhibit *substrate specificity*.
2. Enzymes use the chemical reactivity of amino acid side chains in the vicinity of the substrate binding site to promote the conversion of the substrate to the desired product. Many enzymes also use the chemical properties of a coenzyme or prosthetic group during catalysis.
3. Enzymes act by stabilizing the transition state between substrate and product.

Stabilization of the transition state lowers the energy of activation of the reaction. The energy of activation may be thought of as a barrier or hurdle that impedes the conversion of reactants to products. When the energy of activation is lowered, a larger proportion of reactant molecules have enough energy to be converted to product. Therefore, more reactant molecules per unit time are converted to product, and the reaction rate rises.

ENZYME SPECIFICITY

Enzymes bind substrate at a region called the *active center* or *active site*. The active site has two functions. First, it has the necessary structural and/or hydrophobic/hydrophilic characteristics to recognize and bind the correct substrate. Second, it has the appropriate chemical environment to catalyze the reaction.

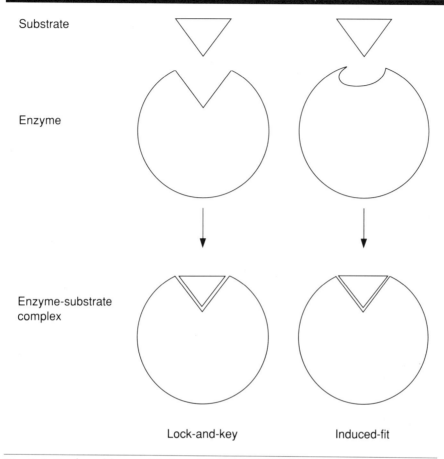

Fig. 2-1. Lock-and-key and induced-fit models of substrate binding.

Two models have been proposed to explain enzyme specificity (Fig. 2-1). In the *lock-and-key model*, the enzyme and substrate have complementary structural features, so that the substrate fits into the active site as a key fits into a lock. In the *induced-fit model*, the substrate does not exactly fit the active site, but binding of the substrate induces a conformational change in the enzyme that causes the active site to fit around the substrate in a lock-and-key manner.

COFACTORS AND COENZYMES

Many enzymes contain non-amino acid groups called *cofactors* that are essential to the function of the enzyme. Both metal ions and organic compounds are used as cofactors. Cofactors that remain tightly bound to the enzyme during and after catalysis are called *prosthetic groups*. Cofactors that easily dissociate from the enzyme before and after catalysis are called *coenzymes* (or *cosubstrates*). Most prosthetic groups are metal ions, whereas most coenzymes are organic compounds derived from vitamins. The complex of protein and cofactor is called the *holoenzyme*; the protein dissociated from the cofactor is called the *apoenzyme*.

HOW ENZYMES AFFECT THE ENERGY OF ACTIVATION

To understand how enzymes catalyze reactions, two features of chemical reactions must be recalled.

1. Reactants are converted to products gradually, by way of a series of *reaction intermediates*.

All reaction intermediates are less stable (have a higher energy content) than either reactants or products. The reaction intermediate with the greatest energy content is called the *transition state intermediate*.

2. For a reactant to be converted to product, enough energy must be available to form the transition state intermediate.

The transition state may be described as the intermediate whose structure and properties most resemble those of both the reactant and the product. After the transition state is formed, there is a decline in energy

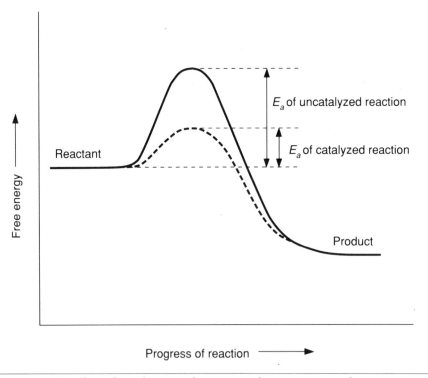

Fig. 2-2. The effect of catalysis on the energy of activation E_a of a reaction.

content to the product. The energy level of the transition state intermediate is called the *energy of activation* (Fig. 2-2). It is the energy required to convert the reactant to the transition state intermediate.

Because the activation energy barrier determines how many substrate molecules have the energy to form product, the activation energy controls the *rate* at which a reaction proceeds. The reaction rate can be increased by either increasing the energy of the reactants or lowering the energy of activation. Enzymes catalyze reactions by lowering the energy of activation (Fig. 2-2).

HOW ENZYMES LOWER THE ENERGY OF ACTIVATION

Enzymes employ at least four different methods for lowering the energy of activation. These are *substrate strain, orientation and proximity effects, acid-base catalysis*, and *covalent catalysis*.

Substrate Strain

Substrate strain is distortion of the substrate molecule that occurs as a result of binding to the enzyme. The distortion involves stretching or weakening of the bond(s) that will be attacked during catalysis. Substrate molecules may undergo distortion in order to bind to a rigid lock-and-key type of active center. Alternatively, distortion can occur when the enzyme undergoes an induced-fit type of conformational change after the substrate is bound.

Orientation and Proximity Effects

Consider the following reaction, which may proceed uncatalyzed or via enzyme catalysis:

$$A + B \rightleftharpoons C + D$$

For the reaction to occur without catalysis, three conditions must be met:

1. A and B must collide.
2. The collision must occur so that the appropriate reacting groups on A and B are correctly oriented with respect to each other.
3. The collision must impart sufficient energy to A and B to allow them to overcome the energy of activation for the reaction.

In other words, a very precise collision between A and B must occur for the reaction to proceed. In solution, collisions occur at random, and the probability of any specific type of collision is small. Consequently, rates of uncatalyzed reactions in solution are often slow.

When the reaction is enzyme catalyzed, both of the substrates are bound to the enzyme in close proximity and in the proper orientation for reaction to occur. As a result, the rate of conversion of A and B into

C and D is significantly enhanced (the reaction rate is increased). In multisubstrate enzyme-catalyzed reactions, the substrates may bind either simultaneously or sequentially (see the section Kinetics of Multisubstrate Enzymic Reactions, below).

Acid-Base Catalysis

Many enzymes employ acid-base catalysis. If the rate of a chemical reaction depends on the concentration of acid, the reaction is *acid-catalyzed*; if the rate depends on the concentration of base, the reaction is *base-catalyzed*. Reactions that depend specifically on H^+ or OH^- exhibit *specific acid-base catalysis*. If a proton donor can be substituted for H^+ and a proton acceptor for OH^-, *general acid-base catalysis* is exhibited. Ionizable amino acid side chains can act as weak acids or bases, and often perform general acid-base catalysis. The imidazole side chain of histidine, which has a pK in the physiologic range, is a particularly common participant in enzyme acid-base catalysis (Chapter 1).

Some enzymes perform *concerted general acid-base catalysis*, in which both an acid and a base participate in the catalytic process. A good example is the hydrolysis of RNA by ribonuclease (Fig. 2-3), in which the histidine residues at positions 12 and 119 alternate as acids and bases.

Covalent Catalysis

In covalent catalysis, the substrate is transiently attached to the enzyme by a covalent bond. The substrate may be attached to an amino acid side chain in the active center or to an enzyme-bound coenzyme. Enzyme mechanisms involving nucleophilic attack usually employ covalent catalysis. An excellent example of covalent catalysis is the hydrolysis of peptide bonds by papain (Fig. 2-4).

Perspective: Enzyme Kinetics

Enzyme kinetics is the study of enzymes during catalysis. Only three questions need to be addressed:

1. How fast is the catalyzed reaction? (i.e., how much product is formed per unit time?)
2. How responsive is the catalytic rate to substrate concentration? (i.e., what is the relationship between catalytic rate and substrate concentration?)
3. Under what conditions can the catalytic rate be modified?

Enzyme kinetics provides the answers to these questions, and generates the parameters that describe the catalytic properties of enzymes.

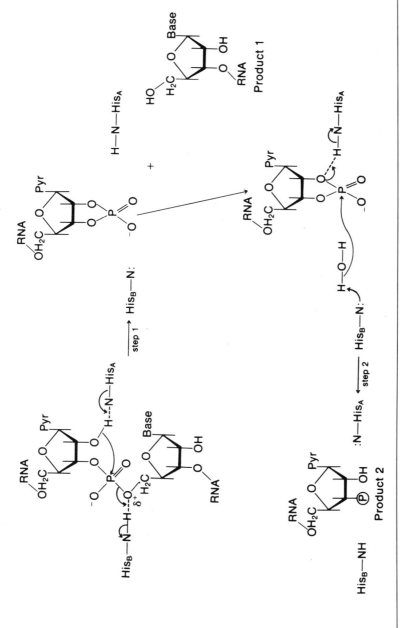

Fig. 2-3. Acid-base catalysis in the cleavage of RNA by ribonuclease. His$_A$ and His$_B$ are histidines 12 and 119, respectively, in the ribonuclease active center. (From Devlin TM: Textbook of Biochemistry with Clinical Correlations. Wiley, New York, 1982, with permission.)

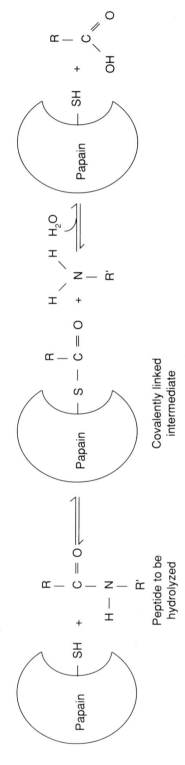

Fig. 2-4. The mechanism by which papain hydrolyzes peptide bonds. A sulfhydryl group on papain forms a thioester linkage with the carboxyl group of the peptide bond.

ENZYME KINETICS

Enzyme-catalyzed reactions occur in two phases (only single-substrate enzyme-catalyzed reactions will be considered in this section). First, substrate and enzyme interact to form an *enzyme-substrate complex*. This complex then dissociates to yield the free enzyme plus product:

$$\text{E} + \text{S} \underset{k_{-1}}{\overset{k_1}{\rightleftharpoons}} \text{ES} \underset{k_{-2}}{\overset{k_2}{\rightleftharpoons}} \text{E} + \text{P} \qquad (1)$$

where k_1, k_2, k_{-1}, and k_{-2} are the *rate constants* of the reactions.

The *Michaelis-Menten* and *Briggs-Haldane* derivations of enzyme rate equations are based on three assumptions:

1. The substrate concentration is large relative to the enzyme concentration, so formation of the enzyme-substrate concentration does not significantly alter the substrate concentration. This assumption is usually valid for enzyme-catalyzed reactions.
2. The rate is measured at the initiation of the reaction, when the concentration of product is zero. Because no product is present, the reverse reaction cannot occur. Therefore only the initial rate will be measured.

The second assumption (called the initial-rate assumption) is required to observe the maximum net forward velocity $V_{\max}$. In light of this assumption, equation (1) can be rewritten

$$\text{E} + \text{S} \underset{k_{-1}}{\overset{k_1}{\rightleftharpoons}} \text{ES} \overset{k_2}{\longrightarrow} \text{E} + \text{P} \qquad (2)$$

Once the reaction is in progress, the exact concentrations of substrate and product are unknown, because these values change constantly during the reaction. However, under the condition of the initial-rate assumption, the substrate concentration is the same as the initial amount of substrate in the reaction mixture.

3. The third assumption has to do with the dynamics of the enzyme-substrate complex. Michaelis and Menten originally assumed that the enzyme-substrate complex is in equilibrium with both the free enzyme and the unbound substrate. For this condition to be satisfied, the formation of product from ES must be slow enough not to disturb the equilibrium of ES with E + S. This assumption is true when $k_{-1} \gg k_2$. In contrast, Briggs and Haldane merely assumed a steady state condition for [ES]; i.e., that the rate of formation of ES is exactly balanced by the rate of its dissociation, whether into S + E or P + E. The relative values of k_{-1} and k_2 are not restricted. The Michaelis-Menten assumption is a special case of this more general formulation.

To apply the steady state assumption to equation (2), we must obtain expressions for the rate of synthesis and dissociation of the ES complex. The rate or velocity V of a chemical reaction A → P is given by the equation

$$V = k[\text{A}]^n \qquad (3)$$

where [A] is the molar concentration of A, n is the number of moles of A required to produce one mole of P, and k is the rate constant. Therefore, from equation (2), the rate of synthesis V_s of the ES complex is constant. Therefore, from equation (2), the rate of synthesis V_s of the ES complex is

$$V_s = k_1 [E][S] \qquad (4)$$

and the rate of dissociation V_d is

$$V_d = k_{-1}[ES] + k_2[ES] \qquad (5)$$

which can be rearranged to give

$$V_d = [ES](k_{-1} + k_2) \qquad (6)$$

The steady state assumption dictates that V_s and V_d must be equal. Therefore,

$$k_1 [E][S] = [ES](k_{-1} + k_2) \qquad (7)$$

Rearrangement of equation (7) yields

$$[E][S] = [ES]\left(\frac{k_{-1} + k_2}{k_1}\right) \qquad (8)$$

Dividing both sides of equation (8) by [ES] yields

$$\frac{[E][S]}{[ES]} = \frac{k_{-1} + k_2}{k_1} \qquad (9)$$

If $(k_{-1} + k_2)/k_1$ is defined as a single constant K_M (the *Michaelis constant*), equation (9) reduces to

$$\frac{[E][S]}{[ES]} = K_M \qquad (10)$$

Note that [E] is the concentration of free enzyme in the reaction mixture, which equals the concentration of total enzyme [E_t] minus the concentration of the ES complex:

$$[E] = [E_t] - [ES] \qquad (11)$$

Substituting [E_t] − [ES] for [E] in equation (10) yields

$$\frac{([E_t] - [ES])[S]}{[ES]} = K_M \qquad (12)$$

Rearrangement yields

$$([E_t] - [ES])[S] = K_M[ES] \qquad (13)$$

Dividing by [ES] and rearranging gives

$$\frac{[E_t]}{[ES]} = \frac{K_M}{[S]} + 1 \qquad (14)$$

From equation (3), we know that the rate of product formation $V = k_2[ES]$. Therefore, as more enzyme is converted to the ES complex, the reaction rate increases. If all the enzyme is in the form of the ES complex, then [ES], [E$_t$S], and [E$_t$] will all be equal ([ES] = [E$_t$S] = [E$_t$]). Under these conditions, the maximum rate V_{max} will be observed. V_{max} may then be expressed as

$$V = V_{max} = k_2[E_t] \tag{15}$$

Solving for [E$_t$], we get

$$[E_t] = \frac{V}{k_2} \tag{16}$$

Using $V = k_2[ES]$, we can also solve for [ES]:

$$[ES] = \frac{V_{max}}{k_2}$$

Substituting V_{max}/k_2 for [E$_t$] and V/k_2 for [ES] into equation (14) and rearranging gives

$$V = \frac{V_{max}[S]}{K_M + [S]} \tag{17}$$

which is the *Michaelis-Menten equation*. The Michaelis-Menten equation relates substrate concentration to the reaction rate or velocity, and makes it possible to calculate the kinetic parameters V_{max} and K_M. The Michaelis constant K_M is the substrate concentration that gives a reaction rate equal to one-half V_{max}. This can be verified by substituting K_M for [S] in equation (17):

$$V = \frac{V_{max}K_M}{2K_M} = \frac{V_{max}}{2} \tag{18}$$

As mentioned above, K_M is defined mathematically as a ratio of rate constants for the synthesis and dissociation of the ES complex:

$$K_M = \frac{k_{-1} + k_2}{k_1} \tag{19}$$

When the substrate concentration equals the numerical value of this ratio, the observed reaction velocity V will be half maximal ($V = V_{max}/2$). If the original [ES] steady state assumption of Michaelis and Menten ($k_{-1} \gg k_2$) holds, then k_2 in equation (19) is negligible, and for all practical purposes K_M reduces to

$$K_M = \frac{k_{-1}}{k_1} \tag{20}$$

Determination of K_M and V_{max}

A plot of V versus [S] using the Michaelis-Menten equation gives a hyperbolic curve (Fig. 2-5) that asymptotically approaches V_{max}. The value of K_M may be estimated as the point where the curve crosses $V_{max}/2$. However, V_{max} is difficult to determine accurately on this plot, because the curve approaches it only asymptotically. K_M and V_{max} are more easily obtained from a plot of the reciprocal of equation (17), which is the equation of a straight line ($y = ax + b$). This is the *Lineweaver-Burk equation*:

$$\frac{1}{V} = \frac{K_M}{V_{max}}\left(\frac{1}{[S]}\right) + \frac{1}{V_{max}} \tag{21}$$

A plot of $1/V$ versus $1/[S]$ yields a straight line that has a slope of K_M/V_{max} and a y-axis intercept of $1/V_{max}$. K_M may also be obtained from the y-intercept, as shown in Figure 2-6. The Lineweaver-Burk equation can be used to determine V_{max} and K_M by measuring the enzyme activity at known substrate concentrations.

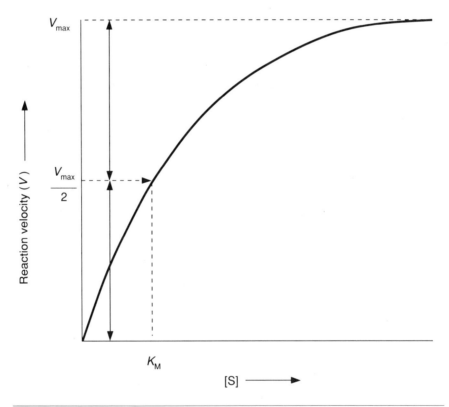

Fig. 2-5. Plot of reaction velocity as a function of substrate concentration according to the Michaelis-Menten equation.

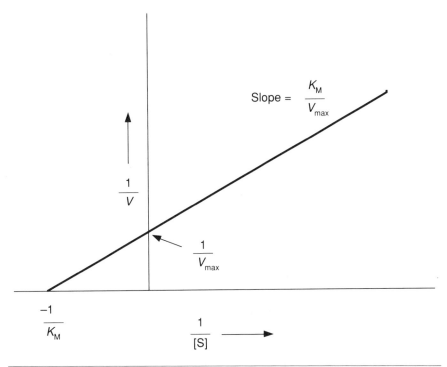

Fig. 2-6. Lineweaver-Burk plot for the determination of K_M and V_{max}.

Enzyme Activity

The *activity* of an enzyme is expressed in three forms: as *enzyme units*, as *specific activity*, and as *turnover number*. One standard unit of enzyme activity converts 1 micromole of substrate to product per minute. If the weight of enzyme in milligrams is known as well as the number of enzyme units in the sample, the specific activity (enzyme units per milligram protein) can be calculated. If the total enzyme concentration in moles is also known, then the turnover number can be calculated. The turnover number is numerically equal to k_2: it is the number of moles of substrate converted to product per minute per mole of enzyme. The turnover number can be used to compare the molecular activities of different enzymes. The turnover number may also be expressed in terms of *katals*. One katal equals the number of moles of substrate converted to product per second per mole of enzyme.

ENZYME INHIBITION

Many drugs and toxic agents exert their effects by inhibiting specific enzymes. Enzyme inhibition can also be used to obtain information on the specificity and mechanism of action of enzymes.

Enzyme inhibition may be *reversible* or *irreversible*. In general, irreversible inhibition permanently destroys enzyme activity. Irreversible inhibition is also called *enzyme inactivation*. In contrast, reversible inhibitors usually decrease enzyme activity, and full activity returns when the inhibitor is removed. The rest of this section is devoted to reversible inhibition. Reversible inhibitors may be classified according to the way they interact with the enzyme:

Competitive inhibitors bind to the same site as the substrate, and therefore compete with the substrate for the enzyme. Competitive inhibitors usually resemble the substrate structurally but cannot react to form product. Competitive inhibition can be overcome by either increasing the concentration of substrate or diluting the inhibitor.

Uncompetitive inhibitors bind only to the enzyme-substrate complex. There is no competition between the inhibitor and the substrate for the enzyme.

Noncompetitive inhibitors can bind to either free enzyme or the enzyme-substrate complex. Again, there is no competition between inhibitor and substrate for the enzyme.

These three types of inhibitor affect the V_{max} and K_M of enzymes differently.

Competitive Inhibition

The molecular events during competitive inhibition may be diagrammed as follows:

$$E + S \underset{k_{-1}}{\overset{k_1}{\rightleftharpoons}} ES \overset{k_2}{\longrightarrow} E + P$$
$$+$$
$$I$$
$$\updownarrow k_i$$
$$EI$$

where k_i is the equilibrium constant for the dissociation of the inhibitor I from the EI complex. The inhibitor and the substrate compete for the same site on the enzyme. The Michaelis-Menten equation for this type of inhibition is

$$V = \frac{V_{max}[S]}{[S] + K_M\left(1 + \frac{[I]}{k_i}\right)} \tag{22}$$

and the Lineweaver-Burk expression is

$$\frac{1}{V} = \frac{K_M}{V_{max}}\left(1 + \frac{[I]}{k_i}\right)\frac{1}{[S]} + \frac{1}{V_{max}} \tag{23}$$

As can be seen from equation 22, V_{max} is unchanged but K_M is increased

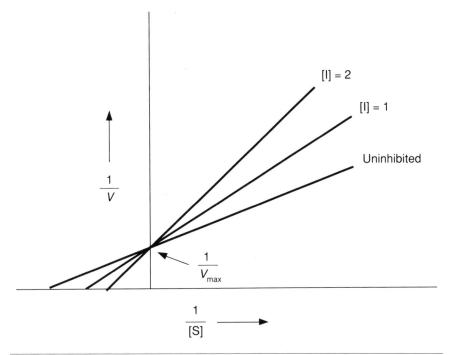

Fig. 2-7. Lineweaver-Burk plot showing the effect of a competitive inhibitor on K_M and slope. $[I] = 1$ and $[I] = 2$ refer to different inhibitior concentrations, where $[I] = 2 > [I] = 1$.

by a factor of $1 + [I]/k_i$. The value of K_M in the presence of inhibitor is known as the *apparent* K_M:

$$\text{Apparent } K_M = K_M \left(1 + \frac{[I]}{k_i}\right) \qquad (24)$$

The effect of competitive inhibition on the Lineweaver-Burk plot is shown in Figure 2-7.

Uncompetitive Inhibition

Uncompetitive inhibition may be diagrammed as follows:

$$E + S \underset{k_{-1}}{\overset{k_1}{\rightleftharpoons}} ES \xrightarrow{k_2} E + P$$
$$+$$
$$I$$
$$\updownarrow k_i$$
$$ESI$$

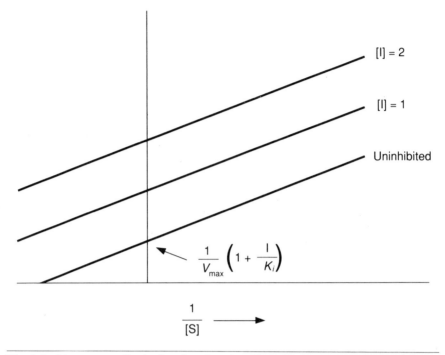

Fig. 2-8. Lineweaver-Burk plot showing the effect of an uncompetitive inhibitor on V_{max} and K_M. $[I] = 1$ and $[I] = 2$ refer to different inhibitor concentrations, where $[I] = 2 > [I] = 1$.

The Michaelis-Menten equation for an uncompetitively inhibited enzyme is

$$V = \frac{\left(\dfrac{V_{max}}{1 + [I]/k_i}\right)[S]}{[S] + \dfrac{K_M}{1 + [I]/k_i}} \tag{26}$$

Both V_{max} and K_M are decreased by the factor $1/(1 + [I]/k_i)$. The corresponding Lineweaver-Burk equation is

$$\frac{1}{V} = \frac{K_M}{V_{max}}\left(\frac{1}{[S]}\right) + \frac{1}{V_{max}}\left(1 + \frac{[I]}{k_i}\right) \tag{27}$$

The effect of uncompetitive inhibition on the Lineweaver-Burk plot is shown in Fig. 2-8.

Noncompetitive Inhibition

Noncompetitive inhibitors bind to both free enzyme and the enzyme-substrate complex:

$$E + S \underset{k_{-1}}{\overset{k_1}{\rightleftharpoons}} ES \overset{k_2}{\longrightarrow} E + P$$
$$+ \qquad\qquad +$$
$$I \qquad\qquad I$$
$$\updownarrow k_i \qquad\qquad \updownarrow k_i$$
$$EI + S \rightleftharpoons ESI$$

The dissociation constant k_i is assumed to be the same for the free enzyme and the enzyme-substrate complex. The resulting Michaelis-Menten equation is

$$V = \left(\frac{V_{max}}{1 + \frac{[I]}{k_i}}\right) \frac{[S]}{[S] + K_M} \qquad (29)$$

In the case of noncompetitive Michaelis-Menten inhibition, K_M is unchanged but V_{max} is decreased by a factor of $1/(1 + [I]/k_i)$. The corresponding Lineweaver-Burk expression is

$$\frac{1}{V} = \frac{K_M}{V_{max}}\left(\frac{1}{[S]}\right)\left(1 + \frac{[I]}{k_i}\right) + \frac{1}{V_{max}}\left(1 + \frac{[I]}{k_i}\right) \qquad (30)$$

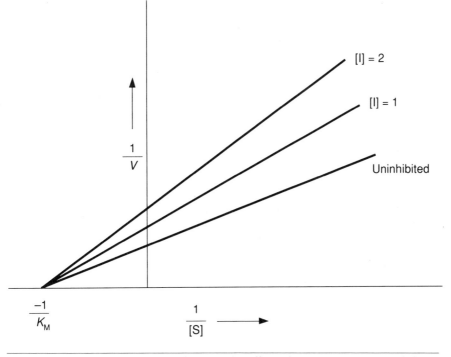

Fig. 2-9. Lineweaver-Burk plot showing the effect of a noncompetitive inhibitor on V_{max} and slope. [I] = 1 and [I] = 2 refer to different inhibitor concentrations, where [I] = 2 > [I] = 1.

Figure 2-9 shows the effect of noncompetitive inhibition on the Lineweaver-Burk plot.

KINETICS OF MULTISUBSTRATE ENZYMIC REACTIONS

The previous sections on enzyme kinetics were only concerned with single-substrate enzyme catalyzed reactions. However, most enzyme catalyzed reactions involve more than one substrate. In a typical bisubstrate enzyme catalyzed reaction

$$A + B \rightleftharpoons C + D$$

both substrates must at some point bind to the enzyme. In the case of single-substrate reactions, the enzyme-substrate complex is binary, consisting of the enzyme and one substrate molecule. Two-substrate reactions introduce the possibility of ternary enzyme-substrate complexes, consisting of the enzyme and both substrate molecules. The interaction of substrates and enzyme in bisubstrate reactions can occur by three mechanisms: *ping-pong* or *double displacement*, *ordered sequential*, and *random sequential*.

The ping-pong or double displacement mechanism may be diagrammed as follows:

$$E \xrightarrow{A} EA \longrightarrow E'C \xrightarrow{C} E' \xrightarrow{B} E'B \longrightarrow ED \xrightarrow{D} E$$

The enzyme preferentially binds one of the substrates, A. After A is bound, it reacts with the enzyme to form product C, which dissociates, and the altered enzyme E', which is able to bind and react with substrate B, yielding the second product D and the regenerated enzyme E. Key features of the ping-pong mechanism are that A and B bind to the enzyme in a preferred order, and no ternary complex is formed. A good example of the ping-pong mechanism is the transamination of amino acids by transaminases (Chapter 9).

The ordered sequential and random sequential mechanisms both involve the formation of complexes of the enzyme with more than one substrate at once. The ordered sequential mechanism may be diagrammed as follows:

$$E \xrightarrow{A} EA \xrightarrow{B} EAB \longrightarrow ECD \xrightarrow{C} ED \xrightarrow{D} E$$

The binding of the substrates and the release of the products both occur in a specific sequence. Presumably, binding of A to the enzyme causes a conformational change that creates a binding site for B.

In a random sequential mechanism, in contrast, the enzyme has binding sites for both A and B, and either substrate can bind first without affecting the reaction rate. The random sequential mechanism may be diagrammed as follows:

$$E \begin{array}{c} A \\ \diagup \\ \diagdown \\ B \end{array} EA + EB \begin{array}{c} B \\ \diagup \\ \diagdown \\ A \end{array} EAB + EBA \longrightarrow$$

$$ECD + EDC \begin{array}{c} C \\ \diagup \\ \diagdown \\ D \end{array} EC + ED \begin{array}{c} D \\ \diagup \\ \diagdown \\ C \end{array} E$$

FACTORS AFFECTING REACTION RATE

Temperature

The reaction velocity of non-enzyme-catalyzed reactions increases with increasing temperature. When the temperature is raised, more reactant molecules have enough kinetic energy to reach the activation energy. The rates of enzyme-catalyzed reactions also increase with temperature up to the temperature at which the enzyme begins to denature. Above that temperature, reaction rate drops precipitously as the enzyme denatures.

pH

The rates of enzyme catalyzed reactions are also influenced by pH. Most enzymes exhibit maximum activity at a particular pH value or range, called the *pH optimum*. The pH optima of enzymes are generally determined by the pKs of key reactive groups at the active site and by the pKs of substrate molecules. The pH optimum usually reflects the pH value(s) at which the active site and/or substrate are in the correct ionization state for the reaction.

PHYSIOLOGIC ENZYME REGULATION

Competitive, uncompetitive, and noncompetitive inhibition are more applicable to the effects of drugs and toxins on enzymes than to the physiologic regulation of enzymes in living systems. In general, cells use much more diverse and sophisticated mechanisms to regulated enzymes and thereby control metabolic pathways. Several major mechanisms—allosterism and cooperativity, covalent modification, and enzyme cascades—are discussed in this section.

Allosteric Regulation

In *allosteric regulation*, the activity of an enzyme is regulated by reversible binding of an *effector* molecule to a site on the enzyme other than the active site, known as the *allosteric site*. In plots of V versus [S], allosteric enzymes typically show sigmoid (S-shaped) curves (Fig. 2-10) rather than hyperbolic curves.

Allosteric effectors can be either positive or negative. Negative effectors decrease the reaction rate, whereas positive effectors increase it. Allosteric effectors act by increasing or decreasing V_{max}, K_M, or both. Figure 2-11 shows the effect of positive and negative effectors on the V versus [S] plot for an allosteric enzyme. A single enzyme may have several allosteric effectors, both positive and negative. Most allosteric enzymes are multisubunit enzyme complexes; allosteric sites are commonly located on separate subunits from the active site.

Feedback Inhibition

Many metabolic pathways are controlled by the mechanism of *feedback inhibition*, in which the end product of the pathway allosterically inhibits the first *committed enzyme* of the pathway. (The first committed enzyme is the first enzyme whose product cannot branch into any other pathway in the cell but must proceed to form the end product of that pathway.) For example, in a pathway converting A to D, the enzyme E1 is likely to be inhibited by the product D:

$$A \xrightarrow{E1} B \longrightarrow C \longrightarrow D$$
$$\underbrace{}_{\text{Inhibition}}$$

When the cell has enough D for its needs, the pathway is inhibited; when the level of D drops, the inhibition is relieved. Feedback inhibition is almost always allosteric, because the end product is unlikely to be similar enough in structure to the first substrate to compete for the active site.

Cooperativity

If the binding of an effector to one subunit influences the binding of the substrate or ligand to another subunit, the enzyme exhibits *cooperativity*. Cooperativity may be positive or negative. If the effector molecule is identical to the substrate or ligand, the cooperativity is *homotrophic*. If the effector and ligand are different, the cooperativity is *heterotrophic*. Many nonenzymic proteins also display cooperativity in the binding of specific ligands; the binding of oxygen by hemoglobin is an excellent example.

Cooperativity and Allosterism

Not all allosteric enzymes display cooperativity. Allosterism denotes regulation of an enzyme by binding of an effector to a site other than the active site. The allosteric site may or may not be located on the

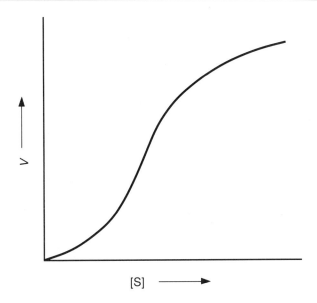

Fig. 2-10. Reaction velocity as a function of substrate concentration for an allosteric enzyme.

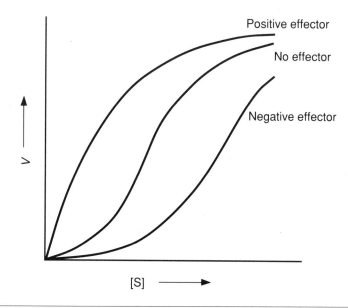

Fig. 2-11. The effects of positive and negative effectors on the plot of reaction velocity V versus substrate concentration [S] for an allosteric enzyme.

same subunit as the active site, and may or may not affect the binding of substrate to the active site. Cooperativity denotes control of substrate binding on one subunit by binding of a ligand or substrate to a site on a different subunit.

Covalent Modification

Enzyme regulation by covalent modification occurs when enzyme activity is modified by either the formation or the lysis of a specific covalent bond in the enzyme. Covalent modification may be reversible or irreversible. In reversible mechanisms, the activity of the enzyme is altered by the addition or removal of a particular group at a specific site on the enzyme. Adding and removing the regulatory group have opposite effects on enzyme activity. Many enzymes are controlled by reversible phosphorylation/dephosphorylation, adenylation/deadenylation, or uridylation/deuridylation. The addition and removal of the regulatory group are catalyzed by different enzymes, which respond to different cellular stimuli. Therefore, covalently modified enzymes are activated and deactivated in response to different cellular conditions.

An example of *irreversible* covalent modification is the activation of zymogen forms of proteolytic enzymes by the cleavage of specific peptide bonds.

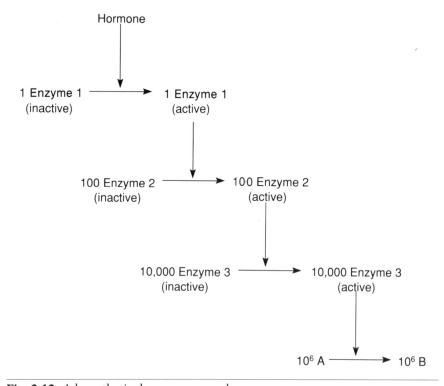

Fig. 2-12. A hypothetical enzyme cascade.

Enzyme Cascades

Enzyme cascades consist of a series of enzymes that sequentially activate each other, usually by covalent modification. Enzyme cascades amplify a weak regulatory signal so that it has a strong effect on a biochemical process or reaction. The first enzyme in the cascade is activated by the initial regulatory signal, and the last enzyme in the cascade controls the regulated process (i.e., catalyzes the reaction). The signal is amplified during the cascade because each enzyme is also the substrate of the preceding enzyme. Consider the hypothetical hormone-controlled conversion of A into B shown in Figure 2-12, in which each enzyme is able to catalyze 100 molecules of substrate to product per unit time. The hormone initially activates enzyme 1, each molecule of which activates 100 molecules of enzyme 2. These 100 molecules then activate 10,000 molecules of enzyme 3, and so forth. At each level, the signal is amplified by a factor of 100. A good example of an enzyme cascade is the hormone control of glycogen synthesis and degradation (Ch. 4).

ISOZYMES

Isozymes (also called *isoenzymes*) are enzymes that catalyze the same reaction but that differ in amino acid sequence, and therefore in structure and physical properties. Isozymes often differ significantly in regulatory properties, and catalyze the same reaction in different tissues that have differing metabolic requirements. Mammalian lactate dehydrogenase, for example, exists as five different isozymes.

CLINICAL CORRELATION: TREATMENT OF LEUKEMIA WITH METHOTREXATE

Methotrexate is a chemotherapeutic agent that is used in the treatment of leukemia and other malignancies. It acts by inhibiting the enzyme dihydrofolate reductase, which catalyzes the conversion of dihydrofolate (FH_2) to tetrahydrofolate (FH_4), a metabolite that is essential for DNA synthesis. Tetrahydrofolates act as one-carbon carriers in a number of biosynthetic pathways, including the biosynthesis of thymidine, which is essential for DNA synthesis. Figure 2-13 shows the FH_4 cycle involved in the synthesis of thymidine 5'-monophosphate (dTMP). Dihydrofolate reductase catalyzes the synthesis of FH_4, which is then converted to N^5,N^{10}-methylenetetrahydrofolate, which donates a methyl group in the conversion of deoxyuridine 5'-monophosphate (dUMP) to dTMP. This reaction also yields FH_2, the substrate of dihydrofolate reductase.

Methotrexate is a structural analog of folate (Fig. 2-14) and acts as a competitive inhibitor of dihydrofolate reductase. It binds to the enzyme about 1,000 times more tightly than FH_2. As a result, the supply of FH_4 is shut off and synthesis of dTMP is inhibited. The de novo synthesis of purine nucleotides is inhibited, because their synthesis also requires

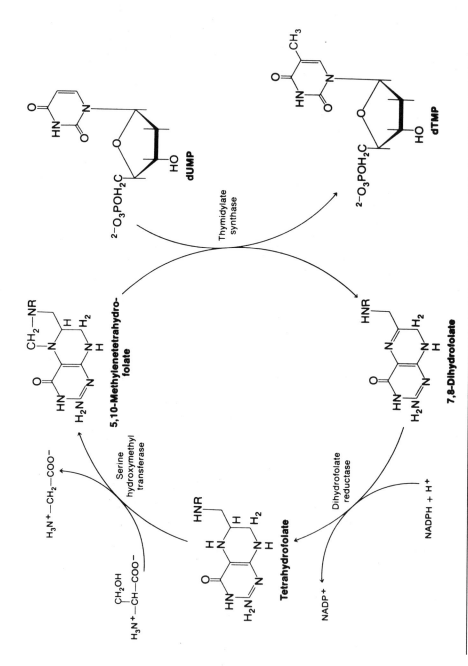

Fig. 2-13. The synthesis of thymidine monophosphate from deoxyuridine monophosphate. (From Zubay G: Biochemistry. 3rd Ed. Macmillan, New York, 1988, with permission.)

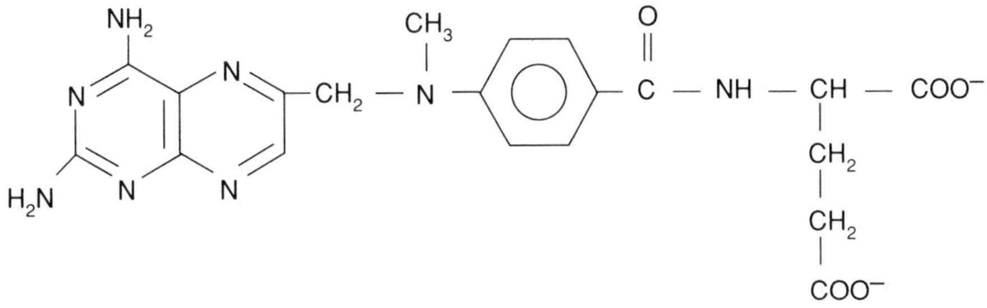

Methotrexate

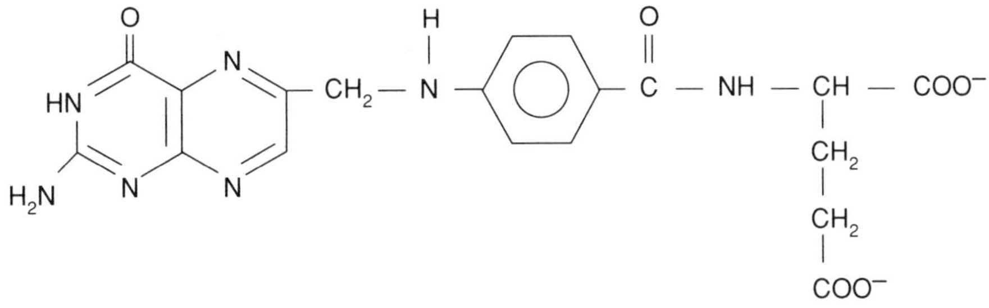

Folic Acid

Fig. 2-14. The structures of methotrexate and folic acid.

FH_4. The cell becomes depleted of nucleotides. The results are inhibition of DNA synthesis, decreased cell growth, and eventual cell death if the deprivation of nucleotides persists. Rapidly dividing cells—such as leukemia cells—succumb first because of their increased demand for DNA precursors.

SUGGESTED READINGS

1. Palmer T: Understanding Enzymes, 2nd Ed. Ellis Horwood Limited, Chichester, England, 1985
2. Cornish-Bowden A: Fundamentals of Enzyme Kinetics. Butterworth (Publishers), London, 1979
3. Engel PC: Enzyme Kinetics: The Steady State Approach, 2nd Ed. Chapman and Hall, New York, 1981
4. Christensen HN, Palmer GA: Enzyme Kinetics: A Learning Program for Students of the Biological and Medical Sciences. WB Saunders, Philadelphia, 1974
5. Fersht A: Enzyme Structure and Mechanism. Freeman, San Francisco, 1977

STUDY QUESTIONS

Directions: For each of the following multiple choice questions, choose the most appropriate answer.

1. Enzymes catalyze reactions by
   A. lowering the activation energy
   B. always using acid-base reactions
   C. shifting the equilibrium to a more favorable position
   D. increasing the energy content of the substrate

2. If NAD^+ readily associates with an enzyme before catalysis and dissociates after catalysis, it is behaving as which of the following?
   A. A cofactor
   B. A coenzyme
   C. A covalent modifier.
   D. More information is to needed to determine the role of NAD^+

3. A noncompetitive inhibitor will affect
   A. K_m
   B. V_{max}
   C. K_M and V_{max}
   D. the apparent K_M

4. Which one of the following statements is *not* true? Allosteric enzymes
   A. show sigmoidal kinetics
   B. are usually multisubunit structures
   C. always show cooperativity
   D. have at least two specific binding sites

5. When considering multisubstrate reaction mechanisms, the term *double displacement* more accurately describes the
   A. ordered sequential mechanism
   B. random sequential mechanism
   C. ping-pong mechanism
   D. none of the above

6. The number of moles of substrate converted to product per minute per mole of enzyme is the
   A. K_M
   B. V_{max}
   C. turnover number
   D. apparent K_M

Directions: Answer the following question(s) using the key outlined below:
- **(A)** if 1, 2, and 3 are correct
- **(B)** if 1 and 3 are correct
- **(C)** if 2 and 4 are correct
- **(D)** if only 4 is correct
- **(E)** if all four are correct

7. Enzymes use which of the following to lower the energy of activation?
   1. General acid-base catalysis
   2. Covalent catalysis
   3. Substrate strain
   4. Temperature variation

3

ENERGY METABOLISM: BIOENERGETICS

Biochemical Thermodynamics/Cellular Energetics/
Phosphate Bond Energy

The student should be able to:

1. Define and distinguish ΔG and $\Delta G^{\circ\prime}$ and describe the relationship between ΔG, $\Delta G^{\circ\prime}$, and the equilibrium constant.
2. Describe the difference between exergonic and endergonic reactions and their relation to ΔG and $\Delta G^{\circ\prime}$.
3. Describe the function of high-energy phosphate compounds, especially ATP, in cellular energetics, and the importance of coupled reactions in energy utilization.

Learning Objectives

Perspective: Bioenergetics

Humans derive the energy they need from food. Food is mostly organic, consisting primarily of carbon, hydrogen, oxygen, nitrogen, sulfur, and an assortment of metal ions. Humans obtain energy by oxidizing food molecules; the major products of complete oxidation are CO_2, H_2O, and energy. In general, *catabolic* or degradative processes produce energy, whereas *anabolic* or synthetic processes consume energy.

The oxidation of food molecules by animals is not an all-or-none phenomenon like a fire or an explosion; instead, it is a gradual, tightly controlled process in which energy is released in increments via many intermediates. The energy released is harvested in the form of chemical energy in certain metabolites, principally *adenosine triphosphate* (ATP).

(continued)

> *Bioenergetics* is the study of energy metabolism. In order to understand bioenergetics, one must be familiar with the principles of reaction thermodynamics. Chemical reactions either consume or release energy, depending on the difference in energy content of the reactants and the products. In this respect, energy is treated as an invisible reactant or product; like matter, it is conserved during the reaction.
> Central to reaction thermodyamics is the determination of the free energy change ΔG of reactions, which is that part of the energy produced or consumed during the reaction that is available to do work. Related to ΔG is the standard free energy change $\Delta G°$, a quantity that is a constant for each reaction.

BIOCHEMICAL THERMODYNAMICS

Consider the biochemical reaction

$$A \rightarrow B$$

In most cases of interest, A and B will have different energy contents. If the energy content of B is less than that of A, the reaction releases energy. Reactions that release energy are called *exergonic*, and proceed spontaneously (although not necessarily at a useful rate). On the other hand, if B contains more energy than A, the reaction will not proceed without input of energy. Reactions that require energy are called *endergonic*.

Not all of the energy produced or consumed in a reaction is available to perform useful work; some is always lost in the form of an increase in entropy (i.e., as heat). A major goal of biochemical thermodynamics is to determine how much of the energy produced by an exergonic reaction will be available for work or, alternatively, how much of the energy supplied to drive an endergonic reaction actually does work. This quantity—the amount of energy in a reaction that is available for doing work under conditions of constant temperature and pressure—is called the *free energy change* ΔG of the reaction. (Biochemical reactions essentially all take place under conditions of constant temperature and pressure.) The free energy change is given by the thermodynamic expression

$$\Delta G = \Delta G° + RT \ln \frac{[\text{product}]^n}{[\text{reactant}]^n} \quad (1)$$

where ΔG is the observed free energy difference between the reactants and the products, $\Delta G°$ is the standard free energy change, R is the gas constant, T is the temperature in degrees kelvin, [reactant] and [product] are the molar concentrations of each species, and n is the number of molecules of each species participating in the reaction.

The *standard free energy change* $\Delta G°$ is the free energy change ΔG when reactants and products are present at 1 M concentrations. It is a constant for any given reaction. $\Delta G°'$ is the same as $\Delta G°$ except that it is measured at pH 7. Free energy and standard free energy are most often expressed in units of kilocalories per mole (kcal/mol).

The absolute value of ΔG for a reaction (the amount of free energy released or consumed during the course of the reaction) depends on the starting concentrations of reactants and products. As the reaction proceeds toward equilibrium, ΔG approaches zero. When equilibrium is achieved, no more net reaction can occur, so ΔG equals zero.

Therefore, at equilibrium,

$$0 = \Delta G° + RT \ln K_{eq} \quad (2)$$

where the *equilibrium constant* K_{eq} is the mole ratio of products to reactants at equilibrium, measured under standard conditions. The above expression can be rewritten

$$\Delta G° = -RT \ln K_{eq} \quad (3)$$

so $\Delta G°$ can be calculated if the equilibrium constant is known. If K_{eq} is greater than 1, $\Delta G°$ is negative, and under standard conditions (with reactants and products both present at 1 M concentrations), the reaction is exergonic and will proceed spontaneously. If K_{eq} is less than 1, $\Delta G°$ is positive and under standard conditions the reaction will not proceed unless sufficient energy is supplied.

Under nonstandard conditions (for example, inside cells), a reaction will be exergonic and will proceed spontaneously if its ΔG is negative, and it will be endergonic and require energy if its ΔG is positive. Therefore, the spontaneity of biologic reactions is determined by ΔG and not by $\Delta G°$. It should be noted, however, that neither ΔG nor $\Delta G°$ provides any information about the *rate* of a reaction. Usually, ΔG for biologic reactions has the same sign as $\Delta G°$, and the two often do not differ greatly in value.

A very important property of ΔG is that the ΔG of a reaction does not depend on the path taken, but only on the concentrations of reactants and products and on $\Delta G°$. Therefore, the ΔGs (and $\Delta G°$s) of the reactions in a pathway are additive, and the spontaneity of the entire pathway may be predicted by the sum of the ΔGs of the individual reactions. Consider the following series of reactions:

$$A \xrightarrow{\Delta G_1} B \xrightarrow{\Delta G_2} C \xrightarrow{\Delta G_3} D \xrightarrow{\Delta G_4} E$$

The entire pathway can proceed spontaneously if the total ΔG is negative, even if one or more of the steps has a positive $\Delta G°$.

To sum up:

1. *If ΔG is negative*, the reaction will proceed spontaneously.
2. *If ΔG is positive*, the reaction will not proceed without input of sufficient energy.

3. The larger the absolute value of ΔG, the farther the reaction is from equilibrium, and the larger the amount of free energy that will be liberated or consumed as the reaction proceeds to equilibrium.

CELLULAR ENERGETICS

The energy used by cells is obtained by the oxidation of fuel molecules, primarily carbohydrates, lipids, and proteins (i.e., amino acids). Lipids release more energy per gram than carbohydrates or proteins because they are more highly reduced. Mammals store energy in the form of glycogen and fat, and release it gradually via oxidation to meet the metabolic needs of the organism (see Chapters 5 through 7).

Energy derived from oxidation of fuel molecules is primarily captured in the phosphoanhydride bonds of ATP. In some reactions (substrate level phosphorylations), ATP is synthesized directly (see Chapter 5). However, during oxidation of fuel molecules, most released energy is immediately captured as the reduced forms of the coenzymes *nicotinamide adenine dinucleotide* (NADH) and *flavin adenine dinucleotide* (FADH$_2$). These reduced coenzymes subsequently drive the synthesis of ATP by the oxidative phosphorylation system (Chapter 6). When carbohydrate (glucose 6-phosphate) is oxidized by an alternative mechanism (the pentose phosphate pathway; Chapter 5), the released energy is captured in the form of reduced *nicotinamide adenine dinucleotide phosphate* (NADPH). Cells use ATP to drive energetically unfavorable enzymic reactions. NADPH is used to supply reducing power for a variety of synthetic reactions.

HIGH ENERGY PHOSPHATE COMPOUNDS

High energy phosphate compounds are phosphate-containing substances that exhibit a large release of free energy when the phosphate group is hydrolyzed. Inside cells, ATP (Fig. 3-1) is the most abundant high energy phosphate compound. The hydrolysis of both its γ and β phosphate groups is associated with the release of a large quantity of free energy (large negative $\Delta G°$):

$$ATP \rightarrow ADP + P_i \qquad \Delta G° \approx -8.4 \text{ kcal/mol}$$

Energy is released both by the hydrolysis of ATP to *adenosine diphosphate* (ADP) and inorganic phosphate (P_i) and by the hydrolysis of ATP to *adenosine monophosphate* (AMP) and pyrophosphate (PP_i). AMP, however, is not a high energy compound. Energy is also released when a phosphate group is transferred from a high energy compound to a low energy compound. For example, in the first step of glycolysis, transfer of the γ phosphate of ATP to glucose yields ADP and glucose 6-phosphate.

A number of other high energy phosphate compounds are involved in energy-transfer reactions in the cell. Some have higher free energies of

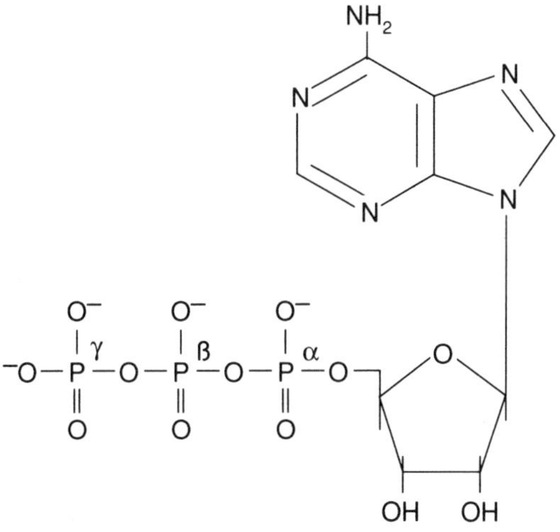

Fig. 3-1. The structure of ATP. The three phosphate groups are designated α, β, and γ.

hydrolysis than ATP, and can be used to regenerate ATP by transferring a phosphate group to ADP. For example, in glycolysis, ADP is regenerated to ATP by accepting phosphate groups from the high energy phosphate compounds phosphoenolpyruvate and 1,3-bisphosphoglycerate. The standard free energy of hydrolysis of the phosphoenolpyruvate high-energy phosphate bond is about -14.8 kcal/mole, obviously more than enough to regenerate ATP.

ATP can also be regenerated by phosphate transfer from *phosphocreatine* (creatine phosphate) (Chapter 19). Phosphocreatine is a compound designed for the short-term storage and intracellular transport of energy. It is abundant in muscle tissue, where it is used as a source of energy to meet the dramatic demands of muscle contraction. Creatine phosphate is synthesized by creatine kinase according to the reaction

$$\text{ATP} + \text{creatine} \rightarrow \text{ADP} + \text{creatine phosphate} + \text{H}^+$$

$$\Delta G^{\circ\prime} = -3 \text{ kcal/mol}$$

Other nucleoside triphosphates—guanosine triphosphate (GTP), cytosine triphosphate (CTP), and uridine triphosphate (UTP)—also participate in various specific energy transfers in the cell. A molecule of GDP is phosphorylated to GTP during the tricarboxylic acid cycle (Chapter 6), and the resulting GTP may be used to regenerate ATP:

$$\text{ADP} + \text{GTP} \rightleftharpoons \text{ATP} + \text{GDP}$$

This reaction is catalyzed by *nucleoside diphosphate kinase*, which catalyzes phosphate transfers between nucleoside diphosphates and nucleoside triphosphates. GTP is also used as an energy source during protein synthesis (Chapter 14).

Coupled Reactions

Exactly how are high-energy phosphate compounds used to drive endergonic processes in the cell? Remember that a pathway will proceed spontaneously if the sum of the ΔGs of its reactions is negative. Therefore, hydrolysis of ATP will pull an endergonic reaction if the two reactions are *coupled* by a common intermediate and the net ΔG is negative. For example, consider the reaction

$$\text{Amino acid} + \text{tRNA} \rightarrow \text{aminoacyl-tRNA}$$

This reaction is endergonic, and is driven by ATP hydrolysis. The energy is supplied via the following coupled reaction mechanism:

$$\text{ATP} + \text{amino acid} \rightarrow \text{aminoacyl-AMP} + \text{PP}_i$$

$$\Delta G° = -2 \text{ kcal/mole}$$

$$\text{Aminoacyl-AMP} + \text{tRNA} \rightarrow \text{aminoacyl-tRNA} + \text{AMP}$$

$$\Delta G° = -2.5 \text{ kcal/mole}$$

Net reaction:

$$\text{ATP} + \text{amino acid} + \text{tRNA} \rightarrow \text{aminoacyl-tRNA} + \text{AMP} + \text{PP}_i$$

$$\Delta G° = -4.5 \text{ kcal/mole}$$

In this reaction series, ATP is used to form an activated (high energy) intermediate, aminoacyl-AMP, which contains enough energy to react with tRNA and form the desired product. In coupled reaction mechanisms, energy is often transferred from ATP to a high-energy reaction intermediate.

SUGGESTED READING

1. Lehinger AL: Bioenergetics. 2nd ed. Benjamin, New York, 1972
2. Ernester L: New Comprehensive Biochemistry. Vol. 9, Bioenergetics. Elsevier, New York, 1985
3. Van Holde KE: Physical Biochemistry. 2nd ed. Prentice-Hall, Englewood Cliffs, NJ, 1985
4. Patton AR: Biochemical Energetics and Kinetics. WB Saunders, Philadelphia, 1965

STUDY QUESTIONS

Directions: For each of the following multiple choice questions, choose the most appropriate answer.

1. The free energy of a reaction
   A. is the energy available to do work
   B. is the heat energy
   C. is only associated with ATP hydrolysis during energy-coupled reactions
   D. cannot be accurately measured

2. When ΔG is negative, the reaction
   A. is endergonic
   B. cannot proceed spontaneously
   C. is exergonic
   D. cannot accomplish useful work

3. All of the following statements are true EXCEPT that
   A. ADP can be regenerated to ATP by phosphate transfer from phosphoenolpyrovate
   B. phosphocreatine is an important energy source for muscle tissues
   C. GTP is an important energy source for protein synthesis
   D. when ATP is used to drive a reaction, it is broken down to either ADP and P_i or AMP and PP_i

4. As the ratio of reactants to products proceeds toward equilibrium in a reaction, the ΔG of the reaction
   A. goes to zero
   B. changes from positive to negative or vice versa
   C. does not change
   D. becomes equal to $-RT \ln K_{eq}$

5. If a cell carries out a hypothetical reaction with a $\Delta G°$ of +40 kcal by coupling it to the hydrolysis of ATP to ADP, what is the minimum number of ATP molecules that must be hydrolyzed? ($\Delta G°$ may be used as an approximation of ΔG.)
   A. 5
   B. 4
   C. 50
   D. 42

Directions: Answer the following questions using the key outlined below:
- **(A)** if 1, 2, and 3 are correct
- **(B)** if 1 and 3 are correct
- **(C)** if 2 and 4 are correct
- **(D)** if only 4 is correct
- **(E)** if all four are correct

6. Reactions that have a K_{eq} greater than 1 are likely to
   1. have a positive ΔG
   2. be exergonic
   3. be endergonic
   4. have a negative $\Delta G°$

7. Consider the following pathway:

$$A \xrightarrow{-\Delta G°} B \xrightarrow{-\Delta G°} C \xrightarrow{+\Delta G°} D \xrightarrow{-\Delta G°} E$$

Which of the following predictions can be made:
   1. The pathway will proceed spontaneously
   2. The pathway will not proceed spontaneously
   3. The pathway has one spontaneous reaction
   4. The pathway has three spontaneous reactions

4

CARBOHYDRATE METABOLISM I

Monosaccharides, Oligosaccharides, and Polysaccharides/Glycogen Metabolism/ Clinical Correlation

Learning Objectives

The student should be able to:

1. Distinguish the structures of aldoses, ketoses, pyranoses, and furanoses; identify the anomeric carbon atom; and diagram glucose mutarotation.
2. Draw the structures of disaccharides on the basis of their systematic names.
3. Diagram the glycosidic linkages of glycogen.
4. Distinguish the metabolic roles of liver and muscle glycogen.
5. Describe the roles of glycogen synthase, the branching enzyme, glycogen phosphorylase, and the debranching enzyme in glycogen synthesis and degradation.
6. Describe both the allosteric regulation of glycogen phosphorylase and glycogen synthase by metabolites and coordinated regulation of glycogen metabolism in liver by the hormones glucagon, epinephrine, and insulin.
7. Distinguish the mechanisms of action of epinephrine binding to α- and β-adrenergic receptors.

Perspective: Carbohydrate Metabolism I

Carbohydrates are polyhydroxy aldehydes and ketones of the empirical formula $(CH_2O)_n$. Carbohydrates are classified as monosaccharides, oligosaccharides, and polysaccharides. Monosaccharides are the smallest carbohydrate units; they cannot be hydrolyzed to yield a carbohydrate of fewer carbon atoms. Oligosaccharides are polymers consisting of from two (in disaccharides) to about ten monosaccharides. Polysaccharides are high-molecular-weight polymers consisting of many monosaccharide units.

MONOSACCHARIDES, OLIGOSACCHARIDES, AND POLYSACCHARIDES

The monosaccharides and disaccharides are also called *sugars*; monosaccharides are sometimes called *simple sugars*. The most important monosaccharide in mammalian cells is the six-carbon sugar glucose. Glucose is an energy source for essentially all tissues. The five-carbon sugars ribose and deoxyribose also play very important roles as components of nucleotides and nucleic acids. Some monosaccharides contain additional functional groups besides the hydroxyl and aldehyde or ketone groups that are normally present; these sugars are called *derived monosaccharides*.

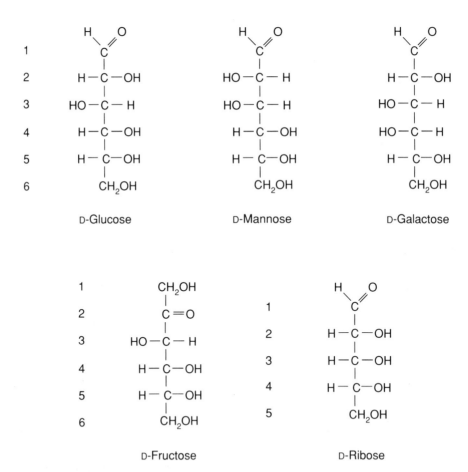

Fig. 4-1. Fisher projections of the aldohexoses D-glucose, D-mannose, and D-galactose, the ketohexose D-fructose, and the aldopentose D-ribose. The conventional numbering of the carbon atoms is indicated. The configuration of the sugar is determined by the configuration about the asymmetric carbon farthest from the carbonyl group. D-Glucose, D-mannose, and D-galactose are stereoisomers of each other; D-glucose and D-mannose are epimers.

Examples of free disaccharides are sucrose (common table sugar), maltose (formed by digestion of dietary starch), and lactose (the sugar of milk). The repeating units of many polysaccharides are disaccharides.

Some oligosaccharides occur primarily as glycoconjugates with proteins or lipids. Both glycoproteins and glycolipids are common components of membranes. The hydrophilic carbohydrate moieties are exposed on the exterior surface of the membrane, and often function in cell-cell recognition and in the recognition of cell surface receptors by some hormones. Several plasma proteins also contain carbohydrate units.

Polysaccharides function as structural molecules or as energy stores. In mammals, structural polysaccharides are usually complexed with protein and are called *acid mucopolysaccharides* or *proteoglycans*. Proteoglycans are found mainly in connective tissue. Members of this group of substances include hyaluronic acid, dermatan sulfate, chondroitin, chondroitin sulfate, keratan sulfate, heparin, and heparin sulfate. The polysaccharide *glycogen* is the energy storage polysaccharide of mammals. Glycogen is phosphorolyzed to glucose 1-phosphate, which can be oxidized to produce energy.

Biochemically Important Monosaccarides

Nomenclature

Aldehyde monosaccharides are called *aldoses*; ketone monosaccharides are called *ketoses*. Aldoses and ketoses are identified according to the number of carbon atoms they contain. Aldoses are called *aldotrioses, aldotetroses, aldopentoses*, etc., and the ketoses are called *ketotrioses, ketotetroses, ketopentoses*, etc. Alternatively, the aldoses can be called simply *trioses, tetroses, pentoses*, etc., and the ketoses *triuloses, tetuloses, pentuloses*, etc. The common sugars are generally referred to by their trivial names, such as glucose, fructose, and galactose. The two three-carbon sugars are *glyceraldehyde* (the aldose) and *dihydroxyacetone* (the ketose).

Structure of Monosaccharides

All the monosaccharides except dihydroxyacetone have at least one asymmetric carbon atom. (Any carbon with four different substituents is asymmetric, and can exist in D and L configurations.) For monosaccharides, the prefixes D and L designate the stereochemical configuration of the asymmetric carbon farthest from the carbonyl carbon.

When carbohydrates are drawn as *Fisher projections* (Fig. 4-1), each asymmetric carbon is in the D configuration if its hydroxyl group lies to the right of the carbon chain, and in the L configuration if its hydroxyl lies to the left of the carbon chain. The sugars D-glucose, D-mannose, and D-galactose shown in Figure 4-1 constitute a series of stereoisomers because they differ only in the configuration of hydroxyl groups about the asymmetric carbons. In addition, glucose and mannose are *epimers*: stereoisomers that differ at only one asymmetric carbon.

Since sugars are aldehydes or ketones, they participate in the chemical reactions typical of these compounds. The reaction of particular interest is the formation of hemiacetals by aldoses and of hemiketals by ketoses. When pentoses and hexoses are put in aqueous media, the carbonyl oxygen attacks the penultimate hydroxyl to spontaneously form a cyclic intramolecular hemiacetal or hemiketal (Fig. 4-2). For example, when D-glucose is placed in an aqueous medium, the hydroxyl on carbon 5 reacts with the carbonyl carbon to form a six-membered hemiacetal ring. This reaction causes the former carbonyl carbon (carbon 1 of aldoses) to become asymmetric. This carbon is called the *anomeric carbon*, and the ring sugar exists as α and β anomers. When the hemiacetal is illustrated as an extension of the Fisher open-chain pro-

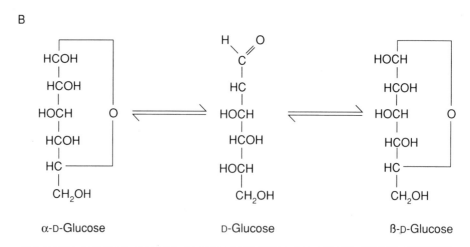

Fig. 4-2. Hemiacetal and hemiketal formation. **(A)** General reactions by which aldehydes and ketones react with alcohols to form hemiacetals and hemiketals. **(B)** In aqueous media the linear form of glucose forms the cyclic hemiacetals α-D-glucose and β-D-glucose. The α and β anomers are distinguished by the configuration about carbon 1, the anomeric carbon. The two anomers undergo mutarotation via the linear chain intermediate.

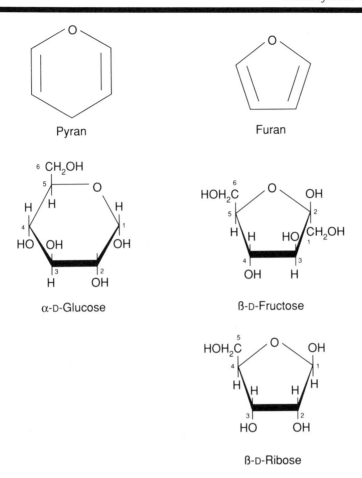

Fig. 4-3. The pyran and furan ring structures, and Haworth projections of some monosaccharides that form pyran- and furan-like rings. The carbon atoms are numbered. Six-membered carbohydrate rings are called pyranoses because they resemble pyran, and five-membered rings are called furanoses because they resemble furan. Aldohexoses cyclize to form pyranoses, whereas ketohexoses and aldopentoses cyclize to form furanoses.

jection (Fig. 4-2), it is considered to be in the α *configuration* (α-D-glucose) if the hydroxyl group on the anomeric carbon lies to the right of the carbon chain, and in the β configuration (β-D-glucose) if it lies to the left of the chain.

α-D-Glucose and β-D-glucose differ significantly in specific optical rotation. The optical rotation of a solution of pure α anomer is +112, whereas that of pure β anomer is +19. However, the optical rotation of a freshly prepared solution of either anomer gradually changes to an intermediate value of +53. This change, called *mutarotation*, is caused by interconversion of the α and β anomers to achieve an equilibrium mixture. The open-chain or linear species is present in solution to only a small extent, and is believed to be a necessary intermediate in the interconversion of the α and β anomers (Fig. 4-2).

The cyclic hemiacetals formed by aldohexoses are six-membered rings that resemble the pyran ring (Fig. 4-3). They are therefore called *pyranoses* or, more accurately, *aldopyranoses*. Aldopentoses, on the other hand, usually form five-membered rings that resemble the furan ring (Fig. 4-3), and are called *furanoses* or *aldofuranoses*. Ketohexoses such as D-fructose also form five-membered furan-like rings, called *ketofuranoses*.

Pyranose and furanose structures are better illustrated using *Haworth projections*, such as those in Figure 4-3, than by Fisher structures, but Haworth projections still do not accurately depict the spatial construction of the ring, which they show as flat. Structural studies have shown that pyranoses assume puckered "chair" or "boat" conformations like those of cyclohexane (Fig. 4-4), with the ring substituents projecting axially (parallel to the ring axis) and equatorially (roughly in the plane of the ring). In pyranose rings, the anomer that has more nonhydrogen substituents in equatorial positions is considered the more stable. Furanose rings also usually exist in a puckered conformation. Both pyranoses and furanoses undergo mutarotation.

Derived Monosaccharides

The most common derivatives of monosaccharides in mammalian cells are carboxylic acid derivatives of glucose (Fig. 4-5). Glucuronic acid

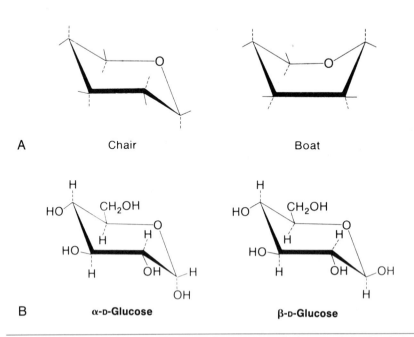

Fig. 4-4. **(A)** Ideal chair and boat configurations of glucose. **(B)** The chair conformations of α-D-glucose and β-D-glucose. β-D-glucose is the more stable anomer because it has more nonhydrogen substituents in equatorial orientation. Dashed lines indicate axial bonds; solid lines indicate equatorial bonds.

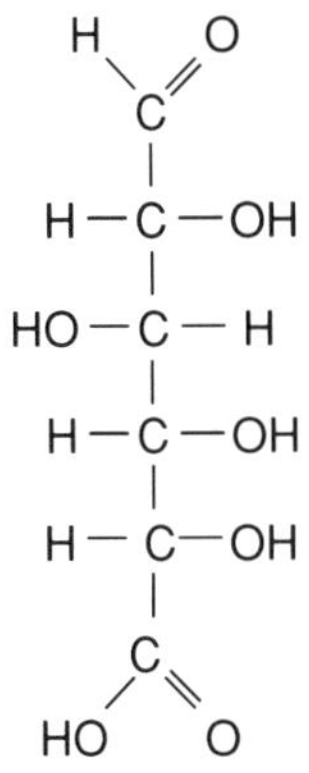

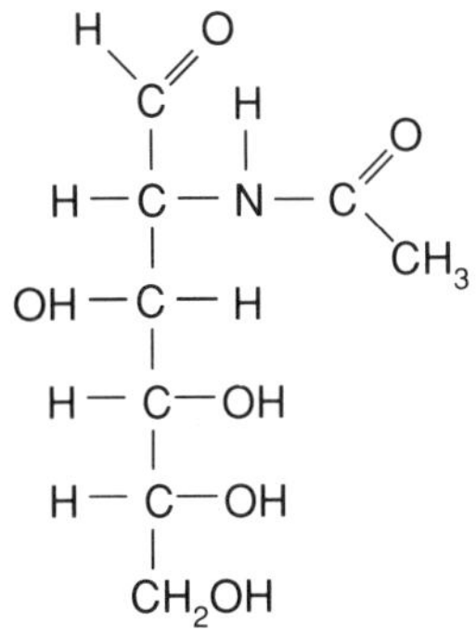

Fig. 4-5. Structures of D-glucuronic acid and D-glucosamine.

acetals are important in the elimination of bilirubin (Chapter 10). Phosphogluconic acid is a key intermediate in the pentose phosphate pathway (Chapter 5), and gluconic acid is also a component of connective tissue polysaccharides (Chapter 22).

Amino sugars (Fig. 4-5) constitute another important group of sugar derivatives. The most common amino sugars are the *N*-acetyl derivatives of D-glucosamine, D-galactosamine, and D-neuraminic acid; these are primarily components of glycoproteins and mucopolysaccharides.

Disaccharides

Disaccharides consist of two monosaccharides joined by a glycosidic bond (Fig. 4-6). They may be generated either de novo or by the breakdown of polysaccharides. To form a glycosidic bond, the hydroxyl attached to the anomeric carbon of one monosaccharide reacts with a different hydroxyl on another monosaccharide, with the elimination of water. Glycosidic bonds are characterized by the configuration of the anomeric carbon that is involved in the bond, and by the number of the involved carbon on the other sugar. Figure 4-6 shows the structure of glucose-$\alpha(1 \rightarrow 4)$-glucose (maltose). This systematic name indicates that maltose is a glucose dimer in which an anomeric carbon in the α configuration is linked to carbon 4 of the other glucose. Note that the anomeric carbon of the second sugar of disaccharides is free, and can act as a reductant for certain inorganic ions. For this reason, the monosaccharide with the free anomeric carbon is referred to as the *reducing end* of the disaccharide, whereas the monosaccharide that does not have a free anomeric carbon is called is the *nonreducing end*. Disaccharides undergo mutarotation at the reducing end.

The most common disaccharides produced by mammals are maltose, isomaltose, and lactose. Maltose and isomaltose are generated during digestion of starch in the gut. Lactose is directly synthesized by the mammary glands and is an important component of milk. Common

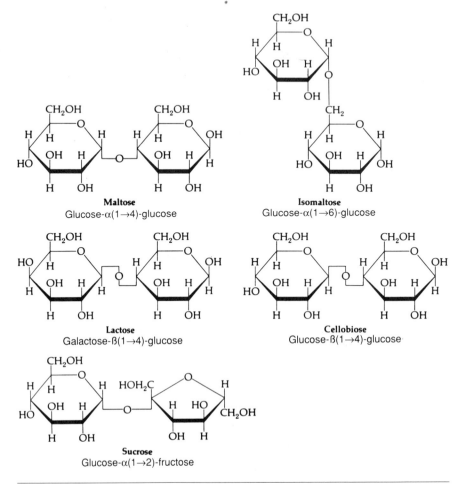

Fig. 4-6. Structures of some of the common disaccharides. (Modified from Smith EL, Hill RL, Lehman, IR, et al: Principles of Biochemistry: General Aspects. 7th Ed. McGraw-Hill, New York, 1983, with permission.)

table sugar, sucrose, is the disaccharide glucose-α(1 → 2)-fructose (Fig. 4-6).

Oligosaccharides and Polysaccharides

Glycoproteins

Oligosaccharides in mammalian cells are usually involved in protein or lipid conjugates. Glycoproteins usually consist of oligosaccharides or small polysaccharides covalently linked to the protein via *N*- or *O*-glycosidic bonds (Fig. 4-7). In *N*-linked glycoproteins, the linkage is usually formed between the anomeric carbon of an *N*-acetylglucosamine and the amide nitrogen of an asparagine residue. Asparagines involved in *N*-glycosidic bonds are usually found in the amino acid sequence

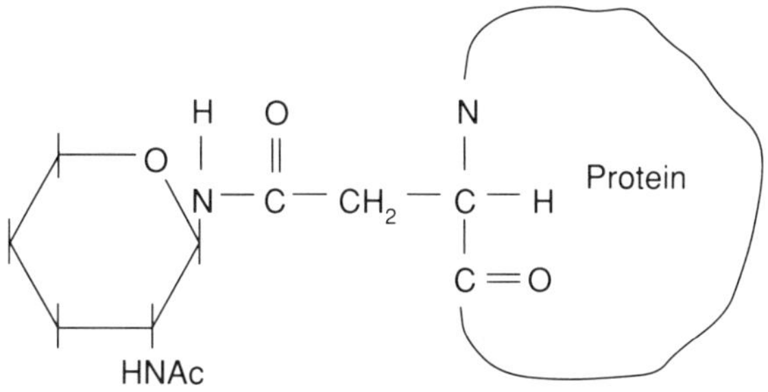

N-Linked carbohydrate

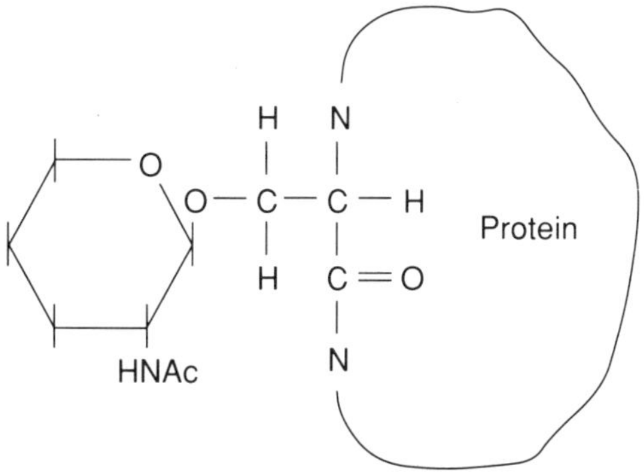

O-Linked carbohydrate

Fig. 4-7. The *N*- and *O*-glycosidic linkages of glycoproteins.

–Asn–X–Thr–, where X can be any amino acid. *O*-glycosidic linkages are formed between the anomeric carbon of *N*-acetylgalactosamine and the hydroxyl group of either serine or threonine. A glycoprotein may bear more than one carbohydrate moiety, but each oligosaccharide unit has only one site of attachment to the polypeptide chain, through either an *N*- or an *O*-glycosidic bond.

O- and *N*-linked oligosaccharides each have a distinctive *core oligosaccharide structure* to which other monosaccharides are usually bound. The core structures are shown in Figure 4-8. Other simple and derived monosaccharides are usually added to the core structure; the resulting oligosaccharide may be quite highly branched. *O*-linked oligosaccharides contain mostly sialic acid (*N*-acetyl neuraminic acid), L-fucose, *N*-acetylgalactosamine, and galactose. Sialic acid, galactose, galactosamine, and mannose are usually added to the core structure of *N*-linked oligosaccharides. Some *N*-linked structures, called *high-mannose structures*, consist of several mannose residues in addition to those present in the core structure. Other *N*-linked oligosaccharides that contain a variety of other sugars are called *complex structures*.

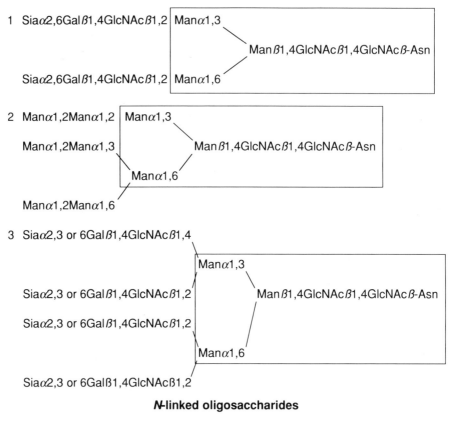

N-linked oligosaccharides

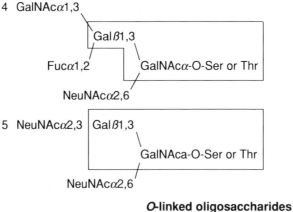

O-linked oligosaccharides

Fig. 4-8. A sampling of *N*- and *O*-linked carbohydrate structures found in glycoproteins. The two types of core structure are boxed. (Modified from Smith EL, Hill RL, Lehman, IR, et al: Principles of Biochemistry: General Aspects. 7th Ed. McGraw-Hill, New York, 1983, with permission.)

The carbohydrate moieties of glycoproteins are often drawn projecting from the protein like antennae. Consequently, carbohydrate structures with two branches are called *biantennary*, those with three are called *triantennary*, etc.

It is not unusual for different molecules of a single type of glycoprotein to carry glycoconjugates that differ only by the presence or absence of one or a few terminal sugar residues. This microheterogeneity is due to incomplete synthesis and/or degradation of the carbohydrate moieties.

Glycoproteins: Key Points
1. Carbohydrate structures are attached to the protein through *N*- or *O*-glycosidic linkages.
2. *N*-glycosidic-linked carbohydrate structures may be high-mannose or complex, depending on the variety of monosaccharides present.
3. Incomplete synthesis and/or degradation of carbohydrate structures contribute to glycoprotein microheterogeneity.

Glycolipids

Sugar residues are found in the sphingolipids, a class of complex membrane lipids, and in the phospholipid phosphatidyl inositol (Chapter 8). Glycosphingolipids may contain one or several sugar residues. The most common sugar components are glucose, galactose, and *N*-acetyl galactosamine. Members of one subclass of glycosphingolipids, the gangliosides, also contain varying amounts of *N*-acetylneuraminic acid. Sphingolipids are especially abundant in the central nervous system.

Polysaccharides

Large polysaccharides have two functions in living systems: as structural molecules and as energy stores. In mammals, two distinct types of polysaccharide, proteoglycans and glycogen, respectively perform these two functions.

Proteoglycans

Proteoglycans are very large protein-polysaccharide conjugates that form the ground substance of connective tissue. Unlike glycoproteins, polysaccharide is the dominant element of proteoglycans: carbohydrate constitutes 5 to 10 percent of glycoproteins but about 95 percent of proteoglycans. The polysaccharides of proteoglycans are high-molecular-weight polyanionic chains called *glycosaminoglycans*. The repeating units of glycosaminoglycans are disaccharides. One sugar of the disaccharide is always an amino sugar—either glucosamine or galactosamine—and the other is usually a uronic acid residue. The sugar residues may also contain an esterified sulfate ligand. The link between the glycosaminoglycan and protein components of the proteoglycan is not always covalent. The polyanionic glycosaminoglycan chains bind a great deal of water, which helps to give the connective tissue ground substance its viscoelastic properties. The structure and function of proteoglycans is discussed in more detail in Chapter 22.

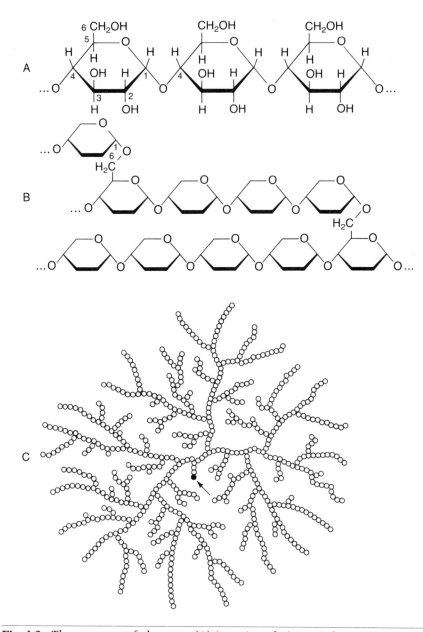

Fig. 4-9. The structure of glycogen. **(A)** A section of α(1 → 4) chain. **(B)** Glycogen chain with α(1 → 6) branches. **(C)** Sketch of a glycogen molecule: The arrow points to the nonreducing end. (From McGilvery RW: Biochemistry: A Functional Approach. 3rd Ed. WB Saunders, Philadelphia, 1983, with permission.)

Glycogen

Glycogen is the energy storage polysaccharide of mammalian cells. It consists entirely of glucose residues. The glucose residues are linked by $\alpha(1 \to 4)$ glycosidic bonds into chains, and the chains branch via $\alpha(1 \to 6)$ linkages which initiate new $\alpha(1 \to 4)$ chains (Fig. 4-9). Intact glycogen molecules are very highly branched, with many nonreducing ends and presumably only one reducing end. The branch points are frequent (about every fourth residue) in the core of the molecule, and less frequent toward the periphery. Glycogen molecules may reach molecular weights into the millions. Glucose residues are added to and released from the nonreducing ends during times of energy storage and utilization, respectively. The highly branched structure of glycogen means that glucose can be added or removed significantly faster than from a linear chain of equal size. Glycogen molecules are stored as aggreagates called glycogen granules.

Most tissues contain some glycogen, but the great bulk of the body's glycogen is stored in the liver and muscles; over half the body's glycogen is in the muscles. These two pools of glycogen have different functions. Muscle cells use their glycogen locally, to meet energy demands of muscle contraction. Liver glycogen, on the other hand, is used to maintain a constant level of glucose in the blood. Blood glucose nourishes other tissues. After a carbohydrate meal, the liver stores glucose as glycogen. When blood glucose falls, the liver degrades glycogen and releases free glucose into the blood. The liver glycogen pool therefore fluctuates much more than the muscle glycogen pool; muscle glycogen is significantly depleted only when muscle works hard. Muscle cells are unable to release glucose from glycogen into the blood.

Perspective: Glycogen Metabolism

Appropriate blood glucose levels are the result of the finely tuned control of glycogen synthesis and degradation in the liver. The processes of glycogen synthesis and degradation are exquisitely regulated by allosteric effectors and by cascades of enzyme phosphorylations and dephosphorylations initiated by hormonal and nervous stimulation. These mechanisms coordinate glycogen synthesis and degradation so that activation of one process switches off the other. The actual enzymic reactions involved in glycogen synthesis and degradation are the same for both muscle and liver, but their regulatory mechanisms differ in some aspects.

GLYCOGEN METABOLISM

Glycogen Degradation

The process of glycogen degradation is also called *glycogenolysis* or *glycogen phosphorolysis* because inorganic phosphate is used in releasing glucose residues. Glycogen is degraded by two enzymes, *glycogen phos-*

phorylase and the *debranching enzyme* (Fig. 4-10). Glycogen phosphorylase catalyzes the following reaction:

$$(\text{Glucose})_n + P_i \rightleftharpoons (\text{glucose})_{n-1} + \text{glucose 1-phosphate}$$

Glycogen phosphorylase is specific for $\alpha(1 \to 4)$ glycosidic linkages. It degrades $\alpha(1 \to 4)$ chains down to a limit of four residues from a branch point. At that point the debranching enzyme takes three of the four residues of the branch and transfers them to the end of another $\alpha(1 \to 4)$ chain (Fig. 4-10). The debranching enzyme then cleaves the $\alpha(1 \to 6)$ glycosidic linkage to yield one free glucose and an unbranched chain.

Glucose 1-phosphate produced from glycogenolysis is converted to glucose 6-phosphate by *phosphoglucomutase*. In the liver, the resulting glucose 6-phosphate is mostly hydrolyzed by *glucose 6-phosphatase* to free glucose and inorganic phosphate:

$$\text{Glucose 6-phosphate} \to \text{glucose} + P_i$$

This free glucose diffuses into the blood. In muscle and other tissues, glucose 6-phosphate is subsequently oxidized to yield energy. Some glucose 6-phosphate produced in the liver is also oxidized for energy production.

Glycogen Synthesis

The synthesis of glycogen from glucose is not the exact reverse of glycogen degradation. In the first step, glucose is phosphorylated:

$$\text{Glucose} + \text{ATP} \to \text{glucose 6-phosphate} + \text{ADP}$$

This reaction is catalyzed mainly by *glucokinase* in the liver and by *hexokinase* in other tissues (see Chapter 5 for a discussion of these two enzymes). Glucose 6-phosphate is then converted to glucose 1-phosphate by phosphoglucomutase, the same enzyme that catalyzes this conversion in glycogen degradation:

$$\text{Glucose 6-phosphate} \rightleftharpoons \text{glucose 1-phosphate}$$

In the next step, catalyzed by *glucose 1-phosphate uridylyl transferase*, glucose 1-phosphate reacts with uridine triphosphate (UTP) to yield uridine diphosphate-glucose (UDP-glucose):

$$\text{Glucose 1-phosphate} + \text{UTP} \rightleftharpoons \text{UDP-glucose} + PP_i$$

UDP-glucose is an activated form of glucose that is commonly used for

Fig. 4-10. Glycogen degradation. Glycogen phosphorylase degrades linear chains to four residues from a branch point. The debranching enzyme then transfers three of the remaining four residues to the end of the other branch, and then cleaves the $\alpha(1 \to 6)$ bond to release the remaining residue as free glucose. Glycogen phosphorylase can now continue to degrade the straight chain.

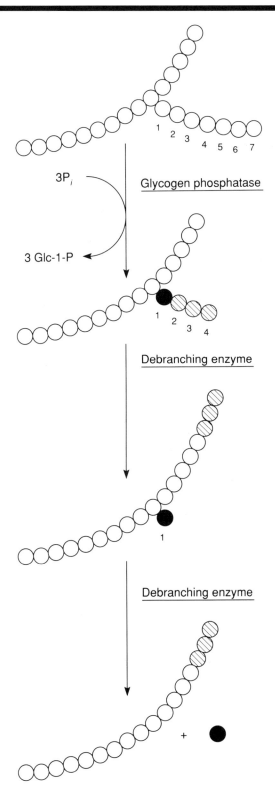

the transfer of a glycosyl residue. This reversible reaction is driven in the direction of UDP-glucose synthesis by the rapid hydrolysis of pyrophosphate by *pyrophosphatase*:

$$PP_i + H_2O \rightarrow 2P_i$$

Glycogen synthase catalyzes the transfer of the glucose to glycogen:

$$\text{Glycogen + UDP-glucose} \rightarrow \text{glycogen-glucose + UDP}$$

Glycogen synthase only synthesizes $\alpha(1-4)$ linkages; $\alpha(1 \rightarrow 6)$ linkages are made by the *branching enzyme*. This enzyme cleaves several residues from a linear $\alpha(1 \rightarrow 4)$ chain and transfers this short segment to an existing $\alpha(1 \rightarrow 4)$ chain that is at least 10 residues long via an $\alpha(1 \rightarrow 6)$ linkage, creating a branch (Fig. 4-11). The new branch is not less than four glucose residues from an adjacent branch point. The combined action of glycogen synthase and the branching enzyme results in the large, highly branched glycogen structure.

UDP is reconverted to UTP, with the hydrolysis of one ATP:

$$UDP + ATP \rightleftharpoons UTP + ADP$$

Thus, glycogen synthesis expends two ATPs per glucose residue.

It should be noted that glycogen synthesis requires a glycogen primer. This need is usually met by pre-existing glycogen molecules or by glucose-protein conjugates. Free glucose and small oligosaccharides do not serve as primers.

Regulation of Glycogen Degradation

Glycogen phosphorylase is the regulated enzyme of glycogen degradation. The enzyme is regulated both by allosteric effectors and by a phosphorylation/dephosphorylation interconversion, which is under hormonal control.

Glycogen phosphorylase exists in two forms: an active, phosphorylated form, *phosphorylase a*, and an unphosphorylated form, *phosphorylase b*. AMP stimulates the unphosphorylated form, whereas ATP and glucose inhibit the phosphorylated form. At cellular AMP and ATP levels, the unphosphorylated enzyme is usually the less active form. Both muscle and liver glycogen phosphorylases are under hormonal control. The liver enzyme is stimulated by glucagon and epinephrine, whereas the muscle enzyme is stimulated primarily by norepinephrine. Both epinephrine and norepinephrine have similar effects on glycogen phosphorylase. In liver cells, epinephrine is the primary catecholamine effector, whereas in muscle cells, norepinephrine is the primary catecholamine effector.

Hormonal Regulation of Glycogen Degradation in the Liver

Liver cells contain both glucagon and catecholamine receptors. In fact, liver cells carry two different types of catecholamine receptors: α- and

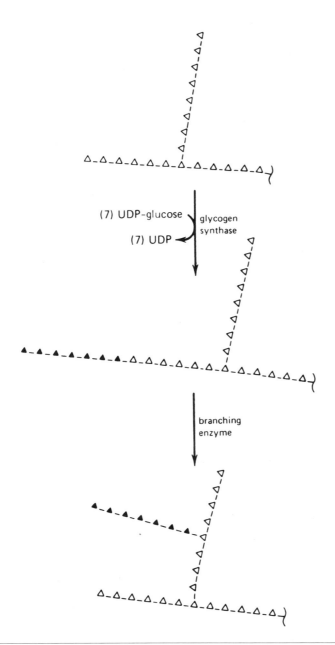

Fig. 4-11. Glycogen synthesis. Glycogen synthase makes α(1 → 4) linear chains. The branching enzyme cleaves off a short terminal segment of an α(1 → 4) chain that is at least 10 residues long, and transfers it to another segment of linear chain via an α(1 → 6) linkage. (From Devlin TM: Textbook of Biochemistry. Wiley, New York, NY 1986.)

β-adrenergic receptors. Binding of glucagon to the glucagon receptor and binding of epinephrine to the β-adrenergic receptor both trigger an identical series of intracellular events mediated by *cyclic adenosine monophosphate* (cAMP), which results in the stimulation of glycogen phosphorylase activity (Fig. 4-12). The events in this enzyme cascade are as follows:

1. Glucagon or epinephrine binds to the membrane receptor.
2. Binding of the hormone activates membrane-bound *adenylate cyclase*, which catalyzes the conversion of ATP to cAMP. Intracellular levels of cAMP rise.
3. cAMP activates *protein kinase A*. Protein kinase A consists of catalytic and regulatory subunits. cAMP binds to the regulatory subunit, causing it to dissociate from the catalytic subunit. Dissociation of the regulatory subunit activates the catalytic subunit.
4. Activated protein kinase A phosphorylates phosphorylase kinase *b*, converting it to the active *a* form.
5. Phosphorylase kinase *a* phosphorylates glycogen phosphorylase *b* (the less active form of the enzyme), converting it to the more active glycogen phosphorylase *a*.
6. Phosphorylase *a* then catalyzes the degradation of glycogen.

Glycogen degradation is inhibited by the inactivation of phosphorylase *a* and phosphorylase kinase by *phosphoprotein phosphatase*. Under the conditions that lead to glycogen degradation, phosphoprotein phosphatase is inhibited as follows by a protein called *inhibitor 1*:

7. The protein kinase activated by cAMP also phosphorylates inhibitor 1, converting it to the active form.
8. Active inhibitor 1 prevents the activation of phosphoprotein phosphatase, in turn preventing the inhibition or shutdown of glycogen degradation.

The binding of epinephrine to its α-adrenergic receptors on the plasma membrane results in the production of intracellular inositol triphosphate and diacylglycerol from membrane-associated phosphatidyl inositol 4,5-bisphosphate in a reaction apparently catalyzed by phospholipase C. Inositol triphosphate stimulates the release of Ca^{2+} from the endoplasmic reticulum. The increased cellular $[Ca^{2+}]$ then functions to activate phosphorylase kinase. The diacylglycerol produced is believed to be active in inhibiting glycogen synthase.

Hormonal Regulation of Glycogen Degradation in Muscle

Muscle cells do not have glucagon receptors. Therefore, glucagon does not influence glycogen metabolism in muscle. Muscle cells do have β-adrenergic receptors, and the effect of norepinephrine on glycogen degradation in the muscle is similar to the effect of epinephrine in the liver.

Calcium released as a result of nervous excitation also stimulates glycogen degradation in muscle tissues by activating phosphorylase kinase.

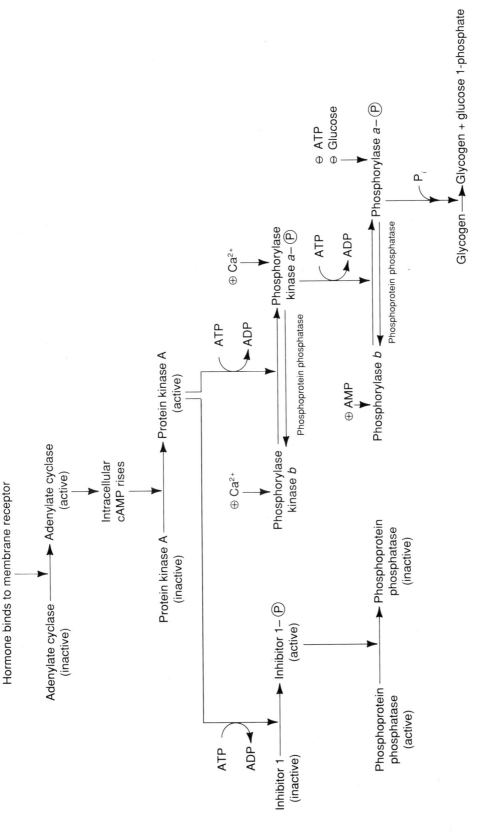

Fig. 4-12. Regulation of glycogen degradation. Phosphorylase kinase a and phosphorylase are inactivated by phosphoprotein phosphatase. However, phosphoprotein phosphatase is inhibited by inhibitor 1. Therefore, glycogen synthesis remains stimulated. Positive and negative allosteric effectors of some enzymes are shown.

The Effect of Insulin

Insulin inhibits glycogen degradation in both muscle and liver cells. The mechanism of insulin action is not well understood, but it is likely that insulin also exerts its intracellular effects through second-messenger molecules.

Regulation of Glycogen Synthesis

The regulated enzyme of glycogen synthesis is glycogen synthase. It exists in two forms, a phosphorylated form, glycogen synthase D, and an unphosphorylated form, glycogen synthase I. Glycogen synthase D requires glucose 6-phosphate as an allosteric activator (hence D for *dependent*), whereas glycogen synthase I does not (I for *independent*). At normal intracellular glucose 6-phosphate levels, glycogen synthase I is the more active form. Glycogen synthase I can be phosphorylated (deactivated) by the cAMP-dependent protein kinase, as well as by several other protein kinases that are also under the control of hormonal second messengers.

Glycogen synthase D is dephosphorylated (activated) to glycogen synthase I by the same phosphoprotein phosphatase that dephosphorylates and thus inactivates glycogen phosphorylase and phosphorylase kinase. As discussed, phosphoprotein phosphatase is inhibited by inhibitor 1, which is activated by the cAMP-dependent protein kinase that also deactivates glycogen synthase and triggers the activation of glycogen phosphorylase. Thus, a rise in intracellular cAMP simultaneously stimulates glycogen degradation and inhibits glycogen synthesis by inhibiting conversion of glycogen synthase D to the active I form.

Glycogen synthesis is also stimulated by the hormone *insulin*, by a mechanism that has not been clarified but that presumably involves a second messenger. Insulin also stimulates the transport of glucose into muscle cells.

Regulation of Glycogen Synthesis: Key Points

1. Glycogen synthase exists in two forms: an unphosphorylated active form, glycogen synthase I, and a phosphorylated inactive form, glycogen synthase D.
2. cAMP-dependent protein kinase and other protein kinases convert glycogen synthase I to glycogen synthase D, thereby inactivating glycogen synthase.
3. Glycogen synthase D may be converted to glycogen synthase I by phosphoprotein phosphatase.
4. However, phosphoprotein phosphatase is inhibited by the active form of inhibitor 1.
5. Inactive inhibitor 1 is converted to active inhibitor 1 by cAMP-dependent protein kinase.
6. Therefore, cAMP-dependent protein kinases both inactivate glycogen synthase and prevent its reactivation.
7. Insulin stimulates glycogen synthesis.

Coordinated Regulation of Glycogen Metabolism

The processes of glycogen synthesis and degradation significantly contribute to the efficient use of glucose as energy. This is achieved because

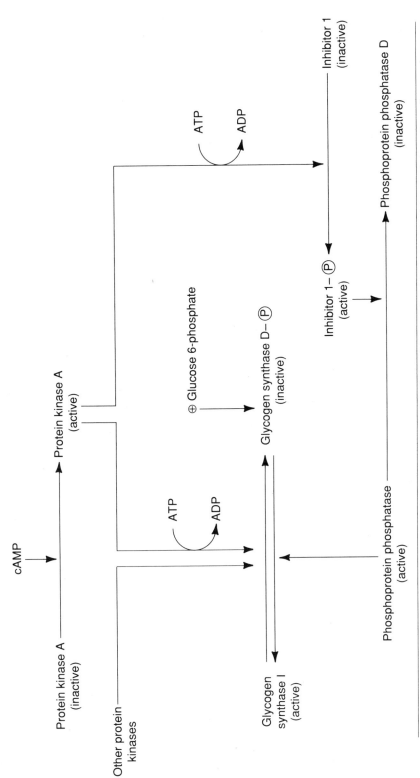

Fig. 4-13. Inhibition of glycogen synthesis. A number of cellular kinases besides protein kinase A can inactivate glycogen synthase. These include calmodulin-dependent protein kinase and protein kinase C. Positive and negative allosteric effectors of some enzymes are shown.

glycogen synthesis and degradation occur maximally under different cellular conditions. When the cAMP-dependent protein kinase is activated, it results in the phosphorylation (activation) of glycogen phosphorylase and in the phosphorylation (inactivation) of glycogen synthase. Therefore, when glycogen degradation is stimulated, glycogen synthesis is inhibited. Conversely, phosphoprotein phosphatase dephosphorylates (activates) glycogen synthase and dephosphorylates (inactivates) glycogen phosphorylase. Therefore, when glycogen synthesis is activated, glycogen degradation is inhibited. This type of reciprocal regulation is called *coordinate control*.

Effect of Hormones on Glycogen Metabolism: Overview

Glucagon is released from the pancreas in response to low blood sugar. The role of this hormone is to maintain adequate blood glucose during times of fasting or starvation. The adrenal medulla, on the other hand, releases epinephrine in response to impending stress or danger. Epinephrine stimulates release of glucagon from the pancreas and interacts directly with muscle and liver cells. High blood glucose levels cause the release of insulin from the β cells of the pancreas. Binding of insulin to cell membranes enhances glycogen synthesis, and also stimulates the uptake of glucose by adipose and muscle cells.

CLINICAL CORRELATION: TYPE III GLYCOGEN STORAGE DISEASE

Type III glycogen storage disease, also known as limit dextrinosis or Cori's disease, is due to a genetic deficiency of the debranching enzyme. Individuals with this disease cannot degrade glycogen beyond the first encountered α(1 → 6) branch points. They usually show an enlarged liver and growth retardation; the spleen may also be moderately enlarged. Partially degraded glycogen structures (limit dextrins) accumulate in muscle, liver, and sometimes the heart. The heart may also be moderately enlarged. Severe hypoglycemia and convulsions occur in some patients, and muscle wasting and weakness in others, indicating an inability to meet tissue glucose needs.

The treatment strategies for type III glycogen storage disease are aimed at increasing the supply of free glucose available to the tissues. Current recommendations are frequent meals in order to maintain a more constant influx of dietary glucose, and a high-protein diet, which allows enhanced conversion of amino acids to glucose via gluconeogenesis. In some cases, surgical diversion of the portal circulation to the tissues before it reaches the liver has also been beneficial. Portal diversion gives the tissues first claim on dietary glucose before the liver sequesters it.

SUGGESTED READING

1. Sharon N: Complex Carbohydrates: Their Chemistry, Biosynthesis, and Function. Addison-Wesley, Reading, MA, 1975
2. Hers HG: The control of glycogen metabolism in the liver. Ann Rev Biochem 45:167, 1976
3. Aspinall GO: Polysaccharides. Pergamon, New York, 1970

STUDY QUESTIONS

Directions: Match the following terms with the most appropriate structure shown below.

1. Ketofuranose

2. Aldofuranose

3. β-D-Glucose

A
$$\text{HOH}_2\text{C} \quad \text{O} \quad \text{OH}$$
H H HO CH₂OH
OH H

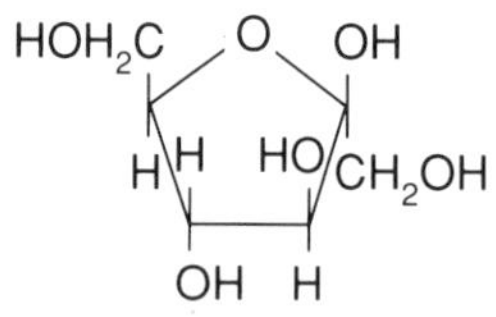

B
CH₂OH
H O H
H
OH H
HO OH
H OH

C
HOH₂C O OH
H H H H
OH OH

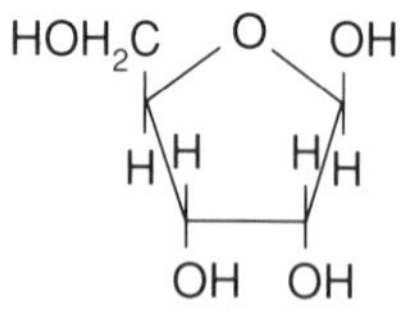

D
CH₂OH
H O OH
H
OH H
HO H
H OH

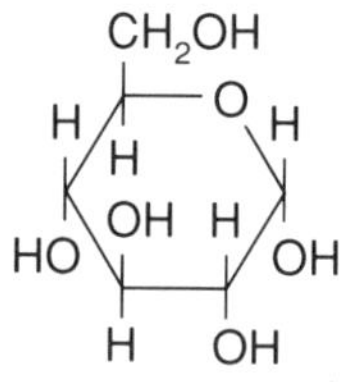

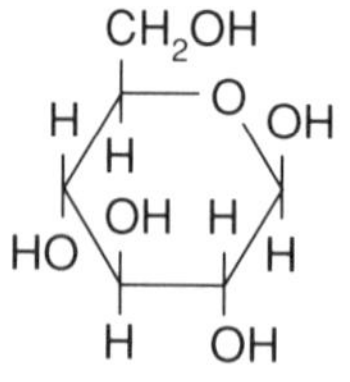

Directions: Answer the following questions using the key outlined below:
- **(A)** if 1, 2, 3 are correct
- **(B)** if 1, 3 are correct
- **(C)** if 2, 4 are correct
- **(D)** if 4 only is correct
- **(E)** if all four are correct

4. All glycosaminoglycans
   1. have disaccharide repeat units
   2. consist of covalently linked protein and carbohydrate components
   3. are polyanions
   4. contain sulfate esters

5. Glycogen degradation is stimulated by the active forms of the enzymes
   1. phosophorylase kinase
   2. inhibitor 1
   3. protein kinase
   4. phosphoprotein phosphatase

6. α-Adrenergic activation of glycogen degradation involves
   1. cAMP
   2. Ca^{2+}
   3. protein kinase A
   4. inositol triphosphate

Directions: For each of the following multiple choice questions, choose the most appropriate answer.

7. The primary role of muscle glycogen is to supply glucose to
   A. meet the energy needs of the liver
   B. supply the brain with glucose during starvation
   C. meet the sudden energy needs associated with muscle contraction
   D. function as a precursor in the synthesis of muscle proteoglycans

8. Which *one* of the following statements *is* true?
   A. Glycogen contains α(1→4) and β(1→6) glycosidic bonds
   B. Glycogen has a high content of hexosamine
   C. Epinephrine functions to maintain adequate blood levels of glucose whenever blood glucose falls
   D. The same phosphorylation process that stimulates glycogen degradation also inhibits glycogen synthesis

9. All of the following statements are true EXCEPT that
    A. insulin triggers glycogen synthesis
    B. insulin and glucagon function in a coordinated manner to stimulate glycogen degradation
    C. glucagon stimulates glycogen degradation
    D. insulin enhances glucose uptake by muscle cells

10. All of the following statements are true EXCEPT that
    A. liver cells contain glucagon and epinephrine receptors
    B. muscle cells contain α and β adrenergic receptors
    C. muscle cells contain norepinephrine receptors
    D. insulin stimulates glucose uptake into muscle cells.

5

CARBOHYDRATE METABOLISM II

Digestion, Absorption, and Cellular Uptake of Dietary Carbohydrates/Glycolysis/Gluconeogenesis/Metabolism of Pyruvate/The Pentose Phosphate Pathway/ Clinical Correlations

Learning Objectives

The student should be able to:

1. List the common digestible polysaccharides of the human diet, describe the role of polysaccharidases, oligosaccharidases, and disaccharidases in the digestion and absorption of dietary carbohydrate, and describe the usual immediate fate of glucose derived from dietary polysaccharides.
2. Describe the differing roles of glycolysis in muscle, liver, brain, adipose tissue, and erythrocytes.
3. Give the gross and net energy yield of glycolysis.
4. List the glycolytic regulatory enzymes and their corresponding effectors.
5. Distinguish the properties and roles of hexokinase and glucokinase.
6. Describe the physiologic role of gluconeogenesis, list the reactions of gluconeogenesis that are not part of glycolysis, and describe the differences in gluconeogenesis depending on whether lactate or pyruvate is the substrate.
7. Describe the function and operation of the mitochondrial NADH shuttle mechanisms.
8. Describe the coordinated regulation of glycolysis and gluconeogenesis.
9. Diagram the pentose phosphate pathway and identify its cellular functions.

Perspective: Digestion, Absorption, and Cellular Uptake of Carbohydrates

Carbohydrate obtained from the diet is an important energy source. Dietary carbohydrate usually consists mostly of polysaccharides, with lesser amounts of disaccharides and monosaccharides. Only monosaccharides can be absorbed into the blood and taken up by tissues; therefore, dietary polysaccharides, oligosaccharides, and disaccharides must be digested to monosaccharides in the intestine. The combined processes of digestion, absorption, and cellular uptake deliver carbohydrate to the cellular pathways where it is used.

DIGESTION, ABSORPTION, AND CELLULAR UPTAKE OF CARBOHYDRATE

Humans can digest only polysaccharides consisting of $\alpha(1 \rightarrow 4)$ glycosidic linkages or $\alpha(1 \rightarrow 4)$ linkages with $\alpha(1 \rightarrow 6)$ branch points—in practice, starch and glycogen. Specific disaccharides and oligosaccharides can be digested if the intestinal lumen contains disaccharidases or oligosaccharidases (glycosidases) capable of hydrolyzing them. This restriction applies to the oligosaccharide and disaccharide products of polysaccharide breakdown as well as to free dietary oligosaccharides and disaccharides.

Nondigestible carbohydrate is acted upon to some extent by intestinal bacteria, but mostly passes unaltered through the intestine and is excreted in the feces.

Starch and Glycogen

Starch and glycogen contain only glucose. They are the energy-storage polysaccharides of plants and animals, respectively. There are two types of starch: *amylose*, which consists of long chains of glucose units in $\alpha(1 \rightarrow 4)$ glycosidic linkage, and *amylopectin*, which consists of $\alpha(1 \rightarrow 4)$ glycosidic chains with periodic $\alpha(1 \rightarrow 6)$ branch points. Amylopectin is identical to glycogen except that glycogen is more densely branched.

Starch and glycogen are digested mainly in the small intestine by pancreatic α-*amylase*. (Saliva also contains α-amylase, but this enzyme accomplishes little starch digestion.) Amylose is digested to maltose and maltotriose, whereas amylopectin and starch digestion yields maltose, maltotriose, and α-*limit dextrins* consisting of five or six glucose residues centered around an $\alpha(1 \rightarrow 6)$ branch point (Fig. 5-1).

Oligosaccharides and Disaccharides

The major dietary disaccharides are lactose from milk and sucrose (table sugar). Along with the disaccharide and oligosaccharide products of starch and glycogen digestion, these compounds are hydrolyzed to monosaccharides by *disaccharidases* and *oligosaccharidases* associated with the luminal membrane of intestinal mucosal cells. The resulting monosaccharides are rapidly transported by carrier-mediated transport systems into the mucosal cells and subsequently into the blood.

Monosaccharides

Dietary monosaccharide consists mainly of fructose and glucose obtained from fruits and honey. These monosaccharides are absorbed directly from the intestinal lumen by specific transport systems.

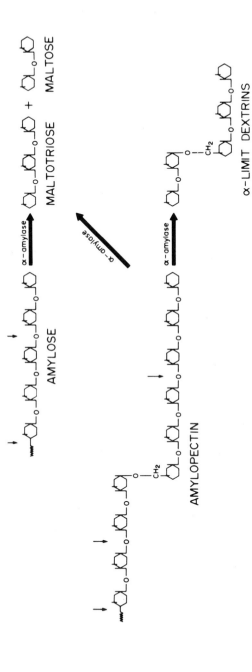

Fig. 5-1. Degradation of amylose and amylopectin. Amylose and amylopectin are degraded in the intestine to yield a mixture of maltotriose, maltose, and the α-limit dextrins that contain the α(1 → 6) branch points of amylopectin. Glycogen degradation is similar to that of amylopectin. (From Smith EL, Hill RL, Lehman IR, et al: Principles of Biochemistry. Mammalian Biochemistry. 7th Ed. McGraw Hill, New York, 1983, with permission.)

Tissue Uptake of Dietary Carbohydrate

Since starch and glycogen are glucose homopolymers, digestion of a carbohydrate meal introduces a bolus of glucose into the portal circulation. This glucose is usually taken up immediately by the liver, where it is stored as glycogen and gradually returned to the blood for distribution to other tissues.

Perspective: Glycolysis

Glycolysis is the pathway by which glucose is degraded to yield either lactate or pyruvate (depending on the availability of oxygen) and energy in the form of ATP.

1. Glycolysis occurs in the cytoplasm of nearly all cells. It does not require oxygen; however, in the presence of low oxygen tension, the primary product of glycolysis is lactate.
2. Different tissues have different metabolic capabilities and energy needs. Therefore, the rate of glycolysis varies with the tissue and the metabolic conditions (i.e., energy demands). Glycolysis adjusts its functional role to meet the needs of the individual tissue.
3. Glycolysis can either function to meet an immediate energy need rapidly or contribute to meeting a sustained energy need. Glycolysis also contributes to the storage of glucose energy as lipid. Depending on the tissue and the current energy needs, glycolysis may perform a combination of these roles.

GLYCOLYSIS

The overall reaction for glycolysis under anaerobic conditions is

$$\text{Glucose} + 2 \text{ ADP} + 2 \text{ P}_i \rightarrow 2 \text{ lactate} + 2 \text{ ATP}$$

Glycolysis yields a net energy gain of two ATPs per glucose. The process by which these ATPs are generated is called *substrate-level phosphorylation* because ADP is phosphorylated to ATP by high-energy phosphate compounds (substrates) generated during the glycolytic pathway.

In the presence of adequate oxygen, the overall reaction of glycolysis is

$$\text{Glucose} + 2 \text{ ADP} + 2 \text{ P}_i + 2 \text{ NAD}^+ \rightarrow$$
$$2 \text{ pyruvate} + 2 \text{ ATP} + 2 \text{ NADH} + 2 \text{ H}^+$$

The NADHs may be used to drive oxidative phosphorylation (Chapter 6), and pyruvate may be metabolized via a number of pathways, including conversion to acetyl CoA, oxaloacetate, or alanine.

The Role of Glycolysis in Different Tissues

Muscle

Contracting muscles have an immediate need for energy, which is met by the ATPs formed during glycolysis. Because NAD^+ is reduced to

NADH during glycolysis, a mechanism to regenerate NAD^+ is necessary if glycolysis is to continue. When the oxygen supply is limited and/or the tissue contains few mitochondria, NAD^+ is regenerated by the reduction of pyruvate to lactate by NADH:

$$\text{Pyruvate} + \text{NADH} + H^+ \rightleftharpoons \text{lactate} + NAD^+$$

The production of lactate from glycolysis is called *lactate fermentation*. Lactate fermentation occurs primarily in white skeletal muscle, which have a limited oxygen supply and relatively few mitochondria. Lactate fermentation is effective at generating short-duration bursts of energy. Most of the resulting lactate enters the circulation and is taken up by the liver.

In dark and cardiac muscle, which have abundant mitochondria and a large oxygen supply, the pyruvate produced during glycolysis is oxidized in the mitochondria to CO_2 and H_2O, and the NADHs produced in glycolysis are shuttled into the mitochondria and oxidized to NAD^+ by the mitochondrial electron transport chain (Chapter 6). In the oxygen-dependent oxidation of glucose to CO_2 and H_2O, glycolysis is coupled to the TCA cycle and cellular respiration. Respiration produces about 10 times as much energy as glycolysis alone, but it responds much more slowly to energy demands.

Brain

Glucose is the preferred fuel of the brain. The brain has both a high demand for energy and an abundant supply of oxygen, and glycolysis in the brain is necessarily coupled to aerobic respiration.

Red Blood Cells

Red blood cells have no mitochondria, and are incapable of aerobic respiration. They depend entirely on lactate fermentation for their energy needs.

Adipose Tissue

In adipose tissue, the pyruvate produced by glycolysis is taken up by the mitochondria and decarboxylated to acetyl CoA. Most of this acetyl CoA is used as the substrate for fatty acid synthesis rather than for aerobic respiration.

Liver

The pyruvate generated by glycolysis in the liver is mostly oxidized to acetyl CoA, which is either oxidized for energy (i.e., coupled to aerobic respiration) or used in the synthesis of fatty acids. In addition, the liver receives most of the lactate produced by the lactate fermentation of erythrocytes and skeletal muscle, and uses it to resynthesize glucose.

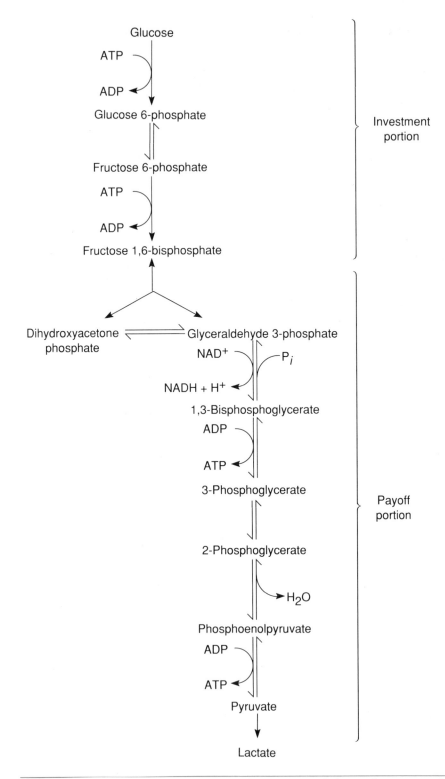

Fig. 5-2. The glycolytic pathway. Note the division of the pathway into the energy investment and payoff portions.

> Figure 5-2 shows the reactions of glycolysis using glucose as substrate. The glycolytic pathway can be divided into two sections. In the first section, the *energy investment portion*, glucose is converted to fructose 1,6-bisphosphate and then split into one molecule of glyceraldehyde 3-phosphate and one of dihydroxyacetone phosphate. Two molecules of ATP are consumed in this process. In the second section of the pathway, the *payoff portion*, the molecule of dihydroxyacetone phosphate is converted to a second molecule of glyceraldehyde 3-phosphate, and the two glyceraldehyde 3-phosphates are then oxidized to pyruvate, with the production of four ATPs per glucose. There is thus a net gain of two ATPs per glucose molecule.

Perspective: The Glycolytic Reactions

THE GLYCOLYSIS PATHWAY

Hexokinase and Glucokinase

The first step in glycolysis is the phosphorylation of glucose to glucose 6-phosphate, at the expense of one ATP:

$$\text{Glucose} + \text{ATP} \xrightarrow{\text{hexokinase or glucokinase}} \text{glucose 6-phosphate} + \text{ADP}$$

Most glucose that enters the cell is phosphorylated by this reaction, which serves to trap glucose inside the cell. This reaction is catalyzed by *hexokinase*, which is present in all tissues. The liver also contains an enzyme called *glucokinase*, which catalyzes the same reaction. The two enzymes have substantially different properties:

1. Hexokinase is relatively nonspecific; it prefers glucose but will catalyze the phosphorylation of a number of aldohexoses and ketohexoses. Glucokinase, on the other hand, is specific for glucose.
2. Hexokinase has a K_M for glucose of approximately 20 μM, whereas glucokinase has a much higher K_M of approximately 12 mM.
3. Hexokinase is inhibited by its product, glucose 6-phosphate, whereas glucokinase is not.

What is the role of glucokinase in the liver? A major function of the liver is to store glucose in the form of glycogen when blood glucose levels are high (after a carbohydrate meal). The effect of glucokinase on this process is illustrated by considering the following points:

1. The concentration of glucose in liver cells is approximately in equilibrium with plasma glucose; therefore, when plasma glucose rises, so does liver cell glucose.
2. Glucose that enters liver cells is prevented from leaving by the phosphorylation reaction. Because of its high K_M, glucokinase traps glucose in liver cells (i.e., for glycogen synthesis) only when plasma glucose is elevated. Therefore, the liver does not accumulate glucose when plasma glucose is in the normal range (approximately 5 mM). If glucokinase had a low K_M, the liver would act as a glucose sink, depleting plasma

glucose. The high K_M of glucokinase, therefore, protects plasma glucose levels.
3. The high K_M of glucokinase also maintains normal plasma glucose levels by allowing the free glucose generated by glycogen degradation to leave the cell.

When liver cell glucose 6-phosphate levels are high and plasma glucose levels are low, glucose 6-phosphate is converted to glucose by *glucose 6-phosphatase*:

$$\text{Glucose 6-phosphate} \xrightarrow{\text{glucose 6-phosphatase}} \text{glucose} + P_i$$

Glucose 6-phosphatase is present in the liver, kidney, and intestine. These tissues release free glucose into the circulation, and collectively contribute to the maintenance of adequate plasma glucose levels.

Glucose Phosphate Isomerase

In the next step of glycolysis, glucose 6-phosphate is converted to fructose 6-phosphate by *glucose phosphate isomerase*.

$$\begin{array}{c}
\text{CH}=\text{O} \\
| \\
\text{H}-\text{C}-\text{OH} \\
| \\
\text{HO}-\text{C}-\text{H} \\
| \\
\text{H}-\text{C}-\text{OH} \\
| \\
\text{H}-\text{C}-\text{OH} \\
| \\
\text{CH}_2-\text{OPO}_3^{2-} \\
\text{Glucose 6-phosphate}
\end{array}
\quad \underset{\text{Glucose phosphate isomerase}}{\rightleftharpoons} \quad
\begin{array}{c}
\text{CH}_2\text{OH} \\
| \\
\text{C}=\text{O} \\
| \\
\text{HO}-\text{C}-\text{H} \\
| \\
\text{H}-\text{C}-\text{OH} \\
| \\
\text{H}-\text{C}-\text{OH} \\
| \\
\text{CH}_2-\text{OPO}_3^{2-} \\
\text{Fructose 6-phosphate}
\end{array}$$

The enzyme catalyzes the rearrangement of the carbonyl carbon atom from the number 1 position to the number 2 position, converting the aldose to a ketose. This reaction is freely reversible, so glucose 6-phosphate and fructose 6-phosphate exist in equilibrium within the cell.

Phosphofructokinase

In the next step, catalyzed by the enzyme *phosphofructokinase* (more accurately called *fructose 6-phosphate-1-kinase*), fructose 6-phosphate is phosphorylated to fructose 1,6-bisphosphate with the consumption of a second ATP:

Carbohydrate Metabolism II

$$\begin{array}{c}CH_2OH\\|\\C=O\\|\\HO-C-H\\|\\H-C-OH\\|\\H-C-OH\\|\\CH_2-OPO_3^{2-}\end{array} + ATP \xrightarrow{\text{phosphofructokinase}} \begin{array}{c}CH_2-OPO_3^{2-}\\|\\C=O\\|\\HO-C-H\\|\\H-C-OH\\|\\H-C-OH\\|\\CH_2-OPO_3^{2-}\end{array} + ADP$$

Fructose 6-phosphate → Fructose 1,6-bisphosphate

This reaction is the rate-limiting step of glycolysis, and it is also *irreversible* and the *committed step* in the pathway. Phosphofructokinase is the major regulated enzyme of glycolysis. This reaction is the second ATP investment of glycolysis.

Aldolase

Aldolase catalyzes the cleavage of fructose 1,6-bisphosphate into the two three-carbon compounds dihydroxyacetone phosphate and glyceraldehyde 3-phosphate:

$$\begin{array}{c}CH_2-OPO_3^{2-}\\|\\C=O\\|\\HO-C-H\\|\\H-C-OH\\|\\H-C-OH\\|\\CH_2-OPO_3^{2-}\end{array} \rightleftharpoons \begin{array}{c}CH_2-OPO_3^{2-}\\|\\C=O\\|\\CH_2OH\end{array} + \begin{array}{c}CH=O\\|\\H-C-OH\\|\\CH_2-OPO_3^{2-}\end{array}$$

Fructose 1,6-bisphosphate   Dihydroxyacetone phosphate   Glyceraldehyde 3-phosphate

Triose Phosphate Isomerase

Dihydroxyacetone phosphate and glyceraldehyde 3-phosphate are freely interconverted by the enzyme *triose phosphate isomerase*:

$$\begin{array}{c}CH_2-PO_3^{2-}\\|\\C=O\\|\\CH_2OH\end{array} \rightleftharpoons \begin{array}{c}CH=O\\|\\H-C-OH\\|\\CH_2-OPO_3^{2-}\end{array}$$

Dihydroxyacetone phosphate   Glyceraldehyde 3-phosphate

This reaction is important because only glyceraldehyde 3-phosphate is metabolized in the subsequent reactions of glycolysis.

Glyceraldehyde 3-Phosphate Dehydrogenase

Glyceraldehyde 3-phosphate is next oxidized to 1,3-bisphosphoglycerate by the enzyme *glyceraldehyde 3-phosphate dehydrogenase*, with the reduction of one molecule of NAD^+:

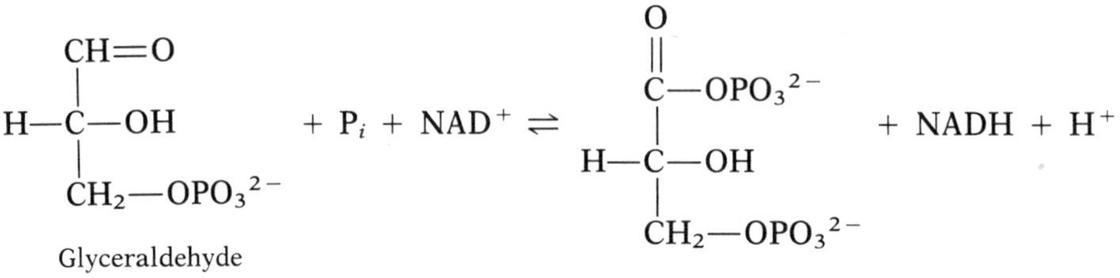

$$\begin{array}{c} CH=O \\ | \\ H-C-OH \\ | \\ CH_2-OPO_3^{2-} \end{array} + P_i + NAD^+ \rightleftharpoons \begin{array}{c} O \\ \| \\ C-OPO_3^{2-} \\ | \\ H-C-OH \\ | \\ CH_2-OPO_3^{2-} \end{array} + NADH + H^+$$

Glyceraldehyde 3-phosphate $\qquad\qquad$ 1,3-Bisphosphoglycerate

This is the first *energy-trapping reaction* of glycolysis: the energy released as a result of the oxidation of the aldehyde group by NAD^+ is trapped in the high-energy phosphate ester.

Phosphoglycerate Kinase

In the next reaction, catalyzed by *phosphoglycerate kinase*, the high energy compound 1,3-bisphosphoglycerate is used to drive the phosphorylation of ADP, yielding the first ATP:

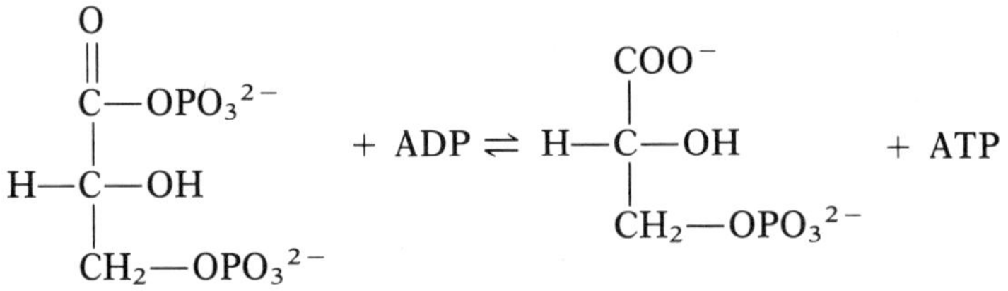

$$\begin{array}{c} O \\ \| \\ C-OPO_3^{2-} \\ | \\ H-C-OH \\ | \\ CH_2-OPO_3^{2-} \end{array} + ADP \rightleftharpoons \begin{array}{c} COO^- \\ | \\ H-C-OH \\ | \\ CH_2-OPO_3^{2-} \end{array} + ATP$$

1,3-Bisphosphoglycerate $\qquad\qquad$ 3-Phosphoglycerate

This reaction is an example of substrate-level phosphorylation.

Phosphoglyceromutase

The enzyme *phosphoglyceromutase* converts 3-phosphoglycerate to 2-phosphoglycerate:

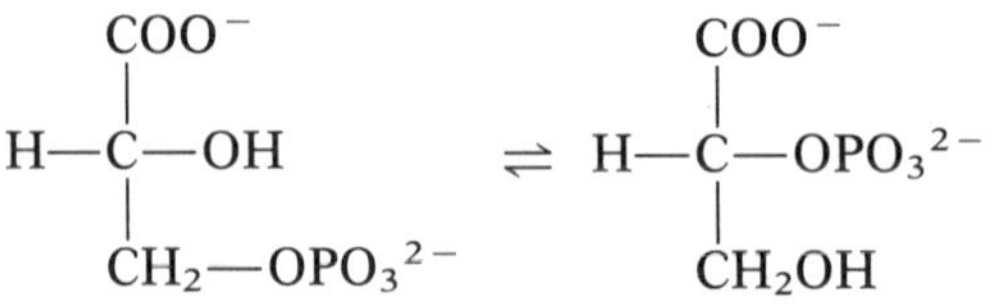

$$\begin{array}{c} COO^- \\ | \\ H-C-OH \\ | \\ CH_2-OPO_3^{2-} \end{array} \rightleftharpoons \begin{array}{c} COO^- \\ | \\ H-C-OPO_3^{2-} \\ | \\ CH_2OH \end{array}$$

3-Phosphoglycerate $\qquad\qquad$ 2-Phosphoglycerate

Enolase

The enzyme *enolase* then catalyzes another energy-trapping reaction: the oxidation of 2-phosphoglycerate to phosphoenolpyruvate (PEP). Again, the energy of oxidation is trapped in the form of a high-energy phosphate bond (in this case, yielding an enolphosphate):

$$\begin{array}{c} COO^- \\ | \\ H-C-OPO_3^{2-} \\ | \\ CH_2OH \end{array} \rightleftarrows \begin{array}{c} COO^- \\ | \\ C-OPO_3^{2-} \\ || \\ CH_2 \end{array} + H_2O$$

    2-Phosphoglycerate   Phosphoenolpyruvate

Pyruvate Kinase

Phosphoenolpyruvate is then converted to pyruvate by *pyruvate kinase*, with the formation of another ATP:

$$\begin{array}{c} COO^- \\ | \\ C-OPO_3^{2-} \\ || \\ CH_2 \end{array} + ADP \rightarrow \begin{array}{c} COO^- \\ | \\ C=O \\ | \\ CH_3 \end{array} + ATP$$

    Phosphoenolpyruvate   Pyruvate

Pyruvate kinase uses the energy of the enolphosphate group to drive the phosphorylation of ADP, with the simultaneous release of pyruvate.

Lactate Dehydrogenase

Pyruvate may then be converted to lactate by *lactate dehydrogenase*, with the regeneration of NAD$^+$:

$$\begin{array}{c} COO^- \\ | \\ C=O \\ | \\ CH_3 \end{array} + NADH + H^+ \rightleftarrows \begin{array}{c} COO^- \\ | \\ H-C-OH \\ | \\ CH_3 \end{array} + NAD^+$$

    Pyruvate   Lactate

METABOLISM OF PYRUVATE

Depending on the tissue involved and the prevailing metabolic conditions, the pyruvate produced by glycolysis undergoes various fates. It may be converted to lactate as described above, or it may enter the mitochondria and be oxidized to acetyl CoA by pyruvate dehydrogenase. Acetyl CoA may be oxidized to CO_2 and H_2O by the TCA cycle (Chapter 6), or it may leave the mitochondria to participate in fatty acid biosynthesis. Alternatively, pyruvate that enters the mitochondrion may be carboxylated by pyruvate carboxylase to oxaloacetate, which is both a TCA cycle intermediate and a precursor for gluconeogenesis. Pyruvate may also be transaminated to alanine (Chapter 9).

ENERGY YIELD OF GLYCOLYSIS

The energy yield from glycolysis depends on whether the final product is pyruvate or lactate.

1. If lactate is the final product, the net energy yield is two ATPs per glucose:

    Glucose + 2 ADP + 2 P_i → 2 lactate + 2 ATP

2. If pyruvate is the final product, two ATPs are produced by substrate-level phosphorylation, as in lactate fermentation. In addition, the NADHs generated during glycolysis are shuttled into the mitochondria, where they are oxidized by the electron transport chain with the production of ATP. Depending on the shuttle mechanism by which the NADH enters the mitochondria, each NADH supports the synthesis of either two or three ATPs. Therefore, when the final product of glycolysis is pyruvate, the net energy yield is from six to eight ATPs:

    Glucose + 6 to 8 ADP + 6 to 8 P_i → 2 pyruvate + 6 to 8 ATP

    If glycolysis starts from the glucose 1-phosphate produced by glycogen degradation rather than from free glucose, the net yield of glycolysis is higher by one ATP, because the first phosphorylation of glucose is accomplished without expending ATP.

NADH Shuttle Mechanisms

NADH shuttle mechanisms function to transport reducing equivalents (reduced nucleotide coenzymes, i.e., NADH) from the cytosol to the mitochondria, where they are oxidized via the electron transport chain for energy production. The inner mitochondrial membrane has two shuttles for NADH: the *glycerol phosphate shuttle* and the *malate-aspartate shuttle*.

NADPH Shuttle Mechanisms: Key Points

1. Some NADH is generated in the cytosol—for example, by glycolysis. The energy demands of most cells require that some of this NADH be oxidized in the mitochondria.
2. The inner mitochondrial membrane is impermeable to NADH and NAD^+, but it is permeable to certain metabolites that can be oxidized and reduced by nucleotide coenzymes.
3. In both the NADH shuttle mechanisms, it is the reducing power (electrons) carried by NADH, rather than NADH itself, that is conveyed across the membrane. The characteristic property of these shuttles is that cytosolic NADH is used to reduce a particular cytosolic metabolite, which in turn reduces a nucleotide coenzyme in the mitochondria and is itself reoxidized.

4. In the case of the malate-aspartate shuttle, the reduced cytosolic metabolite (malate) enters the mitochondrial matrix, where it reduces NAD^+ to NADH. The oxidized metabolite is then returned to the cytosol.
5. In the case of the glycerol phosphate shuttle, the reduced cytosolic metabolite (glycerol 3-phosphate) does not cross the inner mitochondrial membrane, but instead interacts with an enzyme that spans the membrane.

Glycerol Phosphate Shuttle

In the glycerol phosphate shuttle (Fig. 5-3A) NADH first reduces cytosolic dihydroxyacetone phosphate to glycerol 3-phosphate, in a reaction catalyzed by glycerol 3-phosphate dehydrogenase. Glycerol 3-phosphate diffuses to the inner mitochondrial membrane, where it reacts with FAD and glycerol 3-phosphate dehydrogenase present on the inner mitochondrial membrane. The glycerol 3-phosphate is oxidized to dihydroxyacetone phosphate, which is released back into the cytosol, and the mitochondrial FAD is reduced to $FADH_2$. The net reaction of the glycerol phosphate shuttle is

$$NADH_{cytoplasmic} + FAD_{mitochondrial} \rightarrow NAD^+_{cytoplasmic} + FADH_{2\ mitochondrial}$$

Thus, the glycerol phosphate shuttle in effect converts cytosolic NADH into mitochondrial $FADH_2$

Malate-Aspartate Shuttle

In the malate-aspartate shuttle (Fig. 5-3B), cytosolic NADH reduces oxaloacetate to malate via *cytosolic malate dehydrogenase*. Malate is then transported into the mitochondrial matrix, where it reduces NAD^+ to NADH via *mitochondrial malate dehydrogenase*, and is itself reoxidized to oxaloacetate.

At this point, a cytosolic NADH has been exchanged for a mitochondrial NADH. The remainder of the malate-aspartate shuttle is concerned with returning oxaloacetate to the cytosol. First, mitochondrial oxaloacetate transaminates with glutamate to yield α-ketoglutarate and aspartate. Aspartate is transported to the cytosol, where it transaminates with α-ketoglutarate to regenerate oxaloacetate and glutamate. For the shuttle to function properly, the concentrations of α-ketoglutarate and glutamate must be regulated on both sides of the membrane. In fact, a specific inner mitochondrial membrane antiport system (Chapter 21) transports α-ketoglutarate and malate simultaneously in opposite directions across the membrane, while a glutamate-aspartate antiport system simultaneously transports glutamate and aspartate in opposite directions across the membrane.

The net effect of the malate-aspartate shuttle is to use cytosolic NADH to reduce mitochondrial NAD^+:

$$NADH_{cytoplasmic} + NAD^+_{mitochondrial} \rightarrow NADH_{mitochondrial} + NAD^+_{cytoplasmic}$$

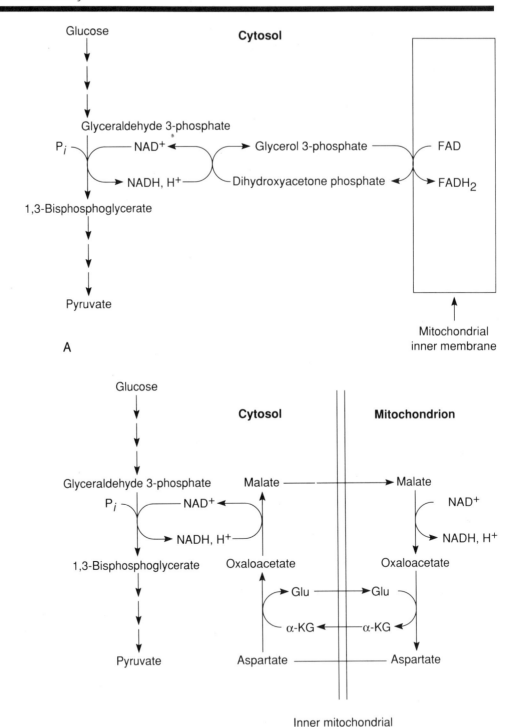

Fig. 5-3. NADH shuttle mechanisms. **(A)** The glycerol phosphate shuttle. **(B)** the malate-aspartate shuttle. Glu, glutamate; α-KG, α-ketoglutarate.

NADH Shuttle Mechanisms: General Points

1. The reducing power of NADH is shuttled into the mitochondria by the cyclic reduction and oxidation of a translocated metabolite. Enzymes to catalyze this oxidation/reduction must therefore exist in both the cytosol and the mitochondria.
2. The cytosolic and mitochondrial pools of the translocated metabolite must be freely interchangeable.

NADH Shuttles and the Glycolysis Energy Yield

As mentioned, the energy yield of glycolysis to pyruvate depends on the NADH shuttle mechanism used. The reason is that the oxidation of NADH and $FADH_2$ via the mitochondrial electron-transport chain (Chapter 6) yield three and two ATPs, respectively.

REGULATION OF GLYCOLYSIS

The regulated enzymes of glycolysis are the ones that catalyze the three irreversible reactions of the pathway: hexokinase/glucokinase, phosphofructokinase, and pyruvate kinase (Fig. 5-4). The reaction catalyzed by phosphofructokinase is both the committed step of the pathway and the rate-limiting step.

Hexokinase and Glucokinase

Hexokinase and glucokinase determine the amount of glucose 6-phosphate that is produced, and therefore the quantity that is available for glycolysis. Hexokinase is inhibited by glucose 6-phosphate, whereas liver glucokinase is not. The level of glucokinase in liver cells varies with hormonal status and with the amount of carbohydrate in the diet. Glucokinase levels rise in response to a high-carbohydrate diet, and fall when dietary carbohydrate is reduced. Glucokinase activity is also lowered by epinephrine and glucagon, and increased by insulin and adrenal corticosteroids.

Phosphofructokinase

Phosphofructokinase, the rate-limiting enzyme of glycolysis, responds to a variety of positive and negative effectors:

Positive effectors: AMP, P_i, fructose 2,6-bisphosphate
Negative effectors: ATP, $[H^+]$, citrate

Lactate produced from glycolysis causes an increase in cellular $[H^+]$, which inhibits phosphofructokinase. This inhibition is at least partially overcome by the transport of lactic acid (lactate and H^+) out of cells to the blood. During vigorous exercise, the flow of blood to the muscle fibers is not sufficient to remove all the lactic acid produced. As a result, H^+ accumulates in the tissue, causing the muscle pain often associated with strenuous physical activity.

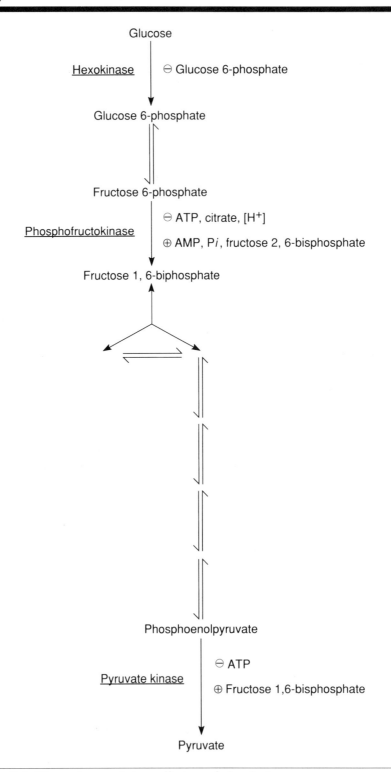

Fig. 5-4. The major regulatory effectors of glycolysis.

The energy status of a cell is determined by its levels of ATP, ADP, and AMP. Under normal conditions, ATP >> ADP >> AMP. The fact that phosphofructokinase is inhibited by ATP and stimulated by AMP and P_i indicates that this enzyme is regulated by the energy status of the cell: phosphofructokinase is activated when the cell needs energy, and inhibited when the cell has enough energy. This type of regulation is not surprising for an enzyme that is involved in the production of ATP.

Citrate, a TCA cycle intermediate, also inhibits phosphofructokinase. As the level of TCA cycle intermediates rises, citrate is transported out of the mitochindria to the cytosol, where it inhibits phosphofructokinase and slows glycolysis. This mechanism sensitizes glycolysis to the status of the TCA cycle.

The most important regulator of glycolysis in the liver is the positive effector fructose 2,6-bisphosphate. This compound is a purely regulatory metabolite, rather than a pathway intermediate. It is synthesized by the enzyme *phosphofructokinase 2* (PFK2):

$$\text{Fructose 6-phosphate} + \text{ATP} \xrightarrow{\text{PFK2}} \text{fructose 2,6-bisphosphate} + \text{ADP}$$

and degraded by *fructose bisphosphatase 2* (FPBase 2):

$$\text{Fructose 2,6-bisphosphate} \xrightarrow{\text{PFBase2}} \text{fructose 6-phosphate} + P_i$$

Both enzyme activities are located on the same polypeptide chain. This protein exists in a phosphorylated and a deposphorylated form. Phosphorylation inhibits PFK2 and stimulates FPBase2, leading to a fall in fructose 2,6-bisphosphate, whereas dephosphorylation activates PFK2 and inhibits FPBase2, leading to a rise in fructose 2,6-bisphosphate. The protein is phosphorylated by the cyclic AMP-dependent protein kinase (page 78), which is activated by a glucagon-triggered enzyme cascade. A fall in blood glucose causes a rise in glucagon, which results in phosphorylation of the protein, leading to a fall in intracellular fructose 2,6-bisphosphate and inhibition of phosphofructokinase.

Pyruvate Kinase

Pyruvate kinase, like phosphofructokinase, is inhibited by ATP. More interesting, however, is the activation of pyruvate kinase by fructose 1,6-bisphosphate. When phosphofructokinase is stimulated, the increase in its product fructose 1,6-bisphosphate in turn stimulates pyruvate kinase downstream. This is an example of an enzyme being kept informed of the activity of its major upstream "feeder" enzyme. Pyruvate kinase is also hormonally inhibited by glucagon via a cAMP-dependent phosphorylation/dephosphorylation system. The phosphorylated enzyme is inactive.

ENTRY OF OTHER SUGARS INTO GLYCOLYSIS

Dietary galactose, fructose, and mannose can be converted into glycolytic intermediates and fed into the glycolysis pathway.

Galactose

Galactose is converted into glucose 6-phosphate as shown in Figure 5-5. The first step is a phosphorylation, catalyzed by *galactokinase*:

Galactose + ATP → galactose 1-phosphate + ADP

In the next step, UDP-glucose transfers UDP to galactose in a reaction catalyzed by a specific uridyltransferase:

Galactose 1-phosphate + UDP-glucose →

UDP-galactose + glucose 1-phosphate

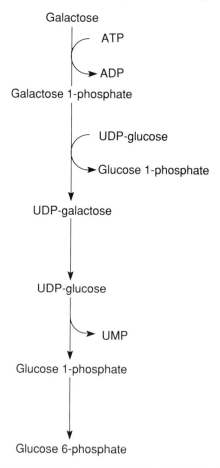

Fig. 5-5. The conversion of galactose to glucose 6-phosphate. This is the major pathway by which galactose enters the pathways of glucose metabolism.

The UDP-galactose from this reaction is converted to UDP-glucose by an epimerase (Fig. 5-5). Glucose 1-phosphate can be released from UDP-glucose by a phosphorylase, and subsequently converted to glucose 6-phosphate by phosphoglucomutase. Glucose 6-phosphate may then enter glycolysis.

Fructose

Most dietary fructose is converted to fructose 1-phosphate by *fructokinase*:

$$\text{Fructose} + \text{ATP} \rightarrow \text{fructose 1-phosphate} + \text{ADP}$$

Fructose 1-phosphate is then cleaved by *fructose 1-phosphate aldolase* to yield dihydroxyacetone phosphate and glyceraldehyde:

Fructose 1-phosphate →

    dihydroxyacetone phosphate + glyceraldehyde

Dihydroxyacetone can proceed down the glycolytic pathway, and glyceraldehyde can be metabolized in various ways to yield glycolytic intermediates (Fig. 5-6).

A portion of dietary fructose is converted directly to fructose 6-phosphate by hexokinase.

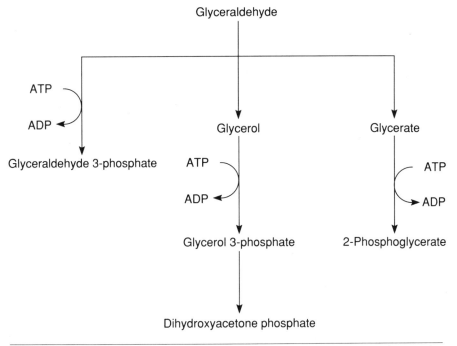

Fig. 5-6. The metabolism of glyceraldehyde. All three pathways end in a glycolysis intermediate.

Mannose

Mannose is first phosphorylated to mannose 6-phosphate by hexokinase. Mannose 6-phosphate is then converted to fructose 6-phosphate by *phosphomannose isomerase*:

$$\text{Mannose} + \text{ATP} \rightarrow \text{mannose 6-phosphate} + \text{ADP}$$

$$\text{Mannose 6-phosphate} \rightarrow \text{fructose 6-phosphate}$$

Perspective: Gluconeogenesis

Gluconeogenesis is the pathway by which glucose is synthesized from noncarbohydrate precursors.

1. Blood glucose levels must be kept constant within certain limits to ensure adequate nourishment for peripheral tissues. Degradation of liver glycogen initially serves this purpose; the liver contains enough glycogen to fuel the body for several hours. The main function of gluconeogenesis in the liver and kidney is to supply the blood with glucose after liver glycogen has been exhausted.
2. Muscle cells also perform gluconeogenesis, but they are unable to release free glucose into the blood. The newly synthesize glucose is used within the muscle cell to synthesize glycogen.
3. The major precursors of gluconeogenesis are lactate, pyruvate, certain amino acids, and the glycerol released by triacylglycerol degradation.

GLUCONEOGENESIS

Glycolysis and gluconeogenesis proceed in opposite directions, and have many reaction steps in common. However, the two pathways are not simply the reverse of each other. The reason is that the glycolytic reactions catalyzed by hexokinase/glucokinase, phosphofructokinase, and pyruvate kinase are irreversible under cellular conditions. Therefore, for gluconeogenesis to occur, these three reactions must be circumvented (Fig. 5-7).

Gluconeogenesis requires energy in the form of ATP, GTP, and reducing power (NADH). The steps that use energy are as follows:

Reactions requiring ATP:

1. Conversion of pyruvate to oxaloacetate by pyruvate carboxylase (the first reaction of the two-step reduction of pyruvate to phosphoenolpyruvate)
2. Conversion of 3-phosphoglycerate to 1,3-bisphosphoglycerate (the reverse of the ATP-requiring phosphoglycerate kinase reaction)

Fig. 5-7. The glycolytic and gluconeogenetic pathways, showing the three steps that are catalyzed by different enzymes in the two pathways (wide arrows).

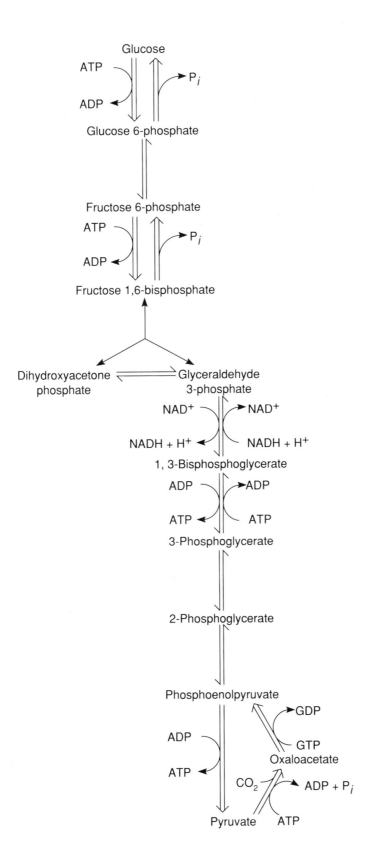

Reaction requiring GTP:

3. Conversion of oxaloacetate to phosphoenolpyruvate by phosphoenolpyruvate kinase (the second reaction of the two-step reduction of pyruvate to phosphoenolpyruvate)

Reaction requiring NADH:

4. Reduction of 1,3-bisphosphoglycerate to glyceraldehyde 3-phosphate (the reversal of the glyceraldehyde 3-phosphate dehydrogenase reaction)

Gluconeogenesis from Lactate

To enter gluconeogenesis, lactate is first converted to pyruvate by lactate dehydrogenase

$$\text{Lactate} + \text{NAD}^+ \xrightleftharpoons[]{\text{lactate dehydrogenase}} \text{pyruvate} + \text{NADH} + \text{H}^+$$

This reaction is also a source of NADH for the glyceraldehyde 3-phosphate reaction.

The following discussion covers only the steps at which gluconeogenesis departs from glycolysis. The first challenge is to convert pyruvate to phosphoenolpyruvate. This is achieved by the combined reactions of *pyruvate carboxylase* and *phosphoenolpyruvate carboxykinase*.

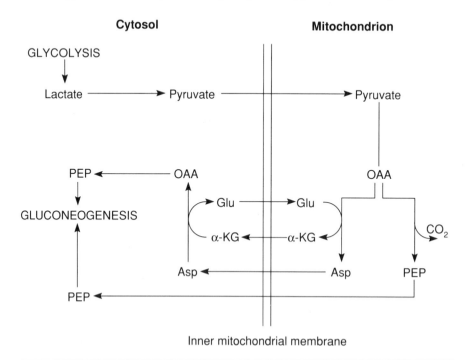

Fig. 5-8. Movement of oxaloacetate as asparte from the mitochondria into the cytosol to support gluconeogenesis from lactate. Asp, aspartate; Glu, glutamate; α-KG, α-ketoglutarate; OAA, oxaloacetate; PEP, phosphoenolpyruvate.

Pyruvate Carboxylase

Pyruvate carboxylase catalyzes the following reaction:

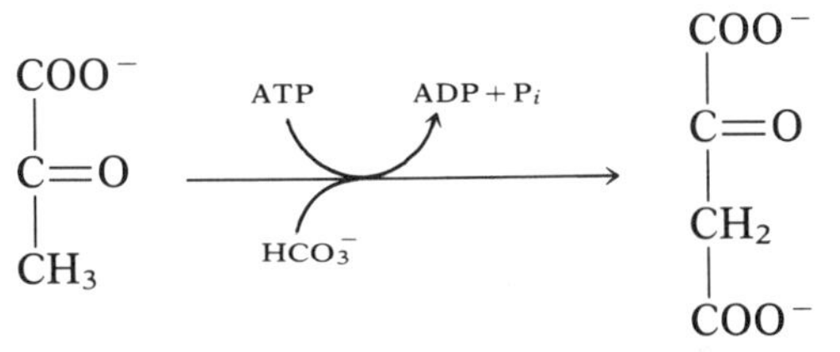

$$\begin{array}{c} COO^- \\ | \\ C=O \\ | \\ CH_3 \end{array} \quad \xrightarrow[HCO_3^-]{ATP \quad ADP+P_i} \quad \begin{array}{c} COO^- \\ | \\ C=O \\ | \\ CH_2 \\ | \\ COO^- \end{array}$$

Pyruvate   Oxaloacetate

Pyruvate carboxylase is located in the mitochondria. Therefore, the pyruvate generated in the cytosol (by lactate dehydrogenase) must be transported into the mitochondria in order to enter gluconeogenesis.

Phosphoenolpyruvate Carboxykinase

Phosphoenolpyruvate carboxykinase catalyzes the reaction

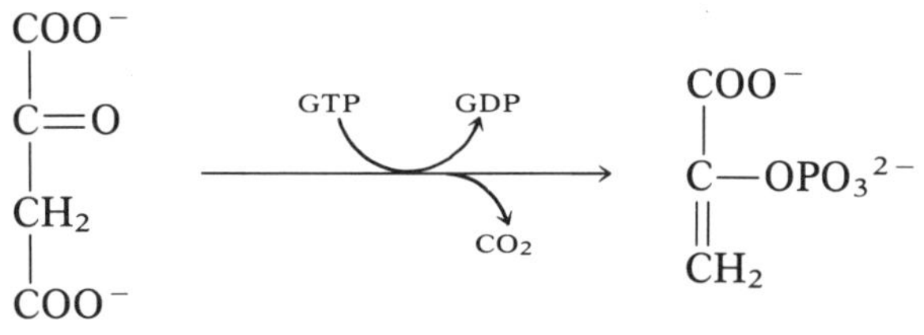

$$\begin{array}{c} COO^- \\ | \\ C=O \\ | \\ CH_2 \\ | \\ COO^- \end{array} \quad \xrightarrow[CO_2]{GTP \quad GDP} \quad \begin{array}{c} COO^- \\ | \\ C-OPO_3^{2-} \\ || \\ CH_2 \end{array}$$

Oxaloacetate   Phosphoenolpyruvate

This enzyme is located in both the mitochondria and the cytosol. Pyruvate carboxylated to oxaloacetate in the mitochondria may be either decarboxylated to phosphoenolpyruvate by mitochondrial phosphoenolpyruvate carboxykinase or transaminated to aspartate. Aspartate is transported to the cytosol, where it undergoes another transamination to yield oxaloacetate. The cytosolic oxaloacetate is converted to phosphoenolpyruvate by the cytosolic phosphoenolpyruvate carboxykinase and used for gluconeogenesis (Fig. 5-8). Phosphoenolpyruvate generated in the mitochondria is also transported in the cytosol, where it is used for gluconeogenesis.

Fructose 1,6-Bisphosphatase

The next step at which gluconeogenesis departs from glycolysis is the dephosphorylation of fructose 1,6-bisphosphate. This reaction is catalyzed by *fructose 1,6-bisphosphatase*:

Fructose 1,6-bisphosphate $\xrightarrow{\text{fructose 1,6-bisphosphatase}}$ fructose 6-phosphate + P_i

Glucose 6-Phosphatase

The final distinct reaction of gluconeogenesis is the dephosphorylation of glucose 6-phosphate to free glucose, catalyzed by *glucose 6-phosphatase*:

$$\text{Glucose 6-phosphate} \xrightarrow{\text{glucose 6-phosphatase}} \text{glucose} + P_i$$

Glucose 6-phosphatase also dephosphorylates the glucose 6-phosphate produced by glycogen breakdown. Tissues that lack glucose 6-phosphatase cannot release free glucose to the blood.

Gluconeogenesis from Pyruvate

The only difference between gluconeogenesis from lactate and gluconeogenesis from pyruvate is the mechanism that generates the cytosolic NADH needed for gluconeogenesis. Pyruvate is formed when the oxygen supply is adequate. Under these conditions, cells tend to shuttle NADH into the mitochondria (as described on page 100), leaving too little NADH in the cytosol to support gluconeogenesis. To solve this problem when pyruvate is the substrate for gluconeogenesis, NADH equivalents are transferred from the mitochondria back to the cytosol by the following process (Fig. 5-9). First, mitochondrial pyruvate is carboxylated

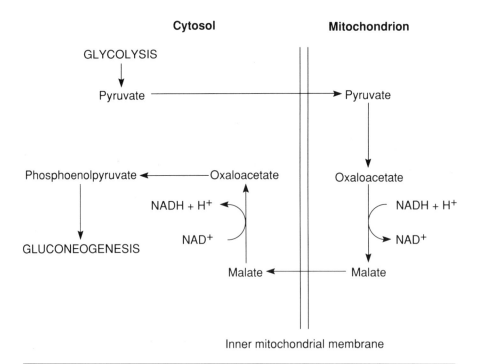

Fig. 5-9. Movement of oxaloacetate as malate from the mitochondria into the cytosol to support gluconeogenesis from pyruvate. This mechanism ensures that the cytosol contains the NADH needed for gluconeogenesis.

by pyruvate carboxylase to oxaloacetate, which is reduced by NADH to malate. Malate is transported out of the mitochondria, and in the cytosol is reoxidized to oxaloacetate, with the production of the required NADH. This cytosolic oxaloacetate is then converted to phosphoenolpyruvate by phosphoenolpyruvate carboxykinase, and gluconeogenesis proceeds as for lactate.

Gluconeogenesis from Amino Acids

Amino acids whose degradation results in net synthesis of pyruvate or oxaloacetate can be used to synthesize glucose. So can amino acids that are degraded to yield α-ketoglutarate or succinate, since these compounds are TCA cycle intermediates and can therefore be transformed to oxaloacetate.

The Contribution of Galactose, Fructose, and Mannose to Gluconeogenesis

The monosaccharides galactose, fructose, and mannose enter gluconeogenesis by the same pathways by which they enter glycolysis (see page 100):

1. *Galactose* is converted to glucose 6-phosphate.
2. *Fructose* from the diet is mostly converted to dihydroxyacetone phosphate and glyceraldehyde. Dihydroxyacetone phosphate can enter gluconeogenesis directly; glyceraldehyde is converted to dihydroxyacetone or phosphoglycerate. Some dietary fructose is phosphorylated by hexokinase to fructose 6-phosphate, which can enter gluconeogenesis.
3. *Mannose* is converted to fructose 6-phosphate.

Energy Considerations

Six ATP equivalents (four ATPs and two GTPs) are required for the synthesis of one molecule of glucose from pyruvate or lactate. The net reaction is

2 Lactate (2 pyruvate) + 6 ATP + 2 NADH + 2 H^+ →

glucose + 6 ADP + 6 P_i + 2 NAD^+

Regulation of Gluconeogenesis

Gluconeogenesis achieves the metabolic opposite of glycolysis. Therefore, the two processes would not be expected to proceed maximally at the same time. When gluconeogenesis is active, glycolysis is inhibited, and vice versa. The reciprocal functioning of these two pathways is an example of *coordinate regulation*.

The intracellular effectors that shut off glycolysis activate gluconeogenesis (Fig. 5-10). For example, AMP and fructose 2,6-bisphosphate activate phosphofructokinase (of glycolysis) and inhibit fructose 1,6-

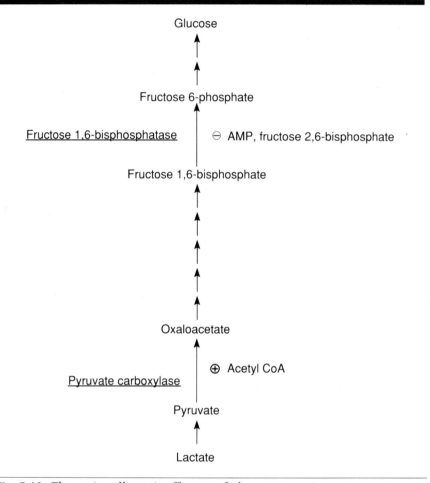

Fig. 5-10. The major allosteric effectors of gluconeogenesis.

bisphosphatase (of gluconeogenesis). A drop in AMP and fructose 2,6-bisphosphate has the opposite effects. When an excess of acetyl CoA is present, pyruvate carboxylase is stimulated and pyruvate kinase is inhibited. In this case, pyruvate kinase is inhibited by the increased level of citrate that accompanies a rise in acetyl CoA. Citrate inhibits phosphofructokinase, which decreases the level of fructose 1,6-bisphosphate, deactivating (or inhibiting) pyruvate kinase. The net effect, again, is inhibition of glycolysis and stimulation of gluconeogenesis.

The two pathways also respond oppositely to hormone regulation:

Glucagon stimulates gluconeogenesis by lowering cellular fructose 2,6-bisphosphate levels. Glucagon also stimulates fatty acid oxidation, resulting in increased acetyl CoA, which stimulates gluconeogenesis.
Insulin, by contrast, stimulates glycolysis and inhibits gluconeogenesis.

> **Perspective: The Pentose Phosphate Pathway**
>
> The pentose phosphate pathway, also known as the *hexose monophosphate shunt* and the *phosphogluconate pathway*, is an additional mechanism of glucose oxidation. It occurs in the cytosol and neither produces nor consumes ATP. The energy product of the pentose phosphate pathway is NADPH.
>
> Both NADH and NADPH are reduced purine nucleotides, but they have distinct functional roles in cells. Most NADH produced is eventually oxidized via the mitochondrial electron transport chain to generate ATP. NADPH, in contrast, is the carrier of the reducing power required for most biosynthetic processes, especially the synthesis of fatty acids and steroids.
>
> In short, the pentose phosphate pathway is:
>
> 1. The major source of NADPH for biosyntheses
> 2. The major source of ribose 5-phosphate for nucleotide synthesis
> 3. An important pathway for the interconversion of hexoses, pentoses, and trioses, which is crucial for the synthesis and utilization of pentoses.

THE PENTOSE PHOSPHATE PATHWAY

The pentose phosphate pathway (Fig. 5-11) is most easily understood if it is divided into two phases. NADPH is produced in the *oxidative phase*, and hexose-pentose-triose interconversions occur in the *nonoxidative phase*.

The Oxidative Phase

The oxidative phase converts glucose 6-phosphate to ribulose 5-phosphate and CO_2 with the production of two NADPHs (Fig. 5-12). In the first reaction, glucose 6-phosphate is oxidized by *glucose 6-phosphate dehydrogenase* to 6-phosphogluconolactone, with the reduction of one $NADP^+$ to NADPH. The 6-phosphogluconolactone is then hydrolyzed by a specific lactonase to 6-phosphogluconate, which is oxidatively decarboxylated by *6-phosphogluconate dehydrogenase* to ribulose 5-phosphate and CO_2, with the reduction of a second $NADP^+$ to NADPH. The overall reaction of the oxidative phase is thus

Glucose 6-phosphate + 2 $NADP^+$ →

$$\text{ribulose 5-phosphate} + CO_2 + 2\ \text{NADPH} + 2\ H^+$$

The Nonoxidative Phase

The nonoxidative phase is primarily concerned with the reorganization of the carbon atoms of ribulose 5-phosphate. It can yield different net products depending on the needs of the cell.

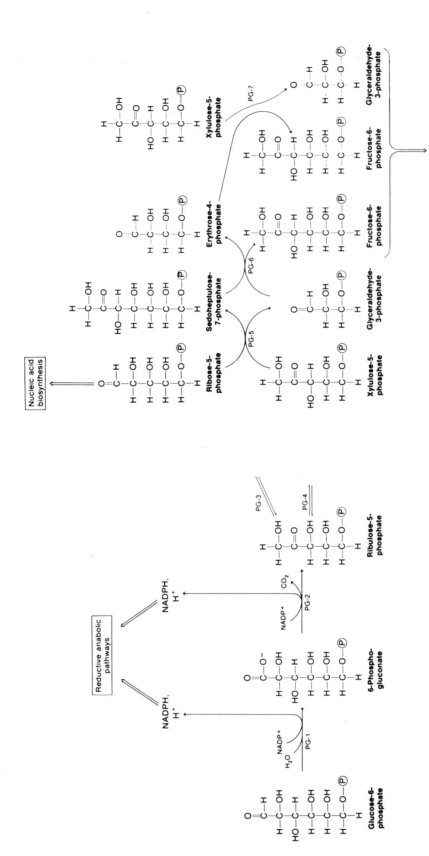

Fig. 5-11. The pentose phosphate pathway. PG-1, glucose 6-phosphate dehydrogenase; PG-2, 6-phosphogluconate decarboxylase; PG-3, phosphopentose isomerase; PG-4, phosphopentose epimerase; PG-5, transketolase; PG-6, transaldolase, PG-7, transketolase. (From Zubay G: Biochemistry. 3rd Ed. Macmillan, New York, 1988, with permission.)

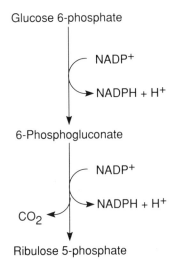

Fig. 5-12. The oxidative phase of the pentose phosphate pathway. Two NADPHs are produced during this phase.

The reactions of the nonoxidative phase are as follows (Fig. 5-13):

1. Ribulose 5-phosphate is converted to both ribose 5-phosphate and xylulose 5-phosphate by an isomerase and an epimerase, respectively.
2. The enzyme *transketolase* transfers the aldehyde group comprising carbon atoms 1 and 2 of xylulose 5-phosphate to the number 1 position of ribose 5-phosphate, yielding the three-carbon compound glyceraldehyde 3-phosphate and the seven-carbon compound sedoheptulose 7-phosphate.
3. The enzyme *transaldolase* transfers the dihydroxyacetone group consisting of the first three carbons of sedoheptulose 7-phosphate to glyceraldehyde 3-phosphate. The products of this reaction are fructose 6-phosphate and the four-carbon compound erythrose 4-phosphate.
4. Erythrose 4-phosphate then reacts with xylulose 5-phosphate to produce another molecule of fructose 6-phosphate and one of glyceraldehyde 3-phosphate. This reaction is also catalyzed by transketolase and involves the transfer of the first two carbons of xylulose 5-phosphate.

All reactions of the nonoxidative phase are reversible.

Cellular Roles of the Pentose Phosphate Pathway

Depending on the current needs of the cell, the nonoxidative phase can yield different net results.

1. If NADPH is needed, the nonoxidative phase results in net production of fructose 6-phosphate, which in turn can be converted to glucose 6-phosphate and reoxidized to yield more NADPH.
2. If the cell needs ribose phosphate for the synthesis of nucleotides, the net product of the nonoxidative phase will be ribose 5-phosphate.
3. If more glycolysis intermediates are needed, the nonoxidative phase results in increased synthesis of fructose 6-phosphate and glyceraldehyde 3-phosphate.

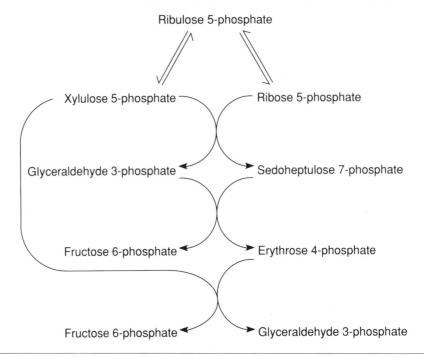

Fig. 5-13. The nonoxidative phase of the pentose phosphate pathway.

Tissue Utilization of the Pentose Phosphate Pathway

Several tissues, including the mammary glands, testis, adipose tissue, erythrocytes, and adrenal cortex, metabolize substantial amounts of glucose via the pentose phosphate pathway. Most of these tissues are active in the synthesis of fatty acids or steroids, and therefore require large quantities of NADPH. Erythrocytes require NADPH in order to maintain glutathione in its reduced state, which is important for normal red cell function.

CLINICAL CORRELATIONS

Fructose 1,6-Bisphosphatase Deficiency

Fructose 1,6-bisphosphatase is essential for gluconeogenesis. Individuals who lack this enzyme rely solely on their liver glycogen stores to maintain blood glucose, and become hypoglycemic when liver glycogen is depleted. Additional complications are accumulation of fructose 1-phosphate, fructose 1,6-bisphosphate, glycerol 1-phosphate, and lactate. Gluconeogenetic precursors are diverted into the glycolytic pathway, causing an increase in lactate production and eventual lactic acidosis.

Typical symptoms of the disorder are lactic acidosis, hypoglycemia, and hepatomegaly. Acidosis and hypoglycemia are often associated with infection. The hypoglycemia is most effectively treated by intravenous administration of glucose.

Fructose Intolerance: Deficiency of Fructose 1-Phosphate Aldolase

As described on page 105, most dietary fructose is phosphorylated to fructose 1-phosphate, which is cleaved by fructose 1-phosphate aldolase to yield dihydroxyacetone phosphate and glyceraldehyde, which can enter either glycolysis or gluconeogenesis. In the absence of fructose 1-phosphate aldolase, fructose 1-phosphate accumulates. Continued ingestion of fructose can lead to a depletion of cellular phosphate and ATP, and ultimately compromise the cell's metabolic functioning.

The clinical symptoms of fructose intolerance are severe hypoglycemia and vomiting after fructose ingestion. In children, fructose intolerance is also associated with hepatomegaly, aminoaciduria, and impaired growth.

SUGGESTED READING

1. Hanson RW, Mehlman MS (eds): Gluconeogenesis. Wiley-Interscience, New York, 1976
2. Stanbury JB, Wyngaarden JB, Fredrickson DS, et al (eds): The Metabolic Basis of Inherited Disease. 4th Ed. McGraw-Hill, New York, 1983
3. Lolowick SP: The Hexokinases. pp. 1–48. In Buyer PD (ed): The Enzymes. 3rd Ed. Vol. 9, Part B. Academic Press. New York, 1976
4. Hers HG, Hue L: Gluconeogenesis and related aspects of glycolysis. Ann Rev Biochem 52:617, 1983
5. Claus TH, El-Maghrabi MR, Regen DM, et al: The role of fructose-2,6-bisphosphate in the regulation of carbohydrate metabolism. Curr Top Cell Regul 23:57, 1984

STUDY QUESTIONS

Directions: Answer the following questions using the key outlined below.
- **(A)** is 1, 2 and 3 are correct
- **(B)** is 1 and 3 are correct
- **(C)** if 2 and 4 are correct
- **(D)** if only 4 is correct
- **(E)** if all four are correct

1. The dietary polysaccharide(s) digested by humans is/are
    1. glycogen
    2. glycogen and amylopectin
    3. amylose
    4. celluloses and glycogen

2. The structure of amylopectin most resembles that of
    1. cellulose
    2. maltotriose
    3. maltose
    4. glycogen

3. All of the following statement(s) about dietary carbohydrate are true EXCEPT
    1. Only monosaccharides are absorbed by the intestine.
    2. Significant starch digestion occurs in the mouth because of the presence of α-amylase in the saliva.
    3. Most dietary glucose is taken up initially by the liver.
    4. Disaccharidases are located inside intestinal mucosal cells.

Directions: For each of the following multiple choice questions, choose the most appropriate answer.

4. The most important allosteric effector of glycolysis in the liver is:
   A. citrate
   B. AMP
   C. fructose 2,6-bisphosphate
   D. glucose 6-phosphate

5. Two enzymes that are important in gluconeogenesis are
   A. glucose 6-phosphatase and pyruvate kinase
   B. pyruvate kinase and pyruvate carboxylase
   C. fructose 1,6-bisphosphatase and pyruvate kinase
   D. pyruvate carboxylase and phosphoenolpyruvate carboxykinase

6. The glycerol phosphate shuttle functions to convert
   A. NADH equivalents from the cytosol to the mitochondria
   B. NADH equivalents in the cytosol to $FADH_2$ equivalents in the mitochondria
   C. glycerol 3-phosphate in the cytosol to dihydroxyacetone phosphate in the mitochondria
   D. NADH equivalents in the mitochondria to $FADH_2$ in cytosol

7. The product of glycolysis may be either lactate or pyruvate, depending on
   A. oxygen availability
   B. the rate of glycolysis
   C. the mitochondrial supply of NADH
   D. the presence of pyruvate dehydrogenase

8. The net ATP production from lactate fermentation of glucose is
   A. 4
   B. 2
   C. 20
   D. 8

9. The high glucose K_M exhibited by glucokinase
   A. causes a much higher rate of glucose synthesis in the liver
   B. makes blood glucose more readily available to liver for glycolysis
   C. ensures that significant liver glycogen synthesis does not occur until blood glucose levels are elevated above normal
   D. ensures a low rate of glycolysis in extrahepatic tissues

10. AMP and fructose 2,6-bisphosphate regulate both glycolysis and gluconeogenesis by
    A. inhibiting phosphofructokinase and activating pyruvate carboxylase
    B. inhibiting phosphofructokinase and activating pyruvate kinase
    C. activating hexokinase and inhibiting fructose 1,6-bisphosphatase
    D. activating phosphofructokinase and inhibiting fructose 1,6-bisphosphatase

11. Which one of the following statements is true?
    A. Dietary glucose is initially stored in the liver as glycogen
    B. Gluconeogenesis occurs to prevent the excessive conversion of glucose to lipid inside the liver
    C. Glucose 6-phosphatase contributes little to the maintenance of blood glucose levels
    D. Citrate allosterically inhibits pyruvate kinase and slows glycolysis

6

MITOCHONDRIAL OXIDATIONS

The Mitochondrion/The Tricarboxylic Acid Cycle/
Electron Transport/Oxidative Phosphorylation/
Clinical Correlation

Learning Objectives

The student should be able to:

1. Describe the structural components of the mitochondrion.
2. Describe the functions of the TCA cycle and its substrates and products.
3. Describe the significance of anaplerotic reactions.
4. Illustrate the sequential arrangement of the components of the mitochondrial electron transport chain.
5. Describe the mechanism of oxidative phosphorylation.
6. Describe the reaction catalyzed by ATP synthetase and its relation to the mitochondrial proton gradient.
7. Define respiratory control and the effect on the electron transport chain and oxidative phosphorylation of uncouplers and of inhibitors of ATP synthetase.

Perspective

Most of the cell's energy comes from the oxidation of acetyl CoA in mitochondria. Glycolysis oxidizes sugars to pyruvate, which is converted to acetyl CoA in mitochondria. Proteins and fatty acids are also broken down to yield acetyl CoA (Fig. 6-1). Acetyl units are oxidized to carbon dioxide by the *tricarboxylic acid (TCA) cycle* in the mitochondrial matrix. The energy released during the oxidation reactions is collected mainly as reducing equivalents—electron pairs—in NADH and FADH$_2$. These compounds are oxidized by the *electron transport chain* on the mitochondrial inner membrane. The electron transport chain consists of a series of electron acceptors of progressively lower reducing potential. The chain terminates with molecular oxygen, which accepts the electrons and is reduced to water. The free energy released as electrons flow

(continued)

down the gradient of reducing potential is used to pump protons out of the mitochondrial matrix, generating a proton gradient across the inner mitochondrial membrane. The controlled flow of these protons back across the membrane drives the synthesis of ATP in the process known as *oxidative phosphorylation*.

THE MITOCHONDRION

The reactions of the TCA cycle and oxidative phosphorylation occur in the mitochondrion, which is thus the premier energy-producing organelle of the cell. Cells with high energy requirements contain more mitochondria per unit cell volume than cells with lower energy requirements.

The mitochondrion consists of an *outer membrane*, an *inner membrane*, an *intermembrane space*, and the *matrix* or *mitosol* inside the inner membrane (Fig. 6-2). The outer membrane is composed of approximately 50 percent protein and 50 percent lipid. This membrane is very permeable, and most cytosolic solutes have free access to the intermembrane space and the outer surface of the inner mitochondrial membrane. Only a few proteins are normally found in the intermembrane space. The most notable of these are adenylate kinase and sulfite oxidase, both of which have been used as marker enzymes in identifying intermembrane space materials.

The inner membrane is structurally and functionally very sophisticated. It consists of about 80 percent protein and 20 percent lipid. The proteins of the inner membrane include the enzymes of electron transport and oxidative phosphorylation as well as the many transport proteins that regulate the flux of ions and metabolites in and out of the mitochondrial matrix. The predominant phospholipids of the membrane are phosphatidylcholine, phosphatidylethanolamine, and car-

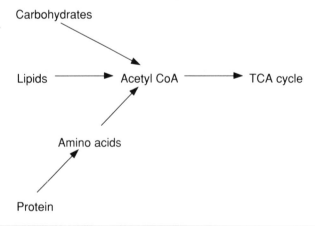

Fig. 6-1. Acetyl CoA is produced by oxidation of carbohydrates, lipids, and proteins and amino acids.

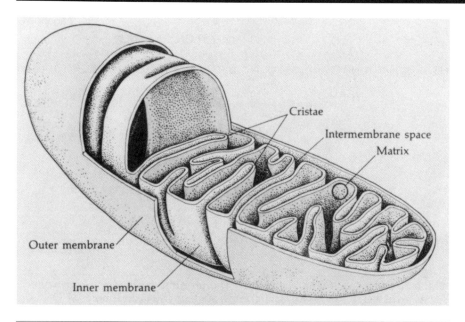

Fig. 6-2. Mitochondrial structure. (From Wolfe S: Introduction to Cell Biology. Wadsworth, Belmont, CA, 1974, with permission.)

diolipin. The membrane is very hydrophobic and very impermeable; polar solutes can cross it only via specific transport systems. The inner membrane is folded into many deep invaginations called *cristae*, which greatly increase its surface area and hence its capacity for transmembrane transport, electron transport, and oxidative phosphorylation.

The matrix contains the enzymes of the TCA cycle, fatty acid β-oxidation (Chapter 7), and some of the enzymes involved in amino acid and nitrogen metabolism. The ion and metabolite content of the matrix is rigidly controlled by the inner mitochondrial membrane transport systems.

PYRUVATE DEHYDROGENASE

The pyruvate produced by glycolysis must be decarboxylated to acetyl CoA before it can enter the TCA cycle. This reaction is catalyzed by *pyruvate dehydrogenase*, a large enzyme complex located in the mitochondrial matrix. The overall reaction is

$$\text{Pyruvate} + \text{CoA} + \text{NAD}^+ \rightarrow \text{acetyl CoA} + \text{CO}_2 + \text{NADH} + \text{H}^+$$

The NADH produced by this reaction feeds its electrons to the electron transport chain. Coenzyme A is a cofactor that is often used in the transfer of activated acyl groups. The terminal sulfhydryl group of the CoA molecule forms a thioester linkage with acyl groups (Fig. 6-3). This thioester bond is weaker than the corresponding oxygen ester bond, so acyl CoA compounds are more reactive.

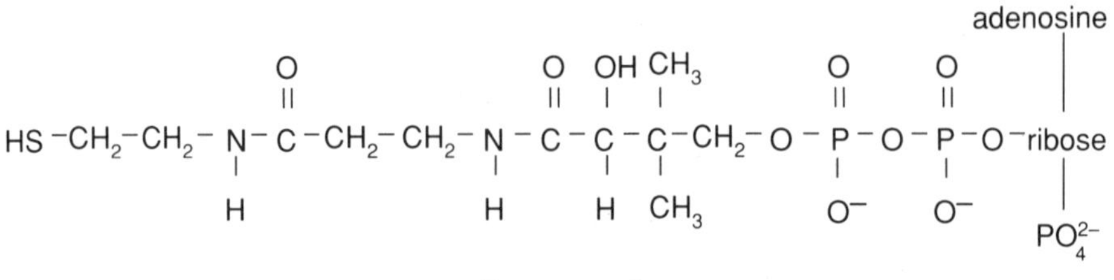

$$\text{HS-CH}_2\text{-CH}_2\text{-N(H)-C(=O)-CH}_2\text{-CH}_2\text{-N(H)-C(=O)-C(OH)(H)-C(CH}_3)_2\text{-CH}_2\text{-O-P(=O)(O}^-)\text{-O-P(=O)(O}^-)\text{-O-ribose-adenosine, PO}_4^{2-}$$

Coenzyme A

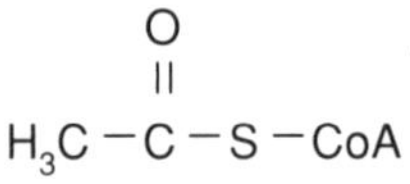

$$\text{H}_3\text{C-C(=O)-S-CoA}$$

Acetyl CoA

Fig. 6-3. Coenzyme A and acetyl CoA. The terminal sulfhydryl of coenzyme A forms a thioester bond with acyl groups, such as the acetyl moiety.

Regulation of Pyruvate Dehydrogenase

Pyruvate dehydrogenase is a very important regulated enzyme because it irreversibly commits pyruvate to a specific range of metabolic fates: animals are unable to convert acetyl CoA back into pyruvate. Like most regulated enzymes of energy metabolism, pyruvate dehydrogenase responds to the energy charge of the cell: it is inhibited by GTP and ATP, and stimulated by ADP and AMP. Like the regulated enzymes of the TCA cycle, it is stimulated by NAD^+ and inhibited by NADH. It is also inhibited by its own product acetyl CoA, and stimulated by free CoA. It can be stimulated by Ca^{2+} and insulin. The regulation of pyruvate dehydrogenase is shown in Figure 6-5.

Perspective: The Tricarboxylic Acid Cycle

The TCA cycle (also called the *citric acid cycle* and the *Krebs cycle*) oxidizes acetyl units to CO_2, and harvests the released free energy in the form of NADH, reduced flavin adenine dinucleotide ($FADH_2$), and GTP (Fig. 6-4). The oxidation of one acetyl group to two molecules of CO_2 requires one complete turn of the cycle, and generates 3 NADH, 1 $FADH_2$, and 1 GTP. Several TCA cycle intermediates are also precursors of gluconeogenesis and amino acid biosynthesis.

THE TRICARBOXYLIC ACID CYCLE

Figure 6-4 shows the reactions of the TCA cycle. The net reaction for the oxidation of one acetyl unit is

Acetyl-CoA + 3 NAD$^+$ + FAD + GDP + P$_i$ →

2 CO$_2$ + 3 NADH + 3 H$^+$ + FADH$_2$ + GTP + CoA

Reactions of the TCA cycle

Citrate Synthase

The first step of the TCA cycle is the condensation of acetyl CoA with oxaloacetate to yield citrate, catalyzed by citrate synthase:

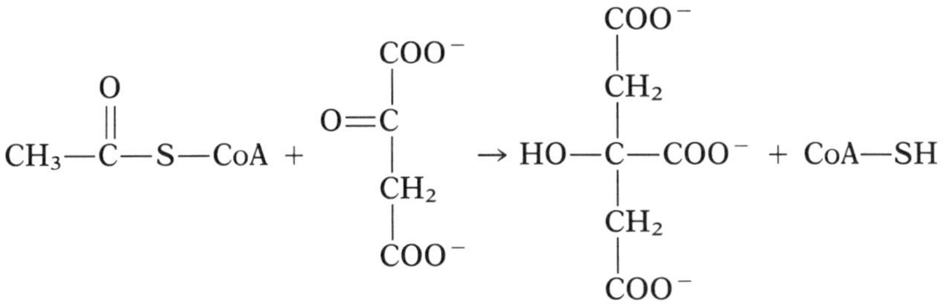

$$\text{CH}_3-\overset{\text{O}}{\underset{\|}{\text{C}}}-\text{S}-\text{CoA} + \begin{array}{c}\text{COO}^-\\|\\\text{O=C}\\|\\\text{CH}_2\\|\\\text{COO}^-\end{array} \rightarrow \begin{array}{c}\text{COO}^-\\|\\\text{CH}_2\\|\\\text{HO}-\text{C}-\text{COO}^-\\|\\\text{CH}_2\\|\\\text{COO}^-\end{array} + \text{CoA}-\text{SH}$$

Acetyl CoA   Oxaloacetate   Citrate

Under cellular conditions, this reaction has a large negative ΔG and is essentially irreversible.

Aconitase

In a reversible reaction, *aconitase* then converts citrate to isocitrate:

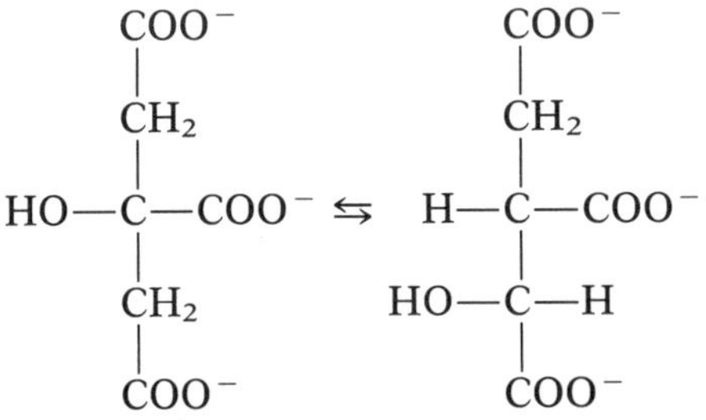

$$\begin{array}{c}\text{COO}^-\\|\\\text{CH}_2\\|\\\text{HO}-\text{C}-\text{COO}^-\\|\\\text{CH}_2\\|\\\text{COO}^-\end{array} \rightleftharpoons \begin{array}{c}\text{COO}^-\\|\\\text{CH}_2\\|\\\text{H}-\text{C}-\text{COO}^-\\|\\\text{HO}-\text{C}-\text{H}\\|\\\text{COO}^-\end{array}$$

Citrate   Isocitrate

The equilibrium for this reaction lies on the side of citrate formation. The reaction is driven in the forward direction by the oxidation of isocitrate in the next reaction.

Isocitrate Dehydrogenase

In the next step, *isocitrate dehydrogenase* decarboxylates isocitrate to α-ketoglutarate and CO$_2$ with the reduction of a molecule of NAD$^+$ to

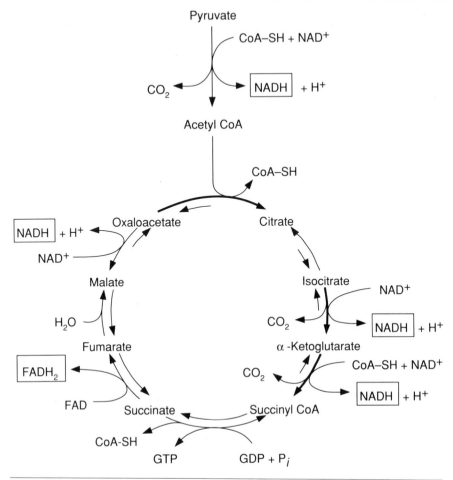

Fig. 6-4. Pyruvate carboxylase and the tricarboxylic acid cycle. The three exergonic reactions of the TCA cycle are indicated by heavy arrows.

NADH. The reaction proceeds via an enzyme-bound oxalosuccinate intermediate, as follows:

Under cellular conditions, the equilibrium of this reaction favors formation of α-ketoglutarate.

α-Ketoglutarate Dehydrogenase

In the next reaction, α-ketoglutarate is oxidized to the four-carbon compound succinyl CoA by *α-ketoglutarate dehydrogenase*, with the release of the second CO_2 and the reduction of another NAD^+ to NADH:

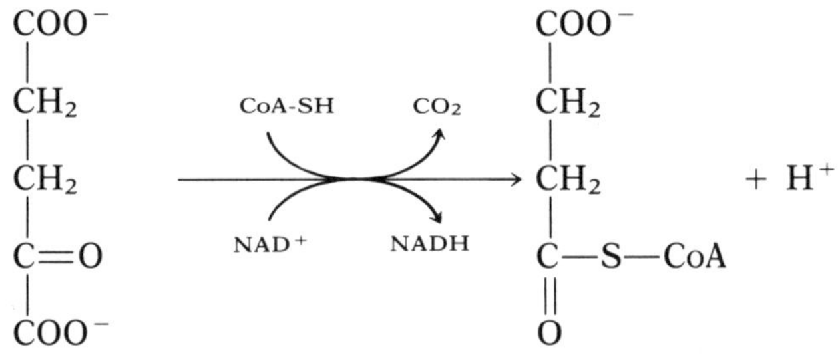

$$\begin{array}{c} COO^- \\ | \\ CH_2 \\ | \\ CH_2 \\ | \\ C=O \\ | \\ COO^- \end{array} \quad \xrightarrow[\text{NAD}^+ \quad \text{NADH}]{\text{CoA-SH} \quad CO_2} \quad \begin{array}{c} COO^- \\ | \\ CH_2 \\ | \\ CH_2 \\ | \\ C-S-CoA \\ \| \\ O \end{array} \quad + \; H^+$$

α-Ketoglutarate   Succinyl CoA

The succinyl CoA produced in this reaction is an energy-rich compound (similar to acetyl CoA). This reaction is irreversible under cellular conditions.

Succinate Thiokinase

The enzyme *succinate thiokinase* (also called *succinyl CoA synthetase*) uses energy-rich succinyl CoA to drive the synthesis of GTP:

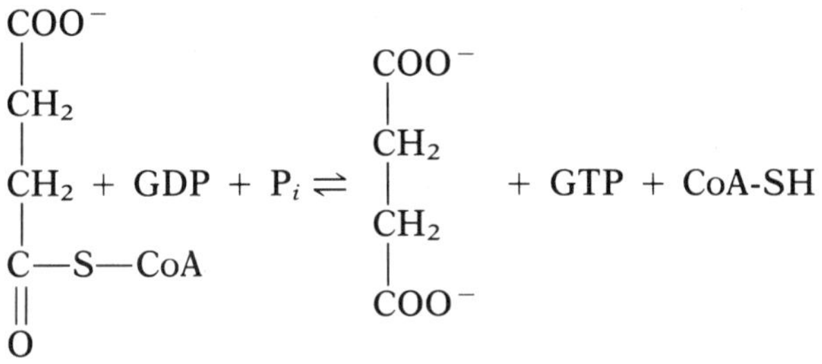

$$\begin{array}{c} COO^- \\ | \\ CH_2 \\ | \\ CH_2 \\ | \\ C-S-CoA \\ \| \\ O \end{array} + GDP + P_i \rightleftharpoons \begin{array}{c} COO^- \\ | \\ CH_2 \\ | \\ CH_2 \\ | \\ COO^- \end{array} + GTP + CoA\text{-}SH$$

Succinyl CoA   Succinate

GTP can phosphorylate ADP, in a reversible reaction catalyzed by *nucleoside diphosphokinase*:

$$GTP + ADP \rightleftharpoons GDP + ATP$$

Succinate Dehydrogenase

In the next reaction, *succinate dehydrogenase* converts succinate to fumarate, with the reduction of FAD to $FADH_2$:

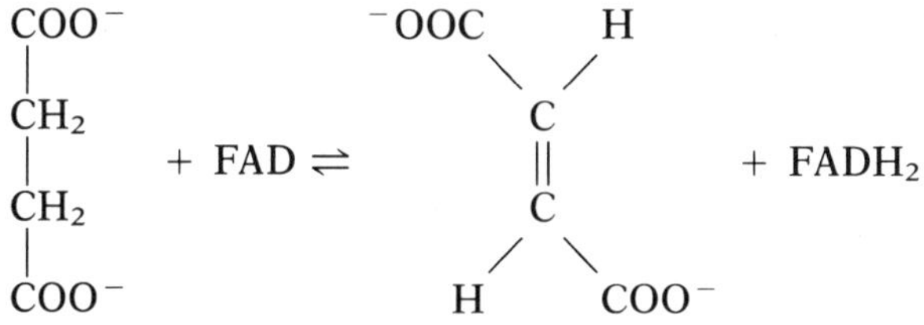

$$\begin{array}{c} COO^- \\ | \\ CH_2 \\ | \\ CH_2 \\ | \\ COO^- \end{array} + FAD \rightleftharpoons \begin{array}{c} {}^-OOC \quad\;\; H \\ \diagdown \;\; \diagup \\ C \\ \| \\ C \\ \diagup \;\; \diagdown \\ H \quad\;\; COO^- \end{array} + FADH_2$$

Succinate   Fumarate

Succinate dehydrogenase is part of an enzyme complex embedded in the inner mitochondrial membrane; the FAD/FADH$_2$ involved in this reaction is a bound cofactor of the enzyme complex. Succinate dehydrogenase is the only enzyme of the TCA cycle that is not in solution in the matrix.

Fumarase

Fumarase then reversibly hydrates fumarate to L-malate:

$$\text{Fumarate} + H_2O \rightleftharpoons \text{L-Malate}$$

Fumarase will only hydrate the *trans* isomer of fumarate.

Malate Dehydrogenase

In the last reaction of the TCA cycle, *malate dehydrogenase* oxidizes malate to oxaloacetate with the production of another NADH:

$$\text{Malate} + NAD^+ \rightleftharpoons \text{Oxaloacetate} + NADH + H^+$$

The regenerated oxaloacetate is ready to condense with another molecule of acetyl CoA.

The malate dehydrogenase reaction is the only reaction of the TCA cycle that is strongly endergonic ($\Delta G° = +7.1$ kcal/mol). It is "pulled" to the right because its product oxaloacetate is continuously consumed by the citrate synthase reaction.

The TCA Cycle Equilibrium

The TCA cycle as a whole has a negative ΔG (see Chapter 3 for a discussion of free energy). Two of its steps (the reactions catalyzed by citrate synthase and α-ketoglutarate dehydrogenase) have substantial negative $\Delta G°$s, one (malate dehydrogenase) has a large positive $\Delta G°$, and the rest have small negative or positive $\Delta G°$s (see Fig. 6-4). The overall direction of the cycle is determined by the sum of positive and negative

$\Delta G°$s. Since the sum of negative $\Delta G°$s is significantly greater than the sum of positive $\Delta G°$s, the cycle proceeds in the forward direction.

REGULATION OF THE TCA CYCLE

The TCA cycle constitutes an intersection of carbohydrate, protein, and lipid metabolism, all three of which supply it with acetyl CoA. In addition, several TCA cycle intermediates are precursors of glucose and amino acid synthesis, and acetyl CoA can be used for lipid biosynthesis. The TCA cycle must respond to these synthetic needs as well as to the cell's requirement for ATP. The regulation of the TCA cycle shows both a coarse and a fine level of control.

Coarse Control of the TCA Cycle

The rate of the TCA cycle is limited by the supply of its various substrates, particularly FAD and NAD^+. Any factor that slows electron transport will cause a buildup of $FADH_2$ and NADH and a corresponding deficit of FAD and NAD^+. Since electron transport is normally coupled to the synthesis of ATP by oxidative phosphorylation, any factor that slows oxidative phosphorylation will also slow the TCA cycle. Oxidative phosphorylation requires both ADP and oxygen. When the cell is using ATP at a low rate, and therefore generating little ADP, the TCA cycle will be inhibited. A diminished oxygen supply also inhibits the TCA cycle. Thus, the coarse control factors that slow the TCA cycle are

1. Decreased [ADP]
2. Decreased oxygen supply

Fine Control of the TCA Cycle

Four enzymes of the TCA cycle are regulated by allosteric effectors. Isocitrate dehydrogenase is the rate-limiting enzyme of the cycle. It is inhibited by ATP and NADH and stimulated by ADP. Figure 6-5 shows the main allosteric effectors of the regulated enzymes of the TCA cycle and of pyruvate decarboxylase. As can be seen, these enzymes are controlled primarily by the $NADH/NAD^+$ and ATP/ADP ratios.

To summarize:

1. The fine regulation of the TCA cycle is provided by allosteric regulation of four TCA cycle enzymes.
2. Isocitrate dehydrogenase is the rate-controlling enzyme.
3. The activity of the regulated enzymes is determined by the $NADH/NAD^+$ and ATP/ADP ratios.

Anaplerotic Reactions

Three TCA cycle intermediates—oxaloacetate, α-ketoglutarate, and succinyl CoA—are substrates of other cellular pathways. When these intermediaries are drawn off into the alternate pathways, they must be

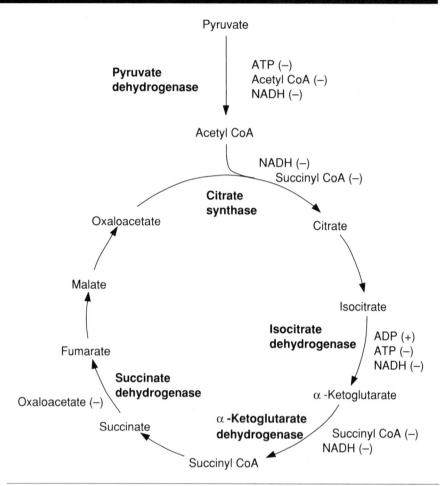

Fig. 6-5. the major allosteric regulators of pyruvate dehydrogenase and the TCA cycle. (+) indicates enzyme activators; (−) indicates enzyme inhibitors. As can be seen by comparing with Figure 6-4, the three exergonic reactions of the TCA cycle are all regulated.

replenished in the mitochondrion in order to maintain TCA cycle activity. The reactions that replenish them are called *anaplerotic reactions*. Three important anaplerotic reactions for the TCA cycle are catalyzed by *pyruvate carboxylase, malic enzyme,* and *glutamate dehydrogenase*:

1. Pyruvate + HCO_3^- + ATP $\xrightarrow{\text{Pyruvate carboxylase}}$ oxaloacetate + ADP + P_i

2. Pyruvate + NADPH + CO_2 + H^+ $\xrightarrow{\text{Malic enzyme}}$ malate + $NADP^+$

3. Glutamate + NAD^+ ($NADP^+$) $\xrightarrow{\text{Glutamate dehydrogenase}}$ α-ketoglutarate + NADH (NADPH) + NH_4^+

The carboxylation of pyruvate to oxaloacetate by pyruvate carboxylase was mentioned in Chapter 5 in connection with gluconeogenesis.

ELECTRON TRANSFER REACTIONS

Electron transfer reactions are by definition oxidation-reduction (redox) reactions. When electrons are transferred from compound A to compound B, compound A is *oxidized* and compound B is *reduced*. The substance that is reduced is called the *oxidant* because it oxidizes the electron donor; the electron donor is called the *reductant* because it reduces the electron acceptor (oxidant). Electron transfer reactions may involve the transfer of electrons alone, or the electrons may be accompanied by protons and be carried in the form of hydrogen atoms. Many biological processes involve electron transfer reactions. As we have seen, most of the energy obtained from the degradation of fuel molecules is initially harvested in the form of reducing power (electrons).

Every redox couple (such as $NAD^+/NADH$) has a *standard reduction potential* E_0' measured in volts (see Appendix 2). Stronger electron donors have more negative reduction potentials. For example, under standard conditions, NADH, which has an $E_0' = -0.32$ V, can reduce pyruvate, which has an $E_0' = -0.19$ V. The change in standard reduction potential, $\Delta E_0'$, for an electron transfer reaction or pathway (the difference between the reduction potentials of the first and last species) is related to the standard free energy change $\Delta G^{o\prime}$ by the following equation:

$$\Delta G^{o\prime} = -nF \Delta E_0'$$

where n is the number of electrons transferred and F is the Faraday constant (see Appendix 2). As is clear from this equation, a redox exchange that is exergonic under standard conditions (negative $\Delta G^{o\prime}$) will have a *positive* $\Delta E_0'$.

Electron Transfer Cofactors

Electron transfer reactions are catalyzed by a wide variety of enzymes, but the electrons themselves are carried by a limited number of specialized cofactors. Some of these cofactors are prosthetic groups, whereas others are coenzymes. Many multisubunit electron-transfer complexes—such as the complexes of the mitochondrial electron-transport chain—contain several cofactors arranged in an internal electron-transport chain. Some electron transfer cofactors are described below.

NAD^+, $NADP^+$, FAD^+, FMN, COENZYME Q

The compounds NAD^+, $NADP^+$, FAD, *flavin mononucleotide* (FMN), and *coenzyme Q* (also called *ubiquinone*) (Fig. 6-6) all participate in reactions involving the transfer of two electrons, which are carried with associ-

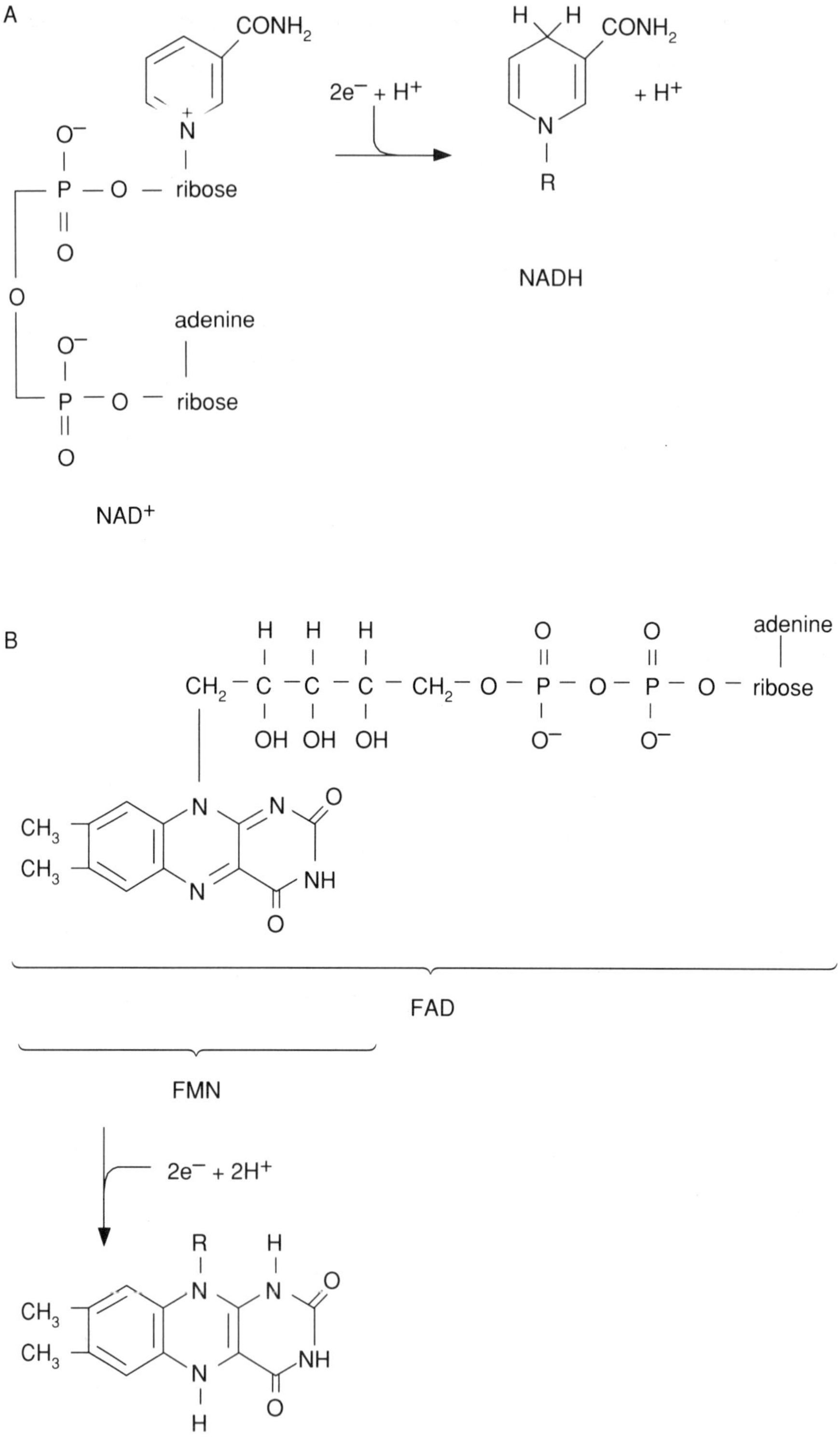

Fig. 6-6. Oxidized and reduced forms of **(A)** NAD^+, **(B)** FMN and FAD (*Figure continues*.)

C

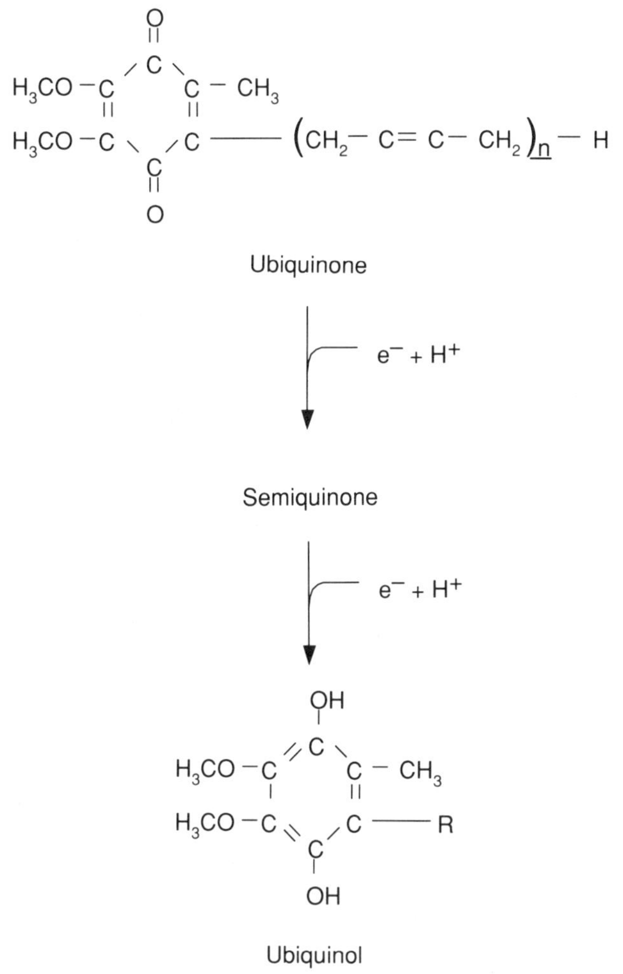

[Structure of Ubiquinone: dimethoxy-dimethyl-benzoquinone with isoprenoid tail $(CH_2-C=C-CH_2)_n-H$]

Ubiquinone

↓ $e^- + H^+$

Semiquinone

↓ $e^- + H^+$

[Structure of Ubiquinol with OH groups and R side chain]

Ubiquinol

Fig. 6-6. (Continued). and **(C)** coenzyme Q (ubiquinone). n in ubiquinone usually varies between 6 and 10. $NADP^+$ is identical to NAD^+ except for a phosphate group esterified to the 2' position of the adenosine moiety.

---

ated protons. The reduced forms of these compounds are NADH, NADPH, $FADH_2$, $FMNH_2$, and ubiquinol (QH_2). Coenzyme Q can alternatively participate in one-electron transfers, in which case the *semiquinone* form (QH·) is generated.

NADH and $FADH_2$ are the immediate donors of electrons to the mitochondrial electron transport chain. FMN and coenzyme Q function as electron carriers in the chain. FMN is associated with an enzyme complex, whereas coenzyme Q is not associated with a protein, but is anchored in the lipid phase of the inner mitochondrial membrane by its long isoprenoid tail.

Non-Heme Iron-Sulfur Centers

Some electron transfer enzymes have a prosthetic group that consists of inorganic iron or iron and sulfur atoms coordinated to cysteine sulfurs on the protein (Fig. 6-7). These *non-heme iron-sulfur centers* all participate in one-electron transfers in which the iron atom oscillates between the ferric (+3) and the ferrous (+2) states. The mitochondrial electron transport chain contains several iron-sulfur proteins.

Cytochromes

Cytochromes are electron-transfer proteins that contain a heme group. The iron of the heme accepts a single electron by changing from the +3 to the +2 state. Different types of cytochromes can be distinguished by their characteristic heme spectra. Cytochromes are present in all aerobic organisms. The mitochondrial electron transport chain also includes several cytochromes.

Metal Ions

Some enzymes use bound metal ions such as Cu, Fe, and Mo to catalyze electron transfers.

Perspective: The Mitochondrial Electron Transport Chain

The mitochondrial electron transport chain consists of four enzyme complexes embedded in the inner mitochondrial membrane, plus two free electron carriers. Electrons enter the chain from NADH and $FADH_2$, and are passed from from carrier to carrier down a gradient of reducing potential, finally reducing oxygen to water. The free energy released as the electrons flow down the chain is used to pump protons out of the mitochondrial matrix. The resulting electrochemical gradient across the inner mitochondrial membrane drives oxidative phosphorylation. The electron transport chain thus acts as a transformer: it takes the energy of oxidation trapped as reducing potential, and converts it to the electrochemical energy of the proton gradient.

In sum, the functions of the electron transport chain are as follows:

1. It oxidizes reduced coenzymes (NADH and $FADH_2$) with the reduction of oxygen to water. This process is associated with a significant release of free energy, which is used to drive the synthesis of ATP.
2. It resupplies the cell with NAD^+ and FAD, allowing the energy-generating processes that use these compounds (glycolysis, the TCA cycle, and fatty acid degradation) to continue.

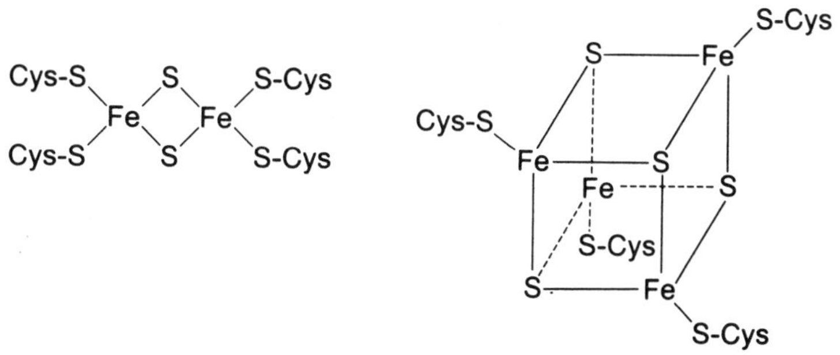

Fig. 6-7. Two examples of iron-sulfur centers. (From Zubay G: Biochemistry. 3rd Ed. Macmillan, New York, 1988, with permission.)

THE MITOCHONDRIAL ELECTRON TRANSPORT CHAIN

Figure 6-8 shows the organization of the mitochondrial electron transport chain. The elements of the mitochondrial electron transport chain are components of the mitochondrial inner membrane. Treatment of mitochondria with detergents yields submitochondrial particles that have been isolated, studied, and assigned functional roles in the electron transport chain. *Complex I* accepts electrons from NADH and transfers them to coenzyme Q. *Complex II* accepts electrons from $FADH_2$ and also reduces coenzyme Q. Coenzyme Q, which is not part of a complex, transfers its electrons to *complex III*. Complex III then reduces cytochrome c, which is also not part of a complex, and which in turn reduces *complex IV*. An additional submitochondrial particle, *complex V*, mediates the phosphorylation of ADP; it will be discussed in the section on oxidative phosphorylation.

Complex I: NADH → Coenzyme Q (NADH-Q Reductase)

Complex I contains bound FMN in association with NADH dehydrogenase and a series of iron-sulfur proteins. The two electrons from NADH first reduce FMN to $FMNH_2$; the iron-sulfur proteins are reduced next, and finally coenzyme Q is reduced to the ubiquinol form.

Complex II: $FADH_2$ → Coenzyme Q

Succinate dehydrogenase and an iron-sulfur protein make up complex II. The $FADH_2$ component of succinate dehydrogenase reduces the iron-sulfur center, which in turn reduces coenzyme Q to ubiquinol.

Complex III: Coenzyme Q → Cytochrome c (Cytochrome Reductase)

Complex III consists of cytochrome b, an iron-sulfur protein, and cytochrome c_1. Electrons are transferred one at a time from coenzyme Q through this complex to cytochrome c, which is only loosely associated

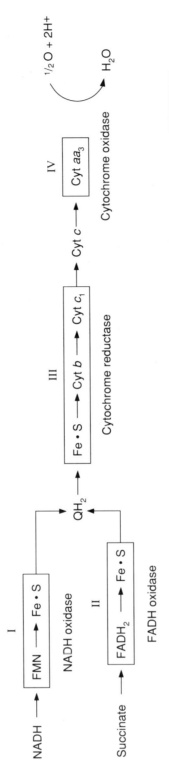

Fig. 6-8. The electron transport chain. The chain represents a gradient of reduction potential, from electron acceptors with stronger (more negative) reduction potentials to those with less strong (more positive) reduction potentials. Complexes I through IV are shown.

with the inner mitochondrial membrane. Coenzyme Q functions as a mediator between the two-electron carriers (complexes I and II) and the one-electron carriers (complexes III and IV).

Complex IV: Cytochrome $c \rightarrow O_2$ (Cytochrome Oxidase)

The cytochrome c oxidase complex (cytochrome oxidase) is the last component in the electron transport chain. It accepts electrons from cytochrome c and catalyzes the four-electron reduction of molecular oxygen to water:

$$4\ e^- + 4\ H^+ + O_2 \rightarrow H_2O$$

Cytochrome oxidase has a very complex quaternary structure. It is believed to exist in the inner mitochondrial membrane as two complexes containing seven subunits each (Fig. 6-9). Subunits I and II both contain a-type hemes. The heme of subunit I is referred to as heme a_3, and the heme of subunit II as heme a (cytochrome oxidase is also called cytochrome aa_3). Subunit II contains two copper atoms, and subunit III is the proton translocase. One of the coppers of subunit II is believed to cooperate with heme a in the oxidation of cytochrome c, and the other copper is believed to be involved with heme a_3 in the reduction of oxygen. The path of electrons through cytochrome oxidase is thus believed to be

$$\text{Cytochrome } c \rightarrow (\text{heme } a \cdot \text{Cu}) \rightarrow (\text{heme } a_3 \cdot \text{Cu}) \rightarrow O_2$$

Inhibitors of Electron Transport

The order of the components of the electron transport chain was deduced by using inhibitors that block electron transfer at specific sites in the chain. When a step in the chain is blocked, the components upstream from the block become fully reduced, whereas the components downstream from it become fully oxidized. Each component in the chain has distinct oxidized and reduced absorption spectra. Thus, the absorption spectrum reveals which components are oxidized and which are reduced after an inhibitor is introduced. Table 6-1 lists some inhibitors and the steps they block.

Table 6-1. Inhibitors of Electron Transport

Inhibitors	Site of Action
Rotenone	Complex I
Amytal	Complex I
Antimycin A	Complex III
Carbon monoxide	Complex IV
Cyanide	Complex IV
Azide	Complex IV

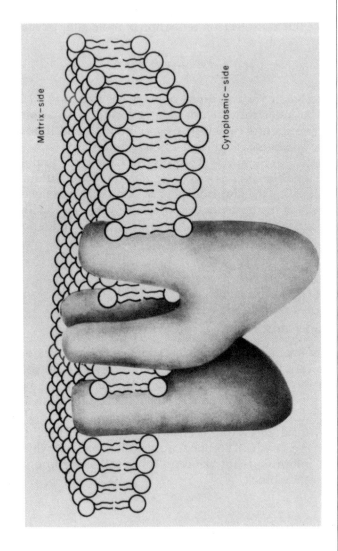

Fig. 6-9. Model of cytochrome oxidase. (From Frey TG, Costello MJ, Karlson B, et al: Structure of the cytochrome c oxidase dimer: Electron microscopy of two-dimensional crystals. J Mol Biol 162:113, 1982, with permission.)

Key Points: The Electron Transport Chain

1. All the components of the chain except unbound cytochrome c are embedded in the inner mitochondrial membrane.
2. The chain consists of four submitochondrial enzyme complexes, each of which is responsible for the transfer of electrons between specific segments of the chain.
3. Electrons flow down a gradient of redox potential (i.e., from compounds with more negative E to compounds with more positive E).
4. Inhibitors can interrupt the flow of electrons at various points in the chain, and the absorption spectra of the chain blocked by various inhibitors can be used to deduce the path of electrons through the chain.

Perspective: Oxidative Phosphorylation

The mitochondrial electron transport chain mediates the reduction of molecular oxygen to water by NADH and $FADH_2$. The transport of electrons through the electron transport chain makes available a significant quantity of free energy, which is used to drive the synthesis of ATP in the process of *oxidative phosphorylation*.

OXIDATIVE PHOSPHORYLATION

The mechanism of oxidative phosphorylation has not been demonstrated unequivocally. The most widely accepted model is the *chemiosmotic hypothesis*. The essential features of this model are as follows (Fig. 6-10):

1. The ATP synthesizing system, like the components of the electron transport chain, is an integral component of the mitochondrial inner membrane.
2. The energy released as a result of the flow of electrons down the chain is used to translocate protons from the mitochondrial matrix to the intermembrane space.
3. This flow of protons establishes an electrochemical gradient across the membrane, with the matrix alkaline and negatively charged relative to the intermembrane space.

Thus, the energy released by the flow of electrons in the electron transport chain is stored in the form of the proton gradient across the membrane.

4. Protons are allowed to flow back into the matrix. The flow of protons down their electrochemical gradient releases energy, which is used to drive the synthesis of ATP.
5. The re-entry of protons into the matrix and the synthesis of ATP are mediated by the *ATP synthetase complex*.

The energy transformations occurring during oxidative phosphorylation may be summarized as follows:

Electron transport → energy → proton gradient → ATP synthesis

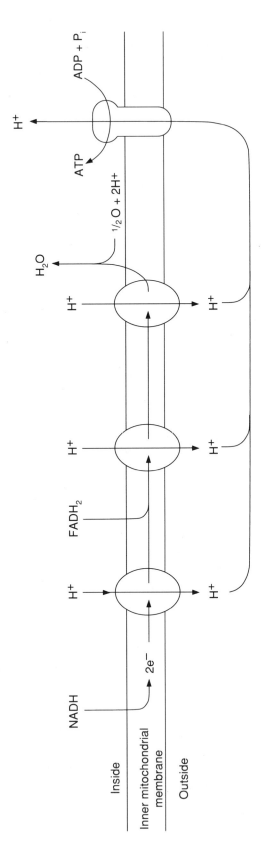

Fig. 6-10. The chemiosmotic model of oxidative phosphorylation.

Several lines of evidence led to the adoption of the chemiosmotic hypothesis. When the components of the electron transport chain are inserted into the membranes of synthetic vesicles and an electron donor is added, electron transport is accompanied by pumping of protons across the vesicle membrane. If ATP synthetase is added to the membrane and ADP and P_i are supplied, ATP synthesis occurs, accompanied by equilibration of protons back across the membrane. Oxidative phosphorylation only occurs when the components of the system are incorporated in a membrane that separates distinct compartments, not when they are free in solution.

In the intact mitochondrion, oxidation of NADH can drive the synthesis of three ATPs, whereas $FADH_2$, which bypasses electron transport complex I, yields only two. Complexes I, III, and IV each contain a proton pump, which is part of the ATP synthetase complex. The passage of a pair of electrons causes each complex to pump enough protons to generate one ATP.

A minimum free energy change of about 13 kcal/mol, corresponding to a change in reduction potential of about 0.25 V, is needed to support the synthesis of one ATP. Complexes I, III, and IV all have large enough ΔEs for the passage of two electrons to support the synthesis of one ATP, and, in fact, flow of electrons through each of these complexes has been shown to support the synthesis of one ATP. Consequently, complexes I, III, and IV are also called *phosphorylation sites I, II, and III*, respectively.

Two electrons are required to reduce one atom of oxygen to water. Therefore, the oxidation of one molecule of NADH or $FADH_2$ corresponds to the synthesis of three or two ATPs, respectively, and to the reduction of one atom of oxygen. As it is usually stated, the oxidation of NADH and $FADH_2$ occur with *P/O ratios* of 3 and 2, respectively.

The ATP Synthetase Complex (Complex V)

The ATP synthetase complex consists of a stem that spans the inner mitochondrial membrane and a stalk and head that protrude into the matrix (Fig. 6-11). The stem is called the F_0 *region*, and the stalk and head constitute the F_1 *region*. The stem contains the proton channel. The head is made up of five kinds of subunits (α, β, γ, δ, and ϵ), and is the catalytic segment. If the F_1 segment is dissociated from the proton channel and solubilized, it acts as a Ca^{2+},Mg^{2+}-dependent reversible ATPase. The proton channel is needed to drive the reaction in the direction of ATP synthesis. The reaction catalyzed by ATP synthetase is

$$ADP + P_i + H^+ \rightarrow ATP + H_2O$$

Although the ATP synthetase complex is called complex V, it is not part of the electron transport chain.

Net Energy Yield from the Complete Oxidation of Glucose

The complete oxidation of a molecule of glucose involves the glycolytic pathway, pyruvate dehydrogenase, the TCA cycle, and electron trans-

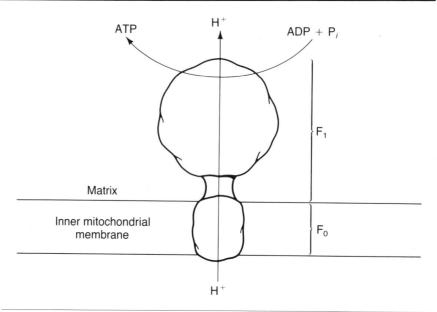

Fig. 6-11. Schematic representation of the ATP synthetase complex of the inner mitochondrial membrane. The F_0 portion (the stem) traverses the membrane and contains the proton channel. The F_1 portion (the head and neck) protrudes into the matrix. The head is the catalytic portion, and is made up of five kinds of subunit.

port coupled to oxidative phosphorylation. Table 6-2 summarizes the net energy yield for the complete oxidation of glucose. Glucose yields two molecules of pyruvate, so two turns of the TCA cycle are required per glucose. The net energy yield of complete glucose oxidation is from 36 to 38 ATP, depending on whether the NADHs from glycolysis enter the mitochondria via the glycerol phosphate shuttle or the malate-aspartate shuttle.

Table 6-2. Net Energy Yield for the Complete Oxidation of One Molecule of Glucose

Source	Substrate-Level Phosphorylation	Oxidative Phosphorylation
Glycolysis		
2 ATP	2 ATP	
2 NADH		
Via malate-aspartate shuttle		6 ATP
or		*or*
Via glycerol phosphate shuttle		4 ATP
Pyruvate dehydrogenase		
2 NADH		6 ATP
TCA cycle		
6 NADH		18 ATP
2 $FADH_2$		4 ATP
2 GTP	2 GTP *or* 2 ATP	
	4 ATP (GTP)	32 to 34 ATP

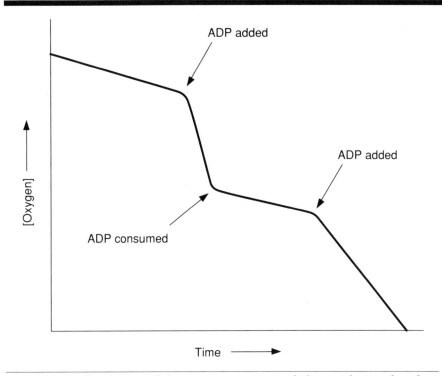

Fig. 6-12. An illustration of electron transport coupled to oxidative phosphorylation. When ADP is added, the rate of oxygen consumption (electron transport) rises until the ADP is used up. If more ADP is added, the rate of oxygen utilization rises again.

COUPLING OF ELECTRON TRANSPORT AND OXIDATIVE PHOSPHORYLATION

Electron transport is normally tightly coupled to oxidative phosphorylation. In order for mitochondria to oxidatively phosphorylate ADP, the electron transport system must have access to

1. A source of electrons (i.e., NADH or $FADH_2$)
2. An adequate supply of O_2

A deficit of either ingredient significantly depresses the rate of oxygen consumption and oxidative phosphorylation. In normally functioning mitochondria, however, the limiting factor is usually [ADP]. The rates of both electron flow (as measured by oxygen consumption) and oxidative phosphorylation vary in direct proportion to [ADP] (Fig. 6-12). Mitochondria in which the rate of oxidative phosphorylation is controlled by the rate of oxygen consumption are said to be *coupled*.

The tightness of the coupling of electron transport and oxidative phosphorylation can be determined by calculating the *respiratory control ratio*:

$$\text{Respiratory control ratio} = \frac{\text{P/O in the presence of ADP}}{\text{P/O in the absence of ADP}}$$

Where P/O is the number of inorganic phosphates incorporated into ATP per oxygen atom consumed. The higher the respiratory control ratio, the tighter the coupling of electron transport and oxidative phosphorylation. Damaged or uncoupled mitochondria show a decreased respiratory control ratio.

Not only electron transport but also the TCA cycle and glycolysis respond to respiratory control:

1. By limiting the rate of the electron transport chain, respiratory control limits the supply of NAD^+ and thus the rate of the TCA cycle. In addition, controlling enzymes of the TCA cycle respond directly to the ATP/ADP and NAD^+/NADH ratios.
2. Some enzymes of glycolysis also respond to the ATP/ADP and NAD^+/NADH ratios.

Uncouplers of Electron Transport and Oxidative Phosphorylation

Uncouplers are compounds that disrupt respiratory control. When an uncoupler is added to mitochondria, electron transport is stimulated and proceeds at a high rate in the absence of ATP synthesis; i.e., the mitochondria are uncoupled. One well-known uncoupler is dinitrophenol:

$$O_2N-\text{C}_6H_3(NO_2)-OH$$

Uncouplers act by dissipating the proton gradient across the inner mitochondrial membrane. Uncouplers are generally both lipophilic and acidic. Because they are lipophilic they can dissolve in the membrane, and because they are acidic they are able to channel protons across the membrane, thus dissipating the transmembrane proton gradient.

Inhibitors of Oxidative Phosphorylation

Some toxins act directly on ATP synthetase, blocking oxidative phosphorylation and thus indirectly inhibiting electron transport and the TCA cycle as well. The antibiotic oligomycin and the biochemical reagent dicyclohexylcarbodiimide are examples of such compounds.

Summary

Inhibitors can act on the electron transport/oxidative phosphorylation system via three mechanisms:

1. *Electron transport inhibitors* block electron-transfer steps in the electron transport chain. By blocking the flow of electrons, they inhibit both oxygen consumption and oxidative phosphorylation. (They also inhibit the TCA cycle by preventing the regeneration of NAD^+/FAD.)
2. *Uncouplers* set the electron transport chain free of respiratory control. Electron transport and oxygen consumption proceed freely in the absence of ATP synthesis.
3. *Oxidative phosphorylation inhibitors* act directly on ATP synthetase. The electron transport flow and the TCA cycle are indirectly inhibited.

CLINICAL CORRELATION: CYANIDE POISONING

Cyanide is probably the most frequently encountered inhibitor of the electron transport chain. It blocks the flow of electrons at cytochrome aa_3 in cytochrome oxidase by rapidly binding to the Fe^{3+} of cytochrome aa_3. Oxidative phosphorylation ceases and cell death ensues.

The strategy of treatment for cyanide poisoning is to chemically divert the cyanide away from cytochrome aa_3. This is achieved first by administering nitrite. Nitrite oxidizes reduced hemoglobin (Fe^{2+}) to methemoglobin (Fe^{3+}). The ferric iron in methemoglobin then binds cyanide, diverting it from cytochrome aa_3.

Treatment of cyanide poisoning also makes use of the enzyme rhodanese in mammalian tissues. This enzyme contains a sulfur/cyanide transferase activity. In the presence of thiosulfate and cyanide, rhodanese releases free sulfite and transfers the sulfur to cyanide, generating thiocyanide, which can be eliminated via the urine. Rhodanese is a very versatile enzyme for eliminating cyanide, because it will react with both free cyanide and cyanide bound to certain other species (such as methemoglobin). Rhodanese is activated for cyanide therapy by administering thiosulfate. Thus, cyanide poisoning treatment involves the administration of nitrite, which diverts the cyanide from cytochrome aa_3, and the administration of thiosulfate, which allows rhodanese to release cyanide from methemoglobin.

SUGGESTED READING

1. Atkinson DE: Cellular Energy Metabolism and its Regulation. Academic Press, New York, 1972
2. Lowenstein JM (ed.): Citric Acid Cycle: Control and Compartmentation. Marcel Dekker, New York, 1969
3. Bahsheffsky H, Baltsheffsky M: Electron transport phosphorylation. Annu Rev Biochem 43:871, 1974
4. Munn EA: The Structure of Mitochondria. Academic Press, New York, 1974
5. Tzagoloff A: Mitochondria. Plenum Press, New York, 1982
6. Boyer PD, Chance B, Grusser L, Mitchell P, et al: Oxidative Phosphorylation and Phosphophorylation. Annu Rev Biochem 46:955, 1977
7. Dagley S, Nicholson DE: Metabolic Pathways. Wiley, New York, 1970

STUDY QUESTIONS

Directions: For each of the following multiple choice questions, choose the most appropriate answer.

1. Which one of the following statements is true?
   A. Oxidative phosphorylation occurs on the outer mitochondrial membrane.
   B. Cristae are found on the outer mitochondrial membrane.
   C. The TCA cycle occurs mainly in the mitochondrial matrix.
   D. The mitochondrial ATP synthetase is located on the outer membrane, and releases ATP into the cytosol.

2. All of the following compounds are intermediates of the TCA cycle EXCEPT
   A. succinate
   B. ketocitrate
   C. malate
   D. oxaloacetate

3. For each cycle turn, the TCA cycle produces
   A. 3 NADH, 1 FADH$_2$, and 1 GTP
   B. 2 FADH$_2$, 3 FADH$_2$, and 1 GTP
   C. 3 NADH, 1 FADH$_2$, and 2 GTP
   D. 1 NADH, 3 FADH$_2$ and 1 GTP

4. Only one reaction of the TCA cycle has a strongly positive $\Delta G°$ value. That reaction is catalyzed by
   A. succinate dehydrogenase
   B. fumarate dehydrogenase
   C. isocitrate dehydrogenase
   D. malate dehydrogenase

5. Which one of the following substances can only participate in electron transfer reactions involving the transfer of a single electron?
   A. NAD$^+$
   B. Cytochrome c
   C. Coenzyme Q
   D. FAD

Directions: Answer the following questions using the key outlined below:
- **(A)** if 1, 2, and 3 are correct
- **(B)** if 1 and 3 are correct
- **(C)** if 2 and 4 are correct
- **(D)** if only 4 is correct
- **(E)** if all four are correct

6. Mitochondria are
   1. the subcellular site of the TCA cycle
   2. the subcellular site of the pentose phosphate pathway
   3. double-membrane organelles
   4. surrounded by a highly impermeable outer membrane

7. Which of the following statements is or are true?
   1. An agent that blocks a step in the electron transport chain will also inhibit oxidative phosphorylation.
   2. An uncoupler will stimulate electron transport but inhibit oxygen consumption.
   3. An uncoupler will stimulate oxygen consumption but inhibit oxidative phosphorylation.
   4. Inhibitors of ATP synthetase do not affect electron transport.

7

LIPID METABOLISM I

Fatty Acids and Triacylglycerols/
Digestion of Dietary Lipids/Lipoproteins/
Fatty Acid Synthesis/Fatty Acid Oxidation/
Clinical Correlations

Learning Objectives

The student should be able to:

1. List four functions of lipids in mammalian organisms.
2. Summarize the chemical nature of lipids and the structure of fatty acids and triacylglycerols.
3. List the events that occur during the digestion and absorption of dietary lipid.
4. List the steps of fatty acid biosynthesis and the factors that regulate it.
5. Describe the shuttle mechanism that transports acetyl CoA for fatty acid synthesis from the mitochondria to the cytosol.
6. Describe the mitochondrial and microsomal mechanisms of fatty acid elongation and desaturation, and the mechanisms for the production of short-chain (less than C_{16}) and methylated fatty acids.
7. Describe the mobilization of lipids from adipose tissue into the blood and from lipoproteins into tissues.
8. Describe the activation of fatty acids and the transport of fatty acyl CoAs into the mitochondria, and list the steps of β-oxidation.
9. Distinguish β, α, and ω oxidation of fatty acids.
10. Describe the factors involved in the regulation of fatty acid oxidation.
11. Describe the synthetic pathway for ketone bodies and their role during long-term starvation.

FUNCTIONS OF BIOLOGIC LIPIDS

Lipids may be defined generally as substances that are soluble in organic solvents and insoluble in water. *Amphipathic lipids*—lipids that contain a hydrophilic group—are sparingly soluble in water. Lipids perform several vital roles in the body, which may be summarized as follows:

1. Energy Storage. The triacylglycerol in adipose tissue is the body's major deposit of stored energy. Energy is stored as fat for use over long time periods (days or weeks). In contrast, energy stored as glycogen is used over short time periods (between meals or overnight). Lipids have two properties that equip them for long-term energy storage. First, they are highly reduced, so that oxidation of lipid releases more energy per unit weight than oxidation of carbohydrate, and more energy is stored in a given weight of lipid. Second, triacylglycerols are relatively nonreactive, so they can be stored for long periods under normal conditions without undergoing unwanted chemical alterations.

Adipose tissue is a type of connective tissue specialized for the synthesis, storage, and degradation of triacylglycerols. Adipose tissue cells are called *adipocytes* or *fat cells* (Fig. 7-1). The cytoplasm of adipocytes contains a large central lipid droplet, which may occupy as much as 90 percent of the cell's volume. The rest of the cytoplasm forms a thin aqueous layer around the lipid droplet. Most adipose tissue is found in subcutaneous and intramuscular deposits.

2. *Membrane Structure.* The fabric of the plasma membrane and the various intracellular membranes is formed of amphipathic lipids (primarily phospholipids) arranged tail-to-tail to form a bilayer structure (Chapter 21).

3. *Surface Active Agents.* Amphipathic lipids are able to interact with both hydrophobic and hydrophilic solvents. When both organic and aqueous phases are present, amphipathic lipid molecules congregate along the interface between the two phases, oriented so that the hydrophilic groups interact with the aqueous phase and the hydrophobic groups interact with the organic phase. If a small amount of an amphipathic lipid is added to an aqueous phase, the lipid molecules form a monolayer, with the hydrophobic portions exposed to the air and the

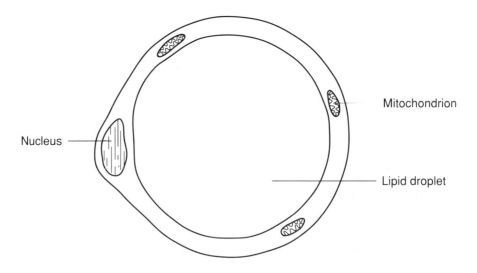

Fig. 7-1. Structure of the adipocyte.

hydrophilic portions dissolved in the water. This amphipathic layer lowers the surface tension of the aqueous phase. Amphipathic lipids therefore act as *surface active agents* (*surfactants*). By lowering the surface tension of an aqueous phase, surfactants can also act as detergents: they make it possible to disperse and emulsify hydrophobic lipids in aqueous environments. Detergent surfactants are important in the digestion of dietary lipids. Surfactants also lower the surface tension of the aqueous film lining the alveoli of the lung so that small alveoli do not collapse during the breathing cycle.

4. Specific Biologic Response. Steroid hormones, prostaglandins, HPETEs, HETEs, and leukotrienes (Chapter 8) are all lipids or lipid derivatives that elicit specific biologic responses in mammalian tissues.

FATTY ACIDS AND TRIACYLGLYCEROLS

This chapter covers the structure of fatty acids and triacylglycerols, the digestion and absorption of dietary lipid, and the synthesis and degradation of fatty acids. The metabolism of other biologic lipids is discussed in Chapter 8.

Fatty Acids

Fatty acids consist of an alkyl chain with a terminal carboxyl group. Some common fatty acids are shown in Table 7-1 and Figure 7-2. The alkyl chains of fatty acids may be either *saturated* or *unsaturated*; unsaturated fatty acids contain one or more double bonds. Saturated fatty acids have the general formula $CH_3(CH_2)_nCOOH$. The carbon atoms in fatty acids are numbered starting with the carboxyl carbon. Carbon atoms 2 and 3 may also be referred to as the α and β carbons, respectively, and the terminal (methyl) carbon is the ω carbon. Many fatty acids also have common or trivial names.

The double bonds in unsaturated fatty acids are usually in the *cis* conformation. When more than one double bond is present, they are always separated by at least one methylene group ($-CH_2-$); i.e., they are nonconjugated. Fatty acids with one double bond are called *monoenoic* fatty acids, and those with more than one double bond are called *polyenoic*. The systematic names of fatty acids indicate how many carbon atoms there are in the fatty acid chain and the conformation and location of any double bonds. For example, as shown in Table 7-1, the systematic name of linoleic acid, *cis,cis*-9,12-octadecenoic acid, indicates that it is an 18-carbon (C_{18}) fatty acid with two *cis* double bonds, one between carbons 9 and 10 and the other between carbons 12 and 13.

Fatty acids are usually linear, although branched-chain (methylated) fatty acids have been found in mammary gland tissue. The function of these branched-chain fatty acids is not known. Most fatty acids in mam-

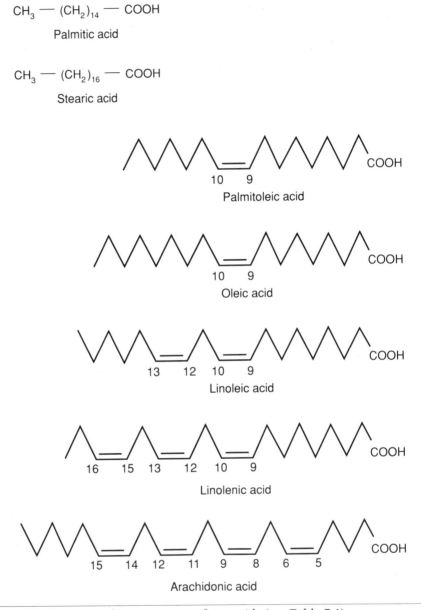

Fig. 7-2. Structure of some common fatty acids (see Table 7-1).

Table 7-1 Trivial Names, Numeral Identifications, and Systematic Names of Some Common Saturated and Unsaturated Fatty Acids

Trivial Name	Numeral Identification[a]	Systematic Name[b]
Palmitic acid	16:0	Hexadecanoic acid
Stearic acid	18:0	Octadecanoic acid
Palmitoleic acid	16:1(9)	cis-9-Hexadecenoic acid
Oleic acid	18:1(9)	cis-9-Octadecenoic acid
Linoleic acid	18:2(9,12)	cis,cis-9,12-Octadecenoic acid
Linolenic acid	18:3(9,12,15)	cis,cis,cis-9,12,15-Octadecenoic acid
Arachidonic acid	20:4(5,8,11,14)	All cis-5,8,11,14-eicosatetraenoic acid

[a] The numeral identification gives, in order, the number of carbons in the fatty acid chain, the number of double bonds, and the number of the first carbon atom in each double bond. Thus, the numeral identification of linoleic acid, 18:2(9,12), indicates that linoleic acid is a C_{18} fatty acid with two double bonds, one between carbons 9 and 10 and the other between carbons 12 and 13.

[b] Systematic names of fatty acids may also be given using Greek deltas to indicate double bonds. In this system, the name of linoleic acid would be cis,cis-Δ^9,Δ^{12}-octadecenoic acid.

mals contain an even number of carbon atoms. The most abundant fatty acids in humans contain 16, 18, or 20 carbon atoms.

Triacylglycerols

The bulk of the body's fatty acids are esterified to glycerol to form *triacylglycerols* (Fig. 7-3), which are found primarily in adipose tissue. The glycerol moiety of a triacylglycerol carries three esterified fatty acid chains. The properties of the triacylglycerol depend significantly on the nature of the fatty acid components. Free fatty acids are ionized at the appropriate pH, but triacylglycerols are neutral, nonpolar, nonionizable, and extremely hydrophobic. They are also chemically quite inert.

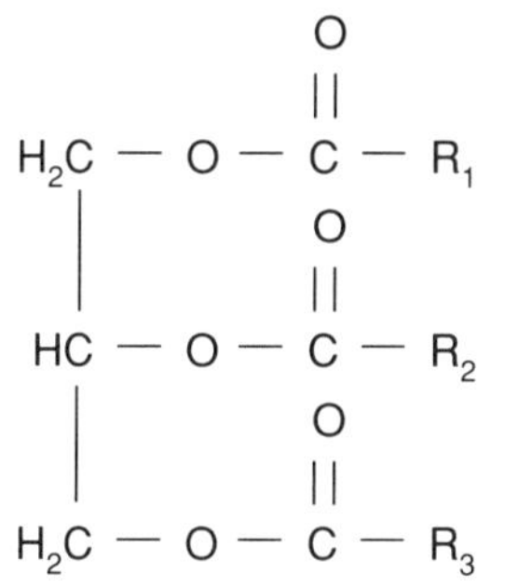

$$\begin{array}{c}
O \\
\parallel \\
H_2C-O-C-R_1 \\
| O \\
\parallel \\
HC-O-C-R_2 \\
| O \\
\parallel \\
H_2C-O-C-R_3
\end{array}$$

Fig. 7-3. The general structure of triacylglycerols. The three R groups are fatty acyl hydrocarbon chains.

Perspective: Lipid Digestion, Absorption, and Metabolism

The individual processes of lipid metabolism are best understood in light of the overall lipid economy, which may be outlined as follows.

1. Mammalian organisms contain a *lipid pool* consisting of stored lipids, newly synthesized lipids, and lipids in transit in the circulation. Most of the body's lipid at any given time is in this pool. Specifically, the pool contains (1) the triacylglycerols stored in adipose tissue, (2) lipids that are being transported in the blood in the form of lipoproteins and chylomicrons, and (3) newly synthesized fatty acids and phospholipids.
2. Lipids enter the lipid pool primarily in the form of dietary lipids and lipids synthesized endogenously from amino acids and carbohydrates.
3. Lipids are removed from the pool for the purpose of energy production and to be used in the synthesis of membranes and complex specialized lipid molecules.

Lipid metabolism is the sum of the processes by which lipids enter and leave the pool (Fig. 7-4). Keep in mind that each process causes lipid to be either added to or removed from the pool.

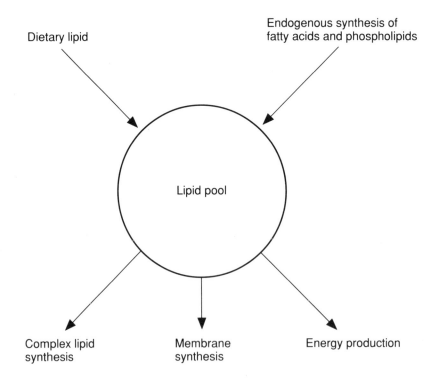

Fig. 7-4. The basic components of lipid metabolism.

DIGESTION AND ABSORPTION OF DIETARY LIPID

Digestion of Dietary Lipid

Most of the lipid in the diet consists of triacylglycerols containing long chain fatty acids (C_{16} or longer). The diet also includes small quantities of phospholipids, triacylglycerols containing short-chain fatty acids, free fatty acids, cholesterol, cholesterol esters, and other specialized lipids. This section focuses on the digestion of triacylglycerols containing long chain fatty acids.

Digestion of dietary lipid starts in the mouth, due to the secretion of *lipases* by the tongue. Lipases are enzymes that hydrolyze the ester bonds of triacylglycerols and phospholipids. In the case of triacylglycerols, complete lipase digestion yields free fatty acids and glycerol. The tongue secretes an acid-stable lipase that is swallowed with the food and is believed to be active in the acidic environment of the stomach. Most lipid digestion, however, is carried out in the small intestine by pancreatic lipase.

Figure 7-5 illustrates the digestion and absorption of lipid in the intestine. Entry of food into the small intestine stimulates the release of the hormone cholecystokinin, which causes the pancreas and gallbladder to release lipase and bile, respectively, into the intestinal lumen. Cholecystokinin also slows the entry of food into the small intestine, presumably to ensure that the available digestive enzymes are not overwhelmed.

Lipid enters the intestine in the form of large lipid droplets. Lipases are water-soluble proteins, so they attack only triacylglycerol molecules exposed on the surface of a droplet. The efficiency of fat digestion is greatly increased by the *bile salts*, which act as powerful detergents, emulsifying the large lipid droplets into very small droplets, thus greatly increasing the surface area available for lipase digestion. Bile salts are amphipathic cholesterol derivatives. Phospholipids present in the intestinal lumen also act as detergents. The mechanical churning of the intestine enhances the emulsifying effect of the bile salts and phospholipids.

Pancreatic lipase specifically hydrolyzes fatty acids from positions 1 and 3 of triacylglycerols, yielding free fatty acids and 2-monoacylglycerols. The enzyme does not discriminate significantly on the basis of fatty acid chain length or degree of unsaturation. The catalytic activity of pancreatic lipase is facilitated by a protein called *colipase*, which forms with the lipase a 1:1 complex that stabilizes the three-dimensional shape of the lipase when it is absorbed by the water-lipid interface at the surface of the lipid droplet.

The pancreas also secretes *esterases*, which hydrolyze a variety of lipid esters, including cholesterol esters and monoacylglycerols. These enzymes are dependent on bile salts for activity. The secretions of the pancreas are also rich in prophospholipase A_2, which is activated by trypsin to *phospholipase A_2*, an enzyme important in the hydrolysis of

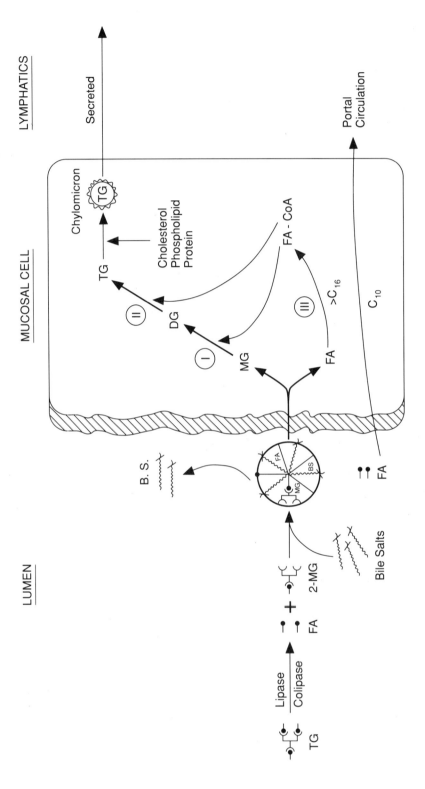

Fig. 7-5. Digestion and absorption of dietary lipid. TG, triacylglycerol; FA, fatty acid; MG, monoacylglycerol; DG, diacylglycerol; BS, bile salt; C_{16}, C_{18}, 16- and 18-carbon fatty acids. Roman numerals refer to different acyltransferases. (From Smith EL, Hill RL, Lehman, RJ, et al: Principles of Biochemistry. Mammalian Biochemistry. 7th Ed. McGraw-Hill, New York, 1983, with permission.)

phospholipids. Phospholipase A_2 requires bile salts and Ca^{2+} for activity. The combined action of pancreatic lipase and the bile salt dependent esterases results in the hydrolysis of essentially all forms of dietary lipid.

Absorption of Dietary Lipid

The products of lipid hydrolysis and nonhydrolyzable lipids are absorbed by the cells of the intestinal mucosa. In order to reach the intestinal mucosal cells, lipids must be soluble (or made to be soluble) in the aqueous medium bathing the mucosal cells. Long-chain fatty acids and 2-monoacylglycerols, the main products of lipid digestion, are not water soluble. They are made soluble, however, by being incorporated into *mixed micelles* (Chapter 21) which contain bile salts. Mixed micelles also contain cholesterol and fat-soluble vitamins.

Lipids enter the mucosal cells by passive diffusion (i.e., by diffusion down a concentration gradient; see Chapter 21). The concentration gradient across the mucosal cell membrane is maintained in two ways:

1. The local concentration of lipids in a mixed micelle is much higher than the concentration in adjacent mucosal cells. Therefore, dietary lipids and fat-soluble vitamins diffuse passively down their concentration gradients from mixed micelles into mucosal cells. Bile salts are ultimately reabsorbed in the distal part of the small intestine and taken to the liver via the enterohepatic circulation.
2. Once inside the mucosal cells, long-chain fatty acids and monoglycerols are resynthesized into triacylglycerols. This process keeps the concentration of long-chain fatty acids and monoglycerols low inside the cell, thus maintaining the concentration gradient across the cell membrane.

In order to be reconverted into triacylglycerols, fatty acids are first converted to their CoA derivatives by *acyl CoA synthetase*. Two fatty acyl CoAs then react successively with each 2-monoacylglycerol to form a triacylglycerol (Fig. 7-5). Cholesterol in the mixed micelles also diffuses into mucosal cells, where it is partially re-esterified.

Transport of Dietary Lipid in Chylomicrons

The resynthesized triacylglycerols are made water-soluble by interacting with phospholipids, cholesterol, cholesterol esters, and certain proteins to form lipoprotein structures called *chylomicrons*. Chylomicrons consist of approximately 80 percent triacylglycerols; the remaining 20 percent is divided among phospholipids, protein, cholesterol, and cholesterol esters. The amphipathic components—protein, cholesterol, and phospholipids—form a hydrophilic layer on the exterior surface of the chylomicron and render it water soluble (Fig. 7-6). The triacylglycerols and cholesterol esters are in the interior, sequestered from the aqueous phase. The protein component of chylomicrons consists of special proteins called *apolipoproteins*. The apolipoproteins are required for the absorption of chylomicrons across the cell membrane and into the lymphatics. Chylomicrons subsequently enter the blood circulation via the thoracic duct.

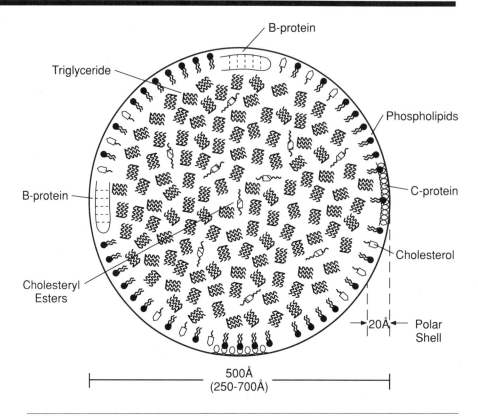

Fig. 7-6. General structure of lipoproteins. (From Morrisett IR, Jackson RL, Gotto AM: Lipid-protein interactions in the plasma lipoproteins. Biochim Biophys Acta 472:93, 1977, with permission.)

Medium- and Short-Chain Fatty Acids

Medium- and short-chain fatty acids (C_{12} and shorter) are water soluble, and are absorbed directly into mucosal cells without the involvement of mixed micelles. From the mucosal cells, they enter the portal circulation and are carried to the liver.

Digestion and Absorption of Dietary Lipid: Key Points

The essential steps of the digestion and absorption of dietary lipid may be summarized as follows:

1. Nonpolar lipids in the intestine are emulsified by the combined action of amphipathic detergent substances (primarily bile salts) and the churning of the intestine.
2. Triacylglycerols are hydrolyzed by pancreatic lipases to long-chain fatty acids and 2-monoacylglycerols.

3. Long-chain fatty acids and monoacylglycerols are incorporated in soluble mixed micelles.
4. Monoacylglycerols, long-chain fatty acids, cholesterol, cholesterol esters, and fat-soluble vitamins diffuse passively from mixed micelles into intestinal mucosal cells. Medium- and short-chain fatty acids are absorbed directly from the aqueous medium into the intestinal mucosal cells.
5. Triacylglycerols containing long-chain fatty acids are resynthesized inside mucosal cells.
6. Chylomicrons are assembled within the mucosal cells and secreted into the lymphatics. From the lymphatics, chylomicrons are carried into the bloodstream and transported to peripheral tissues, particularly adipose tissue.
7. Medium- and short-chain fatty acids enter the portal circulation and are taken up by the liver.

Perspective: Biosynthesis of Fatty Acids

The body has only three primary sources of energy: carbohydrate, protein, and lipid. When the amount of lipid in the diet exceeds the body's metabolic requirements, the excess is stored as triacylglycerols. When the amount of carbohydrate and protein entering the body exceeds what can be stored as glycogen or used otherwise, the excess is also converted to triacylglycerol and stored. This section describes the pathway by which the acetyl CoA from the breakdown of carbohydrates and protein amino acids is used to synthesize fatty acids. In addition, the mechanisms for the modification of fatty acids and the synthesis of short- and medium-chain fatty acids are presented.

BIOSYNTHESIS OF FATTY ACIDS

The major initial product of fatty acid synthesis is the saturated C_{16} fatty acid *palmitate*. The essential features of palmitate synthesis may be summarized as follows:

1. All the carbon atoms in palmitate are derived from acetyl CoA. The fatty acid chain is constructed by the successive addition of two carbons at a time.
2. The first acetyl CoA is used directly, without further activation, and becomes the methyl end of palmitate.
3. All the other acetyl CoA units are activated by carboxylation to malonyl CoA before being used in fatty acid synthesis.
4. After each addition step, NADPH is used in a series of reactions that ultimately reduce the growing chain to the fully saturated state.

Figure 7-7 shows the sources of the carbon atoms of palmitate.

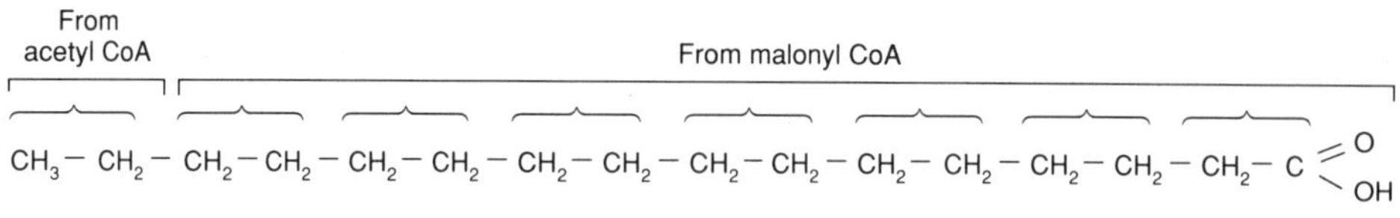

```
         From
       acetyl CoA                      From malonyl CoA
    ⌜‾‾‾‾‾‾⌝ ⌜‾‾‾‾‾‾‾‾‾‾‾‾‾‾‾‾‾‾‾‾‾‾‾‾‾‾‾‾‾‾‾‾‾‾‾‾‾‾‾‾‾‾‾‾‾‾‾‾‾‾‾‾‾‾‾‾‾⌝
    CH₃—CH₂—CH₂—CH₂—CH₂—CH₂—CH₂—CH₂—CH₂—CH₂—CH₂—CH₂—CH₂—CH₂—CH₂—C⟨O/OH
```

Fig. 7-7. The sources of the carbon atoms in palmitate.

Formation of Malonyl CoA

All but the first acetyl units are donated to the growing palmitate chain by *malonyl CoA*. Malonyl CoA is formed by the carboxylation of acetyl CoA in a reaction catalyzed by *acetyl CoA carboxylase*:

$$CH_3-\overset{O}{\underset{\|}{C}}-S-CoA + HCO_3^- + ATP \xrightarrow{\text{Acetyl CoA carboxylase}}$$

Acetyl CoA

$$^-O-\overset{O}{\underset{\|}{C}}-CH_2-\overset{O}{\underset{\|}{C}}-S-CoA + ADP + P_i$$

Malonyl CoA

Acetyl CoA carboxylase is the major regulated enzyme of fatty acid biosynthesis. The level of malonyl CoA determines the quantity of fatty acids synthesized.

Fatty Acid Biosynthesis

The overall reaction of de novo palmitate synthesis is

Acetyl CoA + 7 malonyl CoA + 14 NADPH + 14 H⁺ →

palmitate + 7 CO₂ + 8 CoA + 14 NADP⁺ + 6 H₂O

The reactions in this pathway are all catalyzed by a single enzyme complex, the *fatty acid synthase complex*. The essential features of palmitate biosynthesis are as follows:

1. Acetyl CoA is the primer or initial acceptor molecule required for initiation of fatty acid synthesis.
2. All the rest of the carbons are donated by malonyl CoA, which is used repeatedly to add two-carbon units to the growing fatty acid chain until palmitate is formed.
3. The newly synthesized palmitate is released from the fatty acid synthase complex.

The reactions of chain initiation and elongation are described below.

Step 1: Binding of Acetyl CoA to the Fatty Acid Synthase Complex

In the first step of palmitate biosynthesis, the primer acetyl unit is transferred from acetyl CoA to the correct component of the fatty acid synthase complex. This step occurs in three reactions catalyzed by the *acetyl transacetylase* enzyme activity of the complex. First, acetyl CoA is bound to a specific serine residue:

$$CH_3-\overset{O}{\underset{\|}{C}}-S-CoA + HO-Ser-enzyme \rightleftharpoons$$

$$H_3C-\overset{O}{\underset{\|}{C}}-O-Ser-enzyme + CoA-SH$$

Next, the acetyl group is transferred from the serine to the sulfhydryl group of the *acyl carrier protein* (ACP) component of the complex:

$$CH_3-\overset{O}{\underset{\|}{C}}-O-Ser-enzyme + HS-ACP \rightleftharpoons$$

$$CH_3-\overset{O}{\underset{\|}{C}}-S-ACP + HO-Ser-enzyme$$

Finally, the acetyl group is transferred from the ACP to the *condensing enzyme* (CE), another component of the complex:

$$CH_3-\overset{O}{\underset{\|}{C}}-S-ACP + HS-CE \rightleftharpoons CH_3-\overset{O}{\underset{\|}{C}}-CE + ACP-SH$$

Step 2: Binding of the Malonyl CoA Donor

The malonyl moiety of malonyl CoA is transferred to the ACP by the *malonyl transacetylase* activity of the complex:

$$^{-}OOC-CH_2-\overset{O}{\underset{\|}{C}}-S-CoA + HS-ACP \rightleftharpoons$$

$$^{-}OOC-CH_2-\overset{O}{\underset{\|}{C}}-S-ACP + CoA-SH$$

Step 3: Condensation of Acceptor and Donor (Adduct Formation)

The condensation of the two-carbon unit donated by malonyl with the acetyl acceptor is catalyzed by the condensing enzyme (β-ketoacyl-ACP synthetase):

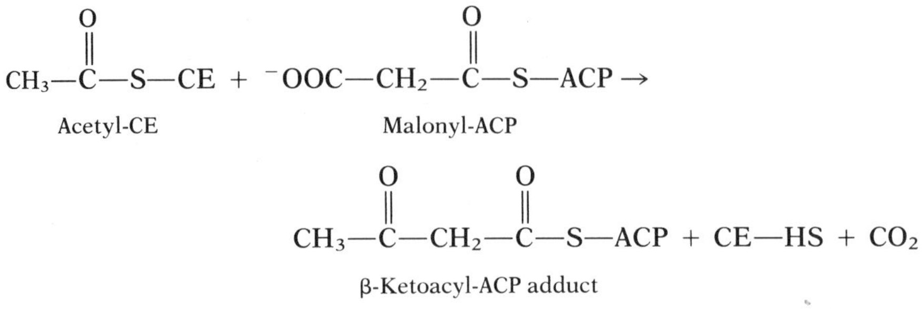

$$CH_3-\overset{O}{\underset{\|}{C}}-S-CE + {}^-OOC-CH_2-\overset{O}{\underset{\|}{C}}-S-ACP \rightarrow$$

Acetyl-CE Malonyl-ACP

$$CH_3-\overset{O}{\underset{\|}{C}}-CH_2-\overset{O}{\underset{\|}{C}}-S-ACP + CE-HS + CO_2$$

β-Ketoacyl-ACP adduct

This reaction involves the release of the CO_2 that was added to acetyl CoA during the formation of malonyl CoA. The acceptor acetyl is transferred from the condensing enzyme to the malonyl group attached to the ACP, and displaces the CO_2. The energy released by this decarboxylation drives the condensation reaction.

Step 4: Formation of the Fully Reduced Alkyl Chain

In the fourth step, the adduct is reduced and dehydrated to a fully saturated hydrocarbon chain. This process occurs in three reactions—a reduction, a dehydration, and another reduction—catalyzed by three different enzymes on the fatty acid synthase complex. These reactions are as follows:

1. The β-ketoacyl is reduced to a β-hydroxyacyl in a reaction catalyzed by β-*ketoacyl-ACP reductase*:

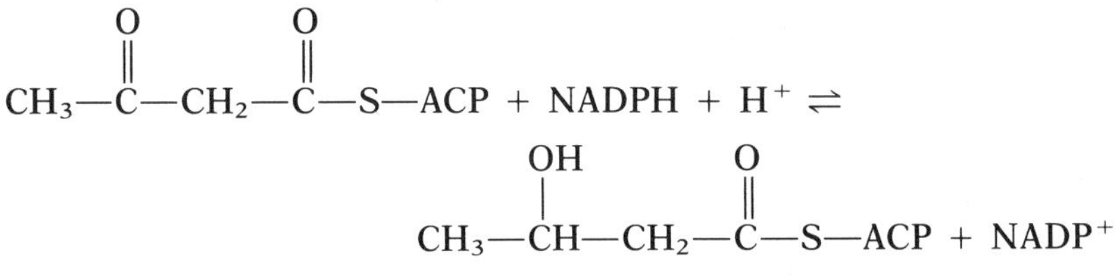

$$CH_3-\overset{O}{\underset{\|}{C}}-CH_2-\overset{O}{\underset{\|}{C}}-S-ACP + NADPH + H^+ \rightleftharpoons$$

$$CH_3-\overset{OH}{\underset{|}{CH}}-CH_2-\overset{O}{\underset{\|}{C}}-S-ACP + NADP^+$$

2. The β-hydroxyacyl is dehydrated by the enzyme β-*hydroxyl-ACP dehydratase*:

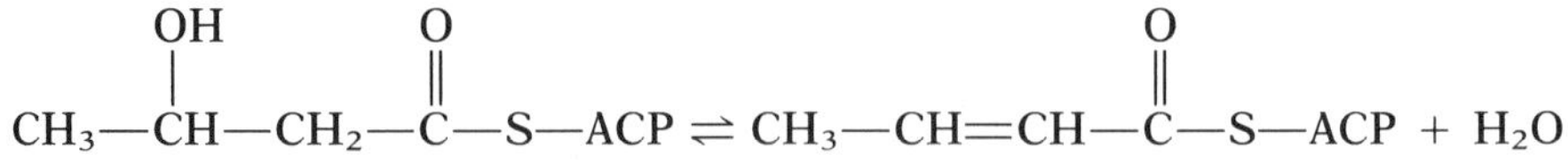

$$CH_3-\overset{OH}{\underset{|}{CH}}-CH_2-\overset{O}{\underset{\|}{C}}-S-ACP \rightleftharpoons CH_3-CH=CH-\overset{O}{\underset{\|}{C}}-S-ACP + H_2O$$

3. The resulting β-enoyl acyl is reduced in a reaction catalyzed by β-*enoyl-ACP reductase*:

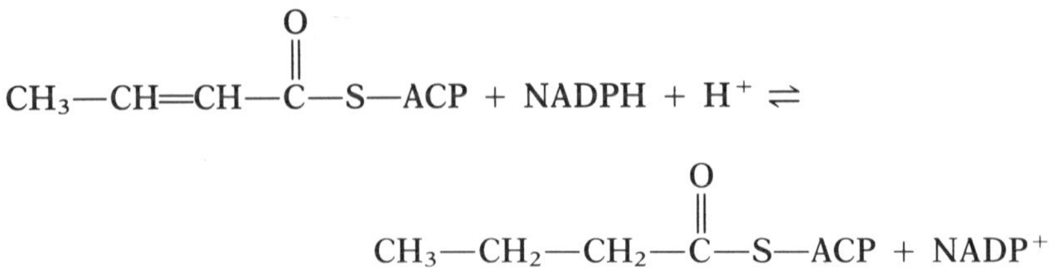

$$CH_3-CH=CH-\overset{O}{\underset{\|}{C}}-S-ACP + NADPH + H^+ \rightleftharpoons$$

$$CH_3-CH_2-CH_2-\overset{O}{\underset{\|}{C}}-S-ACP + NADP^+$$

Lipid Metabolism I 163

Step 5: Transfer of Adduct to the Condensing Enzyme

The fully reduced adduct now becomes the acceptor for the next two-carbon unit from malonyl CoA. To accomplish this, the adduct is transferred by a *transacylase* from the ACP to the acceptor site on the condensing enzyme:

$$CH_3-CH_2-CH_2-\overset{O}{\underset{\|}{C}}-S-ACP + HS-CE \rightleftharpoons$$

$$CH_3-CH_2-CH_2-\overset{O}{\underset{\|}{C}}-S-CE + HS-ACP$$

The acceptor site is now filled and the donor site is available, so the process is ready to begin again. Steps 2 through 5 are repeated until the palmitate chain is complete. The finished chain is released from fatty acid synthase by a *thioesterase*:

$$CH_3-(CH_2)_{14}-\overset{O}{\underset{\|}{C}}-S-ACP \rightleftharpoons CH_3-(CH_2)_{14}-\overset{O}{\underset{\|}{C}}-O^- + ACP$$

The Fatty Acid Synthase Complex

The fatty acid synthase complex from mammalian liver is a dimer of identical subunits. Each subunit is folded into three distinct domains, which collectively possess all the activities required for palmitate synthesis.

The Sources of Acetyl CoA and NADPH for Palmitate Synthesis

The enzymes of palmitate synthesis—acetyl CoA carboxylase and fatty acid synthase—are both cytosolic, but acetyl CoA is produced in the mitochondria. Acetyl CoA is transferred to the cytosol by the *citrate shuttle* (Fig. 7-8). This shuttle also converts cytosolic NADH to NADPH, thus generating some of the NADPH required in palmitate synthesis. The reactions of the citrate shuttle are as follows:

1. Pyruvate inside the mitochondria is converted to oxaloacetate and to acetyl CoA by pyruvate carboxylase and pyruvate dehydrogenase, respectively (see Chapters 5 and 6):

    $Pyruvate_{mito} + ATP + CO_2 \rightarrow oxaloacetate_{mito} + ADP + P_i$

    $Pyruvate_{mito} + CoA + NAD^+ \rightarrow acetyl\ CoA_{mito} + CO_2 + NADH + H^+$

2. Acetyl CoA present in the mitochondria then reacts with oxaloacetate to yield citrate:

    $Acetyl\ CoA_{mito} + oxaloacetate_{mito} \rightarrow citrate_{mito} + CoA_{mito}$

3. Citrate moves to the cytosol via its inner mitochondrial membrane carrier:

    $Citrate_{mito} \rightarrow citrate_{cyt}$

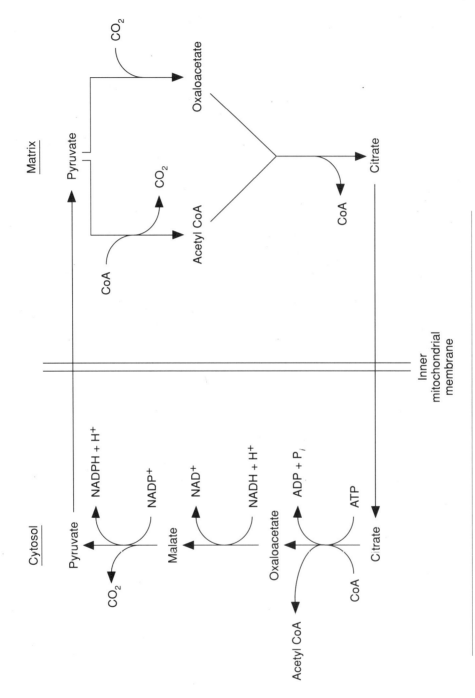

Fig. 7-8. The citrate shuttle.

4. Once in the cytosol, citrate is cleaved by *citrate cleaving enzyme* (*citrate lyase*) to produce oxaloacetate and acetyl CoA. The latter can now be used for fatty acid synthesis.

$$\text{Citrate}_{cyt} + \text{ATP} + \text{CoA} \rightarrow \text{oxaloacetate}_{cyt} + \text{acetyl CoA}_{cyt} + \text{ADP} + P_i$$

5. Oxaloacetate is reduced to malate by *malate dehydrogenase*. This reaction utilizes an NADH.

$$\text{Oxaloacetate}_{cyt} + \text{NADH}_{cyt} + H^+ \rightarrow \text{malate}_{cyt} + \text{NAD}^+_{cyt}$$

6. Malate is then oxidized to pyruvate by *malic enzyme*. During this reaction $NADP^+$ is converted to NADPH.

$$\text{Malate}_{cyt} + \text{NADP}^+_{cyt} \rightarrow \text{pyruvate}_{cyt} + \text{NADPH}_{cyt} + H^+$$

The pyruvate then diffuses back into the mitochondria. The net reaction for the citrate shuttle mechanism is

$$\text{Acetyl CoA}_{mito} + 2\text{ATP} + \text{NADH}_{cyt} + \text{NADP}^+_{cyt} \rightarrow \text{acetyl CoA}_{cyt}$$
$$+ 2\text{ADP} + 2P_i + \text{NAD}^+_{cyt} + \text{NADPH}_{cyt}$$

Eight acetyl CoA molecules are shuttled to the cytosol for each palmitate, with the concurrent production of eight NADPHs. The remaining six NADPHs needed for palmitate synthesis come from the pentose phosphate pathway.

FATTY ACID MODIFICATION

Fatty acids other than palmitate are produced in mammalian cells by modification—lengthening and/or desaturation—of palmitate precursors, or by the premature release of incomplete palmitate chains from fatty acid synthase. In addition, fatty acids obtained from the diet are modified in mammalian cells.

Fatty Acid Elongation

Elongation of fatty acids occurs in both the mitochondria and the endoplasmic reticulum. The mitochondrial and microsomal (endoplasmic reticulum) mechanisms of elongation differ, but in both cases two-carbon units are added to the carboxyl end of existing fatty acid chains.

Mitochondrial Fatty Acid Elongation

In the mitochondrial system, acetyl CoA is used directly as a two-carbon donor, without activation to malonyl CoA (Fig. 7-9). The fatty acid substrates that are elongated by the mitochondrial system are CoA derivatives of medium- and short-chain fatty acids.

Microsomal Fatty Acid Elongation

Microsomal fatty acid elongation uses malonyl CoA as the source of two-carbon units. Palmitoyl CoA is the fatty acid substrate most often elon-

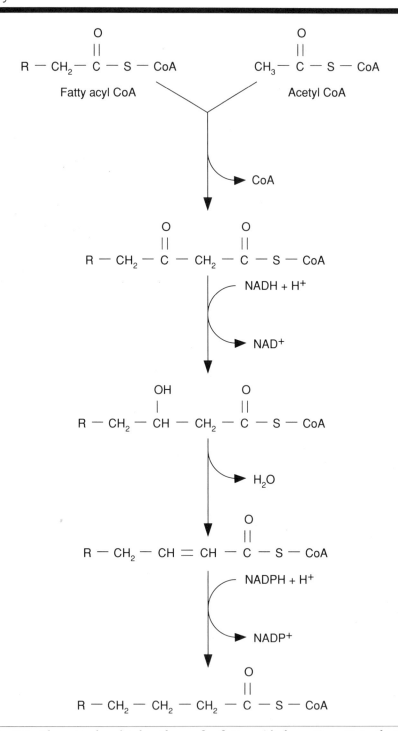

Fig. 7-9. The mitochondrial pathway for fatty acid elongation. Note that the two-carbon donor is acetyl CoA.

gated. The reactions involved are similar to the elongation reactions in the synthesis of palmitate by the fatty acid synthase complex (Fig. 7-10).

Fatty Acid Desaturation

Formation of Monoenoic Fatty Acids

The endoplasmic reticulum contains a microsomal desaturase system that introduces a *cis*-Δ^9 double bond into fatty acids. The principal substrates are palmitate (C_{16}) and stearate (C_{18}), which yield respectively palmitoleate and oleate. The reaction is catalyzed by an enzyme complex, *stearoyl CoA desaturase*, that consists of cytochrome b_5, NADH-cytochrome b_5 reductase, and a desaturase. Molecular oxygen is used as an electron acceptor. The reaction for palmitate is as follows:

$$CH_3-(CH_2)_{14}-\overset{\overset{O}{\|}}{C}-S-CoA + NADH + O_2 + H^+ \rightarrow$$
Palmitoyl CoA

$$CH_3-(CH_2)_5-CH=CH-(CH_2)_7-\overset{\overset{O}{\|}}{C}-S-CoA + NAD^+ + H_2O$$
Palmitoleoyl CoA

Unsaturated fatty acids have lower melting points than their saturated counterparts. Stearoyl desaturase is believed to keep stored triacylglycerols and membrane phospholipids fluid by appropriately regulating the degree of unsaturation of fatty acids. Triacylglycerols that are solid (highly saturated) are not effectively hydrolyzed by lipases. The synthesis of the desaturase complex is repressed by high levels of unsaturated lipids in the diet, and induced by high levels of saturated lipids and/or carbohydrate in the diet. It is also induced by insulin, hydrocortisone, and triiodothyronine.

Formation of Polyenoic Fatty Acids

The polyunsaturated fatty acids of mammalian tissues fall into two classes:

1. Since mammalian tissues lack enzymes to introduce double bonds beyond C_9 in the fatty acid carbon chain, polyunsaturated fatty acids that have *six or fewer* carbon atoms between the terminal (ω) carbon and the nearest double bond cannot be synthesized de novo by mammalian cells. These fatty acids are derived from linoleic or linolenic acid obtained in the diet. Linoleic and linolenic acids are the *nutritionally essential fatty acids*.
2. Polyunsaturated fatty acids that have *seven or more* carbon atoms between the ω-carbon and the nearest double bond can be synthesized de novo in mammalian cells by desaturation of palmitoleic and oleic acids.

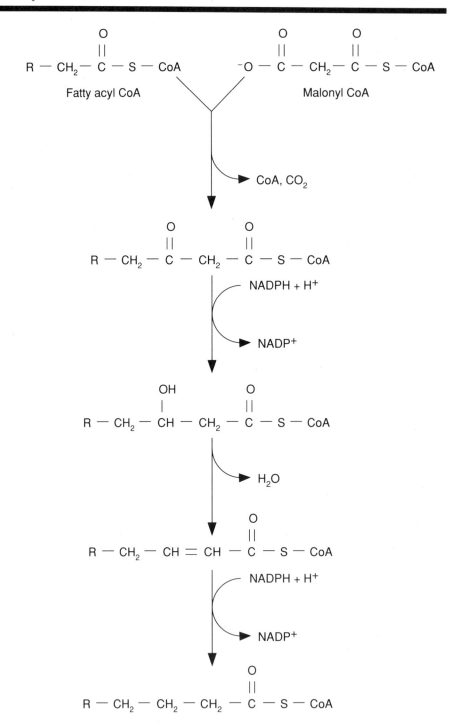

Fig. 7-10. The microsomal (endoplasmic reticulum) pathway for fatty acid elongation. Note that the two-carbon donor is malonyl CoA.

Palmitoleic, oleic, linoleic, and linolenic acids are the precursors of all the polyunsaturated fatty acids generated in mammalian tissues. Numerous polyunsaturated derivatives are created by combinations of chain elongation and desaturation reactions. The desaturase enzymes involved in the synthesis of polyenoic fatty acids are not fully understood; it is believed that specific desaturases introduce double bonds at different positions along the fatty acid chain.

Unsaturated fatty acids are classified according to the position of the double bond nearest the ω-carbon (Fig. 7-11). For example, linolenic acid, which has a double bond between the third and fourth carbons from the ω-carbon, is an ω-3 fatty acid. Therefore, all the fatty acids derived from linolenic acid will also be ω-3 fatty acids. Linoleic acid and its derivatives are all ω-6 fatty acids. Palmitoleate, oleate, and their derivatives constitute the ω-7 and ω-9 families of fatty acids, respectively.

Production of Fatty Acids Shorter than Palmitate

Short- and medium-chain fatty acids (shorter than C_{16}) are synthesized by premature release of the growing fatty acid chain from the fatty acid synthase complex. This cleavage is performed by soluble thioesterases.

Other Fatty Acid Modifications

Fatty acid hydroxylation occurs mainly with certain fatty acids in nerve tissue. The hydroxyl group is placed on the number 2 carbon of these fatty acids. Methylated fatty acids may be produced by replacing malonyl CoA by methylmalonyl CoA during fatty acid synthesis; methylated fatty acids have been found in some secretory glands. Fatty alcohols are formed by an NADPH-dependent reduction of fatty acyl CoAs. Fatty alcohols form the ether-linked substituents of some phospholipids.

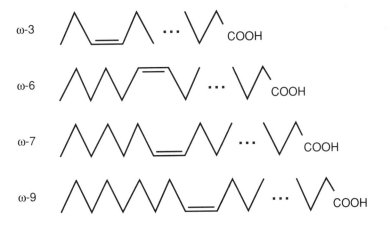

Fig. 7-11. The ω-3, ω-6, ω-7, and ω-9 classes of fatty acids.

Fate of Endogenously Synthesized Fatty Acids

The fate of newly synthesized fatty acids depends on the metabolic state of the organism and on the tissue in which the fatty acid is synthesized. Fatty acids synthesized in adipose tissue are mostly incorporated into triacylglycerols and stored locally. Fatty acids produced in the liver may be used in the synthesis of triacylglycerols, phospholipids, or cholesterol esters. None of these compounds is stored in the liver in significant quantities; instead, they are packaged into lipoproteins and transported in the circulation to the various tissues that use them. Under normal conditions, most of the fatty acid synthesized in the liver is incorporated into triacylglycerols.

Perspective: Mobilization, Transport, and Utilization of Lipid Energy

The utilization of lipids for energy requires that they first be mobilized from the tissue where they are stored (primarily adipose tissue) and transported to the tissues where they will be oxidized. The term *lipid mobilization* refers both to the movement of lipids out of the cells where they have been stored and to the movement of lipids from lipoproteins into the tissue cells. *Lipid transport* refers to the movement of lipids between tissues in the circulation, and also to the movement of lipids across cell membranes (Chapter 21).

The oxidation of fatty acids to yield energy takes place in the mitochondria. Fatty acids are oxidized to acetyl CoA with the production of $FADH_2$ and NADH, which are oxidized via the mitochondrial electron transport chain to produce ATP. The acetyl CoA may enter the TCA cycle and produce even more $FADH_2$ and NADH, which will be oxidized to produce more ATP.

Specific environmental and metabolic stimuli (starvation, cold, exercise, growth, reproduction, and stress) signal the body to mobilize stored (depot) lipid for energy production. Lipid energy is stored in adipose tissue, but the primary site of lipid oxidation is the liver. Triacylglycerols cannot leave adipose tissues, but must first be hydrolyzed to glycerol and fatty acids. Consequently, the utilization of stored lipid for energy production involves:

1. Mobilization of fatty acids from adipose tissue
2. Transport of mobilized fatty acids to the liver and other tissues for oxidation

LIPID MOBILIZATION AND TRANSPORT

Mobilization and Transport of Adipose Tissue Lipids

The mobilization of adipose tissue lipids begins with their hydrolysis to fatty acids and glycerol. This hydrolysis is initiated by a *hormone-sensitive lipase*, which catalyzes the release of a single fatty acid from the 1 or 3 position of the triacylglycerol. The activity of this lipase is controlled by several different cAMP-dependent hormones. The hormones that stimulate this lipase also inhibit triacylglycerol synthesis. After the 1 or 3 fatty acid is liberated by the hormone-sensitive lipase, the remaining diacylglycerol is rapidly hydrolyzed to free fatty acids and glycerol by other lipases.

After hydrolysis, the fatty acids and glycerol diffuse out of the adipose cells. Fatty acids are bound to albumin in the plasma, and glycerol, which is readily soluble in aqueous media, is transported in solution. Both fatty acids and glycerol are taken up mostly by the liver. Most glycerol in the liver enters either the glycolytic or the gluconeogenetic pathway. The major fate of fatty acids is oxidation to acetyl CoA.

Lipoprotein Transport of Triacylglycerols

Lipoproteins function to transport dietary and endogenously synthesized lipids to the sites of lipid storage and/or utilization. The lipoproteins are divided into five different classes on the basis of density. The two classes most important in the transport of triacylglycerols are chylomicrons and *very low density lipoproteins* (*VLDLs*). Chylomicrons and VLDLs contain about 80 percent and 50 percent triacylglycerols, respectively. Chylomicrons are assembled in the intestine and function to transport dietary lipids to the adipose tissue for storage. VLDLs are assembled in the liver and play an important role in the transport of endogenously synthesized triacylglycerols to tissues. Chylomicrons and VLDLs contain several nonidentical polypeptide chains which have been isolated and studied; some of these chains are shared by chylomicrons and VLDLs.

Mobilization of Lipoprotein Triacylglycerols

Lipoprotein-bound triacylglycerols are mobilized by *lipoprotein lipases* that are present on the surface of capillary endothelial cells and possibly on the surface of tissue cells. These lipases hydrolyze fatty acids from the 1 and 3 positions of lipoprotein triacylglycerols to yield free fatty acids and 2-monoacylglycerols. The free fatty acids diffuse into tissue cells; the monoacylglycerols may either diffuse into the cells or be degraded to glycerol and free fatty acid by *serum monoacylglycerol hydrolase*.

Most of the triacylglycerols in chylomicrons are hydrolyzed at or near the surface of adipose tissue cells. The resulting fatty acids and monoacylglycerols diffuse into the adipose cells, where they are reconverted to triacylglycerols for storage. In contrast, the fatty acids from triacylglycerols in VLDLs usually are oxidized in tissue cells to yield energy.

FATTY ACID OXIDATION: ENERGY PRODUCTION FROM LIPIDS

The extent to which a cell obtains energy by oxidation of fatty acids varies from tissue to tissue and with the metabolic status of the individual. Fatty acids that are to be oxidized for energy are first activated in the cytosol, then shuttled into the mitochondria for oxidation.

Fatty Acid Activation

Fatty acids are initially activated to the corresponding CoA derivatives.

$$R-COOH + CoA + ATP \rightarrow R-\overset{O}{\underset{\|}{C}}-S-CoA + AMP + PP_i$$

Note that the ATP is hydrolyzed to AMP and pyrophosphate. The pyrophosphate is subsequently hydrolyzed to 2 P_i. Therefore, the activation of a fatty acid consumes two high energy phosphate bonds.

The enzymes of fatty acid oxidation are located in the mitochondrial matrix. Therefore, fatty acyl CoAs generated in the cytosol must be transported into the mitochondrial matrix. The inner mitochondrial membrane is impermeable to CoA and its derivatives, so fatty acyl CoA enters the mitochondria via a special mechanism.

Entry of Fatty Acyl CoAs into Mitochondria

Fatty acyl groups enter the mitochondria by the *carnitine fatty acyl carrier system* (Fig. 7-12). First, *carnitine acyltransferase I* located on the outside of the inner mitochondrial membrane catalyzes the transacylation reaction

$$\text{Acyl CoA} + \text{carnitine} \rightleftharpoons \text{acyl carnitine} + \text{CoA-SH}$$

A translocase transports the acyl carnitine across the inner mitochondrial membrane into the matrix, and simultaneously transports free carnitine to the cytosol. In the matrix, carnitine acyltransferase II re-

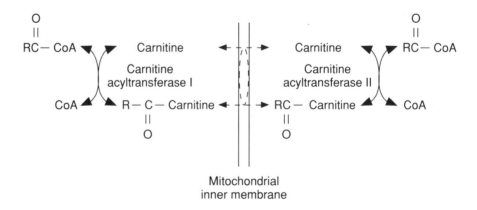

Fig. 7-12. The carnitine carrier system that transports fatty acyls into the mitochondrial matrix. (From Zubay G: Biochemistry. Macmillan, New York, 1988, with permission.)

synthesizes the fatty acyl CoA and releases free carnitine:

$$\text{Acyl carnitine} + \text{CoA} \rightleftharpoons \text{acyl CoA} + \text{carnitine}$$

The carnitine fatty acyl carrier system depends on the presence of CoA on both sides of the inner mitochondrial membrane.

β-Oxidation of Fatty Acids

In the mitochondrial matrix, fatty acyl CoAs are oxidized to acetyl CoA by a recurring reaction sequence that cleaves successive two-carbon units off of the fatty acid chain. This process is known as β-*oxidation*. The reactions of β-oxidation are as follows:

1. Oxidation

The fatty acyl CoA is oxidized by the appropriate *acyl CoA dehydrogenase*. FAD is reduced in the process:

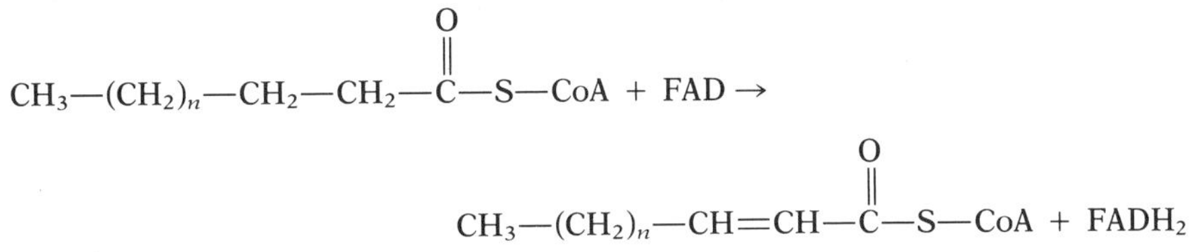

$$CH_3-(CH_2)_n-CH_2-CH_2-\underset{\underset{O}{\|}}{C}-S-CoA + FAD \rightarrow$$

$$CH_3-(CH_2)_n-CH=CH-\underset{\underset{O}{\|}}{C}-S-CoA + FADH_2$$

The mitochondrion contains various dehydrogenases specific for fatty acyl CoAs of different chain lengths.

2. Hydration

The unsaturated fatty acyl CoA is hydrated by an *enoyl CoA hydratase* to yield the β-hydroxyacyl derivative:

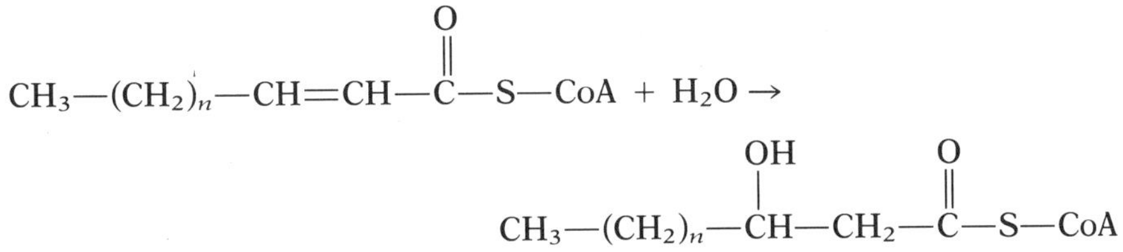

$$CH_3-(CH_2)_n-CH=CH-\underset{\underset{O}{\|}}{C}-S-CoA + H_2O \rightarrow$$

$$CH_3-(CH_2)_n-\underset{\underset{OH}{|}}{CH}-CH_2-\underset{\underset{O}{\|}}{C}-S-CoA$$

The hydratases also show chain length specificity.

3. Oxidation

The β-hydroxy derivative is oxidized by β-*hydroxyacyl CoA dehydrogenase* to the corresponding β-ketoacyl CoA, with the reduction of an NAD$^+$:

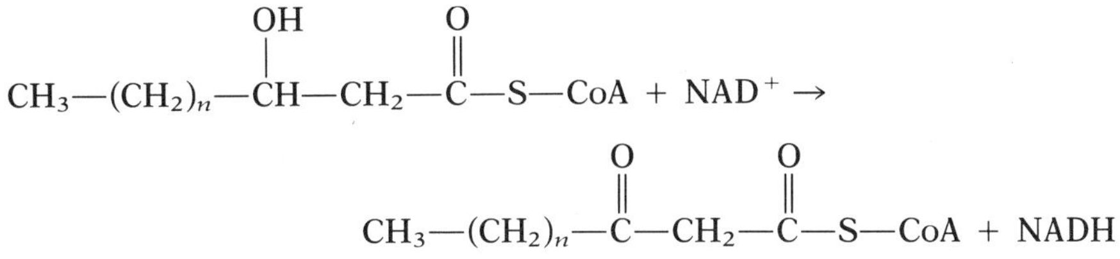

$$CH_3-(CH_2)_n-\underset{\underset{OH}{|}}{CH}-CH_2-\underset{\underset{O}{\|}}{C}-S-CoA + NAD^+ \rightarrow$$

$$CH_3-(CH_2)_n-\underset{\underset{O}{\|}}{C}-CH_2-\underset{\underset{O}{\|}}{C}-S-CoA + NADH$$

4. Thiolysis

The final reaction is the thiolytic cleavage of the bond in the β-keto derivative by an incoming CoA to yield acetyl CoA and the shortened fatty acyl CoA. This reaction is catalyzed by *thiolase*:

$$CH_3-(CH_2)_n-\overset{O}{\underset{\|}{C}}-CH_2-\overset{O}{\underset{\|}{C}}-S-CoA + HS-CoA \rightarrow$$

$$CH_3-\overset{O}{\underset{\|}{C}}-S-CoA + CH_3-(CH_2)_n-\overset{O}{\underset{\|}{C}}-S-CoA$$

The shortened fatty acid chain is now ready for the next cycle of β-oxidation.

Net ATP Yield from Palmitate Oxidation

Each cycle of β-oxidation produces one $FADH_2$, one NADH, and one acetyl CoA. During the last β-oxidation cycle, two acetyl CoAs are formed. Thus, the products of complete β-oxidation of palmitate are

   8 acetyl CoA
   7 $FADH_2$
   7 NADH

As discussed in Chapters 5 and 6, oxidation of $FADH_2$ and NADH by electron transport and oxidative phosphorylation yield respectively 2 and 3 ATPs, and oxidation of acetyl CoA by the TCA cycle coupled to electron transport and oxidative phosphorylation yields 12 ATPs. Therefore, the total yield of oxidation of palmitate to CO_2 and H_2O is 131 ATPs. However, two high-energy phosphate bonds (the equivalent of two ATPs) are consumed in the activation of palmitate. Thus, the net yield of palmitate oxidation is 129 ATPs.

Other Mechanisms of Fatty Acid Oxidation

Most fatty acids (especially saturated fatty acids) can be oxidized by the β-oxidation pathway described above. However, fatty acids that contain an odd number of carbon atoms, certain unsaturated fatty acids, and methylated fatty acids require modifications of the β-oxidation sequence.

Oxidation of Unsaturated Fatty Acids

The double bond that is generated between the α and β carbons in the first step of β-oxidation is in the *trans* configuration. Hydration of this bond by the hydratase yields the L-hydroxy derivative, which is the required substrate of the next enzyme, β-hydroxyacyl CoA dehydrogenase. The double bonds in most unsaturated fatty acids, however, are *cis*, and yield the D stereoisomer when hydrated by the hydratase. These D isomers are converted to the L isomers by a *racemase* (Fig. 7-13). β-Oxidation then proceeds normally.

If a double bond in an unsaturated fatty acid is located between the β and γ carbon atoms instead of between the α and β carbons, the fatty acid cannot directly enter β-oxidation. Instead, an *isomerase* moves the double bond to the correct position, and β-oxidation proceeds (Fig. 7-14).

Oxidation of Fatty Acids Containing an Odd Number of Carbons

Fatty acids that contain an odd number of carbons are oxidized by β-oxidation, with the successive removal of two-carbon units, until a final three-carbon propionyl CoA is obtained. Propionyl CoA is utilized as shown in Figure 7-15. First, propionyl CoA is carboxylated to D-methylmalonyl CoA by *propionyl CoA carboxylase*. Methylmalonyl CoA

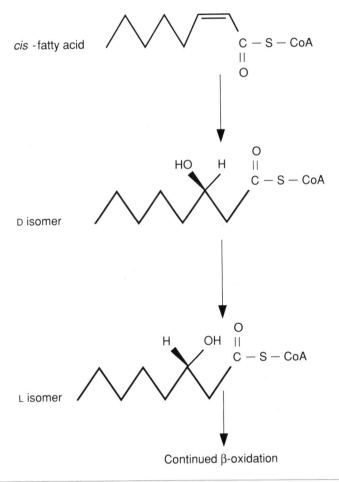

Fig. 7-13. Pathway by which fatty acids containing *cis* double bonds enter β-oxidation.

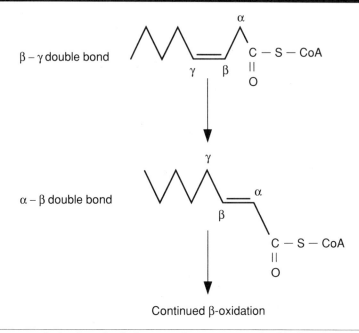

Fig. 7-14. Pathway by which fatty acids containing double bonds between the β and γ carbons enter β-oxidation.

racemase then converts D-methylmalonyl CoA to L-methylmalonyl CoA. Finally, L-methylmalonyl CoA undergoes rearrangement by *methylmalonyl CoA mutase* to yield succinyl CoA. As a result of entering the TCA cycle, succinyl CoA may be oxidized to CO_2 and H_2O or used as a precursor for gluconeogenesis.

α- and ω-Oxidation of Fatty Acids

Animal tissues contain minor pathways that involve oxidation of fatty acids at the α- and ω-carbons. The products of α and ω oxidation can enter β-oxidation.

α-Oxidation

During their metabolism, some fatty acids are hydroxylated on C-2 (the α carbon). The resulting α-hydroxy derivatives may be further oxidized to yield CO_2 and fatty acids consisting of one less carbon atom, which may then be metabolized by β-oxidation. α-Oxidation, which occurs in the endoplasmic reticulum, is especially important in the oxidation of methylated fatty acids. It is a method of generating odd-chain fatty acids.

ω-Oxidation

Fatty acids that are hydroxylated on the terminal carbon can undergo ω-oxidation. The ω-hydroxy group is converted to an ω-carboxy group, yielding an α,ω-dicarboxylic fatty acid. If this dicarboxylic fatty acid

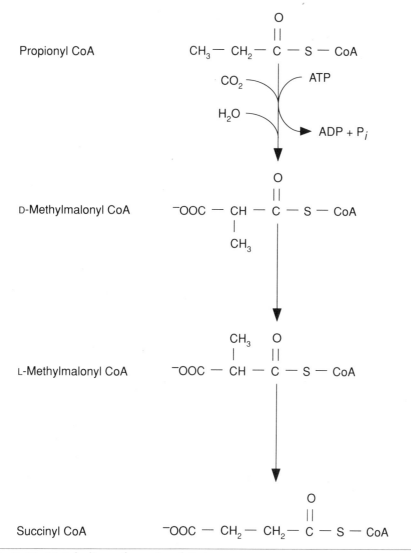

Fig. 7-15. Metabolism of propionyl CoA.

enters β-oxidation, it can be oxidized from both ends. ω-Oxidation of fatty acids has been detected on the endoplasmic reticulum of liver cells.

REGULATION OF FATTY ACID SYNTHESIS AND DEGRADATION

The activity of the TCA cycle and the availability of fatty acids are major factors determining whether fatty acids will be degraded or synthesized. When the TCA cycle is active—under conditions of a low energy charge in the cell (low ratio of ATP to ADP + AMP)—fatty acid synthesis is inhibited and fatty acid degradation is stimulated, so acetyl CoA is channeled into the TCA cycle. When the energy charge is high, the TCA cycle is inhibited, and acetyl CoA is shunted from the TCA cycle into fatty acid synthesis. Two enzymes—isocitrate dehydrogenase and acetyl CoA carboxylase—are important in regulating fatty acid synthesis:

1. *Isocitrate dehydrogenase* is the rate-limiting enzyme of the TCA cycle. As discussed in Chapter 6, it is inhibited by ATP and NADH and is activated by ADP. Inhibition of isocitrate dehydrogenase inhibits the TCA cycle and leads to activation of fatty acid synthesis.
2. *Acetyl CoA carboxylase* catalyzes the committed step of fatty acid synthesis, and is the rate limiting enzyme of the pathway. It is activated by citrate and inhibited by palmitoyl CoA.

Three enzymes are important in the regulation of fatty acid degradation:

1. *Carnitine acyltransferase I* catalyzes the synthesis of fatty acylcarnitine derivatives, which are required for the transport of fatty acids into the mitochondria. This enzyme is inhibited by malonyl CoA.
2. *β-Hydroxyacyl CoA dehydrogenase* catalyzes the formation of β-keto intermediates during fatty acid degradation. It is inhibited by NADH.
3. *Thiolase* catalyzes the release of acetyl CoA units. It is inhibited by acetyl CoA.

Regulation of Fatty Acid Synthesis

When isocitrate dehydrogenase is inhibited by a high energy charge in the cell, little isocitrate is converted to α-ketoglutarate in the TCA cycle. Due to the citrate synthase reaction, citrate accumulates in the mitochondrion and diffuses out into the cytosol. Cytosolic citrate activates acetyl CoA carboxylase and turns on fatty acid synthesis. The accumulation of citrate in the cytosol will also generate acetyl CoA via the citrate lyase reaction, where it can enter fatty acid synthesis. As fatty acid synthesis proceeds,

1. Palmitoyl CoA accumulates
2. ATP is hydrolyzed to ADP and P_i

Palmitoyl CoA is an inhibitor of acetyl CoA carboxylase. A rise in the ADP/ATP ratio activates isocitrate dehydrogenase and thus stimulates the TCA cycle, leading to a fall in cytosolic citrate and therefore to deactivation of acetyl CoA carboxylase. Thus, a buildup of palmitoyl CoA and ADP shuts down fatty acid synthesis and activates the TCA cycle.

The activity of acetyl CoA carboxylase is also controlled by a cAMP-mediated phosphorylation/dephosphorylation. The phosphorylated enzyme is the less active form.

1. *Glucagon* promotes phosphorylation, inhibiting fatty acid synthesis.
2. *Insulin* promotes dephosphorylation, stimulating fatty acid synthesis.

The levels of both acetyl CoA carboxylase and the fatty acid synthase complex in cells are increased by high-carbohydrate or fat-free diets, and decreased by high-fat diets. The rate at which fatty acid synthetase is synthesized is also increased by high-carbohydrate and fat-free diets and decreased by high-fat diets. Moreover, it is decreased by fasting and glucagon.

Regulation of Fatty Acid Degradation

When fatty acid synthesis is occurring (at times of high energy charge in the cell), the concentration of malonyl CoA is increased, which inhibits carnitine acyltransferase I. β-Hydroxyacyl CoA dehydrogenase and thiolase are also inhibited under these conditions. When the energy charge is low, there is an accompanying decrease in the concentrations of malonyl CoA (because fatty acid synthesis is inhibited), NADH, and acetyl CoA, leading to deinhibition of fatty acid degradation. The result is that fatty acid synthesis is inhibited when fatty acid oxidation is stimulated, and vice versa.

KETONE BODIES

The major ketone bodies in humans are acetoacetate and β-hydroxybutyrate. They are synthesized from acetyl CoA (Fig. 7-16), and are important sources of energy for the brain during extended starvation. They may also serve as energy sources for other tissues, especially muscle, when available glucose is low.

During starvation, the body relies largely on lipid oxidation to meet its energy needs. More acetyl CoA is consequently produced, which leads to increased blood levels of acetoacetate and β-hydroxybutyrate. The presence of ketone bodies eases the demand on blood glucose, which is in short supply during starvation, thus reserving it for the central nervous system. Ketone bodies become a major source of energy for the central nervous system only during extended starvation, when the supply of glucose is significantly depressed.

Liver cells produce ketone bodies but are unable to metabolize them. Nonhepatic tissues utilize ketone bodies by degrading them to acetyl CoA. First, β-hydroxybutyrate is re-oxidized to acetoacetate, which is converted to acetoacetyl CoA by *acetoacetate-succinyl CoA CoA transferase*:

Acetoacetate + succinyl CoA → acetoacetyl CoA + succinate

Acetoacetyl CoA is then converted to two molecules of acetyl CoA by a thiolase.

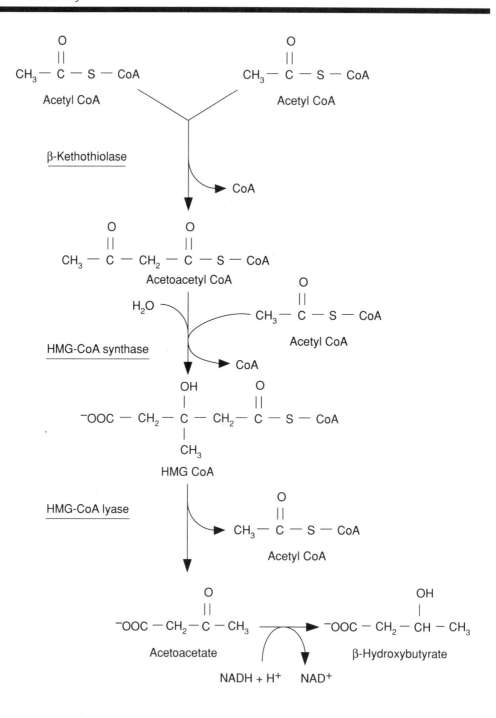

Fig. 7-16. Synthesis of ketone bodies. HMG, hydroxymethylglutaryl.

CLINICAL CORRELATIONS

Abetalipoproteinemia: Apolipoprotein B Deficiency

Dietary and endogenously synthesized lipids are transported to the tissues by chylomicrons and VLDLs. The protein components of these lipoproteins consist of several different polypeptide chains. One type, *apolipoprotein B,* is a major component of both chylomicrons and VLDLs. In patients with abetalipoproteinemia, a genetic deficiency of this protein results in the absence of chylomicrons, VLDLs, and low density lipoproteins (LDLs). LDLs are a metabolic product of VLDLs.

Abetaliproteinemia is characterized clinically by deformed red blood cells (acanthocytosis), retinal degeneration, neuromuscular disorders, and malabsorption. The plasma is usually very low in cholesterol and triacylglycerols. The diagnosis of abetaliproteinemia is confirmed by the absence of apolipoprotein B from the plasma.

The most obvious treatment for abetalipoproteinemia is to restrict the dietary intake of long-chain fatty acids. However, some lipid is essential in the diet. It has therefore been suggested that long-chain fatty acids be replaced in the diet by short- and medium-chain fatty acids. Liver function is carefully monitored. The rationale for this treatment is that short- and medium-chain fatty acids are soluble and do not require lipoproteins for transport and cellular uptake. The diet is also supplemented with the fat-soluble vitamins.

Refsum's Disease: Phytanic Acid Hydroxylase Deficiency

Phytanic acid and phytol are dietary lipids derived from chlorophyll. Phytol is readily oxidized to phytanic acid. Phytanic acid is normally metabolized by initial hydroxylation of the α-carbon by phytanic acid hydroxylase, followed by conversion to pristanic acid via α-oxidation. Pristanic acid is metabolized by β-oxidation. A genetic deficiency of phytanic acid hydroxylase prevents α-oxidation of phytanic acid, resulting in the accumulation of phytanic acid in adipose tissue and plasma. Affected individuals may display retinitis pigmentosa, diminished deep tendon reflexes, and incoordination. The palms and soles may also become scaly and thickened. These characteristic symptoms usually appear gradually. The underlying biochemical defect is shown in Figure 7-17.

Refsum's disease is treated by eliminating phytanic acid and phytol from the diet. Once phytanic acid intake is reduced, the phytanic acid accumulated in the plasma and adipose tissue is gradually eliminated. This approach can successfully arrest progression of the disease; in some cases, regression of symptoms has been observed. If the diet is not followed strictly, the patient relapses. Some patients lose weight rapidly on the diet, which causes an increase in plasma levels of phytanic acid because of the influx of phytanic acid previously stored in the adipose

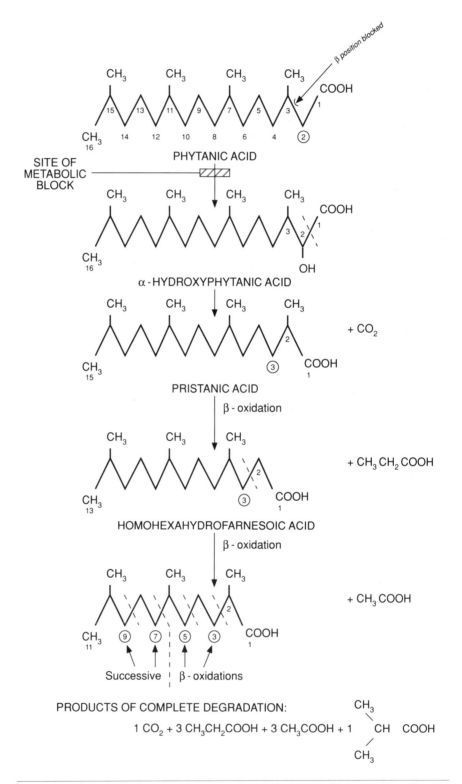

Fig. 7-17. The metabolic block associated with Refsum's disease (phytanic acid hydroxylase deficiency). (From Stanbury JB, Wyngaarden JB, Fredrickson DS: The Metabolic Basis of Inherited Disease. McGraw-Hill, New York, 1978, with permission.)

tissue. These patients experience a temporary worsening of symptoms until the plasma levels of phytanic acid finally decrease.

Ketosis

At normal blood levels of ketone bodies, a small amount of acetoacetate is spontaneously oxidized to acetone:

$$CH_3-\overset{O}{\underset{\|}{C}}-CH_2-\overset{O}{\underset{\|}{C}}-O^- + H^+ \rightarrow CH_3-\overset{O}{\underset{\|}{C}}-CH_3 + CO_2$$

If the blood level of ketone bodies increases (ketonemia), the capacity of the kidney to resorb them may be exceeded, so that ketone bodies appear in the urine (ketonuria). An additional consequence of ketonemia is increased decarboxylation of acetoacetate to acetone, which is expired in the breath. The combination of ketonuria, ketonemia, and acetone breath is referred to as ketosis. Any circumstance leading to excessive oxidation of fatty acids can cause ketosis, for example, prolonged starvation and prolonged overconsumption of lipid-rich foods. The underutilization of glucose in diabetics can also cause ketosis.

SUGGESTED READINGS

1. Galton DJ: The Human Adipose Cell. Appleton-Century-Crofts, East Norwalk, CT, 1971
2. Cryer A, Van RLR: New Perspectives in Adipose Tissue. Butterworth (Publisher), London, 1985
3. Stanbury JB, Wyngaarden JB, Fredrickson DS, et al (eds.): The Metabolic Basis of Inherited Disease. 5th Ed. McGraw-Hill, New York, 1983
4. Masono EJ: Lipids and lipid metabolism. Annu Rev Physiol 39:301, 1977
5. Gurr MI, James AT: Lipid Biochemistry: An Introduction. 3rd Ed. Chapman and Hall, London, 1980
6. McGarry JD, Foster DW: Regulation of hepatic fatty acid oxidation and ketone body production. Annu Rev Biochem 49:395, 1980
7. Wakil SJ, Stoops JK, Joshi VC: Fatty acid synthesis and its regulation. Ann Res Biochem 56:537, 1983

STUDY QUESTIONS

Directions: Answer the following questions using the key outlined below.
- **(A)** if 1, 2, and 3 are correct
- **(B)** if 1 and 3 are correct
- **(C)** if 2 and 4 are correct
- **(D)** if only 4 is correct
- **(E)** if all four are correct

1. Lipids are especially suited for energy storage because they
   1. are highly reduced
   2. generally represent a large proportion of the dietary intake
   3. are nonreactive substances
   4. are usually less reduced than carbohydrate

2. Triacylglycerols
   1. contain fatty acids esterified to glycerol
   2. are the major form of stored lipid
   3. are neutral lipids
   4. are amphipathic lipids

3. The effect of a high carbohydrate diet on the intracellular levels of the fatty acid synthase complex is believed to be mediated by
   1. increased degradation of fatty acid synthase
   2. feedback inhibition
   3. a cAMP-mediated phosphorylation/dephosphorylation regulatory mechanism
   4. increased synthesis of fatty acid synthase

Directions: For each of the following multiple choice questions, choose the most appropriate answer.

4. Endogenously synthesized lipid is transported from the liver to the adipose tissue by
   A. chylomicrons
   B. HDLs
   C. LDLs
   D. VLDLs

5. Lipoprotein lipase cleaves fatty acids from positions
   A. 1 and 3 of triacylglycerols
   B. 1, 2, and 3 of triacylglycerols
   C. 2 and 3 of triacylglycerols
   D. 1 and 2 of triacylglycerols

6. The number of NADPHs required for addition of one acetyl unit to a growing fatty acid chain is
   A. 7
   B. 2
   C. 1
   D. 8

7. The citrate shuttle ultimately generates
   A. citrate and NADPH in the cytosol
   B. acetyl CoA and NADH in the cytosol
   C. acetyl CoA in the cytosol
   D. acetyl CoA and NADPH in the cytosol

8. Fatty acids are transported from adipose tissue to liver while bound to
   A. HDLs
   B. albumin
   C. chylomicrons
   D. LDLs

9. Each cycle of β-oxidation in the degradation of palmitate yields
   A. 2 NADPHs
   B. 1 NADH and 1 $FADH_2$
   C. 2 $FADH_2$s
   D. 1 NADPH and 1 $FADH_2$

10. Regulation of fatty acid β-oxidation is largely mediated by
    A. feedback inhibition by acetyl CoA
    B. the extent of acetyl CoA utilization by fatty acid synthesis
    C. the extent of acetyl CoA utilization by the TCA cycle
    D. Repression of acyl CoA dehydrogenase by ketone bodies

8

LIPID METABOLISM II

Metabolism of Complex Lipids: Triacylglycerols/ Phosphoglycerides/Sphingolipids/Cholesterol/ Prostaglandins/Leukotrienes/Clinical Correlation

The student should be able to:

1. Distinguish the synthesis of triacylgycerols in liver, adipose tissue, and the intestinal mucosa.
2. Describe the synthesis of phosphoglycerides starting from either glycerol 3-phosphate or dihydroxyacetone phosphate.
3. Define the structural features of plasmalogens.
4. Describe phosphoglyceride fatty acid rearrangement and its impact on the fatty acid composition of tissue phospholipids.
5. Describe the roles of the various phospholipases in phospholipid degradation.
6. Distinguish ceramides from cerebrosides.
7. Name three functions of cholesterol in the body, outline the synthesis of cholesterol, and identify the regulated enzyme of cholesterol synthesis and the mechanisms of its regulation.
8. Describe the transport and cellular uptake of cholesterol, and the metabolic fates of cholesterol in the liver.
9. Describe the pathway of prostaglandin synthesis and distinguish prostaglandins and thromboxanes; distinguish HETEs and HPETEs; and understand the role of lipoxygenase in leukotriene synthesis.

Learning Objectives

The term *complex lipids* usually refers to lipids that are structurally complex. The structures of these lipids may consist of two or more components that can be separated by hydrolysis, as in the case of triacylglycerols, phosphoglycerides, sphingolipids, and cholesterol esters. In addition, nonhydrolyzable lipids of sufficient structural complexity, such as cholesterol and the prostaglandins, are also considered complex lipids. This chapter covers the function and metabolism of the complex lipids.

Perspective: Complex Lipids

TRIACYLGLYCEROLS

As described in the last chapter, the main function of triacylglycerols is to store energy in adipose tissue. Triacylglycerols are synthesized in the liver, adipose tissue, and the intestinal mucosa. The synthetic pathway is slightly different in each tissue.

Triacylglycerol Synthesis

Liver

In the liver, triacylglycerols are synthesized from fatty acyl CoAs and either glycerol, glycerol 3-phosphate, or dihydroxyacetone phosphate. If free glycerol is used, it is first converted to glycerol 3-phosphate by *glycerol kinase*:

$$\text{Glycerol} + \text{ATP} \rightarrow \text{glycerol 3-phosphate} + \text{ADP}$$

Dihydroxyacetone phosphate can also be converted to glycerol 3-phosphate by reduction of the keto group by a dehydrogenase:

$$\text{Dihydroxyacetone phosphate} + \text{NADH} + \text{H}^+ \rightleftharpoons$$
$$\text{glycerol 3-phosphate} + \text{NAD}^+$$

Triacylglycerols are synthesized using glycerol 3-phosphate by the pathway shown in Figure 8-1. Glycerol 3-phosphate is acylated at the 1 and 2 positions by two fatty acyl CoAs to yield *phosphatidic acid* (phosphatidate). The phosphate group is then hydrolyzed to yield a 1,2-diacylglycerol and phosphate. The diacylglycerol is subsequently acylated with another fatty acyl CoA to yield a complete triacylglycerol. Each of the three acylation reactions is catalyzed by a specific acyltransferase.

Figure 8-1 also shows an alternative pathway that uses dihydroxyacetone phosphate rather than glycerol 3-phosphate as the substrate. This pathway involves initial acylation at the 1 position followed by reduction at the 2 position to yield lysophosphatidate. The lysophosphatidate is then acylated at position 2 to yield phosphatidic acid, which proceeds to form triacylglycerol as described above.

Adipose Tissue

Adipose tissue does not contain glycerol kinase (the enzyme that catalyzes the conversion of glycerol to glycerol 3-phosphate), and therefore depends on glycerol 3-phosphate and dihydroxyacetone phosphate for triacylglycerol synthesis. Triacylglycerol synthesis in adipose tissue is otherwise similar to that in the liver.

Intestinal Mucosa

As mentioned in Chapter 7, the intestinal mucosa synthesizes triacylglycerols primarily from the 2-monoacylglycerols and fatty acids released by the digestion of dietary lipid. Phosphatidate is not involved as an intermediate. The fatty acids are activated to fatty acyl CoAs and

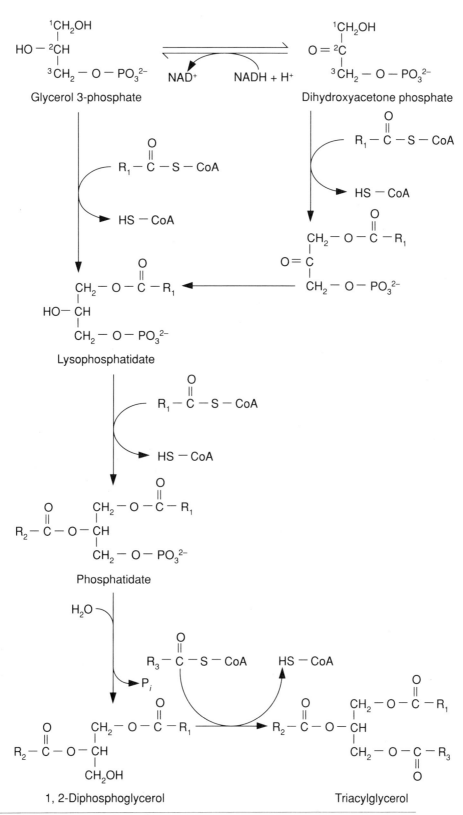

Fig. 8-1. Triacylglycerol synthesis.

used to acylate the 2-monoacylglycerols to form 1,2-diacylglycerols, which are in turn acylated to yield triacylglycerols (see Fig. 7-5).

Degradation of Triacylglycerols

As mentioned in Chapter 7, triacylglycerols are degraded by lipases during the mobilization of fatty acids from adipose tissue and from lipoproteins. The products of triacylglycerol degradation are free fatty acids, glycerol, and mono- and diacylglycerols.

PHOSPHOGLYCERIDES

Phosphoglycerides (sometimes called phosphoglycerols) are amphipathic organic lipids consisting of a glycerol moiety with fatty acids esterified to the 1 and 2 positions and a phosphate group on the 3 position. In almost all cases, the phosphate group is esterified to a polar organic molecule, which usually contains a nitrogenous base (choline, ethanolamine, or serine). Notable exceptions are the non-nitrogenous head groups inositol and glycerol (see Fig. 8-8). Figure 8-2 shows the general structure of a phosphoglyceride. Phosphoglycerides are often called *phospholipids*, but that term is somewhat ambiguous, as it also covers the phosphosphingolipid sphingomyelin.

Phosphoglycerides are named according to the identity of the organic polar head group. For example, if the polar head group is choline, the phosphoglyceride is called phosphatidylcholine. The term *phosphatidylcholine* actually refers to the family of phosphoglycerides that contain choline as the polar head group. The fatty acid components may be specified in order to distinguish members of a phosphoglyceride family (for example, to distinguish different phosphatidylcholines).

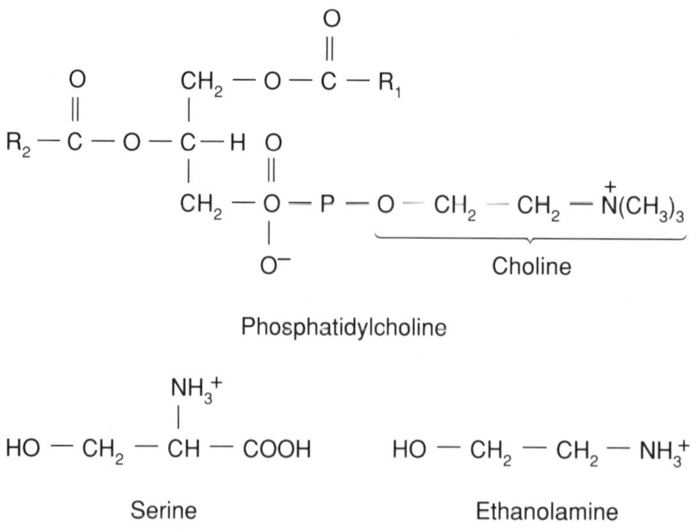

Fig. 8-2. Phosphatidylcholine, illustrating the structure of phosphatidylglycerides. The nitrogenous bases serine and ethanolamine are also shown.

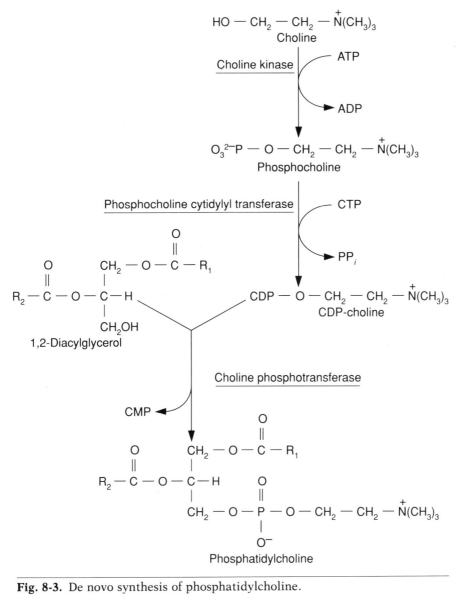

Fig. 8-3. De novo synthesis of phosphatidylcholine.

Most phosphoglycerides are membrane components. In addition, some phosphoglycerides play important roles in specific physiologic and biochemical processes. For example, dipalmitoylphosphatidylcholine is the major component of alveolar surfactant, and lysophosphatidylcholine is important in solubilizing dietary lipid in the intestine. Phosphatidylcholine and phosphatidylinositol serve as sources of arachidonic acid in the synthesis of prostaglandins.

Synthesis of Phosphoglycerides

Phosphoglycerides are synthesized in most tissues. The synthesis of phosphoglycerides occurs in two phases. The first phase is the synthesis of 1,2-diacylglycerol by the pathway shown in Figure 8-1. In the second phase, a specific polar group is attached to carbon 3 via a phosphate ester.

Synthesis of Phosphatidylcholine

Phosphatidylcholine synthesis is shown in Figure 8-3. Choline is first converted to phosphocholine by *choline kinase* and ATP. Phosphocholine then reacts with cytidine triphosphate (CTP) in a reaction catalyzed by *phosphocholine cytidylyl transferase* to yield cytidine diphosphocholine (CDP-choline) and pyrophosphate. CDP-choline then reacts with 1,2-diacylglycerol to yield phosphatidylcholine and cytidine monophosphate (CMP). This reaction is catalyzed by *choline phosphotransferase*. The liver is also capable of synthesizing phosphatidylcholine by using S-adenosylmethionine to successively methylate phosphatidylethanolamine (Fig. 8-4).

Synthesis of Phosphatidylethanolamine

The major pathway of phosphatidylethanolamine synthesis is via CDP-ethanolamine in reactions analogous to those for phosphatidylcholine (Fig. 8-5). Liver mitochondria can also synthesize phosphatidylethanolamine by decarboxylation of phosphatidylserine (Fig. 8-6).

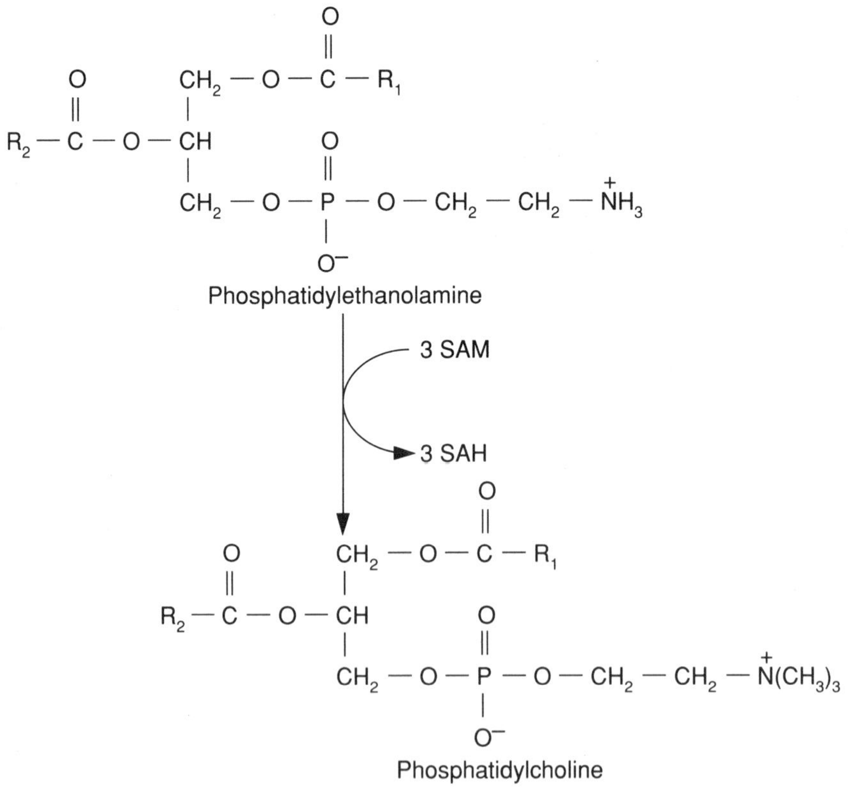

Fig. 8-4. Synthesis of phosphatidylcholine from phosphatidylethanolamine. SAM, S-adenosylmethionine; SAH, S-adenosylhomocysteine.

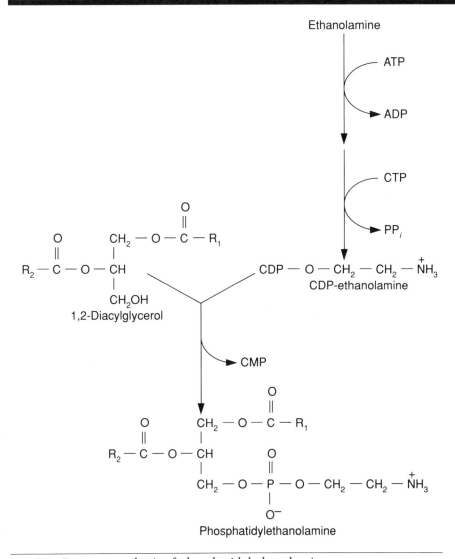

Fig. 8-5. De novo synthesis of phosphatidylethanolamine.

Synthesis of Phosphatidylserine

Phosphatidylserine is formed via a base exchange reaction involving phosphatidylethanolamine and serine, catalyzed by *Phosphatidylethanolamine transferase*:

Phosphatidylethanolamine + serine ⇌
　　　　　　　　　　　phosphatidylserine + ethanolamine

CDP-Diacylglycerols: Synthesis of Phosphatidylglycerol and Phosphatidylinositol

CDP-diacylglycerols are formed by the reaction of phosphatidate with CTP (Fig. 8-7). CDP diacylglycerols are intermediates in the synthesis

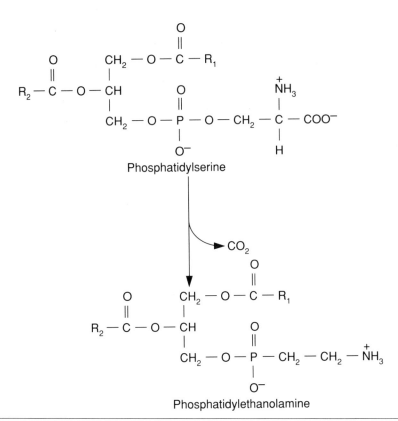

Fig. 8-6. Synthesis of phosphatidylethanolamine via decarboxylation of phosphatidylserine.

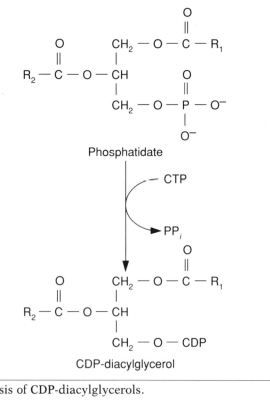

Fig. 8-7. Synthesis of CDP-diacylglycerols.

of phosphatidylglycerol and phosphatidylinositol. The reactions involved in the synthesis of these phosphoglycerides are shown in Figure 8-8.

Plasmalogens

Plasmalogens are a special class of phosphoglyceride in which the fatty acid on the number 1 position is replaced by an α-β unsaturated fatty alcohol. Plasmalogens are significant constituents of nerve and muscle cell membranes. Ethanolamine plasmalogen is the most abundant type in mammals. Plasmalogens are synthesized from dihydroxyacetone phosphate as shown in Figure 8-9. Dihydroxyacetone phosphate is acylated at the 1 position to yield 1-fatty acyldihydroxyacetone. The fatty acyl in the 1 position is then displaced by a fatty alcohol to yield 1-alkyl dihydroxyacetone phosphate. Next, the keto group is reduced and acylated to yield a 1-alkyl-2-acylphosphatidate, which is dephosphorylated to a 1-alkyl-2-acylglycerol. The 1-alkyl-2-acylglycerol reacts with CDP-ethanolamine to yield the phosphoethanolamine derivative. In the final reaction, a double bond is introduced in the α-β position of the fatty alcohol, forming ethanolamine plasmalogen.

Phosphoglyceride Fatty Acid Exchange Reactions

Examination of tissue phosphoglycerides reveals that in most cases the 1 position of the glycerol moiety carries a saturated fatty acid and the 2 position carries an unsaturated fatty acid. These fatty acids are not necessarily the ones incorporated during phosphoglyceride synthesis, but are the result of fatty acid exchange reactions. Two types of fatty acid exchange reactions occur: *direct exchange* and *phospholipid exchange*. The essential features of these two mechanisms are as follows:

Direct Exchange (Fig. 8-10)

1. A fatty acid is hydrolyzed from either position 1 or 2 of the phosphoglyceride. Hydrolysis at the 1 and 2 positions is catalyzed by *phospholipases A_1 and A_2*, respectively.
2. The hydrolyzed fatty acid is replaced by direct acylation using the desired fatty acyl CoA.

Phospholipid Exchange (Fig. 8-11)

1. A fatty acid is exchanged between one phosphoglyceride and one lysophosphoglyceride (or between two lysophosphoglycerides).

Both mechanisms employ specific acyltransferases.

Degradation of Phosphoglycerides

Phosphoglycerides are degraded by phospholipases. A different phospholipase catalyzes the hydrolysis of each hydrolyzable bond in phos-

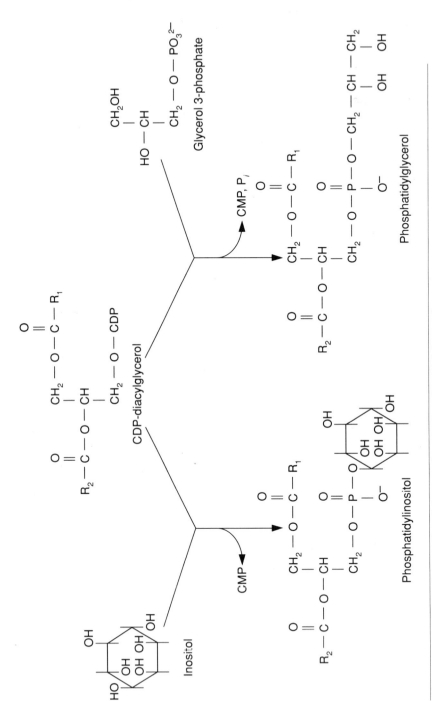

Fig. 8-8. Synthesis of phosphatidylinositol and phosphatidylglycerol.

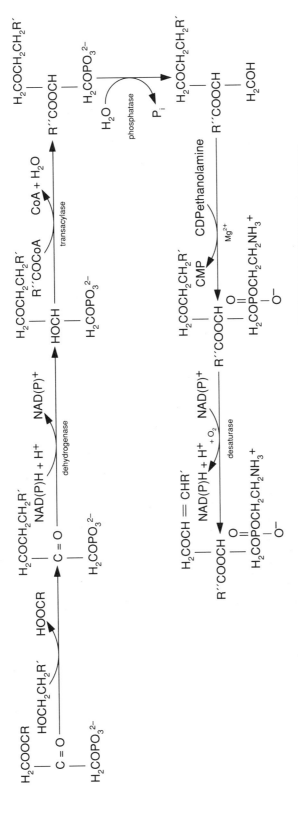

Fig. 8-9. Synthesis of ethanolamine plasmalogen. (Redrawn from Smith EL, Hill RL, Lehman IR, et al: Principles of Biochemistry: General Aspects. 7th Ed. McGraw-Hill, New York, 1983, with permission.)

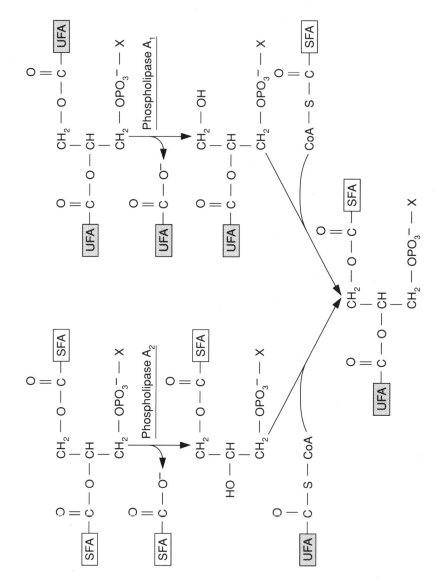

Fig. 8-10. Phosphoglyceride fatty acid exchange, showing the exchanges catalyzed by phospholipases A_1 and A_2. **SFA**, saturated fatty acid; **UFA**, unsaturated fatty acid.

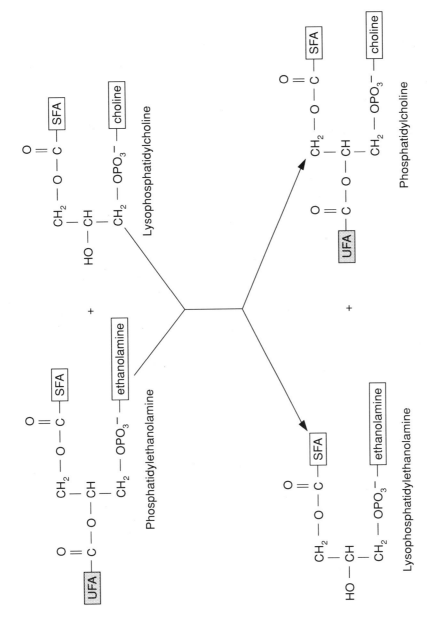

Fig. 8-11. The lysophosphoglyceride exchange reaction. SFA, saturated fatty acid; UFA, unsaturated fatty acid.

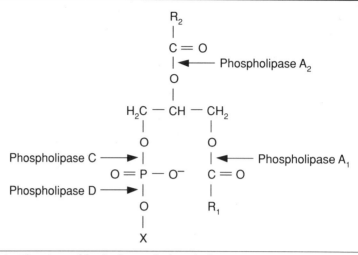

Fig. 8-12. The sites of hydrolysis of phospholipases A_1, A_2, C, and D. X, polar head group.

phoglycerides (Fig. 8-12). These enzymes are present in most mammalian cells. Phospholipases A_1 and A_2 function in direct phospholipid exchange, as discussed above. Phospholipase C hydrolyzes the glycerol phosphate bond, and phospholipase D hydrolyzes the bond between the polar head group and the phosphate. There are also lysophospholipases that hydrolyze the single fatty acid from lysophosphoglycerides. Some phospholipases show additional specificity by attacking only certain phosphoglycerides.

Triacylglycerols and Phosphoglycerides: Key Points

1. The glycerol moiety for the synthesis of triacylglycerols and phosphoglycerides can be derived from free glycerol (except in adipose tissue), glycerol 3-phosphate, or dihydroxyacetone phosphate.
2. Phosphatidate and 1,2-diacylglycerols are common intermediates in the de novo synthesis of phospholipids and triacylglycerols.
3. Phosphoglycerides can also be synthesized by base exchange and base modification reactions involving existing phospholipids (as in the synthesis of phosphatidylserine and the methylation of phosphatidylethanolamine, respectively).
4. Phosphoglycerides often undergo fatty acid exchange reactions.
5. Phosphoglycerides are degraded by a variety of phospholipases, which attack specific ester linkages in the molecule.

SPHINGOLIPIDS

Structure of Sphingolipids

Sphingolipids take their name from the long-chain (C_{18}) base *sphingosine*. As can be seen from Figure 8-13, the carbon chain of sphingosine has a *trans* double bond between carbons 4 and 5, hydroxyl groups on carbons 1 and 3, and an amino group on carbon 2. In mature sphingolipids the hydroxyl group on carbon 3 is always free, and the amino group is always in an amide linkage with a long-chain fatty acid (C_{18}

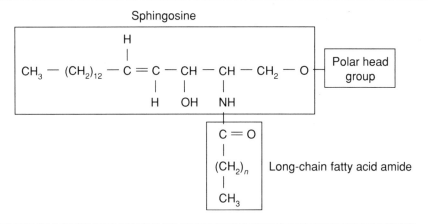

Fig. 8-13. The general structure of sphingolipids.

to C_{24}). The hydroxyl group on carbon 1 is always esterified to a polar head group, which may be either phosphocholine or a carbohydrate. When the polar group is phosphocholine, the resulting sphingolipid is a *sphingomyelin*. When it is a carbohydrate, the sphingolipid is one of an array of *glycosphingolipids*. The sphingosine-fatty acid amide adduct is called a *ceramide*. The ceramide structure is common to all sphingolipids.

The glycosphingolipids are divided into *neutral* and *acidic* glycosphingolipids, depending on whether the carbohydrate polar group is neutral or negatively charged at physiologic pH. The major types of acidic glycosphingolipids are *sulfatide* and the *gangliosides*. The polar group of sulfatide consists of a galactose moiety carrying a charged sulfate group on carbon 3. Gangliosides are acidic because they contain *N*-acetylneuraminic acid (sialic acid) (Fig. 8-14).

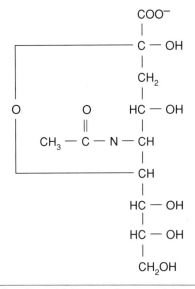

Fig. 8-14. Structure of sialic acid (*N*-acetylneuraminic acid).

Biologic Function of Sphingolipids

Sphingolipids are found in most tissues, especially nerve tissue. Most glycosphingolipids are structural components of cell membranes. In addition, some glycosphingolipids have been identified as blood group determinants, and are believed to be active as cell surface receptor sites for certain toxins. Sphingomyelin is found in high concentration in the myelin sheath of nerve fibers. Gangliosides are abundant in the nerve endings of the ganglion cells of nerve tissue.

Synthesis of Sphingolipids

Formation of Sphingosine and Ceramide

The starting materials for the synthesis of sphingosine are palmitoyl CoA and serine (Fig. 8-15). First, *3-ketodihydrosphingosine synthetase* catalyzes the condensation of palmitoyl CoA and serine to yield 3-ketodihydrosphingosine (dehydrosphinganine), CO_2, and CoA. 3-Ketodihydrosphingosine is then reduced to 3-hydroxydihydrosphingosine. In the next reaction, a long-chain fatty acid forms an amide bond with the amino group of dihydrosphingosine to yield dihydroceramide. Dihydroceramide is then dehydrogenated by a flavin-requiring desaturase to yield ceramide. Alternatively, dihydrosphingosine may be dehydrogenated to sphingosine before formation of the fatty acid amide. Free ceramide is not a mature sphingolipid, but it is an intermediate in the synthesis and degradation of all sphingolipids.

Synthesis of Sphingomyelin

As mentioned above, the phosphosphingolipid sphingomyelin is an important structural component of myelin sheaths. The fatty acid of sphingomyelin is usually palmitic acid, stearic acid, or a C_{24} fatty acid (saturated or unsaturated). The major pathway of sphingomyelin synthesis involves the transfer of phosphocholine to the hydroxyl on carbon 1 of ceramide, catalyzed by *CDP-choline:ceramide cholinephosphotransferase*:

$$\text{Ceramide} + \text{CDP-choline} \rightarrow \text{sphingomyelin} + \text{CMP}$$

A minor pathway of sphingomyelin synthesis employs phosphatidylcholine as the phosphocholine donor:

$$\text{Phosphatidylcholine} + \text{ceramide} \rightarrow \text{sphingomyelin} + \text{1,2-diacylglycerol}$$

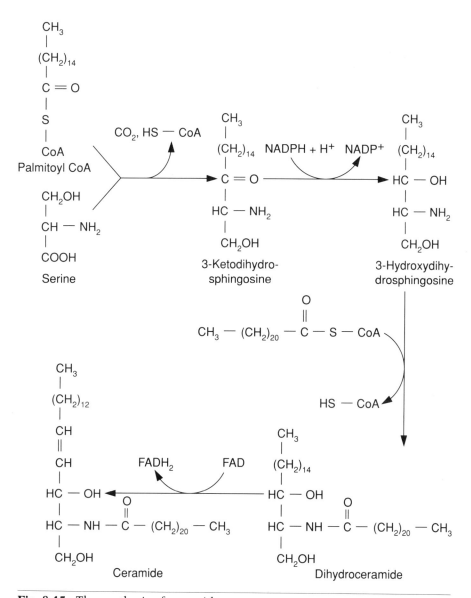

Fig. 8-15. The synthesis of ceramide.

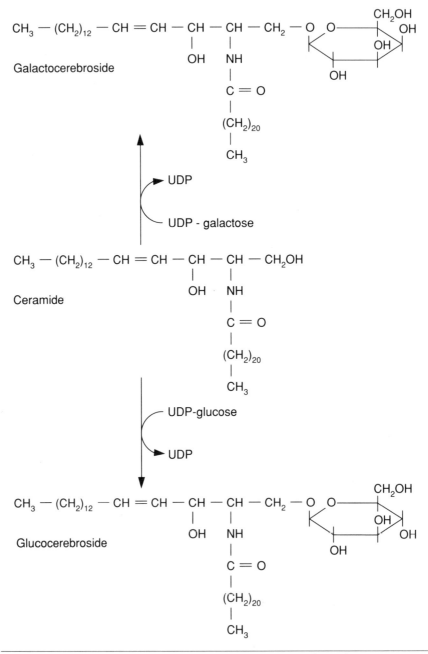

Fig. 8-16. Synthesis of galactocerebroside and glucocerebroside.

Formation of Glycosphingolipids

The polar head group of glycosphingolipids may consist of either a single sugar or an oligosaccharide. Glycosphingolipids with a single sugar residue are called *cerebrosides* (*ceramide monosaccharides*), whereas glycosphingolipids with an oligosaccharide unit are called *globosides* (*ceramide oligosaccharides*). The *gangliosides*, which contain sialic acid (Fig. 8-14) in the oligosaccharide head group, are a special type of ceramide oligosaccharide. To sum up, the different types of glycosphingolipids are

1. Cerebrosides (single-sugar head group)
2. Globosides (oligosaccharide head group)
3. Gangliosides (oligosaccharide head group containing sialic acid)

The most abundant sugar residues in all types of glycosphingolipid are glucose and galactose.

The enzymes responsible for synthesizing glycosphingolipids from ceramide are *glycosyltransferases*, which catalyze reactions of the type

Acceptor + UDP-monosaccharide → acceptor-monosaccharide + UDP

The acceptor may be a ceramide, a cerebroside, or a globoside. The more complicated oligosaccharide structures, containing a variety of different monosaccharides, require the action of many different glycosyltransferases. The glycosyltransferases involved in the transfer of sialic acid residues use CMP-sialic acid in a reaction analogous to the preceding one:

Acceptor + CMP-sialic acid → acceptor-sialic acid + CMP

Cerebrosides and Sulfatide. The synthesis of cerebrosides involves the formation of an α-*glycosidic linkage* between carbon 1 of the ceramide and carbon 1 of a monosaccharide. The synthesis of the two most common cerebrosides—glucocerebroside and galactocerebroside—is shown in Figure 8-16.

Sulfatide is a derivative of galactocerebroside that has a sulfate ester on carbon 3 of the galactose moiety (Fig. 8-17). The source of the sulfate

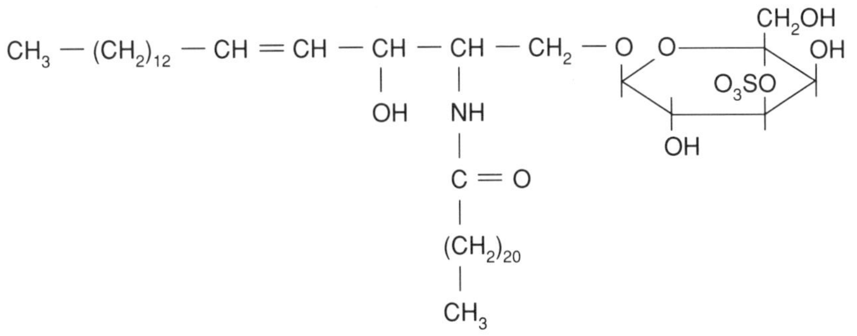

Fig. 8-17. Sulfatide.

in sulfatide synthesis is 3-phosphoadenosine-5-phosphosulfate (PAPS). The sulfate transfer is catalyzed by a microsomal sulfotransferase:

Galactocerebroside + PAPS →
       PAP + galactocerebroside-3-sulfate (sulfatide)

Globosides and Gangliosides. Figure 8-18 shows the synthesis of selected ceramide oligosaccharides, including some gangliosides. The sialic acid component(s) of gangliosides are usually linked to the carbohydrate chain via an α-glycosidic linkages involving carbon 2 of the sialic acid moiety.

Gangliosides are described by a special nomenclature code that indicates the number of sialic acid residues present and the monosaccharide sequence of the oligosaccharide unit attached to the ceramide. For example, in the case of the ganglioside G_{M3} shown in Figure 8-18,

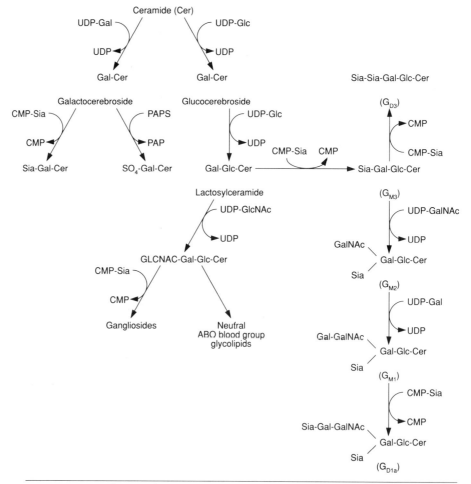

Fig. 8-18. Synthesis of some common glycosphingolipids. (Redrawn from Smith EL, Hill RL, Lehman IR, et al: Principles of Biochemistry. General Aspects. 7th Ed. McGraw-Hill, New York, 1983, with permission.)

Table 8-1. Some Oligosaccharide Sequences Found in Gangliosides

Sequence	Number Designation
Gal-Glc-Cer	3
GalNAc-Gal-Glc	2
Gal-GalNAc-Gal-Glc	1

G indicates that the glycosphingolipid is a ganglioside; M (standing for "mono") indicates that there is only one sialic acid residue; and 3 refers to the particular sequence of sugars attached to the ceramide. Table 8-1 shows some common sugar sequences associated with gangliosides. The subscripts D, T, and Q respectively indicate structures containing two, three, and four sialic acid residues.

Perspective: Degradation of Sphingolipids

In normal tissues, the concentration of sphingolipids is kept approximately constant by a balance between sphingolipid synthesis and degradation. Sphingolipids are degraded in lysosomes by a series of hydrolases. The initial degradation products are ceramide and the component sugar moieties or phosphocholine. Ceramide is then degraded further to a fatty acid and sphingosine; sphingosine is degraded to palmitaldehyde and ethanolamine phosphate.

Degradation of Sphingolipids to Ceramide

Each sugar molecule of the oligosaccharide moiety is removed sequentially by a specific hydrolytic enzyme (Fig. 8-19). The removal of all the sugar residues thus requires the action of a precise sequence of hydrolases: the product of each enzyme is the substrate for the next. A deficiency in any of the enzymes causes an accumulation of the substrate of that enzyme. Genetic deficiencies of several of the enzymes involved in the degradation of sphingolipids have been identified. These deficiencies cause the diseases known as sphingolipidoses, which are discussed in the Clinical Correlations at the end of this chapter.

Degradation of Ceramide and Sphingosine

Ceramide is cleaved by a specific hydrolase (*ceramidase*) to release sphingosine and the fatty acid:

$$\text{Ceramide} \rightarrow \text{fatty acid} + \text{sphingosine}$$

The resulting sphingosine is phosphorylated:

$$\text{Sphingosine} + \text{ATP} \rightarrow \text{sphingosine 1-phosphate} + \text{ADP}$$

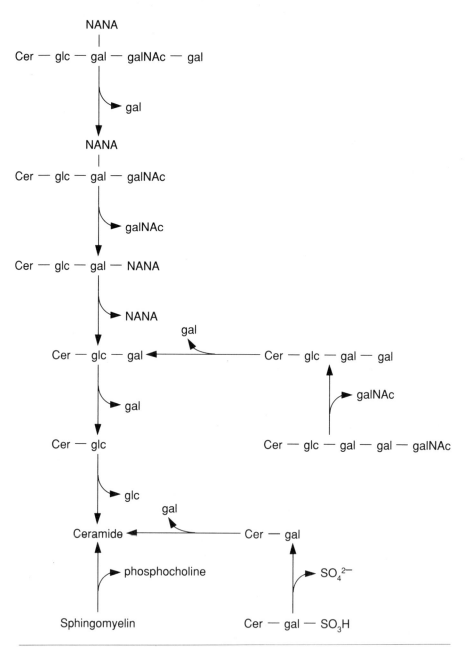

Fig. 8-19. An example of sphingolipid degradation. Sugar residues are removed from the oligosaccharide moiety one by one and in a precise order.

and sphingosine 1-phosphate is hydrolyzed by a pyridoxal phosphate-dependent enzyme:

Sphingosine 1-phosphate → palmitaldehyde + ethanolamine phosphate

Ethanolamine phosphate may be used in the synthesis of phosphoglycerides. Palmitaldehyde may be oxidized to palmitate or further reduced to palmitol, depending on the level of reducing power prevailing in the cell.

Sphingolipids: Key Points

1. Sphingolipids are membrane lipids, and are especially prevalent in nerve tissue.
2. One sphingolipid, sphingomyelin, contains a noncarbohydrate polar head group (phosphocholine). All other sphingolipids have a carbohydrate or carbohydrate derivative as a polar head group. Cerebrosides have a monosaccharide, globosides have an oligosaccharide, and gangliosides have an oligosaccharide polar head group that contains sialic acid.
3. Ceramide (sphingosine plus a fatty acid unit) is an intermediate in the synthesis and degradation of all sphingolipids.
4. Sphingolipids are degraded by series of specific lysosomal hydrolases. The carbohydrate polar head groups of glycosphingolipids are degraded by the sequential removal of monosaccharides. Specific hydrolases also remove phosphocholine from sphingomyelin and convert ceramide to sphingosine and fatty acid.
5. Sphingosine may be further degraded to yield a long-chain fatty aldehyde (palmitaldehyde) and ethanolamine phosphate.

CHOLESTEROL

Cholesterol (Fig. 8-20) is present in all tissues. Most tissues synthesize some cholesterol, but the bulk of endogenous cholesterol is manufactured in the liver and intestine. Cholesterol plays several important roles:

1. It is a component of nearly all cell membranes.
2. It is the precursor of the bile salts, which are crucial in the emulsification of dietary lipid.
3. It is the precursor of steroid hormones.

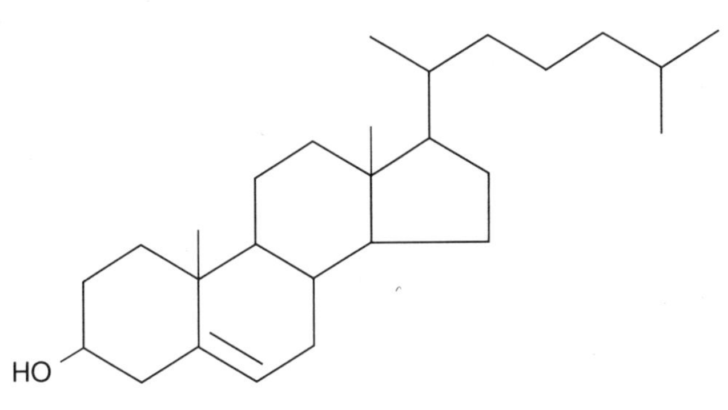

Fig. 8-20. Structure of cholesterol.

Synthesis of Cholesterol

Cholesterol biosynthesis occurs in the cytoplasm. All carbon atoms present in cholesterol are derived from acetyl CoA. Cholesterol synthesis may best be understood by considering it as consisting of two phases: first, a series of condensation reactions that build the C_{30} polyunsaturated hydrocarbon *squalene*, and, second, the reactions that transform squalene into cholesterol.

Figure 8-21 shows the key steps in the reaction sequence that generates squalene. In general terms, acetyl CoA is used to generate a mix-

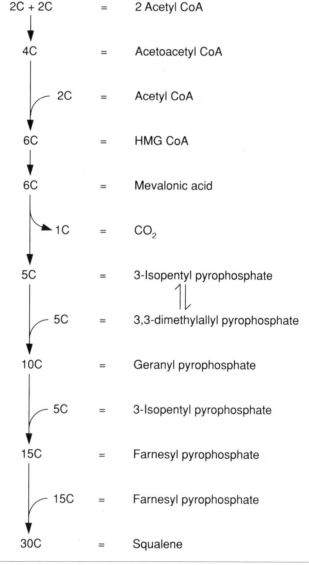

Fig. 8-21. The synthesis of squalene from acetyl CoA. HMG CoA, hydroxymethylglutaryl CoA.

ture of two five-carbon compounds, which condense to yield a C_{15} compound, two molecules of which condense to yield squalene. The principle steps in this synthetic pathway are as follows:

1. Three molecules of acetyl CoA are condensed to yield the six-carbon compound mevalonic acid via the intermediate hydroxymethylglutaryl CoA (HMG-CoA). Reduction of HMG-CoA to mevalonic acid is the rate-limiting step of cholesterol biosynthesis; it is catalyzed by *HMG CoA reductase*.
2. Mevalonic acid is converted to the five-carbon compound 3-isopentyl pyrophosphate via two phosphorylations and a decarboxylation.
3. 3-Isopentyl pyrophosphate exists in equilibrium with 3,3-dimethylallyl pyrophosphate. These two five-carbon compounds condense to form geranyl pyrophosphate, a ten-carbon intermediate.
4. Geranyl pyrophosphate condenses with another 3-isopentyl pyrophosphate to yield the 15-carbon intermediate farnesyl pyrophosphate.
5. Two molecules of farnesyl pyrophosphate condense to form squalene.

Squalene is converted to cholesterol with no further condensations (Fig. 8-22). First, squalene is oxidized to squalene 2,3-epoxide, which is converted to lanosterol. Lanosterol is converted (via several reaction steps) to cholesterol.

Regulation of Cholesterol Metabolism

The rate-limiting enzyme of cholesterol biosynthesis is HMG-CoA reductase. This enzyme is subject to feedback repression by cholesterol. In addition, it is controlled by the opposing influences of glucagon and insulin: insulin stimulates the enzyme, and glucagon antagonizes the effect of insulin. Thyroid hormone induces the synthesis of HMG-CoA reductase.

The total amount of cholesterol in the body is the sum of the cholesterol obtained from the diet plus the cholesterol synthesized endogenously. The amount of cholesterol in the diet influences the rate of endogenous cholesterol synthesis. When dietary cholesterol is high, cholesterol biosynthesis decreases in the liver and intestine; when dietary cholesterol is low, cholesterol biosynthesis increases.

Transport and Cellular Uptake of Cholesterol

The cholesterol needs of most tissues are met by cholesterol obtained from the diet or synthesized in the liver and intestine. Intestinal cholesterol (dietary and endogenous) is packaged for transport in *chylomicrons* (Chapter 7), whereas cholesterol synthesized in the liver is packaged in *very low density lipoproteins (VLDLs)*. These lipoproteins enter the circulation and transport triacylglycerols and other lipids to the tissues. Removal of the triacylglycerols leaves lipoprotein remnants that have a higher percentage of cholesterol by weight and therefore a higher density. The removal of triacylglycerols from chylomicrons and VLDLs yields, respectively, *chylomicron remnants* and *intermediate density lipoproteins (IDLs)*. Some IDLs are further converted into *low density li-*

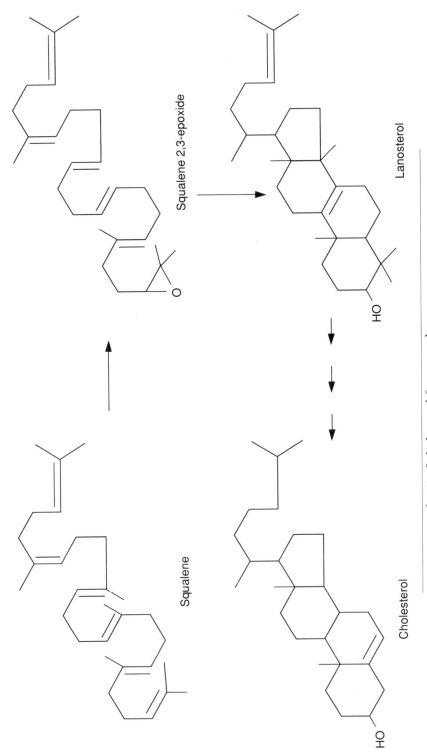

Fig. 8-22. Synthesis of cholesterol from squalene.

poproteins (*LDLs*), whereas the others are taken up by the liver. The liver also takes up chylomicron remnants. LDLs are also assembled in the liver and released directly into the circulation.

LDLs are the primary carriers of cholesterol to the peripheral tissues. Cellular uptake of cholesterol from lipoproteins requires the presence on the cell surface of specific receptors that recognize LDLs. Binding of the lipoprotein to the receptor initiates a series of events that result in cholesterol uptake, and also in the regulation of both intracellular cholesterol synthesis and further cholesterol uptake. The sequence of events is as follows:

1. The lipoprotein binds to the cell surface receptor.
2. The lipoprotein-receptor complex is internalized and taken to the lysosomes.
3. In the lysosomes, the lipoprotein apoproteins are degraded to amino acids and cholesterol esters are hydrolyzed.
4. Free cholesterol leaves the lysosomes. It is incorporated into membranes or used for intracellular needs.
5. As excess free cholesterol accumulates it inhibits and represses HMG CoA reductase, thus shutting down cholesterol production.
6. The accumulation of free cholesterol activates *fatty acyl CoA cholesterol acyltransferase*, which converts excess free cholesterol to cholesterol esters for storage.
7. Accumulation of cholesterol esters inhibits the synthesis of the LDL receptor; cholesterol uptake is inhibited.
8. Excess free cholesterol that continues to accumulate diffuses out of the cell and is bound to *high density lipoproteins* (*HDLs*), which transport it to the liver.

HDLs act as cholesterol scavengers. Free cholesterol becomes associated with the surface of HDLs. If this cholesterol were to diffuse away from the HDL surface, it could either be incorporated into cell membranes or other lipoproteins or collect on blood vessel walls and contribute to the formation of atherosclerotic plaques. To prevent the release of free cholesterol from HDLs, the plasma enzyme *lecithin:cholesterol acyltransferase* (*LCAT*) esterifies free cholesterol associated with HDLs:

Phosphatidylcholine (lecithin) + cholesterol →

$$\text{cholesterol ester + lysophosphatidylcholine}$$

The resulting cholesterol ester diffuses from the surface of the HDL into the hydrophobic core, in which it is transported to the liver.

Cholesterol that enters the liver can be:

1. stored as cholesterol esters
2. Incorporated into lipoproteins
3. Used to synthesize bile acids which are excreted in the bile
4. Excreted directly into the bile

Which combination of these alternatives actually occurs depends on the metabolic needs of the individual.

FORMATION OF BILE ACIDS AND CHOLESTEROL EXCRETION

Bile acids are synthesized in the liver, stored in the gallbladder, and released into the intestine when needed to emulsify dietary fat and fat-soluble vitamins. Moreover, because mammalian cells are unable to degrade the cholesterol ring structure, bile acids are the end products of cholesterol metabolism. Cholesterol is eliminated from the body mainly via the intestine in the form of bile acids.

The liver is incapable of synthesizing de novo enough bile acids to meet the needs of dietary fat digestion. Most of the bile salts excreted

Cholic acid

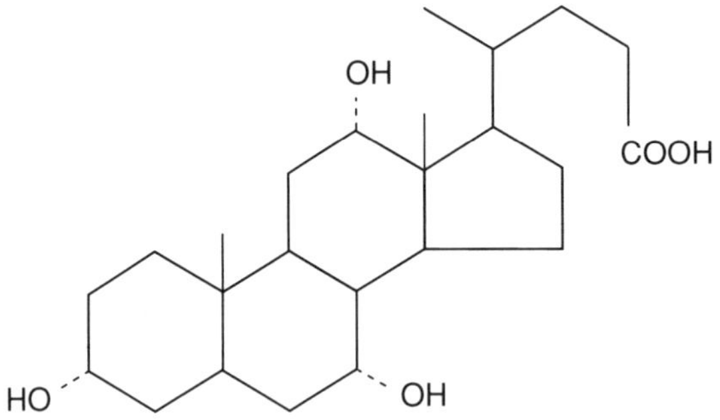

Chenodeoxycholic acid

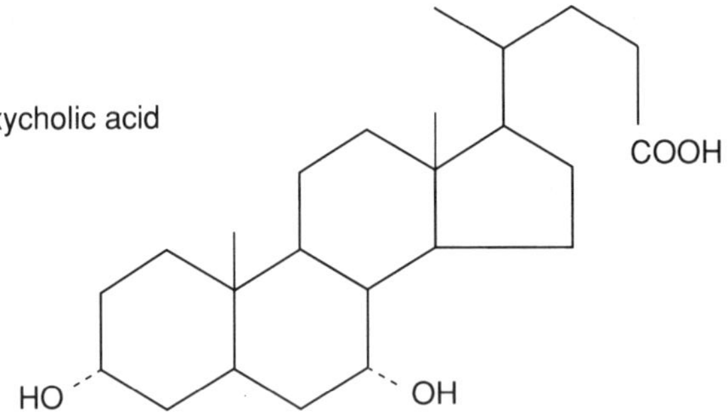

Glycocholic acid

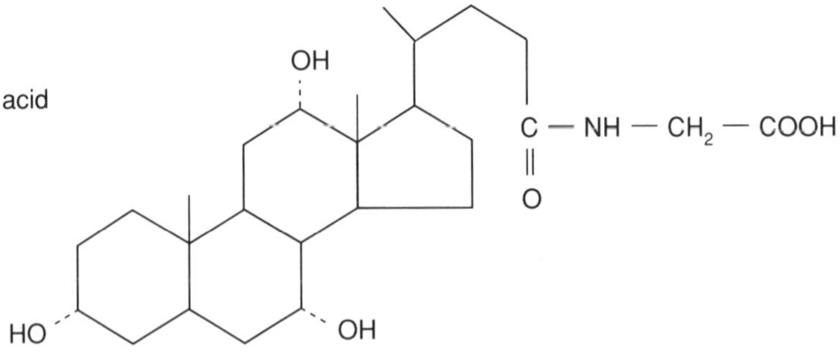

Fig. 8-23. The structures of the primary bile salts (cholic and chenodeoxycholic acids) and the conjugated primary bile salt glycocholic acid.

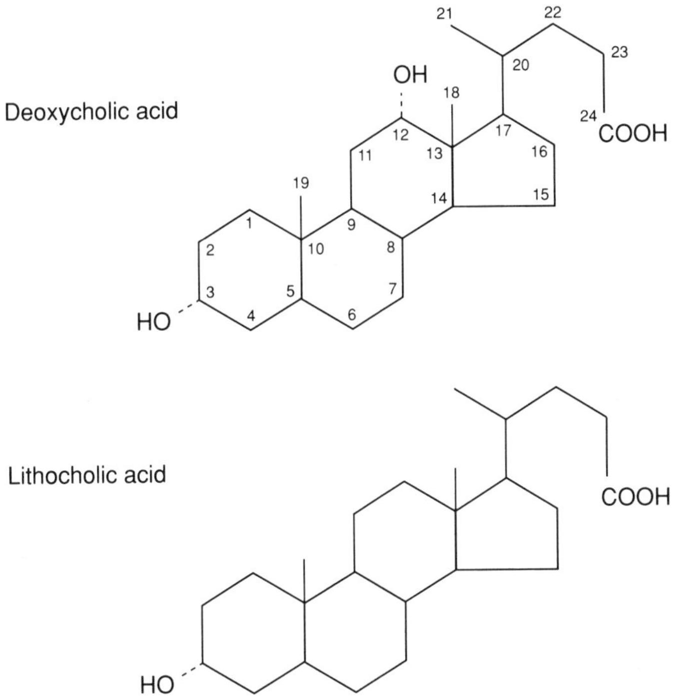

Fig. 8-24. Structures of the secondary bile salts (deoxycholic acid and lithocholic acid).

into the intestine are reabsorbed from the intestine, returned to the liver via the hepatic portal system, and used again.

Synthesis of Bile Acids

The liver converts cholesterol into *primary bile acids*. Two very abundant primary bile acids are *cholic acid* and *chenodeoxycholic acid* (Fig. 8-23). The amino group of glycine or taurine is often found in amide linkage with the carboxylic acid group of bile acids; when this occurs, the bile acid is said to be *conjugated* (Fig. 8-23). Bile salts that are not reabsorbed in the small intestine are converted to *secondary bile salts* by bacteria in the intestinal lumen. Secondary bile salts have fewer hydroxyl groups than the corresponding primary bile salts (Fig. 8-24). Most secondary bile salts are eliminated in the feces. However, small quantities of secondary bile salts (deoxycholate and lithocholate) are reabsorbed and taken back to the liver.

PROSTAGLANDINS AND THROMBOXANES

Prostaglandins are a very important class of lipid signal molecules that elicit a variety of responses. For example, some prostaglandins are believed to be involved in the mediation of inflammatory responses and the regulation of blood pressure. Prostaglandins are synthesized and secreted by most mammalian tissues, and exert their effects near the site of secretion. They have very short half lives ($t_{1/2} \sim$ minutes).

The thromboxanes closely resemble prostaglandins and share the biosynthetic pathway of one group of prostaglandins. Thromboxanes are released by platelets and are potent inducers of thrombus formation.

Structure and Nomenclature

Prostaglandins are 20-carbon carboxylic acids that contain a 5-carbon ring. They may be considered as derivatives of the hypothetical compound prostanoic acid:

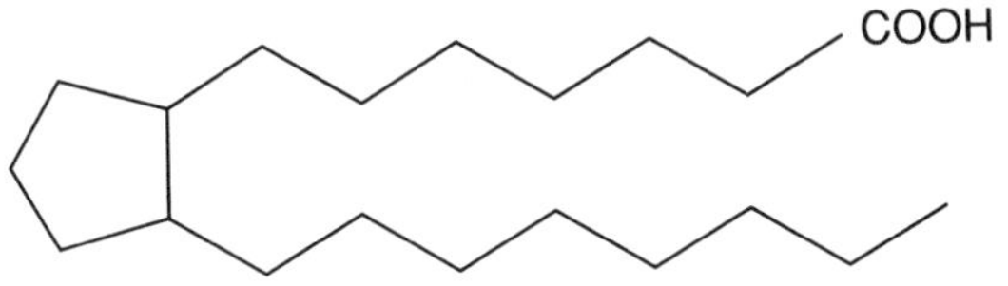

Prostaglandins are classified by the substituents on the cyclopentane ring. The major classes are *prostaglandins A, E, and F* (PGA, PGE, and PGF, respectively). The ring substituents of these classes are as follows:

> PGA: one keto group
> one α-β double bond
> PGE: one keto group
> one β-hydroxyl
> PGF: two hydroxyls that are β to each other

The prostaglandin classes are subdivided according to the number of double bonds in the side chains, which is indicated by a subscript number. For example, PGAs may contain a total of one, two, or three double bonds in the side chains; these subclasses are called PGA_1, PGA_2, and PGA_3. All the prostaglandins with one double bond in the side chains comprise the 1 series (e.g., PGA_1, PGE_1, and PGF_1). Similarly, all the prostaglandins with two and three double bonds in the side chains comprise the 2 and 3 series, respectively.

Some prostaglandins also carry a subscript α (e.g., $PGF_{2\alpha}$). This subscript refers to the configuration of the hydroxyl on carbon 9 of the chain: if the hydroxyl projects below the plane of the ring, the prostaglandin is in the α-configuration.

Perspective: Synthesis of Prostaglandins

Prostaglandins are synthesized from C_{20} polyunsaturated fatty acids. The most abundant prostaglandins (the 2 series) are synthesized from *arachidonic acid* (Fig. 8-25). The synthetic pathway of these prostaglandins may be divided into two phases: (1) the *cyclization* of arachidonic acid to yield the endoperoxide PGG_2, which is converted to PGH_2, and (2) the conversion of PGH_2 to the individual series 2 prostaglandins.

Synthesis of Prostaglandins

The polyunsaturated C_{20} fatty acid arachidonic acid is generated by elongation and desaturation of the essential fatty acid linoleate. Arachidonic acid for prostaglandin synthesis is released from phosphoglycerides or from cholesterol esters by the action of *phospholipase A_2* or a cholesterol esterase, respectively. The phospholipase A_2 reaction is the rate-limiting step in prostaglandin synthesis. Free arachidonate is acted on by the *microsomal cyclooxygenase* to yield the cyclic intermediate PGG_2, which is converted to PGH_2 by a *glutathione-dependent peroxidase* (Fig. 8-25). Both the cyclooxygenase and the glutathione-dependent peroxidase are components of a *prostaglandin synthetase complex*. PGH_2 is converted to the individual series 2 prostaglandins by separate enzymes (Fig. 8-26).

The 1 and 3 prostaglandin series are synthesized, respectively, from dihomo-γ-linoleic acid and eicospentaenoic acid (Fig. 8-27).

Thromboxanes

PGH_2 is also the precursor of the two thromboxanes, *thromboxane A_2* and *thromboxane B_2* (TXA_2 and TXB_2). The thromboxanes resemble series 2 prostaglandins except that they have a six-membered oxane ring

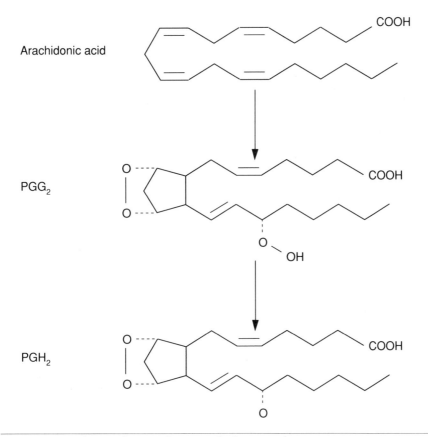

Fig. 8-25. Synthesis of PGH_2 from arachidonic acid.

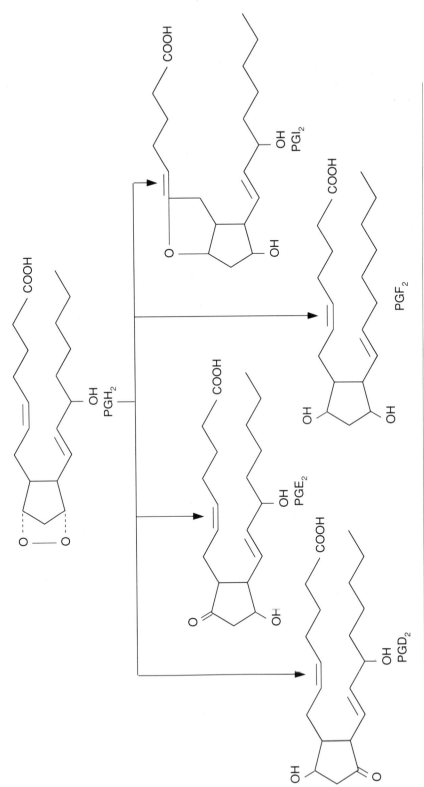

Fig. 8-26. Conversion of PGH$_2$ to other prostaglandins (PGD$_2$, PGE$_2$, PGF$_2$, and PGI$_2$).

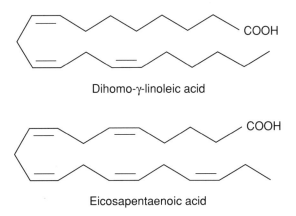

Fig. 8-27. Structures of dihomo-γ-linoleic and eicosapentaenoic acids.

(Fig. 8-28). TXA_2 is released by stimulated platelets, and quickly (≤ 1 min) breaks down to the inactive form TXB_2.

Regulation of Prostaglandin and Thromboxane Synthesis

The controlling factor that determines which of the series 2 prostaglandins or thromboxanes will be produced from PGH_2 in the various tissues is the relative levels of PGH_2 metabolizing enzymes in each tissue. Different tissues therefore produce different prostaglandins. It is believed that the release of prostaglandins from cells is hormone-controlled and is mediated by cyclic AMP and phospholipase A_2. Steroidal anti-inflammatory drugs act by inhibiting phospholipase A_2, whereas the nonsteroidal anti-inflammatory drugs aspirin and indomethacin act by inhibiting cyclooxygenase.

Degradation of Prostaglandins

Prostaglandins are degraded by a series of oxidations and reductions to a wide variety of compounds which are mostly excreted in the urine. The following is a typical urinary product of prostaglandin metabolism:

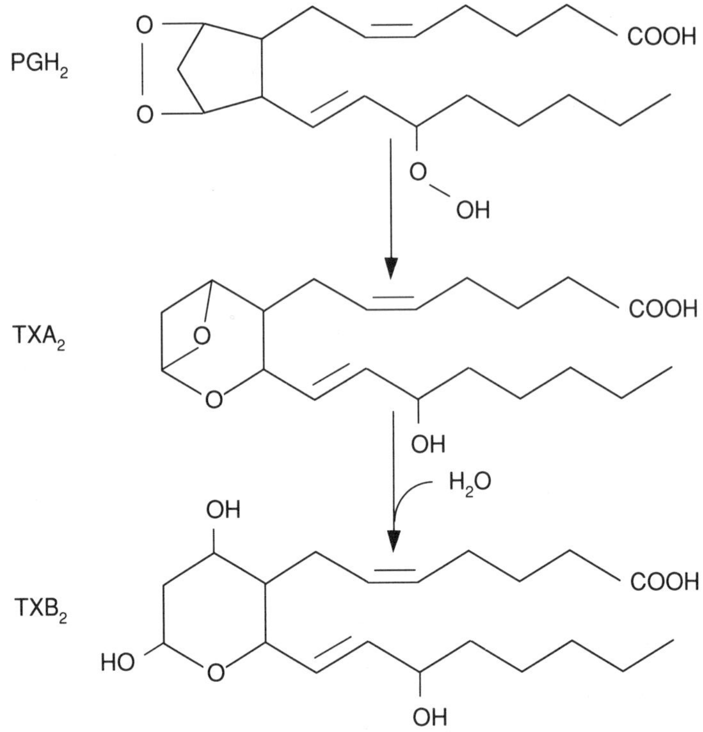

Fig. 8-28. Synthesis of thromboxanes from PGH_2.

LEUKOTRIENES AND HETEs

Leukotrienes and hydroxyeicosatetraenoic acids (HETEs), like thromboxanes and series 2 prostaglandins, are synthesized from arachidonic acid. As a group, leukotrienes and HETEs induce a wide variety of biologic effects, including mediation of chemotaxis, promotion of smooth muscle contraction, airway constriction, mediation of allergic inflammatory responses, and stimulation of adenyl cyclase.

Synthesis of Leukotrienes and HETEs

The initial reaction in the conversion of arachidonic acid to leukotrienes and HETEs is catalyzed by *lipoxygenases* present in certain leukocytes, mast cells, platelets, and lung cells. The product of this reaction is *hydroperoxyeicosatetraenoic acid* (*HPETE*), which is made up of arachidonic acid with a hydroperoxy group (—O—OH) covalently bound in an allylic position. The hydroperoxy group is located at different positions in different HPETEs; Figure 8-29 shows the synthesis of the 5-hydroperoxy derivative.

5-HETE is formed by reduction of 5-HPETE. 5-HPETE is also a precursor of the leukotrienes. Figure 8-30 shows the production of three leukotrienes from 5-HPETE: 5-HPETE is converted to leukotriene A_4

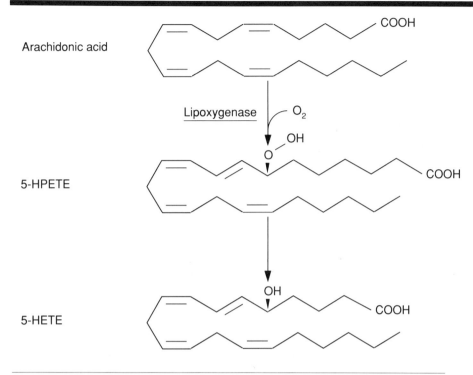

Fig. 8-29. The synthesis of 5-HPETE and 5-HETE from arachidonic acid.

(LTA$_4$), which may be converted to LTB$_4$ (5,12-di-HETE) or to LTC$_4$ (by substitution of reduced glutathione at carbon 6). LTD$_4$ and LTE$_4$ are produced by the hydrolysis of a glutamate and a glycine, respectively, from LTC$_4$.

Prostaglandins, Thromboxanes, and Leukotrienes: Key Points

1. The rate-limiting step of prostaglandin synthesis is the reaction catalyzed by phospholipase A$_2$ (the release of arachidonic acid from phosphoglycerides).
2. Prostaglandins are grouped into classes (A, E, F, etc) on the basis of the functional groups on the cyclopentane ring.
3. The prostaglandin classes are subdivided into series according to the number of double bonds in the side chains. The 2 series (two double bonds in the side chains) is the most abundant class.
4. The series 2 prostaglandins (as well as the thromboxanes, leukotrienes, and HETEs) are synthesized from arachidonic acid, a C$_{20}$ polyunsaturated fatty acid.
5. The different prostaglandin classes are synthesized by different enzymes, which are present in different amounts in different tissues. Different tissues therefore usually produce different mixtures of prostaglandins.
6. Thromboxanes are derived from the prostaglandin intermediate PGH$_2$. Thromboxanes differ structurally from prostaglandins in having a six-membered oxane ring.

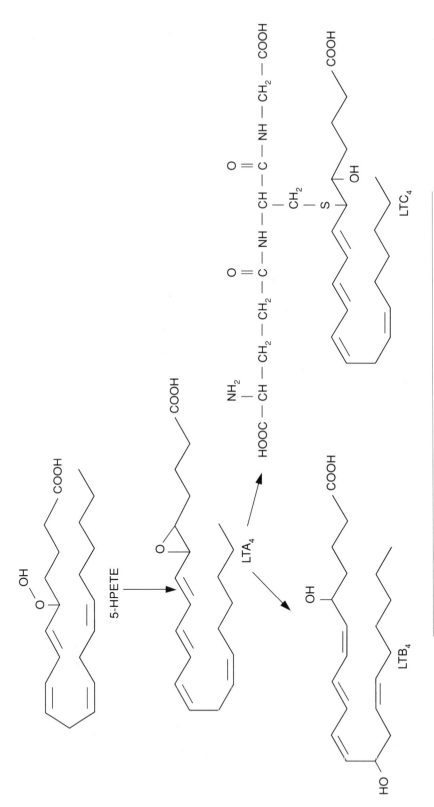

Fig. 8-30. The synthesis of LTA$_4$, LTB$_4$, and LTC$_4$ from 5-HPETE.

7. Lipoxygenase enzymes catalyze the conversion of arachidonic acid to HPETEs, the precursors of HETEs and leukotrienes.

CLINICAL CORRELATION: LECITHIN-CHOLESTEROL ACYLTRANSFERASE (LCAT) DEFICIENCY

The clinical observations of anemia, corneal opacities, proteinuria, and very low plasma levels of cholesterol esters in conjunction with the biochemical abnormality of significantly reduced or absent LCAT activity characterize LCAT deficiency. Under normal conditions, LCAT controls the plasma level of free cholesterol by synthesizing cholesterol esters. A deficiency of LCAT leads to increased plasma free cholesterol, which causes excessive incorporation of cholesterol into membranes. High plasma free cholesterol also results in changes in the lipid composition of lipoproteins and alterations in lipoprotein structure. Clinical symptoms may appear gradually over a period of years. The combined consequences of high plasma cholesterol are likely contributors to the renal abnormalities and premature atherosclerosis observed in LCAT-deficient individuals. In some LCAT-deficient patients, renal function deteriorates to the point that a transplant is required.

Treatment

The current treatment for LCAT deficiency is to restrict the dietary intake of fat. This treatment has been observed to result in partial reversal of the structural abnormalities of plasma LDLs.

SUGGESTED READING

1. Ansell GB, Hawthorne JN, Dawson RMC (eds): Form and Function of Phospholipids. 2nd Ed. Elsevier, New York, 1973
2. Stanbury JB, Wyngaarden JB, Fredrickson DS, et al (eds): The Metabolic Basis of Inherited Disease. 5th Ed. McGraw-Hill, New York, 1983
3. Bell RM, Coleman R: Enzymes of glycerol lipid synthesis in eukaryotes. Annu Rev Biochem 49:459, 1980
4. Bloch K: The biological synthesis of cholesterol. Science 150:19, 1965
5. Danielsson H, Sjovall J: Bile acid metabolism. Annu Rev Biochem 44:233, 1975
6. Curtis-Prior BP: Prostaglandins: An Introduction to Their Biochemistry, Physiology and Pharmacology. North Holland, New York, 1976
7. Piper PJ: Formation and action of leukotrienes. Physiol Rev 64:744, 1984
8. Samuelson B: Prostaglandins and thromboxanes. In Greep RO (ed): Recent Progress in Hormone Research. Vol. 34. Academic Press, New York, 1978

STUDY QUESTIONS

Directions: For each of the following multiple choice questions, choose the most appropriate answer.

1. Triacylglycerol synthesis in adipose tissue
   A. requires free glycerol
   B. cannot use free glycerol
   C. involves 2-monoacyl intermediates
   D. does not require activated fatty acids

2. The synthesis of phosphatidylcholine involves
   A. phosphocholine
   B. CMP-choline
   C. UDP-choline
   D. 2-monoacylglycerol

3. All of the following statements concerning phospholipid and triacylglycerol synthesis are true EXCEPT that
   A. phosphatidylcholine can be synthesized from phosphatidylethanolamine.
   B. phosphatidylethanolamine can be synthesized from phosphatidylserine.
   C. phosphatidylethanolamine cannot be synthesized from phosphatidylcholine.
   D. triacylglycerol synthesis in the liver and adipose tissue involves phosphatidate but not 1,2-diacylglycerol.

4. Phosphocholine is released from phosphatidylcholine by the action of
   A. phospholipase A_1
   B. phospholipase A_2
   C. phospholipase C
   D. phospholipase D

5. Globosides are
   A. ceramide oligosaccharides
   B. ceramide monosaccharides
   C. sphingomyelin derivatives
   D. degradation products of ceramides

6. Cholesterol synthesized in the liver is distributed to the tissues in
   A. LDLs
   B. HDLs
   C. chylomicrons
   D. VLDLs

7. Lipoxygenase enzymes are involved in the synthesis of
   A. prostaglandins and thromboxanes
   B. prostaglandins and HPETES
   C. thromboxanes and HETES
   D. HPETES

Directions: Answer the following questions using the key outlined below.
(A) if 1, 2 and 3 are correct
(B) if 1 and 3 are correct
(C) if 2 and 4 are correct
(D) if only 4 is correct
(E) if all four are correct

---

8. HMG CoA reductase is
   1. inhibited by free cholesterol
   2. repressed by free cholesterol
   3. regulated by glucagon and insulin
   4. regulated by glucagon but not insulin

9. Bile salts are
   1. synthesized from cholesterol
   2. synthesized from dietary lipids by intestinal bacteria
   3. important for the intestinal absorption of dietary lipids and fat-soluble vitamins
   4. only important as a means of eliminating cholesterol from the body.

10. Immediate precursors to prostaglandins include:
    1. arachidonic acid
    2. linoleic acid
    3. dihomo-γ-linoleic acid
    4. arachidic acid

11. Cholesterol synthesis
    1. may be considered as a series of condensation reactions
    2. starts with acetyl CoA
    3. involves isopentyl pyrophosphate intermediates
    4. occurs in the mitochondria

9

AMINO ACID METABOLISM I

Digestion of Proteins and Absorption and Tissue Distribution of Amino Acids/Ammonia Metabolism/ The Urea Cycle/Metabolism of Amino Acid Carbon Chains/One-Carbon Chemistry/Clinical Correlations

Learning Objectives

The student should be able to:

1. Name the major source of free amino acids in mammalian cells.
2. Describe the major mechanism for the production of free ammonia in the liver.
3. Describe the metabolic fate of free ammonia.
4. List the metabolic products of amino acid carbon chain degradation.
5. List the amino acids that contribute to the intracellular pool of one-carbon units.
6. Distinguish the essential and nonessential amino acids.

DIGESTION OF PROTEINS AND ABSORPTION AND TISSUE DISTRIBUTION OF AMINO ACIDS

The primary source of free amino acids in the body is the breakdown of dietary protein. Free amino acids are used mainly to synthesize proteins, but they also serve as precursors to a variety of nitrogenous compounds (Fig. 9-1). Amino acids present in amounts exceeding the quantities needed for the synthesis of proteins and other nitrogenous compounds cannot be stored. Instead, they are degraded to yield ammonia and intermediates of carbohydrate and lipid metabolism. The latter can be converted to glucose or fatty acids or oxidized to yield energy. Ammonia is used to synthesize glutamine (the principal fate of free ammonia) and glutamate. Excess ammonia is used to make urea, which is excreted in the urine.

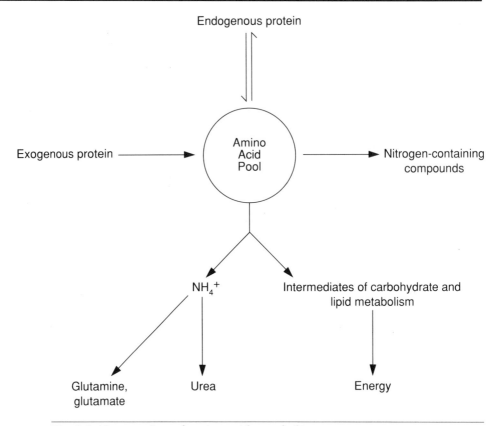

Fig. 9-1. An overview of amino acid metabolism.

Digestion of Dietary Protein

Protein digestion begins in the stomach (Fig. 9-2). Gastric mucosal cells secrete HCl and *pepsinogen*, the zymogen precursor of the proteolytic enzyme *pepsin*. Pepsinogen is activated to pepsin by the acidity of the gastric secretions and by the proteolytic activity of acid-activated pepsin molecules. Pepsin requires the acid environment of the stomach (pH 2 to 3) for optimal activity. In addition, the acidity of gastric juice probably partially denatures the proteins in some uncooked foods, rendering them more susceptible to proteolytic attack.

Pepsin rapidly attacks peptide bonds involving the carbonyl groups of phenylalanine, tyrosine, and tryptophan. It also shows some activity toward peptide bonds involving aliphatic and acidic residues. Pepsin activity usually results in the degradation of proteins to a mixture of oligopeptides, the degradation of which continues in the small intestine.

The pancreas secretes into the small intestine an alkaline mixture that contains a number of proteases in inactive zymogen form: *trypsinogen*, several *chymotrypsinogens*, *proelastase*, and *procarboxypeptidases A* and *B*. Trypsinogen is converted to active trypsin by enteropeptidase, an enzyme produced by the intestinal brush border cells. The resulting trypsin then activates the other zymogens to the corresponding active enzymes.

The pancreatic enzymes degrade oligopeptides and polypeptides to free amino acids and short oligopeptides (usually di- and tripeptides), which are transported into the intestinal mucosal cells. *Trypsin* cleaves peptide bonds involving the carboxyl group of lysine or arginine. The *chymotrypsins* are most active on peptide bonds involving the carboxyl groups of phenylalanine, tyrosine, and tryptophan. *Elastase* is active toward peptide bonds involving neutral aliphatic amino acids. *Carboxypeptidase A* releases aliphatic and aromatic amino acids one at a time from the C-terminus. *Carboxypeptidase B* releases C-terminal lysine or arginine residues.

The digestion of small peptides is completed within the intestinal mucosal cells, which contain aminopeptidases and dipeptidases, including prolidase (Fig. 9-2). *Aminopeptidases* release one amino acid at a time from the N-terminus. *Dipeptidases* hydrolyze dipeptides. *Prolidase* is a dipeptidase specific for dipeptides containing proline in the carboxyl position.

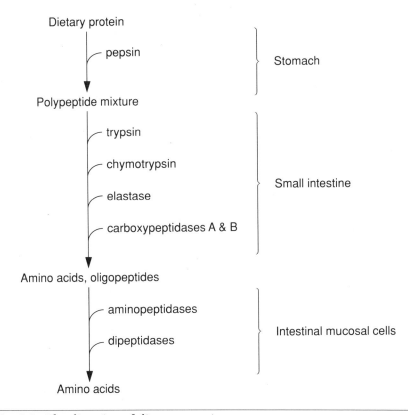

Fig. 9-2. The digestion of dietary protein.

Absorption of Amino Acids and Oligopeptides

The cells of the intestinal mucosa have carrier-mediated active transport systems for the absorption of neutral amino acids, dibasic amino acids, dicarboxylic amino acids, and proline. These systems are sodium-dependent, and are capable of transporting L-amino acids against a concentration gradient. Oligopeptides are also transported into the intestinal mucosal cells by a sodium-dependent active transport mechanism.

Another mechanism that may be important in the transport of some amino acids is the γ-*glutamyl cycle* (Fig. 9-3). This cycle involves the transfer of extracellular amino acids to the γ-glutamyl moiety of intracellular glutathione. The resulting intracellular γ-glutamyl amino acid is subsequently hydrolyzed to yield the free amino acid and 5-oxoproline. 5-oxoproline is hydrolyzed to glutamate, which reacts with cysteine to yield γ-glutamyl cysteine. The resulting γ-glutamyl cysteinyl dipeptide then reacts with glycine to regenerate glutathione. As can be seen from Figure 9-3, three ATPs are consumed in the transport of one amino acid. The brain and kidney, as well as the intestine, possess the enzymes of the γ-glutamyl cycle.

Digestion of Endogenous Protein

Endogenous protein contributes to the supply of free amino acids in two ways. Proteins are degraded in cells by lysosomal proteases. In

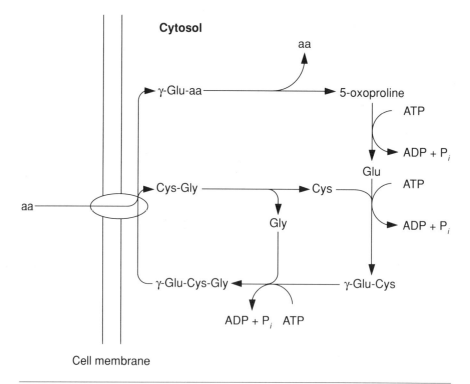

Fig. 9-3. Transport of amino acids into cells by the γ-glutamyl cycle. aa, amino acid.

addition, proteins in the gastric and pancreatic secretions and proteins sloughed off from the lining of the intestinal mucosa are digested and absorbed in the same way as dietary protein.

TISSUE DISTRIBUTION OF AMINO ACID METABOLISM

Figure 9-4 shows the distribution of amino acid metabolism in the body. The principal site of amino acid metabolism is the liver, which is able to take up all of the amino acids from the circulation. Muscle shows a

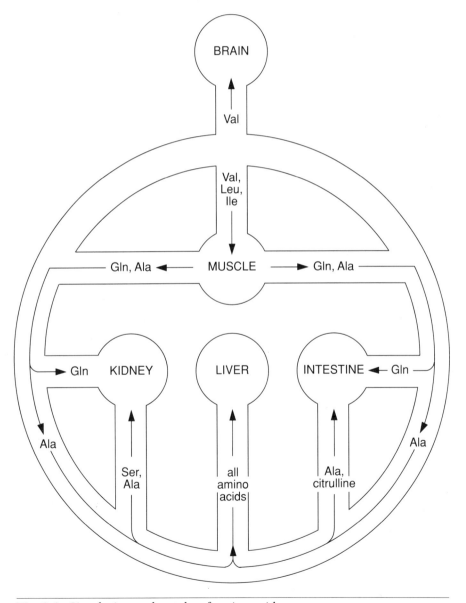

Fig. 9-4. Circulation and uptake of amino acids.

preference for the uptake of valine, isoleucine, and leucine. Brain preferentially takes up valine. Therefore, muscle and brain contribute significantly to the metabolism of branched chain amino acids. Ammonia produced as a result of amino acid degradation in the muscles is transported in the form of alanine and glutamine. Alanine is taken up by the liver, whereas glutamine is preferentially taken up by the small intestine and kidney. The kidney also preferentially takes up proline and glycine. The nitrogen from glutamine degraded in the intestine is transported to the liver in the form of alanine, citrulline, and a small quantity of free ammonia. Nitrogenous products released by the kidney include alanine, serine, and ammonia. Ammonia is excreted in the urine, whereas alanine and serine are transported to the liver. In sum, the products of extrahepatic amino acid degradation are mostly transported to the liver for further metabolism.

Perspective: Ammonia Metabolism

Amino acids are degraded in two steps:

1. Removal of the α-amino group
2. Conversion of the carbon chain to an intermediate of lipid and/or carbohydrate metabolism

The ammonia generated by amino group removal is used in two ways: (1) as much as required is used in the synthesis of nitrogenous compounds; and (2) any excess is fixed in the form of nontoxic urea for elimination via the urine. These two processes make up *ammonia metabolism*.

AMMONIA METABOLISM

Removal of the Amino Group

Removal of the amino groups from amino acids is accomplished by a considerable variety of processes: transamination and oxidative deamination; nonoxidative deamination; hydrolytic deamination; direct deamination; and, in muscle, the purine nucleotide/aspartate cycle.

Transamination and Oxidative Deamination

Transamination followed by oxidative deamination is the major pathway for the production of free ammonia from amino acids. Transamination reactions are catalyzed by *aminotransferases* that are dependent on the cofactor *pyridoxal phosphate*, a derivative of vitamin B_6. These enzymes transfer an amino group from an amino acid to an α-keto acid, yielding a different α-keto acid and a different amino acid:

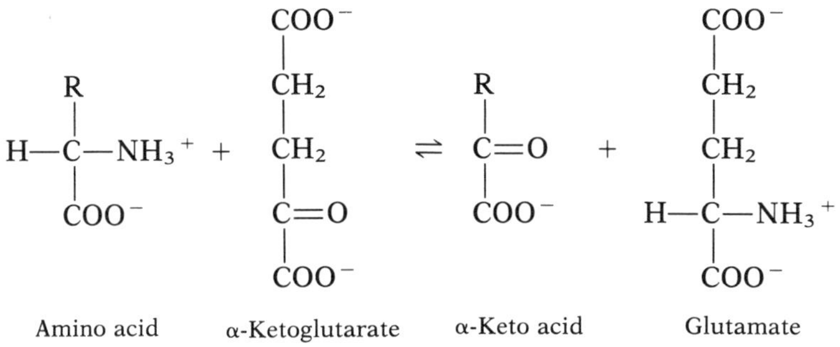

$$\underset{\text{Amino acid}}{\overset{R}{\underset{COO^-}{\overset{|}{H-C-NH_3^+}}}} + \underset{\alpha\text{-Ketoglutarate}}{\overset{COO^-}{\underset{COO^-}{\overset{|}{\underset{|}{\overset{CH_2}{\underset{C=O}{\overset{|}{CH_2}}}}}}}} \rightleftharpoons \underset{\alpha\text{-Keto acid}}{\overset{R}{\underset{COO^-}{\overset{|}{C=O}}}} + \underset{\text{Glutamate}}{\overset{COO^-}{\underset{COO^-}{\overset{|}{\underset{|}{\overset{CH_2}{\underset{H-C-NH_3^+}{\overset{|}{CH_2}}}}}}}}$$

The glutamate/α-ketoglutarate couple usually participates in all transamination reactions.

Transaminations are important in the removal of amino groups (rather than merely the transfer of amino groups among amino acids) because glutamate undergoes oxidative deamination to yield α-ketoglutarate and free ammonia. This reaction is catalyzed by *glutamate dehydrogenase*:

Glutamate + NAD$^+$(NADP$^+$) $\xrightleftharpoons{\text{Glutamate dehydrogenase}}$

α-ketoglutarate + NH$_4^+$ + NADH (NADPH)

The glutamate dehydrogenase reaction is the major mechanism of oxidative deamination in mammalian cells. α-Ketoglutarate acts as a sink for amino groups by accepting amino groups from various amino acids, with the resultant formation of glutamate.

In mammals, all α-amino acids except lysine and threonine can be transaminated. The imino acid proline, which has a secondary rather than a primary amino group, cannot participate in pyridoxal phosphate dependent reactions including transamination and decarboxylation. It is degraded to glutamate by a pathway that will be described later.

L-Amino Acid Oxidase. A minor role in the oxidative removal of amino groups from amino acids is played by *L-amino acid oxidase*, which catalyzes the following reaction:

$$\overset{NH_3^+}{\underset{|}{R-CH-COO^-}} + FMN + H_2O \rightleftharpoons \overset{O}{\underset{||}{R-C-COO^-}} + NH_4^+ + FMNH_2$$

$$FMNH_2 + O_2 \rightarrow FMN + H_2O_2$$

The peroxide formed is converted to oxygen and water by *catalase*.

L-Amino acid oxidase can catalyze the oxidation of all the L-amino acids normally found in proteins except serine, threonine, aspartate, and glutamate.

D-Amino Acid Oxidase. The enzyme *D-amino acid oxidase*, which is present in the liver, kidney, and brain, oxidizes D-amino acids to the corresponding α-keto acids and ammonia. D-Amino acids present in the cell walls of the intestinal bacteria may occasionally be absorbed. The enzyme D-amino acid oxidase prevents any absorbed D-amino acids from becoming incorporated into proteins. This reaction occurs as follows:

$$\underset{\underset{R-CH-COO^-}{|}}{NH_3^+} + O_2 + H_2O \rightarrow R-\underset{\underset{}{\overset{\overset{O}{\|}}{C}}}{}-COO^- + NH_4^+ + H_2O_2$$

Glycine, which is not stereospecific, is also highly active with D-amino acid oxidase.

Nonoxidative Deamination

The enzyme *serine-threonine dehydratase* catalyzes the deamination of serine and threonine (and, to a lesser extent, of cysteine and homoserine). The reactions for serine and threonine are as follows:

$$\text{Serine} \rightarrow \text{pyruvate} + NH_4^+$$

$$\text{Threonine} \rightarrow \alpha\text{-ketobutyrate} + NH_4^+$$

This enzyme requires pyridoxal phosphate.

Hydrolytic Deamination

The hydrolysis of the amide groups of glutamine and asparagine are examples of hydrolytic deamination:

$$\text{Glutamine} \xrightarrow{\text{Glutaminase}} \text{glutamate} + NH_4^+$$

$$\text{Asparagine} \xrightarrow{\text{Asparaginase}} \text{aspartate} + NH_4^+$$

The renal glutaminase reaction is an important source of ammonia to neutralize acidic urine.

Direct Deamination

Direct deamination occurs in the first step of histidine degradation. The amino group is directly cleaved out by *histidine-ammonia lyase*, leaving urocanic acid:

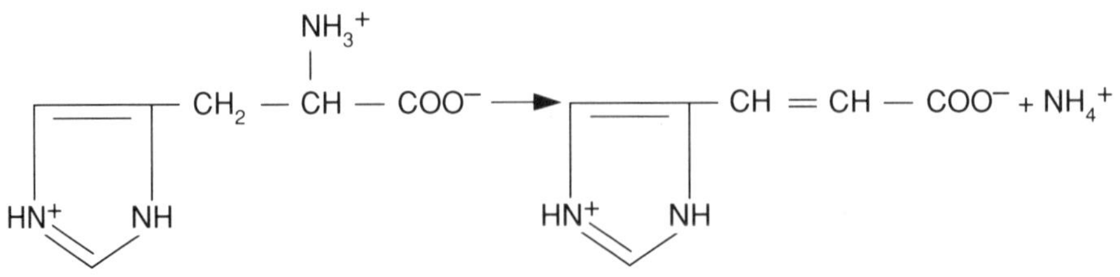

[Imidazole ring]—CH_2—$\underset{\underset{}{|}}{\overset{NH_3^+}{CH}}$—$COO^-$ $\rightarrow$ [Imidazole ring]—$CH=CH$—COO^- + NH_4^+

Histidine   Urocanate

Purine Nucleotide/Aspartate Cycle

The purine nucleotide/aspartate cycle is an important deamination mechanism in skeletal muscle. As shown in Figure 9-5, inosine monophosphate (IMP) accepts an amino group from aspartate, yielding adenosine monophosphate (AMP). AMP is then hydrolyzed to IMP and ammonia.

Utilization of Ammonia

Free ammonia may be metabolized by three different routes:

1. Synthesis of glutamate
2. Synthesis of glutamine
3. Synthesis of urea

The principal fate of ammonia is the synthesis of glutamine. This reaction is catalyzed by *glutamine synthetase*, and occurs in two steps. ATP is used to form an activated enzyme-bound ADP-γ-glutamyl phosphate, which then reacts with ammonia to yield glutamine:

1. Enzyme + glutamate + ATP →

    Enzyme-ADP-γ-glutamyl phosphate

2. Enzyme-ADP-γ-glutamyl phosphate + NH_4^+ →

    Enzyme + glutamine + ADP + P_i

Glutamine is the amino group donor in the biosynthesis of several nitrogenous metabolites. Ammonia may be used to synthesize glutamate by the reversal of the glutamate dehydrogenase reaction.

Ammonia that is not used to synthesize glutamate or glutamine is converted to the nontoxic compound *urea*, which is eliminated in the urine. Most of the nitrogen in the urine is in the form of urea. Urea is synthesized by the pathway known as the *urea cycle*.

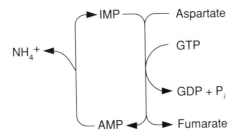

Fig. 9-5. The purine nucleotide/aspartate cycle.

Perspective: The Urea Cycle

In the liver, urea is synthesized via the urea cycle from free ammonia, bicarbonate, ATP, and the α-amino group of aspartate. Tissues outside the liver contain only small amounts of arginase (the urea producing enzyme), and therefore release very little urea. In these tissues the truncated pathway is used primarily to synthesize arginine.

THE UREA CYCLE

Figures 9-6 and 9-7 show the full urea cycle as it occurs in the liver. As can be seen, part of the pathway occurs in the cytosol and part in the mitochondria. The pathway proceeds as follows:

1. The enzyme *carbamoyl phosphate synthetase I* catalyzes the reaction of ammonia with CO_2 and ATP to form *carbamoyl phosphate*.
2. Carbamoyl phosphate reacts with the cycle intermediate ornithine to form citrulline in a reaction catalyzed by *ornithine transcarbamoylase*.
3. Citrulline reacts with aspartate to form argininosuccinate. This reaction is catalyzed by *argininosuccinate synthetase*, and consumes one ATP.
4. *Argininosuccinate lyase* cleaves argininosuccinate to arginine and fumarate. The fumarate is oxidized via the TCA cycle to oxaloacetate, which can be transaminated to regenerate aspartate.
5. In the final reaction of the urea cycle, arginine is cleaved by *arginase* to urea and ornithine:

$$\begin{array}{c}COO^-\\|\\HC-NH_3^+\\|\\CH_2\\|\\CH_2\\|\\CH_2\\|\\NH\\|\\C=NH\\|\\NH_3^+\end{array} \rightarrow \begin{array}{c}COO^-\\|\\HC-NH_3^+\\|\\CH_2\\|\\CH_2\\|\\CH_2\\|\\NH_3^+\end{array} + \begin{array}{c}NH_2\\|\\C=O\\|\\NH_2\end{array}$$

Arginine Ornithine Urea

Arginase is primarily a liver enzyme, so most urea is synthesized in the liver.

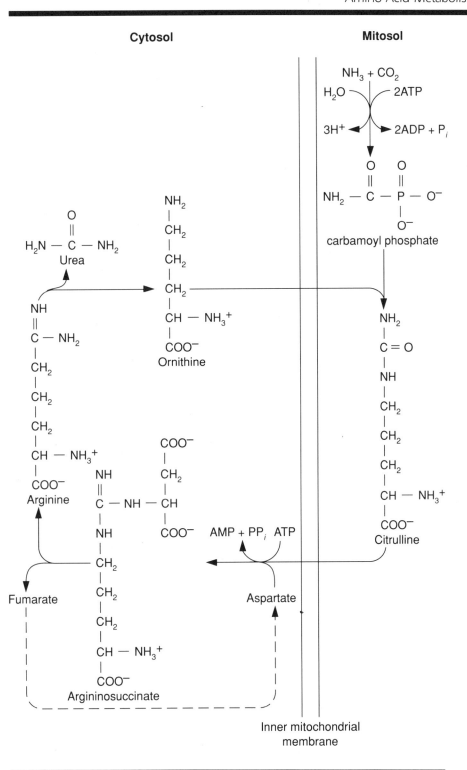

Fig. 9-6. The urea cycle.

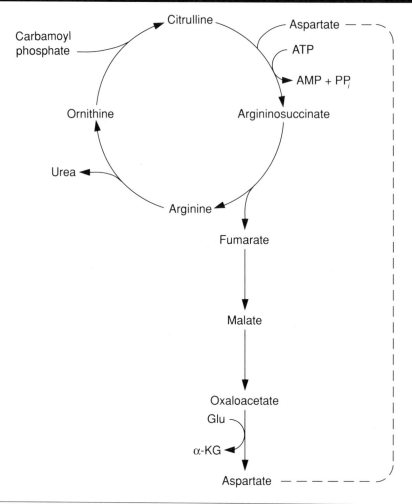

Fig. 9-7. The link between the urea cycle and the citric acid cycle. α-KG, α-ketoglutarate.

Regulation of the Urea Cycle

The regulated enzyme of the urea cycle is carbamoyl phosphate synthetase I. This enzyme responds to the level of amino acid transamination. A rise in amino acid transamination causes a rise in glutamate, leading to a rise in *N*-acetylglutamate, which is an allosteric activator of carbamoyl phosphate synthetase I.

GLYCOGENIC AND KETOGENIC AMINO ACIDS

The second phase of amino acid degradation is the conversion of the carbon chains to intermediates of carbohydrate and/or lipid metabolism. All amino acids give rise to one or more of the following substances: pyruvate, oxaloacetate, fumarate, succinyl CoA, α-ketoglutarate, acetyl CoA, and acetoacetyl CoA.

Table 9-1. Glycogenic and Ketogenic Amino Acids

Glycogenic	Ketogenic	Glycogenic and Ketogenic
Alanine	Leucine	Isoleucine
Asparagine	Lysine	Phenylalanine
Aspartate		Tryptophan
Arginine		Tyrosine
Cysteine		
Glutamate		
Glutamine		
Histidine		
Isoleucine		
Methionine		
Phenylalanine		
Proline		
Serine		
Threonine		
Tyrosine		
Tryptophan		
Valine		

Amino acids whose degradation products make possible a net synthesis of glucose are called *glycogenic* or *glucogenic*, whereas amino acids whose products can be used for a net synthesis of fatty acids or ketone bodies are called *ketogenic*:

1. *Products of glycogenic amino acids:* pyruvate, oxaloacetate, fumarate, succinyl CoA, α-ketoglutarate
2. *Products of ketogenic amino acids:* acetyl CoA, acetoacetyl CoA

Some amino acids generate products of both groups, and are therefore *glycogenic and ketogenic*. Table 9-1 lists the amino acids in each group. Historically, an amino acid was classified as glycogenic or ketogenic depending on whether its administration to diabetic animals resulted in an increase in urinary excretion of glucose or of ketone bodies, respectively. In animals that are neither fasted nor diabetic, intermediates of amino acid degradation are oxidized by the TCA cycle to produce energy.

Perspective: Metabolism of Individual Amino Acid Carbon Chains

The amino acids may be divided into five groups on the basis of their degradation products:

1. Alanine, serine, glycine, cysteine, and threonine are converted to *pyruvate*.
2. Asparagine and aspartate are converted to *oxaloacetate*.
3. Histidine, arginine, proline, glutamine, and glutamate are converted to α-ketoglutarate.
4. Methionine, threonine, valine, and isoleucine are converted to *propionyl CoA*. Isoleucine yields *acetyl CoA* as well.
5. Tryptophan, lysine, leucine, tyrosine, and phenylalanine are converted to *acetyl CoA*, *acetoacetyl CoA*, or *acetic acid*. Phenylalanine and tyrosine also produce *fumarate*, and tryptophan degradation also produces alanine.

METABOLISM OF INDIVIDUAL AMINO ACID CARBON CHAINS

Alanine. Alanine is converted to pyruvate via transamination.

Serine. Serine is converted to pyruvate, ammonia, and water by *serine-threonine dehydratase*. Serine is also converted to glycine and N^5,N^{10}-methylenetetrahydrofolate (see Fig. 9-16) by *serine transhydroxymethylase*.

Glycine. Glycine is converted to CO_2, NH_4^+, and N^5,N^{10}-methylenetetrahydrofolate by reversal of the *glycine synthase* reaction:

Glycine + tetrahydrofolate $\rightleftharpoons$

$$CO_2 + NH_4^+ + N^5,N^{10}\text{-methylenetetrahydrofolate}$$

Glycine is also converted to serine by reversal of the serine transhydroxymethylase reaction:

Glycine + N^5,N^{10}-methylenetetrahydrofolate $\rightleftharpoons$

serine + tetrahydrofolate

Cysteine. The major products of cysteine degradation are pyruvate and sulfate. The major pathway for cysteine degradation is shown in Figure 9-8.

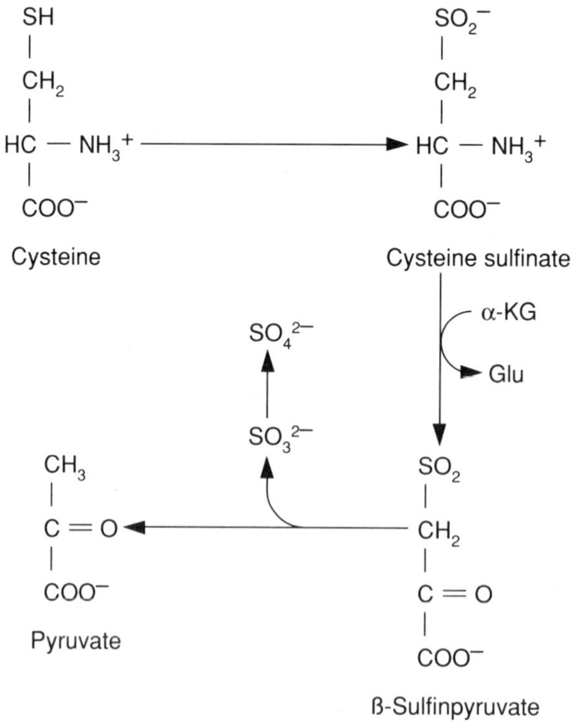

Fig. 9-8. Cysteine degradation. α-KG, α-ketoglutarate.

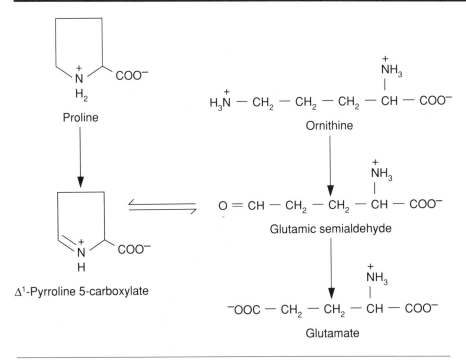

Fig. 9-9. Degradation of proline and ornithine.

Asparagine and Aspartate. The enzyme *asparaginase* converts asparagine to aspartate by hydrolysis of the amide group. Aspartate then undergoes transamination to yield oxaloacetate. Because of the involvement of aspartate in the urea cycle, fumarate may be formed from aspartate.

Proline and Ornithine. Proline and ornithine are converted to glutamate (Fig. 9-9). *Proline oxidase* converts proline to Δ^1-pyrroline 5-carboxylic acid, which undergoes immediate hydrolysis to glutamate semialdehyde. Oxidation of glutamate semialdehyde results in the formation of glutamate. Transamination of ornithine also produces glutamate semialdehyde, which is oxidized to glutamate.

Arginine. As in the urea cycle, arginine is cleaved to urea and ornithine by *arginase*. Urea is excreted in the urine. Ornithine may be reutilized in the synthesis of urea or transaminated to yield glutamate semialdehyde, which is oxidized to yield glutamate.

Histidine. Histidine is degraded to glutamate and N^5,N^{10}-methenyltetrahydrofolate (Fig. 9-10). The initial product of histidine degradation is *N*-formiminoglutamate. The formimino group is transferred to tetrahydrofolate to yield glutamate and N^5-formiminotetrahydrofolate. The latter is hydrolyzed to ammonia and N^5,N^{10}-methenyltetrahydrofolate.

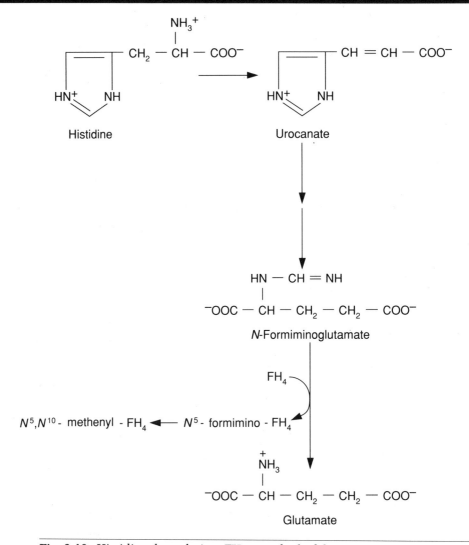

Fig. 9-10. Histidine degradation. FH_4, tetrahydrofolate.

Glutamine and Glutamate. Glutamine is converted to glutamate by hydrolysis of the amide group by *glutaminase*. Glutamate is transaminated to yield α-ketoglutarate. Glutamate may also be oxidized to α-ketoglutarate and ammonia by glutamate dehydrogenase.

Methionine. Methionine is converted to *S-adenosylmethionine*, the primary methylating agent in mammalian cells (Fig. 9-11). Reaction of *S*-adenosylmethionine with an appropriate methyl group acceptor results in the formation of *S*-adenosylhomocysteine, which is subsequently hydrolyzed to adenosine and homocysteine. Homocysteine combines with serine to yield cystathionine. Cystathionine is then degraded to yield cysteine and α-ketobutyrate, which is converted to propionyl

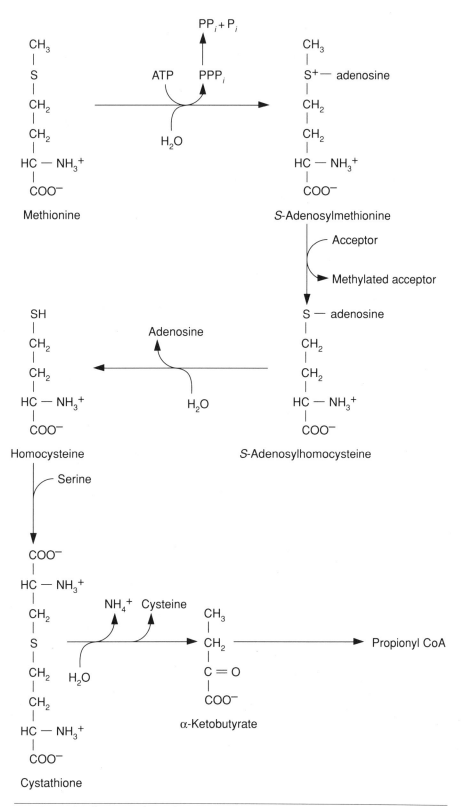

Fig. 9-11. Methionine degradation.

CoA. The methionine in mammalian cells originally comes from the diet. However, the mammalian enzyme *homocysteine methyltransferase* (methionine synthetase) and the remethylation of homocysteine by betaine allows the regeneration of methionine from homocysteine. The quantity of methionine formed by the regeneration of homocysteine is not sufficient to meet the needs of the cell, however, since most homocysteine is degraded to yield cysteine and α-ketobutyrate.

Threonine. Threonine is converted to α-ketobutyrate by *serine-threonine dehydratase* and to pyruvate via the aminoacetone pathway:

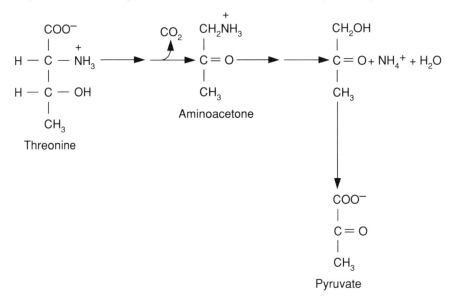

α-Ketobutyrate produced from threonine is also converted to propionyl CoA.

The Branched Chain Amino Acids: Valine, Leucine, Isoleucine. The degradative pathways of the branched chain amino acids have certain features in common (Fig. 9-12). In each case, the amino acid undergoes an initial transamination to yield the corresponding α-keto acid, which is then oxidatively decarboxylated in the presence of CoA to yield the CoA derivative containing one less carbon atom. The degradation of these CoA derivatives is similar to β-oxidation of fatty acids in that it consists of sequential dehydrogenation, hydration, and oxidation steps.

   Valine. Valine is ultimately degraded to propionyl CoA.
   Leucine. The products of leucine catabolism are acetyl CoA and acetoacetate.
   Isoleucine. Isoleucine degradation results in the formation of acetyl CoA and propionyl CoA.

The Metabolism of Propionyl CoA. Propionyl CoA is generated in the degradation of methionine, threonine, valine, and isoleucine. This pro-

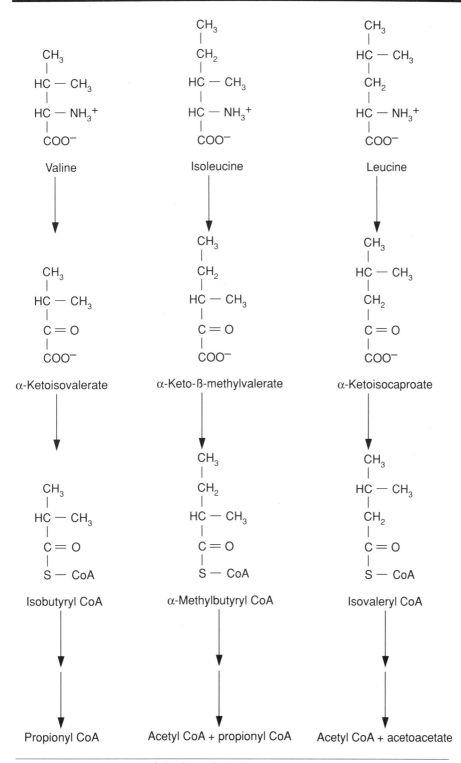

Fig. 9-12. Degradation of the branched chain amino acids.

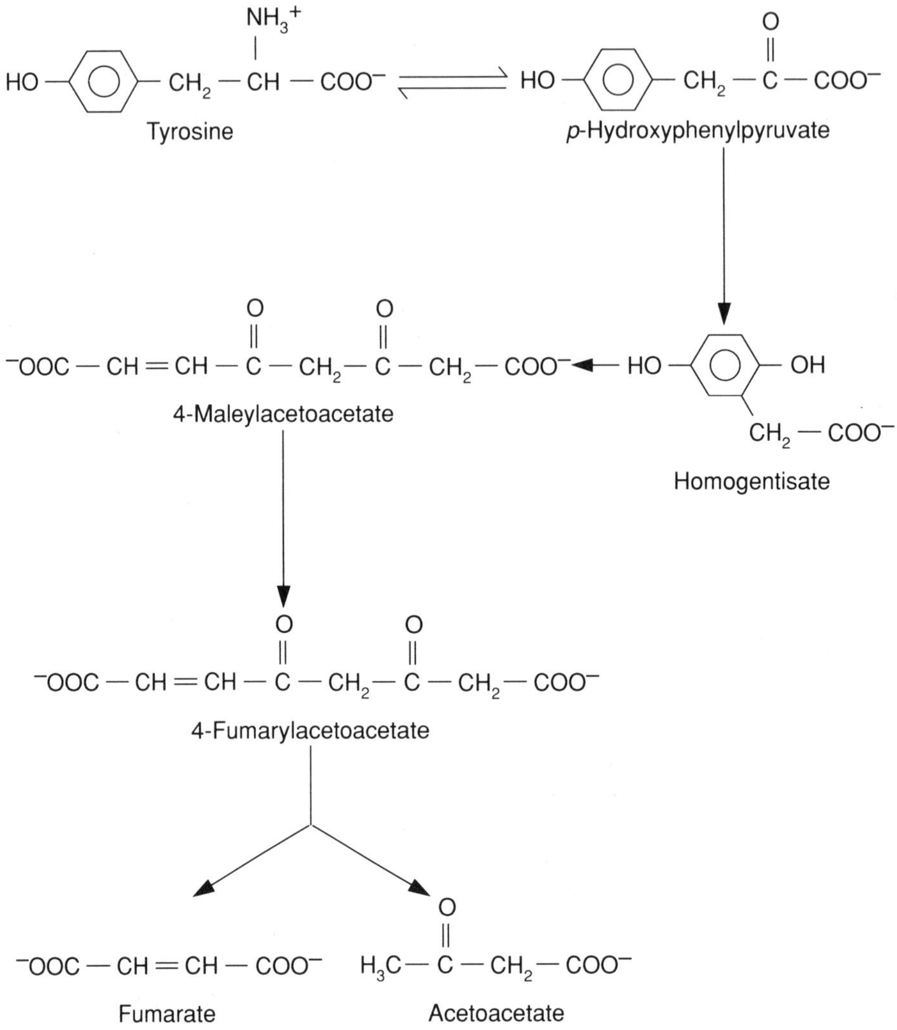

Fig. 9-13. Tyrosine degradation.

pionyl CoA is metabolized to succinyl CoA by the same pathway used for propionyl CoA generated by oxidation of fatty acids containing an odd number of carbons (Chapter 7; see Fig. 7-15).

Phenylalanine. Phenylalanine is converted to tyrosine by the *phenylalanine hydroxylase* system in a two-step process. First, phenylalanine is converted to tyrosine, and the electron donor tetrahydrobiopterin is concomitantly oxidized to dihydrobiopterin. *Dihydrobiopterin reductase* then catalyzes the regeneration of tetrahydrobiopterin:

1.
$$\text{Phenylalanine} \left(\begin{array}{c} \overset{+}{NH_3} \\ | \\ H-C-COO^- \\ | \\ CH_2-C_6H_5 \end{array}\right) + \text{tetrahydrobiopterin} + O_2 \xrightarrow{\text{Phenylalanine hydroxylase}}$$

$$\text{Tyrosine} \left(\begin{array}{c} \overset{+}{NH_3} \\ | \\ H-C-COO^- \\ | \\ CH_2-C_6H_4-OH \end{array}\right) + \text{dihydrobiopterin}$$

2. Dihydrobiopterin + NADPH + H$^+$ $\xrightarrow{\text{Dihydrobiopterin reductase}}$ tetrahydrobiopterin + NADP$^+$

Tyrosine. Tyrosine is degraded to fumarate and acetoacetate. After an initial transamination, the aromatic ring is opened by the successive actions of *p-hydroxyphenylpyruvate* and *homogentisate dioxygenases*. The resulting maleylacetoacetate is isomerized to fumarylacetoacetate, which is subsequently hydrolyzed to yield fumarate and acetoacetate (Fig. 9-13).

Tryptophan. Tryptophan is degraded to formate, alanine, and acetoacetyl CoA (Fig. 9-14). The first intermediate is *N*-formylkynurenine, which is hydrolyzed to formate and kynurenine. Kynurenine is hydroxylated to yield 3-hydroxykynurenine, which is cleaved by kynureninase to alanine and 3-hydroxyanthranilate. The aromatic ring of 3-hydroxyanthranilate is opened, and the resulting 2-amino-3-carboxymuconic semialdehyde is degraded to acetoacetyl CoA (two acetyl CoAs). A limited amount of this 2-amino-3-carboxymuconic semialdehyde is used to synthesize nicotinamide.

Lysine. Lysine is degraded to acetoacetyl CoA. The major pathway of lysine degradation (Fig. 9-15) proceeds via the initial intermediate saccharopine, which is formed by a reductive amination of α-ketoglutarate with the ε-amino group of lysine. Saccharopine is cleaved to yield glutamate and α-aminoadipic semialdehyde. The semialdehyde is con-

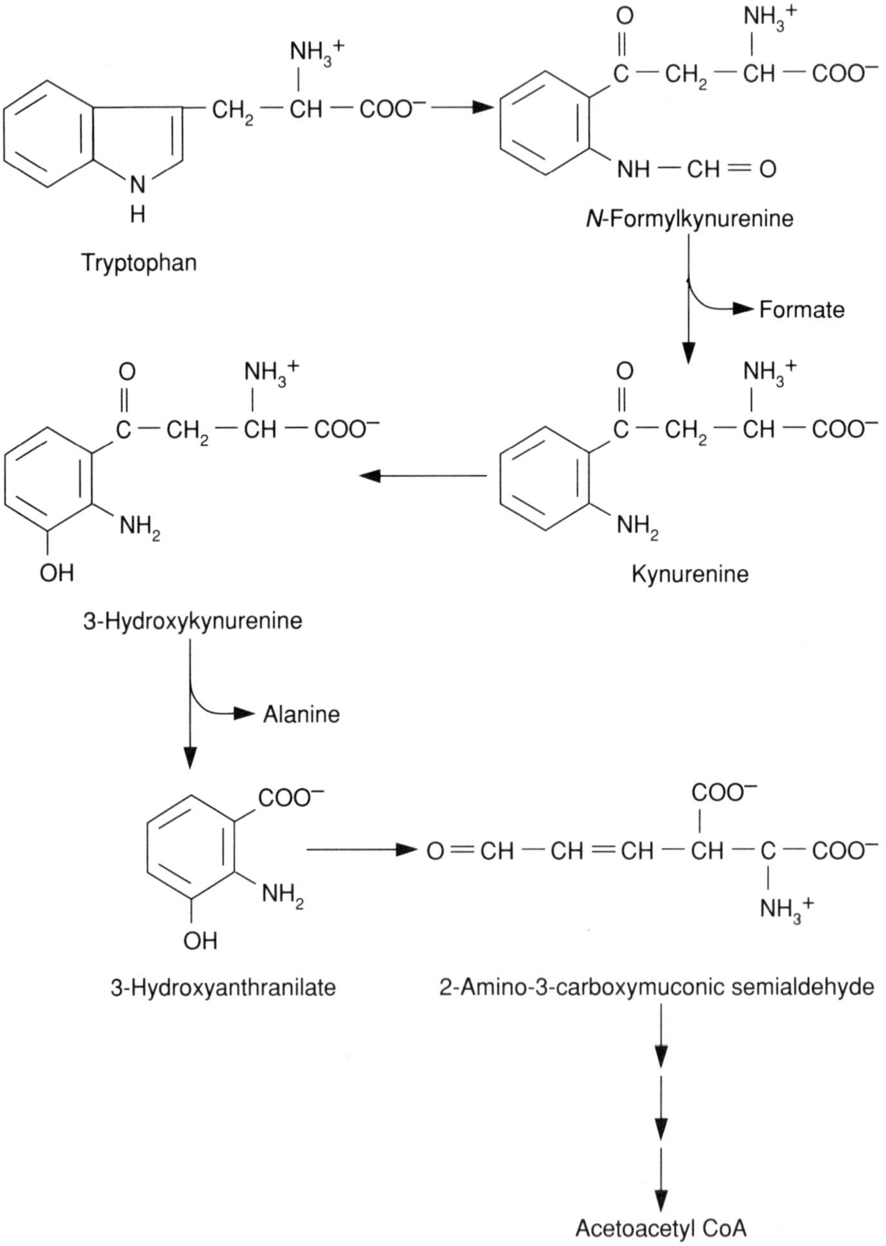

Fig. 9-14. Tryptophan degradation.

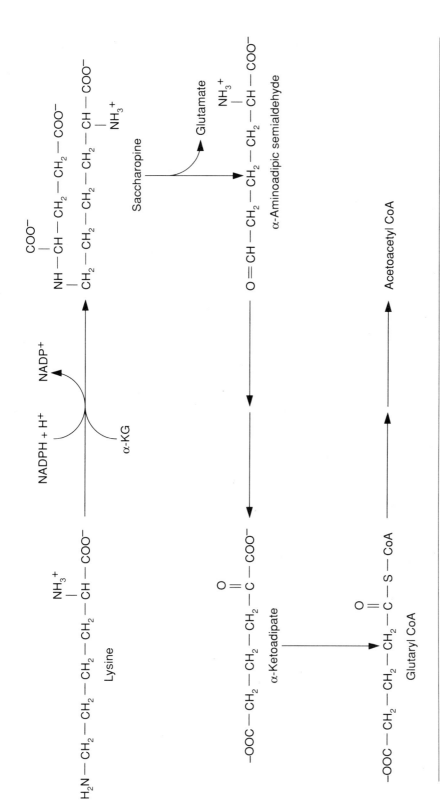

Fig. 9-15. Lysine degradation. α-KG, α-ketoglutarate.

verted to α-ketoadipic acid, which is subsequently decarboxylated to yield glutaryl CoA. Glutaryl CoA is ultimately converted to acetoacetyl CoA.

A second pathway of lysine degradation proceeds via an initial oxidative deamination of the α-amino group to yield α-keto-ε-aminocaproic acid, which is converted to the cyclic compound pipecolic acid. The pipecolic acid ring is opened to yield α-aminoadipic semialdehyde, which is degraded as in the saccharopine pathway.

Perspective: One-Carbon Chemistry

The degradation of certain amino acid carbon chains produces one-carbon units. Serine is the primary source of endogenously generated one-carbon units. One-carbon units generated during metabolism are transferred to tetrahydrofolate, which serves as a carrier for one-carbon units.

ONE-CARBON CHEMISTRY

One-carbon units are carried by tetrahydrofolate at the N^5 or N^{10} position. Alternatively, a one-carbon unit may be carried as a one-carbon bridge between N^5 and N^{10}. Figure 9-16 shows some of the oxidation states of one-carbon units carried by tetrahydrofolate.

In mammalian cells, one-carbon units are used in the synthesis of purines and thymidylate and in the remethylation of homocysteine to methionine. They are also used in the interconversion of serine and glycine.

Source of One-Carbon Units

Serine. Serine is the most important source of endogenous one-carbon units. Serine transhydroxymethylase catalyzes the transfer of the hydroxymethyl group of serine to tetrahydrofolate to form glycine and N^5,N^{10}-methylenetetrahydrofolate. Serine may also be synthesized by the reversal of this reaction. Serine transhydroxylmethylase is a pyridoxal phosphate dependent enzyme.

Glycine. Glycine contributes one-carbon units via the reversal of the glycine synthase reaction (p. 240).

Histidine. Degradation of histidine also ultimately produces N^5,N^{10}-methenyltetrahydrofolate (see Fig. 9-10).

ESSENTIAL AND NONESSENTIAL AMINO ACIDS

Amino acids can be classified as nutritionally *essential* and nutritionally *nonessential*. Essential amino acids either cannot be synthesized in

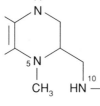

Fig. 9-16. (A) Tetrahydrofolate. The boxed region is shown in Fig. B. **(B)** Oxidation states of one-carbon units carried by tetrahydrofolate.

Table 9-2 Nutritionally Essential and Nonessential Amino Acids

Essential	Nonessential
Arginine	Alanine
Histidine	Asparagine
Isoleucine	Aspartic acid
Leucine	Cysteine
Lysine	Glutamic acid
Methionine	Glutamine
Phenylalanine	Glycine
Threonine	Proline
Tryptophan	Serine
Valine	Tyrosine

mammalian cells or cannot be synthesized at a rate adequate for normal growth, and must be obtained from the diet. Nonessential amino acids can be synthesized by mammalian cells in sufficient quantities for normal growth. Table 9-2 lists the essential and nonessential amino acids.

Tyrosine, cysteine, and arginine merit further discussion. Tyrosine can be synthesized in the cell, but only from the essential amino acid phenylalanine. Similarly, the sulfur atom of cysteine is obtained only from the essential amino acid methionine. Tyrosine and cysteine are classified as nonessential, but their synthesis depends on the availability of the essential amino acids phenylalanine and methionine, respectively. Arginine can be synthesized in human tissues at a rate sufficient to meet the needs of an adult but not those of a growing child. Arginine may therefore be considered either essential or nonessential.

SYNTHESIS OF NONESSENTIAL AMINO ACIDS

Alanine. Alanine is synthesized by transamination of pyruvate. The alanine formed during tryptophan degradation is not likely to contribute significantly to the amino acid pool.

Aspartate. Aspartate is synthesized by the transamination of oxaloacetate and by hydrolysis of asparagine by asparaginase.

Asparagine. Asparagine is synthesized from aspartate by the glutamine-dependent asparagine synthetase:

$$\text{Aspartate} + \text{glutamine} + \text{ATP} + \text{H}_2\text{O} \xrightarrow{\text{Asparagine synthetase}} \text{asparagine} + \text{glutamate} + \text{AMP} + \text{PP}_i$$

Glutamate. Glutamate is synthesized by transamination of α-ketoglutarate, by the reversal of the glutamate dehydrogenase reaction, and by the hydrolysis of glutamine by glutaminase. Glutamate is also produced in the degradation of proline, ornithine, and histidine.

Glutamine. Glutamine is synthesized from glutamate and ammonia by glutamine synthetase.

Cysteine. Cysteine is produced by the cystathionine-γ-lyase reaction during methionine degradation.

Glycine. Glycine is produced by the glycine synthase reaction and from serine by the serine transhydroxymethylase reaction.

Proline. To synthesize proline, glutamate is reduced to glutamate semialdehyde. The semialdehyde undergoes a cyclization reaction to form Δ^1-pyrroline-5-carboxylic acid. Reduction of the Δ^1-pyrroline-5-carboxylic acid yields proline (Fig. 9-17).

Serine. Serine is formed from 3-phosphoglycerate, an intermediate of glycolysis. 3-Phosphoglycerate is converted to glycerate via a 2-phosphoglycerate intermediate. The resulting glycerate is oxidized to 3-hydroxypyruvate, which is transaminated to yield serine (Fig. 9-18). Serine is also generated from glycine and N^5,N^{10}-methylenetetrahydrofolate by reversal of the serine transhydroxymethylase reaction.

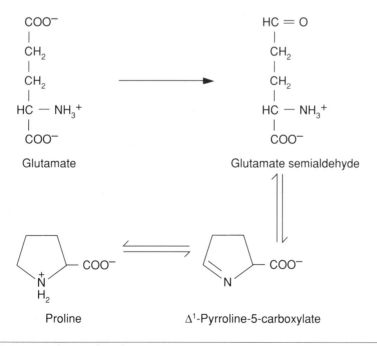

Fig. 9-17. Synthesis of proline.

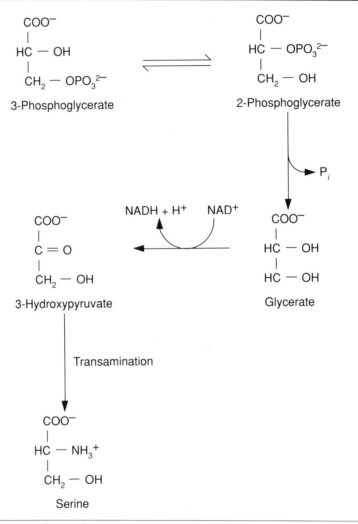

Fig. 9-18. Synthesis of serine. α-KG, α-ketoglutarate.

Tyrosine. Phenylalanine is hydroxylated by the phenylalanine hydroxylase system to form tyrosine.

CLINICAL CORRELATIONS

Deficiencies of Urea Cycle Enzymes and Ammonia Toxicity

Deficiencies of each of the urea cycle enzymes have been described—carbamoyl phosphate synthetase deficiency, ornithine transcarbamoylase deficiency, argininosuccinate synthetase deficiency (citrullinemia), argininosuccinate lyase deficiency (argininosuccinic aciduria), and arginase deficiency (hyperargininemia). All these deficiencies are associated with the accumulation and excretion of increased quantities of a particular urea cycle intermediate and of ammonia. Clinical findings include vomiting, lethargy, irritability, and mental retardation. These symptoms may lead to seizures, coma, and death during the first two weeks of life. The blood urea concentration usually remains in the normal range; consequently, the deficiency in each case is probably only partial, and the clinical symptoms are most likely due to the associated rise in ammonia concentration.

Treatment

Improvement of symptoms has been achieved by putting the patients on a low-protein diet, and sometimes also administering the α-keto acids of the essential amino acids. Another treatment that has been used, benzoate therapy, acts by stimulating the elimination of some excess ammonium by a mechanism independent of the urea cycle. Benzoate is converted to benzoyl CoA, which reacts with glycine to form hippurate:

$$\text{Benzoate} + \text{HS-CoA} \xrightarrow{\text{ATP} \rightarrow \text{AMP} + PP_i} \text{Benzoyl CoA}$$

$$\text{Benzoyl CoA} + \text{Glycine (}{}^+\!NH_3\text{-}CH_2\text{-}COO^-\text{)} \longrightarrow \text{Hippurate} + \text{HS-CoA}$$

Hippurate is excreted in the urine; one nitrogen atom is eliminated for each hippurate excreted. A similar reaction occurs in the synthesis of phenylacetylglutamine from phenylacetic acid and glutamine; in this case, two nitrogens would be eliminated.

The toxicity of ammonia is believed to be due in part to the ability of excess ammonia to drive the glutamate dehydrogenase reaction in the reverse direction to the extent that α-ketoglutarate is drained from the TCA cycle and the activity of the TCA cycle is depressed.

Disorders Affecting Aromatic Amino Acid Metabolism

Phenylketonuria

Phenylketonuria (PKU) occurs when phenylalanine hydroxylase activity is absent or significantly reduced. Phenylalanine accumulates, which stimulates the metabolism of phenylalanine by alternative pathways, resulting in urinary excretion of high amounts of phenyllactate, phenylpyruvate, o-hydroxyphenylacetate, and phenylacetylglutamine, as well as phenylalanine (Fig. 9-19). The concentration of tyrosine in patients with phenylketonuria is usually low to normal. PKU is treated by restricting the intake of phenylalanine.

Tyrosinase Deficiency

Melanin, the pigment of human skin, is synthesized inside melanocytes from tyrosine. In the first two steps of melanin formation, tyrosine is converted to dopa and then to dopaquinone by tyrosinase (see Fig. 10-

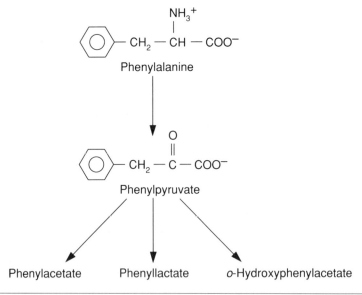

Fig. 9-19. The alternative routes of phenylalanine degradation that are increased in phenylketonuria.

2). A tyrosinase deficiency, which blocks the conversion of tyrosine to dopa, is the cause of certain types of albinism.

Alkaptonuria

In alkaptonuria, the enzyme homogentisate oxidase of tyrosine metabolism is deficient. High amounts of homogentisate are excreted in the urine, which causes the urine to darken gradually or rapidly due to oxidation of homogentisate. Clinical conditions resulting from alkaptonuria are ochronosis and, in later years, arthritis. Ochronosis is pigmentation of connective tissues due to deposition of products of homogentisate metabolism. The ochronotic pigment is believed to be a polymer derived from homogentisate, but its structure is not known.

Disorders Affecting the Metabolism of Sulfur-Containing Amino Acids

Cystathionuria

Cystathionuria, the accumulation and excretion of cystathione, may be due to either underutilization or increased production of cystathione (an intermediate in the breakdown of methionine). Underutilization is due to a genetic deficiency of γ-cystathionase, the enzyme that converts cystathione to α-ketobutyrate. Increased production occurs when an abnormally high amount of homocysteine is available for the cystathione-β-synthase reaction, a situation that arises when the methionine synthetase reaction is inhibited by a lack of cobalamin or 5-methyltetrahydrofolate.

A deficiency in vitamin B_6 (pyridoxine) also produces cystathionuria. Both γ-cystathionase and cystathione-β-synthase are dependent on pyridoxal phosphate. However, the activity of γ-cystathionase is more severely depressed by a deficit of pyridoxal phosphate than that of cystathione-β-synthase, so vitamin B_6 deficiency results in a buildup of cystathione. No clinical abnormalities are usually associated with this condition.

Sulfite Oxidase Deficiency

Sulfite oxidase catalyzes the oxidation of sulfite to sulfate (the terminal step in cysteine degradation). A deficiency of this enzyme is responsible for severe neurologic problems, mental retardation, and the excretion of high amounts of sulfite, thiosulfate, and S-sulfocysteine in the urine.

Deficiencies of Enzymes of Branched Chain Amino Acids

The first two steps in the degradation of the branched chain amino acids (valine, leucine, isoleucine) are identical: transamination followed by oxidative decarboxylation of the resulting keto acids to yield the corresponding CoA derivatives containing one less carbon atom. The decarboxylation reaction is catalyzed by the branched chain α-keto acid dehydrogenase complex. A genetic deficiency of this enzyme causes elevated levels of valine, isoleucine, and leucine and their α-keto acids in blood and urine. The high concentration of α-keto acids gives the urine of these patients a maple syrup odor, leading to the name "maple syrup urine disease" for the disorder. Clinical symptoms are poor feeding after the first week of life, vomiting, lethargy, muscular hypertonicity, and, sometimes, convulsions. Variation in clinical symptoms is likely due to variation in the extent of the disorder. The disorder is treated by restricting the dietary intake of branched chain amino acids.

SUGGESTED READING

1. Meister A: Biochemistry of the Amino Acids, 2nd Ed. Academic Press, Orlando, FL, 1965
2. Cooper AJL: Biochemistry of sulfur-containing amino acids. Annu Rev Biochem 56:187, 1983
3. Grisiola S, Baguener R, and Major F: The Urea Cycle. Wiley, New York, 1976
4. Stanbury JB, Wyngaarden JB, Fredrickson DS, et al: The Metabolic Basis of Inherited Disease, 4th Ed. McGraw-Hill, New York, 1986
5. Wellner D, and Meister A: A survey of inborn errors of amino acid metabolism and transport. Annu Rev Biochem 50:911, 1987

STUDY QUESTIONS

Directions: For each of the following multiple choice questions, choose the most appropriate answer.

1. The principal fate of free ammonia is the synthesis of
   A. glutamate
   B. glutamine
   C. asparagine
   D. urea

2. Although all amino acids are taken up and degraded by the liver, the muscle and brain also significantly contribute to the degradation of
   A. the branched chain amino acids
   B. proline and arginine
   C. the aromatic amino acids
   D. cysteine and methionine

3. All of the following enzymes catalyze the direct production of free ammonium ion EXCEPT
   A. glutamate dehydrogenase
   B. glutaminase
   C. tyrosine-glutamate transaminase
   D. L-amino acid oxidase

4. Which of the following enzymes is NOT a part of the urea cycle?
   A. Arginase
   B. Ornithine transcarbamoylase
   C. Argininosuccinyl synthetase
   D. Ornithine decarboxylase

5. Which of the following groups of amino acids consists of purely glycogenic amino acids?
   A. Serine, alanine, tyrosine
   B. Aspartate, cysteine, histidine
   C. Isoleucine, asparagine, threonine
   D. Glutamine, proline, lysine

6. All of the following acids are converted to α-ketoglutarate EXCEPT
   A. histidine
   B. asparagine
   C. arginine
   D. proline

7. Which of the following amino acids is nutritionally essential and is capable of producing pyruvate on degradation?
   A. Threonine
   B. Alanine
   C. Serine
   D. Isoleucine

8. Genetic deficiency of any of the urea cycle enzymes usually results in elevated blood levels of
   A. citrulline
   B. arginine
   C. NH_4^+
   D. ornithine

259

Directions: Answer the following questions using the key outlined below:
- **(A)** if 1, 2, and 3 are correct
- **(B)** if 1 and 3 are correct
- **(C)** if 2 and 4 are correct
- **(D)** if only 4 is correct
- **(E)** if all four are correct

9. Proteolytic enzymes that are active in the degradation of dietary protein include
   1. trypsin
   2. elastase
   3. chymotrypsin
   4. carboxypeptidase A

10. One-carbon units are synthesized de novo during the degradation of
    1. serine
    2. glycine
    3. histidine
    4. methionine

10

AMINO ACID METABOLISM II

Reactions of Amino Acids/Heme Metabolism/ Clinical Correlations

The student should be able to:

1. Write reactions that illustrate amino acid decarboxylation and hydroxylation, *N*-acylamino acid formation, and transamidination.
2. List the amino acid precursors of polyamines.
3. Describe the reactions of heme biosynthesis and their intracellular location, and list the regulated enzymes of heme biosynthesis.
4. Detail the reactions and enzymes involved in the conversion of heme to bilirubin.
5. Understand the mechanism for the circulatory transport of bilirubin, its solubilization in liver cells, and its elimination.

REACTIONS OF AMINO ACIDS

As indicated in the previous chapter, most amino acids are either incorporated into proteins or degraded to intermediates of lipid and carbohydrate metabolism. However, small amounts of amino acids are used as precursors in the synthesis of various nitrogen-containing compounds. Nitrogenous compounds derived from amino acids are ubiquitous in cells and perform diverse functions. This section is concerned with the synthesis of the nitrogenous compounds that are derived from amino acids via decarboxylation, hydroxylation, acylation, and transamidination. In addition, the synthesis of polyamines and the synthesis of oligopeptides not directed by messenger RNA are covered.

Decarboxylation and Hydroxylation

GABA

γ-Aminobutyric acid (GABA), an inhibitory neurotransmitter of the brain, is synthesized by decarboxylation of the α-carboxyl group of glutamic acid, catalyzed by *glutamate α-decarboxylase*:

$$\begin{array}{c} NH_3^+ \\ | \\ HC-COO^- \\ | \\ CH_2 \\ | \\ CH_2 \\ | \\ COO^- \\ \text{Glutamate} \end{array} \rightarrow \begin{array}{c} NH_3^+ \\ | \\ CH_2 \\ | \\ CH_2 \\ | \\ CH_2 \\ | \\ COO^- \\ \text{GABA} \end{array} + CO_2$$

GABA is metabolized by transamination to succinate semialdehyde, followed by oxidation to succinate, which enters the TCA cycle of nervous tissue:

$$\begin{array}{c} NH_3^+ \\ | \\ CH_2 \\ | \\ CH_2 \\ | \\ CH_2 \\ | \\ COO^- \\ \text{GABA} \end{array} \xrightarrow[\text{Glutamate}]{\alpha\text{-Ketoglutarate}} \begin{array}{c} HC=O \\ | \\ CH_2 \\ | \\ CH_2 \\ | \\ COO^- \\ \text{Succinate semialdehyde} \end{array} \longrightarrow \begin{array}{c} COO^- \\ | \\ CH_2 \\ | \\ CH_2 \\ | \\ COO^- \\ \text{Succinate} \end{array}$$

Histamine

Histamine is found in connective tissue mast cells, gastric mucosa, muscle, lung, and liver. Histamine has a variety of functions:

1. Vasodilation
2. Stimulation of gastric secretion
3. Action as a neurohumoral agent in nerve tissue (in general, histamine is released at times of traumatic shock and at sites of inflammation)

Histamine is synthesized from histidine, in a reaction catalyzed by *histidine decarboxylase*:

$$\text{Imidazole-}CH_2-\underset{\underset{NH_3^+}{|}}{CH}-COO^- \longrightarrow \text{Imidazole-}CH_2-CH_2-NH_3^+ + CO_2$$

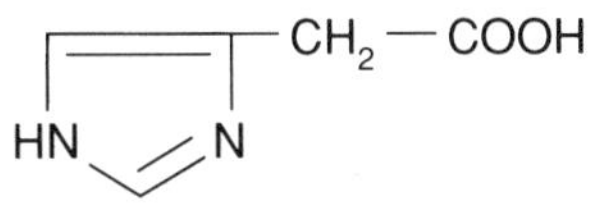

Amino Acid Metabolism II 263

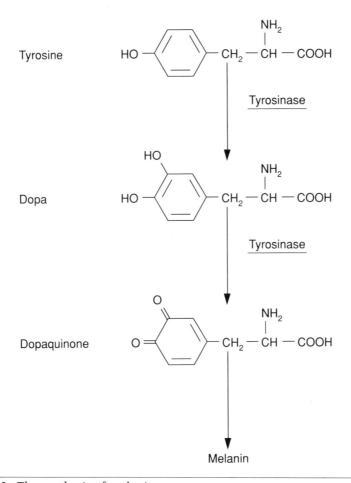

Fig. 10-1. Imidazoleacetic acid.

Histamine degradation may occur by methylation at the N^1 position using *S*-adenosylmethionine (SAM) followed by oxidation to the corresponding aldehyde by diamine oxidase. The aldehyde is subsequently oxidized to the N^1-methylimidazoleacetic acid. An alternative pathway to histamine degradation bypasses the methylation step; the amine is oxidized directly by diamine oxidase to yield imidazoleacetic acid (Fig. 10-1). Both N^1-methylimidazoleacetic acid and imidazoleacetic acid are excreted in the urine.

Fig. 10-2. The synthesis of melanin.

Tyrosine Derivatives

Tyrosine is the precursor of *melanin* and of *dopamine, norepinephrine,* and *epinephrine*. Melanin, the pigment of skin, is a polymer of unspecified structure. Dopamine, norepinephrine, and epinephrine all function as neurotransmitters, and are collectively referred to as the *catecholamines*.

Melanin is synthesized in the melanoblasts of the epidermis by the process summarized in Figure 10-2. The rate-limiting enzyme is *tyrosinase*, which catalyzes the conversion of tyrosine to dopa and of dopa to dopaquinone. The conversion of tyrosine to dopa is enhanced by ultraviolet light.

Catecholamines are synthesized in nerve tissue and the adrenal medulla by the pathway shown in Figure 10-3. The rate-limiting step is again the conversion of tyrosine to dopa, catalyzed in this case by *tyrosine hydroxylase*. Dopa is then converted to dopamine by decarboxylation. Dopamine is hydroxylated to yield norepinephrine, which can be methylated to yield epinephrine. After synthesis, the catecholamines are packaged into cellular vesicles or granules.

Most catecholamines are degraded in the liver or kidney, and the resulting products are excreted in the urine. Unmetabolized catecholamine is also excreted. Catecholamine degradation can occur via methylation or oxidation pathways, and a single catecholamine molecule may undergo both processes (Fig. 10-4). The products excreted in the urine may also be conjugated at the 4-hydroxy position to either sulfate or glucuronic acid. The flexibility of catecholamine metabolism explains the wide variety of epinephrine, norepinephrine, and metabolites of these compounds found in urine. The catecholamine products appearing in the urine may thus be grouped into four classes, each of which may be conjugated or unconjugated:

1. Unmetabolized
2. Methylated
3. Oxidized
4. Methylated and oxidized

Methylation of catecholamines occurs at the 3-hydroxy position (Fig. 10-4), and is catalyzed by *catechol-O-methyltransferase* (*COMT*). Oxidation is initiated by the enzyme *monoamine oxidase*, which yields the corresponding aldehydes; these may be oxidized further to carboxylic acids or reduced to alcohols.

Serotonin

Serotonin is a neurotransmitter and vasoconstrictor found in brain cells, platelets, mast cells, and intestinal tissue. It is synthesized from tryptophan in two steps (Fig. 10-5): tryptophan is hydroxylated by *tryptophan hydroxylase* to 5-hydroxytryptophan, which is decarboxylated to 5-hydroxytryptamine (serotonin). Serotonin is metabolized primarily

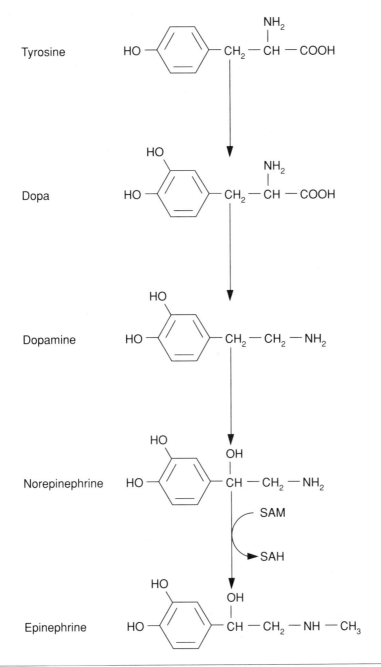

Fig. 10-3. The synthesis of catecholomines. SAM, S-adenosylmethionine; SAH, S-adenosylhomocysteine.

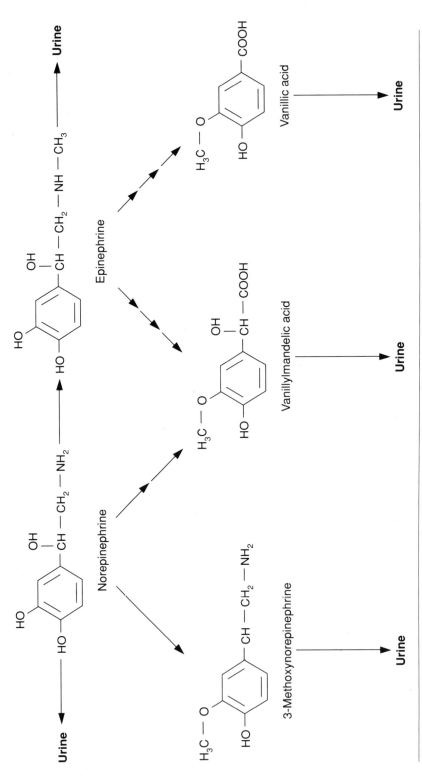

Fig. 10-4. Catecholamine metabolism. The products excreted in the urine may or may not be conjugated.

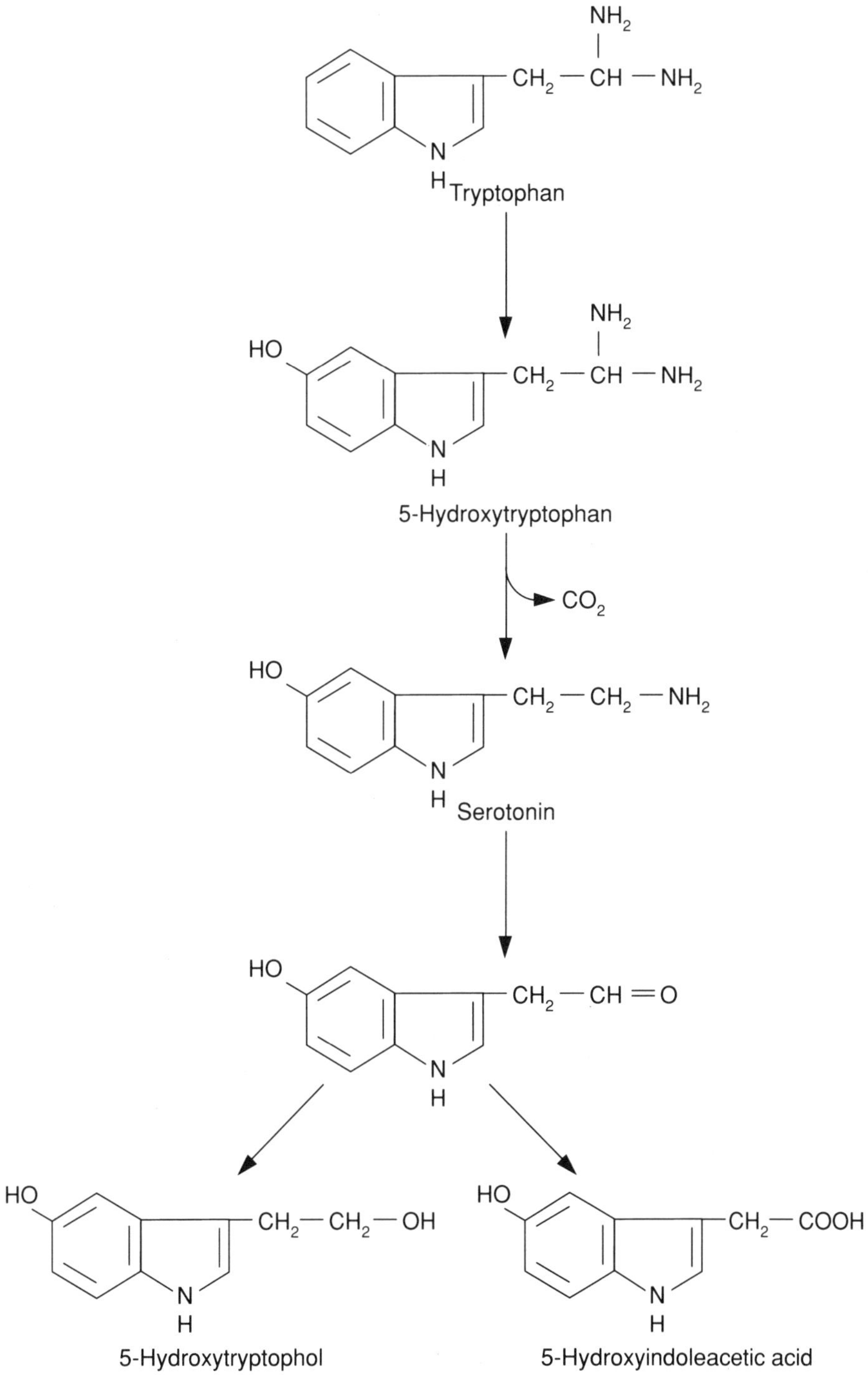

Fig. 10-5. The synthesis and degradation of serotonin.

by N-methylation, N-acetylation, or by initial oxidation via monoamine oxidase. The latter pathway results in the production of mostly 5-hydroxyindoleacetic acid and a small quantity of 5-hydroxytryptophol (Fig. 10-5). The major serotonin metabolite in the urine is normally 5-hydroxyindoleacetic acid; administration of monoamine oxidase inhibitors causes an increase in the urinary excretion of the N-methylated and N-acetylated derivatives.

Formation of N-Acylamino Acids

Amino acids are capable of N-acylating the carboxyl group of aromatic carboxylic acids. The resulting carboxylic acid-amino acid adducts are eliminated in the urine. These reactions normally serve to detoxify and enhance the excretion of aromatic carboxylic acids derived from the diet or from metabolism of aromatic amino acids. As discussed in the Clinical Correlations in Chapter 9, the pathway has been utilized clinically to enhance the elimination of excess ammonia from the body. The reactions involved in the synthesis of N-acylamino acids are described in Chapter 9 (p. 256)

Transamidination: Phosphacreatine

As described in Chapter 3, cells (especially muscle cells) store phosphate bond energy in the form of *phosphocreatine*. Phosphocreatine is capable of phosphorylating ADP. The existence of phosphocreatine allows the cell to store phosphate bond energy without raising the concentration of ATP. Large changes in [ADP] would unbalance the many cellular processes controlled by the adenine nucleotides. Phosphocreatine is also believed to function in the transport of energy (ATP equivalents) among cellular organelles and locations.

The phosphate exchange between ATP/ADP and phosphocreatine is catalyzed by *creatine kinase*:

$$ATP + \begin{array}{c} NH_2 \\ | \\ C=\overset{+}{N}H_2 \\ | \\ N-CH_3 \\ | \\ CH_2 \\ | \\ COO^- \end{array} \rightleftharpoons \begin{array}{c} PO_3^{2-} \\ | \\ {}^+NH_2 \\ | \\ C=\overset{+}{N}H_2 \\ | \\ N-CH_3 \\ | \\ CH_2 \\ | \\ COO^- \end{array} + ADP$$

Creatine   Phosphocreatine

The phosphate-nitrogen bond of phosphocreatine releases approximately as much energy on hydrolysis as an ATP phosphodiester bond.

Synthesis of Creatine

Creatine is synthesized in the liver and pancreas by a pathway involving amino acid transamidination (Fig. 10-6). First, the amidine group of arginine is transferred to the amino group of glycine to produce guanidineoacetic acid and ornithine. This transamidination is catalyzed by *arginine-glycine amidotransferase*. Guanidinoacetic acid is methylated on the α-nitrogen to produce creatine, which is transported in the circulation to the tissues that use it (primarily muscle and brain). An excess of creatine represses the amidotransferase. The formation of *creatinine* (Fig. 10-7) helps to maintain appropriate levels of creatine and creatine phosphate.

Polyamines

Putrescine, spermidine, and *spermine* (Fig. 10-8) are polyamines synthesized in mammalian cells. The physiologic roles of polyamines have not been established. It has been suggested that they play a role in various

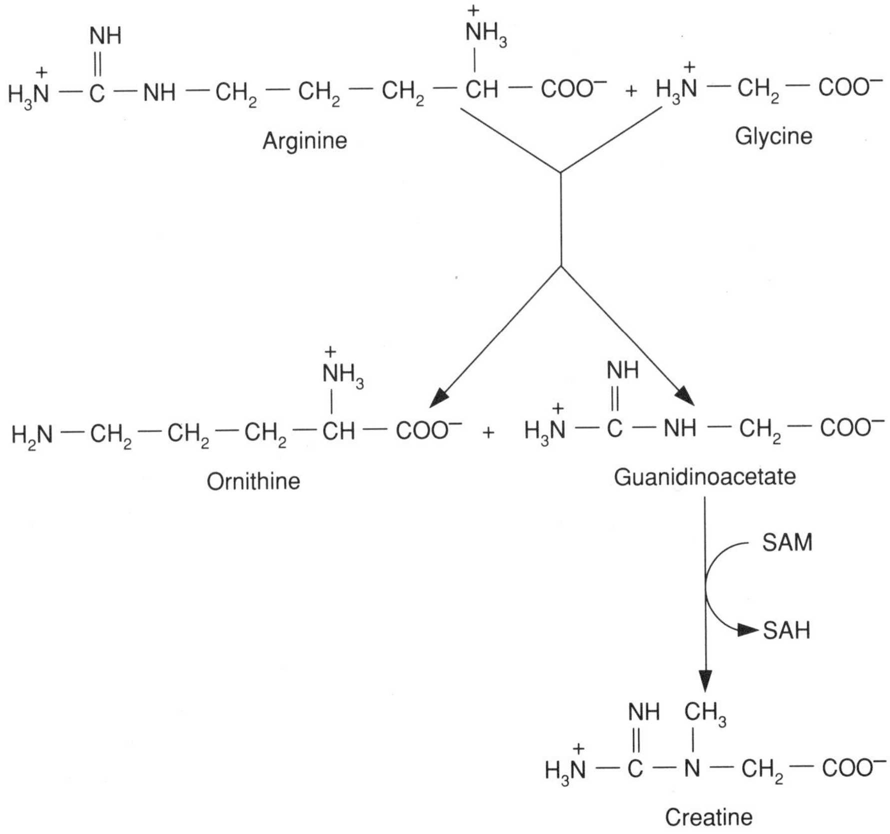

Fig. 10-6. The synthesis of creatine. SAM, S-adenosylmethionine; SAH, S-adenosylhomocysteine.

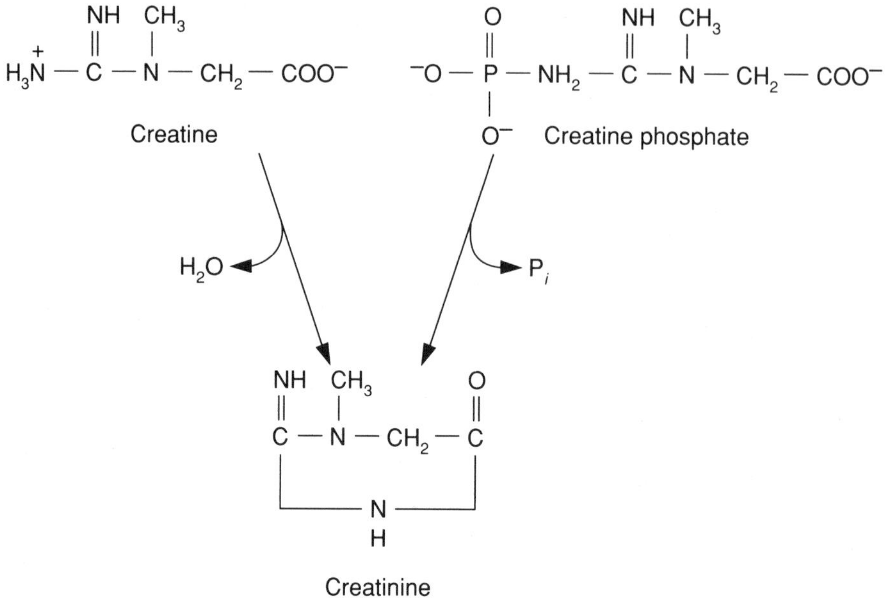

Fig. 10-7. The synthesis of creatinine from creatine and creatine phosphate.

cellular functions, including cell growth and proliferation, platelet aggregation, and modification of protein phosphorylation. In support of their role in cell growth and proliferation, polyamine levels increase in rapidly dividing cells prior to DNA synthesis. Spermidine and spermine are abundant in seminal fluid and the prostate glands.

Synthesis of Polyamines

Figure 10-8 shows the synthesis of putrescine, spermidine, and spermine. Putrescine is formed by the decarboxylation of ornithine by *ornithine decarboxylase*. Decarboxylation of the methionine moiety of S-adenosylmethionine yields a decarboxylated S-adenosylmethionine, the propylamine group of which can be transferred to putrescine to form spermidine. Spermidine can react with another decarboxylated S-adenosylmethionine to yield spermine.

Degradation of Polyamines

As shown in Figure 10-9, spermine and spermidine can be acted upon by either *diamine oxidase* or *polyamine oxidase*. Polyamines may also undergo acetylation by *N*-acetyltransferases.

Non-mRNA-Dependent Synthesis of Oligopeptides

A number of oligopeptides are generated by direct enzymatic synthesis not involving messenger RNA. An example is the tripeptide *glutathione*

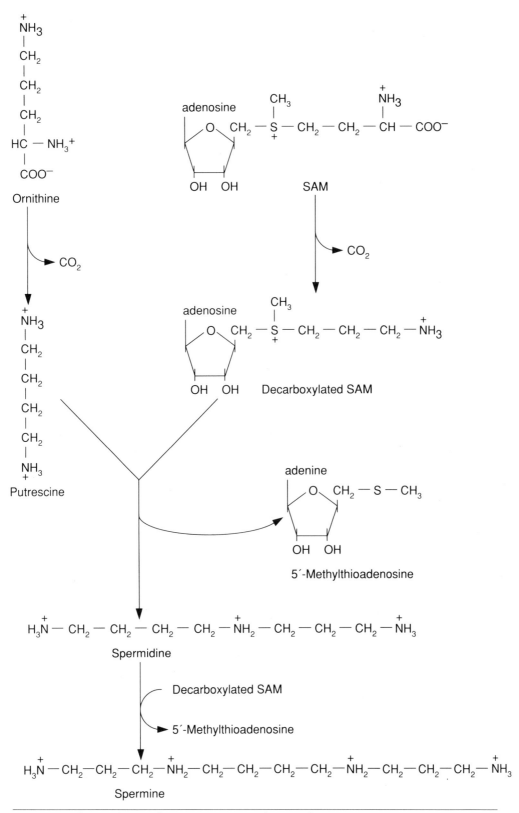

Fig. 10-8. The synthesis of putescine, spermidine, and spermine.

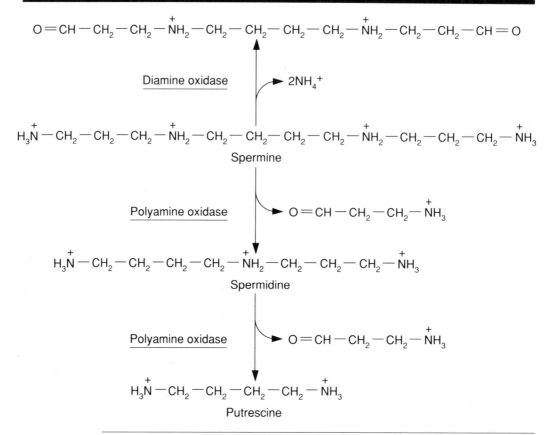

Fig. 10-9. The degradation of polyamines via diamine oxidase and polyamine oxidase, using spermine as an example. The diamine oxidase reaction for spermidine is analogous.

(γ-glutamylcysteinylglycine). Glutathione functions in a variety of capacities:

1. As a cellular reductant
2. In the stabilization of disulfites in proteins
3. As an enzyme cofactor
4. In the detoxification of xenobiotics
5. In the transport of amino acids across cell membranes (the γ-glutamyl cycle; see Fig. 9-3)

Reduced glutathione (GSH) is particularly important as a cellular reductant in erythrocytes, where it functions to reduce methemoglobin (MetHb) to hemoglobin (Hb). In the process, glutathione is oxidized to the disulfide (GSSG):

$$2 \text{ GSH} + 2 \text{ MetHb} \rightleftharpoons \text{GSSG} + 2 \text{ Hb}$$

Glutathione is also important in scavenging toxic peroxides and free radicals. For example:

$$2 \text{ GSH} + \text{ROOH} \rightarrow \text{GSSG} + H_2O + \text{ROH}$$

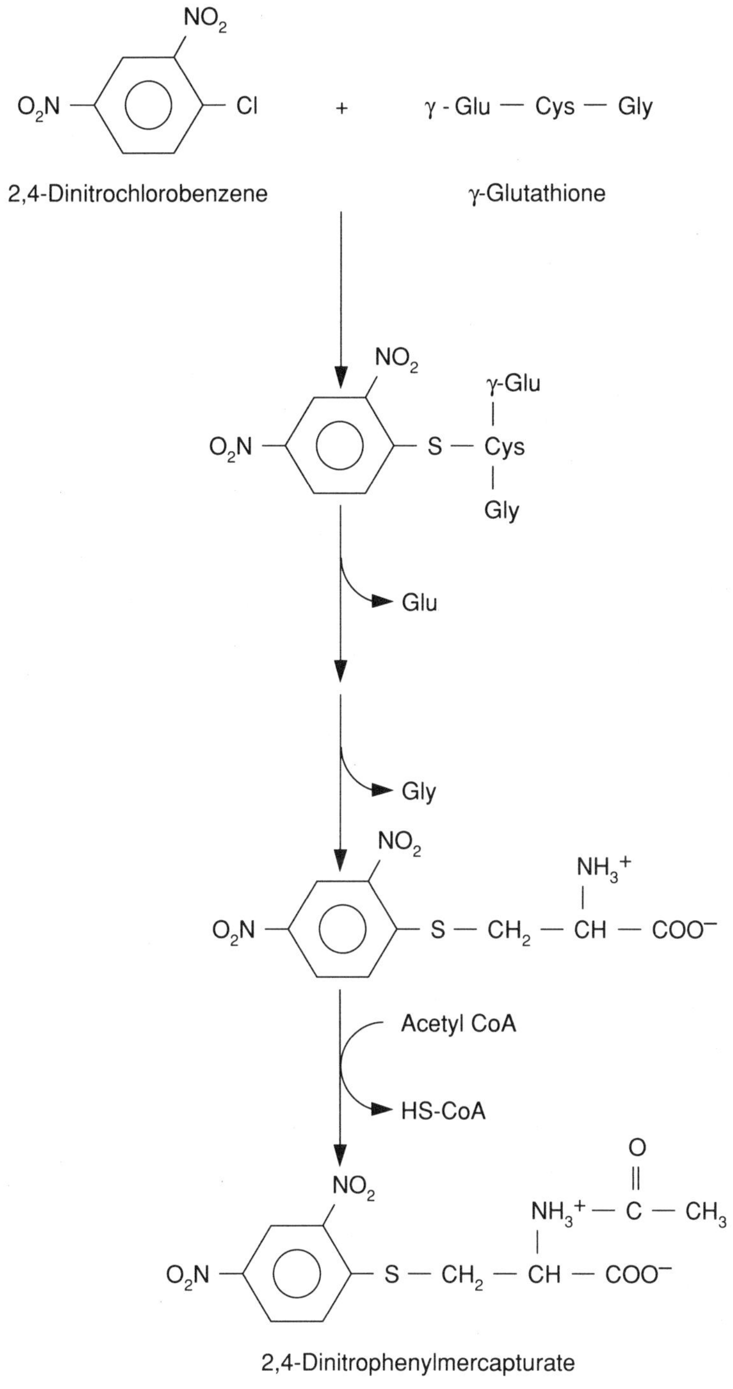

Fig. 10-10. An example of formation of a mercapturic acid.

Figure 10-10 shows an example of the detoxification of foreign substances by glutathione, a process referred to as *mercapturic acid formation*. The cysteine moiety of glutathione first forms a covalent adduct with the foreign substance (3,4-dinitrochlorobenzene in this example) in a reaction catalyzed by *glutathione S-transferase*. The γ-glutamyl and glycyl moieties are then successively cleaved off, and the cysteine moiety is *N*-acetylated by acetyl CoA to form a mercapturic acid. Mercapturic acids are water soluble and are eliminated in the urine.

Synthesis and Degradation of Glutathione

Glutathione is synthesized from its component amino acids in two steps:

$$\text{Glutamate} + \text{cysteine} + \text{ATP} \rightarrow \gamma\text{-glutamylcysteine} + \text{ADP} + P_i$$

$$\gamma\text{-Glutamylcysteine} + \text{glycine} + \text{ATP} \rightarrow \text{glutathione} + \text{ADP} + P_i$$

Glutathione is partially degraded and resynthesized in the operation of the γ-glutamyl cycle (see Fig. 9-3).

Perspective: Heme Biosynthesis

Hemes consist of a porphyrin ring containing a complexed iron molecule. As mentioned in earlier chapters, hemes act as prosthetic groups in heme proteins such as hemoglobin, myoglobin, and the cytochromes. The porphyrin ring of heme is synthesized from glycine and succinyl CoA. The synthesis of heme in mammalian cells can be divided into four phases:

1. Synthesis of porphobilinogen
2. Formation of the tetrapyrrole ring
3. Modification of the tetrapyrrole ring
4. Incorporation of iron to form herme

HEME BIOSYNTHESIS

Reactions of Heme Synthesis

Most tissues synthesize heme. The bulk of the body's heme, however, is synthesized in the bone marrow and liver. As mentioned above, heme biosynthesis may be divided into four phases.

Phase 1: Synthesis of Porphobilinogen

The porphyrin ring of heme is constructed from four molecules of porphobilinogen, a pyrrole derivative. Porphobilinogen is synthesized in two steps from glycine and succinyl CoA. In the first reaction, glycine and succinyl CoA combine to yield δ-aminolevulinic acid. This reaction is catalyzed by δ-*aminolevulinic acid synthetase*, the major regulated enzyme of heme synthesis:

$$\text{Glycine} + \text{Succinyl CoA} \xrightarrow{\text{∂-Aminolevulinic acid synthetase}} \text{∂-Aminolevulinic acid}$$

Next, *δ-aminolevulinic acid dehydrase* catalyzes the condensation of two molecules of δ-aminolevulinic acid to form porphobilinogen, with the elimination of two molecules of water:

$$2 \; \text{∂-Aminolevulinic acid} \xrightarrow{-2H_2O} \text{Porphobilinogen}$$

Phase 2: Formation of the Tetrapyrrole Ring

Four porphobilinogens are joined together to yield the cyclic tetrapyrrole uroporphyrinogen III:

$$4 \; \text{Porphobilinogen} \longrightarrow \longrightarrow \text{Uroporphyrinogen III}$$

where A = acetyl and P = propionyl. Two enzymes are involved in this reaction: *uroporphyrinogen I synthase* and *uroporphyrinogen III cosynthase*. Uroporphyrinogen III cosynthase does not have enzymatic activity. Instead, it directs the activity of uroporphyrinogen I synthase so that the correct product is synthesized. In the absence of uroporphyrinogen III cosynthase, uroporphyrinogen I synthase produces a different tetrapyrrole isomer.

The reactions of phase 2 involve the formation of a linear tetrapyrrole, followed by deamination of the terminal amino groups to yield a cyclic product. To produce the arrangement of acetyl and propionyl groups in uroporphyrinogen III, one of the pyrrole rings must be inverted prior to closure of the tetrapyrrole ring. Uroporphyrinogen III cosynthase plays a role in this inversion.

Phase 3: Modification of the Tetrapyrrole Ring

There are two differences between uroporphyrinogen III and *protoporphyrin IX*, the intermediate into which iron is inserted to form heme:

1. Protoporphyrin IX has methyl groups in place of the acetyl groups on uroporphyrinogen III, and vinyl groups in place of two of the propionyl groups on uroporphyrinogen III.
2. In uroporphyrinogen III, the pyrrole rings are joined by full reduced methylene bridges ($-CH_2-$). In protoporphyrin IX, these methylenes are replaced by methenyl bridges ($=CH-$). The result is a resonance-stabilized system of conjugated double bonds that permeates the entire tetrapyrrole structure.

In the conversion of uroporphyrinogen III to protoporphyrin IX, the four acetyl groups are first decarboxylated to methyl groups by *uroporphyrinogen decarboxylase*, yielding coproporphyrinogen III. Next, the enzyme *coproporphyrinogen oxidase* oxidizes two of the propionyl groups to vinyls, yielding protoporphyrinogen IX. This product is converted to protoporphyrin IX by *protoporphyrinogen oxidase*, which oxidizes the methylene bridging groups to methenyls. The overall reaction is

Uroporphyrinogen III → Protoporphyrin IX

where M = methyl and V = vinyl.

Phase 4: Iron Incorporation

The enzyme *ferrochelatase* (*heme synthetase*) inserts ferrous iron into the protoporphyrin ring to form heme:

[Structures: Protoporphyrin IX + Fe²⁺ → Heme]

The iron atom is coordinated to a nitrogen from each pyrrole ring; the fifth and sixth coordination sites of the iron project on either side of the ring plane.

Intracellular Compartmentation of Heme Synthesis

As shown in Figure 10-11, heme synthesis begins and ends in the mitochondria, but the reactions involving the conversion of δ-aminolevulinic acid to coproporphyrinogen III occur in the cytosol. δ-Aminolevulinic acid and coproporphyrinogen III, therefore, move across the inner mitochondrial membrane in the process of heme synthesis.

Regulation of Heme Synthesis

The rate-controlling enzyme of heme synthesis is δ-aminolevulinic acid synthetase. Heme exerts feedback inhibition and repression on this enzyme. In addition, the enzyme has an absolute requirement for the cofactor pyridoxal phosphate, a derivative of vitamin B_6. Any circumstance that causes a deficiency of pyridoxal phosphate will affect heme synthesis. Likewise, ferrochelatase requires reducing agents (i.e., glutathione), so any interruption in the supply of reducing power to this enzyme will inhibit heme synthesis.

It should be noted that both δ-aminolevulinic acid dehydrase and ferrochelatase are very sensitive to inhibition by heavy metals. In fact, inhibition of heme synthesis is a characteristic effect of lead poisoning.

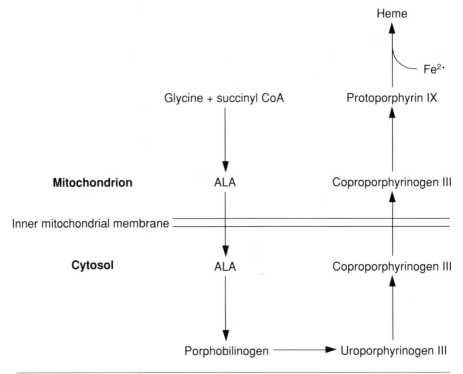

Fig. 10-11. The intracellular compartmentation of heme synthesis.

DEGRADATION AND ELIMINATION OF HEME

Most heme is degraded in the reticuloendothelial cells of bone marrow and liver. The bulk of this heme comes from hemoglobin and myoglobin, with lesser amounts derived from cytochromes and other heme-containing enzymes.

Heme is degraded to *bilirubin* by the pathway shown in Figure 10-12. First, the microsomal enzyme complex *heme oxygenase* cleaves the α-methenyl bridge to yield the linear tetrapyrrole *biliverdin*. This reaction frees the iron atom and also produces carbon monoxide. In fact, this is the only mammalian biochemical reaction that normally produces carbon monoxide. Biliverdin is then reduced to bilirubin by *biliverdin reductase*.

A significant amount of bilirubin is generated outside the liver. It is transported to the liver complexed to plasma albumin (free bilirubin is not very soluble in water). In the liver, bilirubin is rendered water soluble by conjugation with glucuronic acid (Fig. 10-13). A glucuronate molecule is esterified to each of the two propionate groups of bilirubin, resulting in bilirubin diglucuronide.

Bilirubin diglucuronide is excreted into the bile and enters the intestine. In the large intestine, bacterial enzymes hydrolyze the glucuronic acid moieties and release free bilirubin, most of which is converted

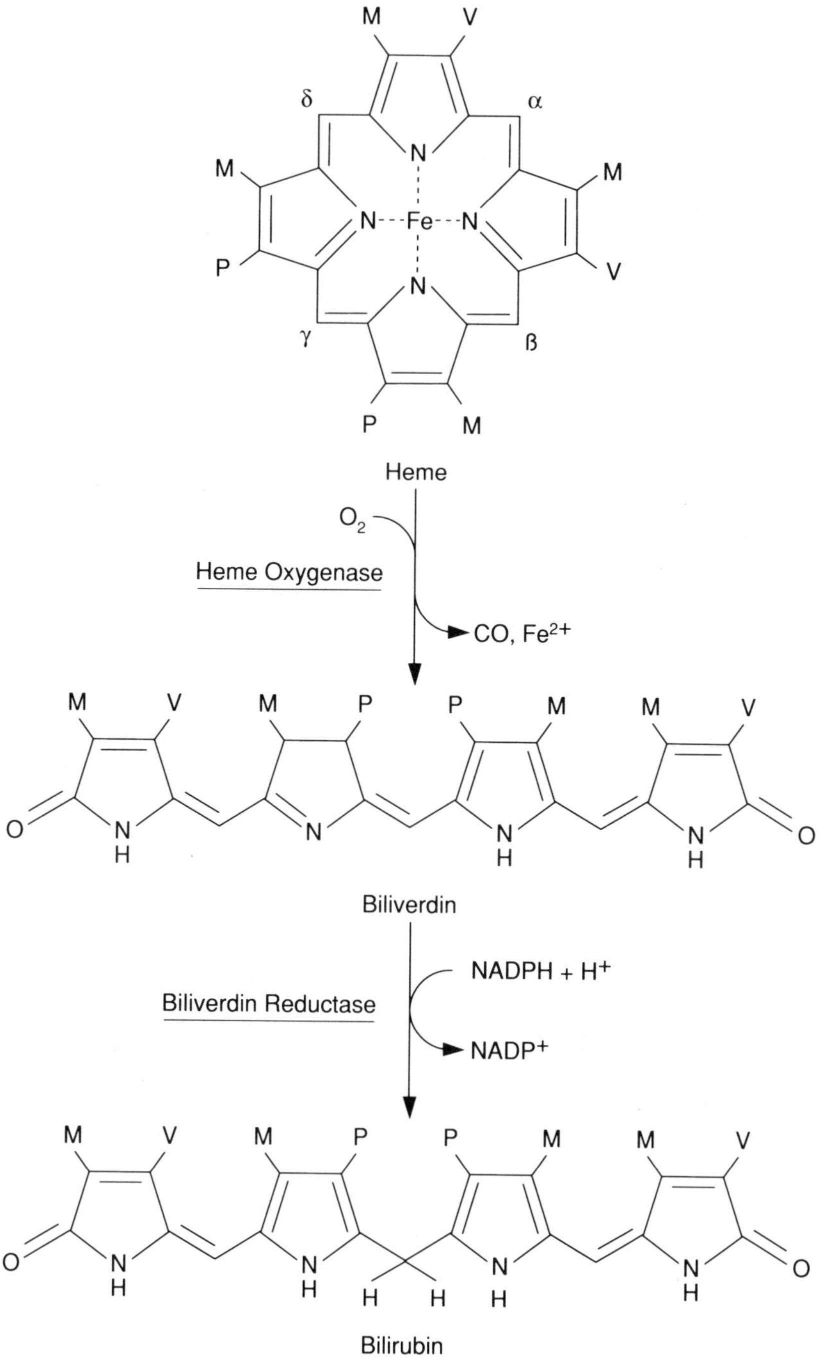

Fig. 10-12. Heme degradation.

by bacteria to a mixture of colorless linear tetrapyrroles called *urobilinogens*, which may in turn be converted to colored substances. Urobilinogens and their reaction products are mostly eliminated in the feces. A very small percentage of urobilinogen is absorbed from the large intestine and enters the blood. Part of it is taken up by the liver, reconjugated with glucuronic acid, and re-excreted into the bile; the remainder is eliminated in the urine.

Heme Degradation and Elimination: Key Points

1. Heme is broken down to the linear tetrapyrrole bilirubin, with release of the iron atom.
2. Extrahepatic bilirubin is transported to the liver complexed to plasma albumin.
3. In the liver, bilirubin is rendered water soluble by conjugation with glucuronide, and is excreted in the bile.

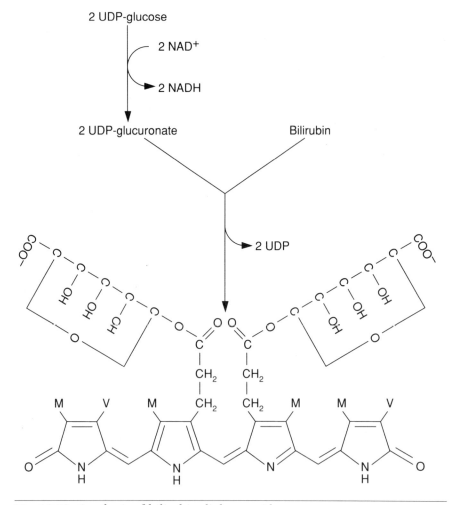

Fig. 10-13. Synthesis of bilirubin diglucuronide.

4. In the intestine, the glucuronide moieties are removed, and bilirubin is converted to urobilinogens and further metabolites by intestinal bacteria.
5. Urobilinogen and subsequent metabolites are eliminated in the feces.

CLINICAL CORRELATIONS: THE PORPHYRIAS

The porphyrias are rare disorders involving defects in heme synthesis. They are characterized by the accumulation and excretion of large amounts of porphyrins in the urine (and sometimes the feces) and by sensitivity to sunlight. Specific porphyrias may exhibit additional symptoms. In general, the porphyrias are classified into two groups, erythropoietic and hepatic. In the erythropoietic porphyrias, porphyrins accumulate in red blood cells, whereas in the hepatic porphyrias, porphyrins accumulate in the liver. This section presents the clinical findings and biochemical defects for selected porphyrias (Fig. 10-14).

Protoporphyria

Protoporphyria is characterized by

1. Increased protoporphyrin concentration in erythrocytes and plasma
2. Elimination of large quantities of protoporphyrin in the feces
3. An acute or chronic skin condition caused by sensitivity to sunlight

The skin condition varies in severity, and is characterized by itching, burning, stinging, and the formation of smooth, usually red patches. In some cases, the biliary excretion of excess protoporphyrin causes the formation of protoporphyrin gallstones and/or progressive liver deterioration. The biochemical defect in protoporphyria is believed to be a deficiency of ferrochelatase, the enzyme that converts protoporphyrin to heme.

Congenital Erythropoietic Porphyria

Congenital erythropoietic porphyria is characterized by an overproduction of uroporphyrinogen I and its metabolites, which accumulate in red blood cells. Symptoms associated with the disorder are chronic skin sensitivity to sunlight, excretion of excessive amounts of porphyrins in the urine, and, in many cases, hemolytic anemia. The biochemical defect is an increase in the ratio of uroporphyrinogen I synthase to uroporphyrinogen III cosynthase, which leads to an accumulation of uroporphyrinogen I in red blood cells.

Toxic Aquired Porphyrias

Certain toxic chemicals can cause porphyria symptoms similar to those observed with the genetic diseases. The chemically induced porphyrias are hepatic rather than erythropoietic. Allylisopropylacetamide, dicarbethoxydihydrocollidine, griseofulvin, and hexachlorobenzene have

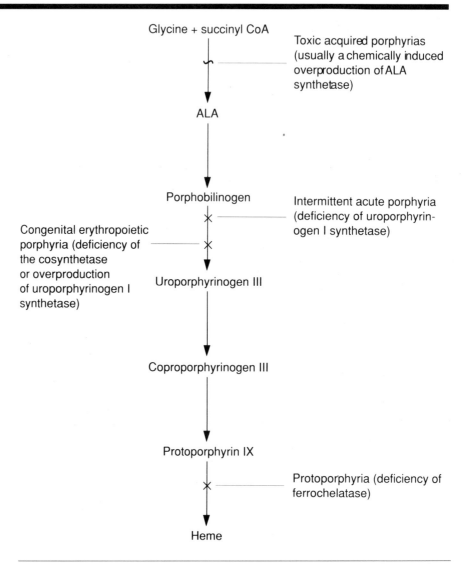

Fig. 10-14. Some interruptions in the heme synthesis pathway that result in known porphyrias.

been observed to cause toxic aquired porphyrias. These substances act by inducing the synthesis of δ-aminolevulinic acid synthase, thus leading to overproduction of porphyrins.

Intermittent Acute Porphyria

Intermittent acute porphyria is characterized by abdominal pain, hypertension, peripheral and central neuropathy, and psychosis. Symptoms occur periodically and vary greatly in severity. Attacks may be precipitated by a variety of stimuli, including certain drugs, hormones, and nutritional status. Individuals with this disorder also excrete large quantities of δ-aminolevulinic acid and porphobilinogen in the urine. This is the only porphyria that is not associated with sensitivity to sunlight. The enzyme defect in this case is a deficiency of uroporphyrinogen I synthetase, which results in the accumulation of δ-aminolevulinic acid and porphobilinogen in the liver.

SUGGESTED READING

1. With TK: Bile Pigments; Chemical, Biological and Clinical Aspects. Academic Press, Orlando, FL, 1968
2. Schmid R, McDonagh AF: The enzymatic formation of filirubin. Ann NY Acad Sci 244:533, 1975
3. Tabor CW, Tabor H: Polyamines. Annu Rev Biochem 53:749, 1984
4. Walker JB: Creatinine: biosynthesis, regulation and function. Adv Enzymol 50:177, 1979
5. Stanbury JB, Wyngaarden JB, Fredrickson DS, et al (eds): The Metabolic Basis of Inherited Disease. 5th Ed. McGraw-Hill, New York, 1983

STUDY QUESTIONS

Directions: Answer the following questions using the key outlined below:
- **(A)** if 1, 2 and 3 are correct
- **(B)** if 1 and 3 are correct
- **(C)** if 2 and 4 are correct
- **(D)** if only 4 is correct
- **(E)** if all four are correct

1. Which of the following substances is directly formed by the decarboxylation of an amino acid?
   1. Serotonin
   2. Histamine
   3. Dopamine
   4. γ-Aminobutyric acid

2. Although some unmetabolized catecholamine is found in the urine, the pathways of catecholamine degradation have steps that involve
   1. conjugation
   2. methylation
   3. oxidation
   4. decarboxylation

3. All of the following substances influence the activity or intracellular levels of δ-aminolevulinic acid synthetase EXCEPT
   1. heme
   2. succinyl CoA
   3. allylisopropylacetamide
   4. iron

4. Substances associated with heme degradation and/or elimination include
   1. carbon monoxide and protoporphyrin
   2. albumin and carbon monoxide
   3. carbon dioxide and albumin
   4. biliverdin and heme oxygenase

5. Heme synthesis intermediates synthesized in the cytosol include
   1. ALA
   2. heme
   3. protoporphyrin
   4. porphobilinogen

Directions: For each of the following multiple choice questions, choose the most appropriate answer.

6. Which of the following reaction types is involved in the synthesis of creatine?
   A. N-acylamino acid formation and transamidination
   B. Decarboxylation and methylation
   C. Transamination and decarboxylation
   D. Methylation and transamidination

7. All of the following are involved in polyamine metabolism EXCEPT
   A. ornithine
   B. monoamine oxidase
   C. S-adenosylmethionine
   D. diamine oxidase

8. All of the following sequences of intermediates occur in the metabolism (synthesis or degradation) of heme EXCEPT
   A. heme → biliverdin → bilirubin
   B. δ-aminolevulinic acid → porphobilinogen → → → protoporphyrin
   C. protoporphyrin → heme → biliverdin
   D. δ-aminolevulinic acid → protoporphyrin → porphobilinogen

11

NUCLEOTIDE METABOLISM

Structure and Function of Nucleotides/De Novo Synthesis/
Deoxyribonucleotides/Nucleotide Degradation/
Salvage Pathways/Coenzyme Synthesis/Clinical Correlation

Learning Objectives

The student should be able to:

1. List three functions of nucleotides in mammalian cells.
2. Distinguish purines from pyrimidines, nucleosides from nucleotides, and ribonucleotides from deoxyribonucleotides.
3. List the starting materials and the regulated steps in the de novo synthesis of purine and pyrimidine nucleotides.
4. Write reactions illustrating the salvage pathways of purine and pyrimidine nucleotide synthesis.
5. Describe the roles of AMP deaminase and GMP reductase in nucleotide interconversion.
6. Write the reaction catalyzed by nucleotide diphosphate reductase.
7. Write the reaction for the synthesis of TMP.
8. List the products of the degradation of purine and pyrimidine bases.
9. Identify the nucleoside moiety in the structures of NAD^+, FAD^+, and coenzyme A.

STRUCTURE AND FUNCTION OF NUCLEOTIDES

The purine and pyrimidine nucleotides are the precursors of DNA and RNA. Specific nucleotides are also important participants in a great variety of metabolic processes. For example:

1. ATP is the primary source of energy for energy-requiring cellular processes.
2. cAMP and cGMP are mediators of physiologic processes.
3. Certain nucleotides are precursors of coenzymes, glycoproteins, glycolipids, polysaccharides, and phospholipids.
4. Many enzymes are allosterically regulated by nucleotide effectors.

Figure 11-1 and Table 11-1 show the common purine and pyrimidine bases and their relation to nucleoside and nucleotide derivatives.

1. The *purine* and *pyrimidine* bases in cells may be considered as substituted derivatives of the nitrogenous purine and pyrimidine ring structures (see Fig. 11-2).

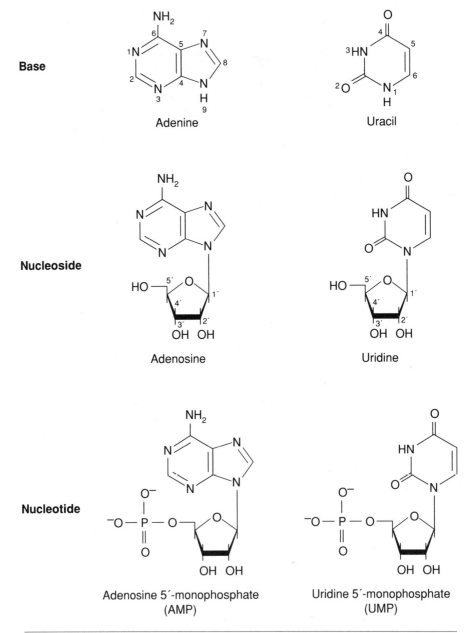

Fig. 11-1. Examples of a purine and a pyrimidine, and their nucleoside and nucleotide derivatives.

Table 11-1. The Purine and Pyrimidine Bases Occurring in Nucleic Acids and Their Nucleoside and Nucleotide Monophosphate Derivatives

Free Base		Nucleoside and Deoxynucleoside Derivatives	Nucleotide Derivatives
Purines			
Adenine	A	Adenosine	Adenosine 5'-monophosphate
		Deoxyadenosine	Deoxyadenosine 5'-monophosphate
Guanine	G	Guanosine	Guanosine 5'-monophosphate
		Deoxyguanosine	Deoxyguanosine 5'-monophosphate
Pyrimidines			
Uracil	U	Uridine	Uridine 5'-monophosphate
		Deoxyuridine	Deoxyuridine 5'-monophosphate
Cytosine	C	Cytidine	Cytidine 5'-monophosphate
		Deoxycytidine	Deoxycytidine 5'-monophosphate
Thymine	T	Deoxythymidine	Deoxythymidine 5'-monophosphate

2. *Nucleosides* consist of a purine or pyrimidine base linked to a sugar residue (ribose or deoxyribose). The linkage is a β-*N*-glycosidic bond to C-1 of the sugar. Nucleosides are thus *ribosyl* or *deoxyribosyl derivatives* of purine and pyrimidine bases.
3. *Nucleotides* are phosphate esters of nucleosides. The phosphate group is usually esterified to the 5' group of the sugar, yielding a *5'-phosphoribosyl derivative* of the purine or pyrimidine base. Nucleotides are called *mononucleotides*, *dinucleotides*, and *trinucleotides*, depending on whether a mono-, di-, or triphosphate chain is esterified to the sugar.

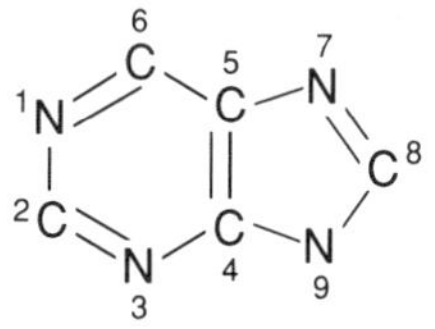

Purine ring

Atom	Source
N-1	Aspartate
N-3, N-9	Amide nitrogen of glutamine
C-2, C-8	Formate
C-6	CO_2
C-4, C-5, N-7	Glycine

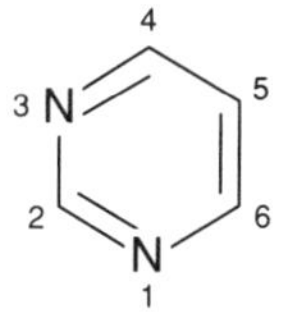

Pyrimidine ring

Atom	Source
N-1, C-4, C-6	Aspartate
N-3	Amide nitrogen of glutamine
C-2	CO_2

Fig. 11-2. The sources of the atoms in the purine and pyrimidine rings.

Notice in Figure 11-1 that the atoms of the base are numbered 1, 2, 3, etc., whereas the carbons of the sugar moiety are indicated by primed numerals (1', 2', 3', etc). The sugar moiety in nucleosides and nucleotides is either ribose or *deoxyribose* (in which case the hydroxyl on carbon 2' is replaced by a hydrogen). The ribonucleoside 5'-monophosphates may also be referred to as *adenylate (AMP), guanylate (GMP), uridylate (UMP),* and *cytidylate (CMP),* and the corresponding deoxyribonucleoside 5'-monophosphates as *deoxyadenylate, deoxyguanylate,* etc.

Perspective: Nucleotide Metabolism

Both purine and pyrimidine nucleotides may be synthesized either de novo or via salvage pathways that re-use purine and pyrimidine bases obtained from the diet or from degradation of endogenous nucleotides. As shown in Figure 11-2, the precursors for de novo synthesis of the purine ring are aspartate, glutamine, glycine, formate, and CO_2, whereas the precursors for de novo synthesis of the pyrimidine ring are aspartate, glutamine, and CO_2. In both cases, the sugar moiety for nucleoside synthesis is supplied by 5-phosphoribosyl pyrophosphate.

The following preview provides an additional perspective on nucleotide metabolism:

1. In the de novo synthesis of purine nucleotides, the phosphoribosyl moiety is introduced in the first reaction, and the purine ring is attached to the phosphoribosyl group.
2. In the de novo synthesis of pyrimidine nucleotides, the pyrimidine ring is completely synthesized before being linked to the phosphoribosyl.
3. The initial nucleotides formed in purine and pyrimidine synthesis are, respectively, *inosine monophosphate* (IMP) and *orotidine monophosphate* (OMP). Neither of these nucleotides is commonly used in cellular metabolism. Inosine monophosphate is converted to AMP and GMP. Orotidine monophosphate is converted to UMP, which can be converted to UTP or indirectly to TMP. UTP is the precursor of CTP.
4. In the degradation of purine nucleotides, the purine bases are degraded to xanthine, which is oxidized to uric acid and ultimately eliminated in the urine. Therefore, the purine ring structure is not disrupted during catabolism.
5. In the degradation of pyrimidines, uracil and thymine are the only free bases produced. Uracil is degraded to β-alanine, and thymine is degraded to β-aminoisobutyrate.

The de novo synthesis of nucleotides requires a significant quantity of energy, mostly for the synthesis of the purine and pyrimidine bases. It is therefore to the cell's advantage to synthesize nucleotides from preformed bases via the salvage pathways. Some cells (red blood cells) are incapable of de novo nucleotide synthesis and depend entirely on the salvage pathways.

Nucleotide Metabolism

SYNTHESIS OF 5-PHOSPHORIBOSYL PYROPHOSPHATE

In the de novo and salvage synthesis of all nucleotides, the sugar-phosphate moiety is supplied by *5-phosphoribosyl pyrophospate* (PRPP). PRPP is synthesized from ribose 5-phosphate by *5-phosphoribosyl pyrophosphate synthetase* (PRPP synthetase):

Ribose 5-phosphate → 5-Phosphoribosyl pyrophosphate (PRPP)
(ATP → AMP)

The ribose 5-phosphate for this reaction is derived from the pentose phosphate pathway or from the degradation of nucleotides. PRPP synthetase requires inorganic phosphate for activity, and is inhibited by ADP. Since the pentose phosphate shunt is an important source of ribose 5-phosphate, factors that stimulate that pathway increase the synthesis of PRPP.

DE NOVO SYNTHESIS OF PURINE NUCLEOTIDES

The purine ring structure is assembled on the phosphoribosyl moiety in 10 steps (Fig. 11-3). Except for the addition of glycine in step 2, the structure is assembled one ring atom at a time. The enzymes of de novo purine synthesis are located in the cytosol.

Reaction 1

In the first step of purine synthesis, 5-phosphoribosyl-1-amine is formed from PRPP and glutamine:

PRPP → 5-Phosphoribosyl-1-amine
(Gln → Glu, PP_i)

In this reaction, the amido group of glutamine displaces the pyrophosphate group from PRPP. In the process, the configuration on C-1 of ribose changes from α to β. The amino group is the starting point for

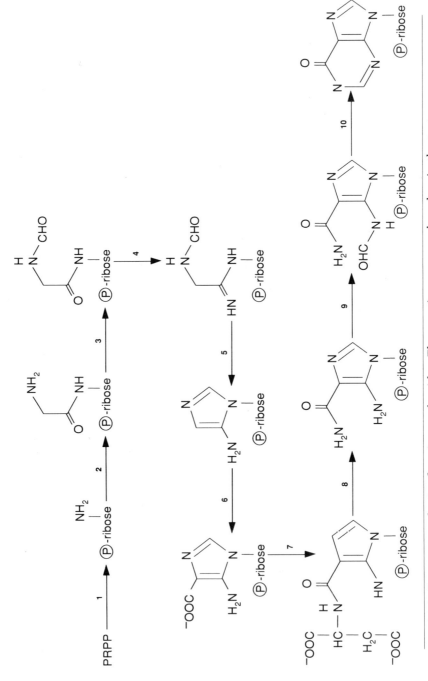

Fig. 11-3. De novo synthesis of purine nucleotides. The reactions are numbered as in the text.

the synthesis of the purine base, and the bond between the nitrogen and C-1 is the *N*-glycosidic bond that will join the base to the ribosyl moiety.

This reaction is the committed step in purine nucleotide synthesis. It is catalyzed by *PRPP amidotransferase*, which is a major regulated enzyme of purine synthesis.

Reaction 2

In the second reaction, glycine is added to the amino group to yield 5'-phosphoribosylglycinamide:

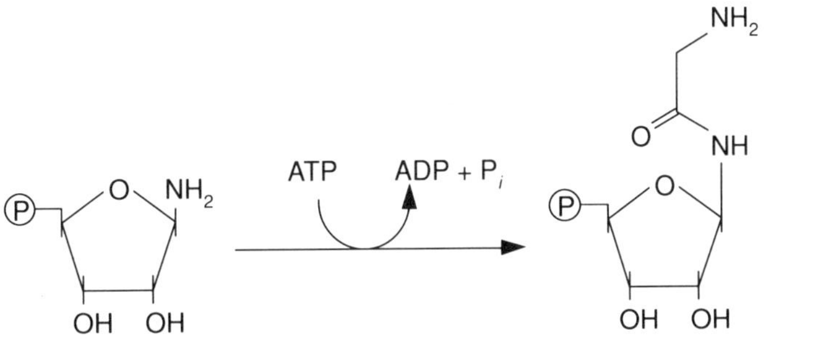

5-Phosphoribosylamine → (ATP → ADP + P_i) → 5'-Phosphoribosylglycinamide

Reaction 3

In the third reaction, the first of two formyl groups is added. Both of the formyl groups involved in purine synthesis are delivered to the growing structure by tetrahydrofolate (FH_4) (see Fig. 9-16).

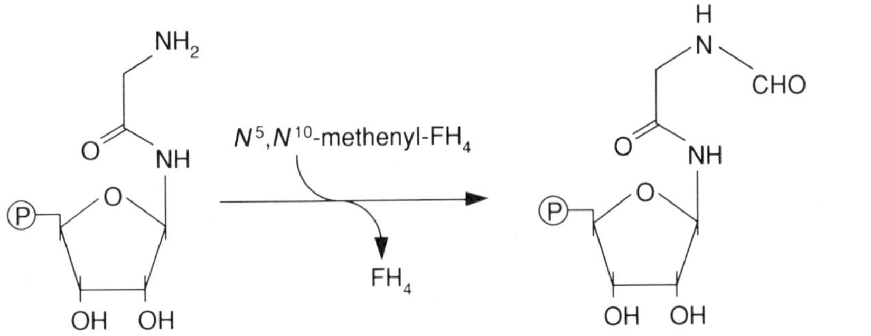

5'-Phosphoribosylglycinamide → (N^5,N^{10}-methenyl-FH_4 → FH_4) → 5'-Phosphoribosylformylglycinamide

Reaction 4

A nitrogen from glutamine is added next:

5′-Phosphoribosylformylglycinamide → 5′-Phosphoribosylformylglycinamidine (Gln → Glu; ATP → ADP + P_i)

Reaction 5

Step 5 is a ring closure reaction:

5′-Phosphoribosylformylglycinamidine → 5′-Phosphoribosyl-5-aminoimidazole (ATP → ADP + P_i)

Reaction 6

Carbon dioxide is added:

5′-Phosphoribosyl-5-aminoimidazole + CO_2 → 5′-Phosphoribosyl-5-aminoimidazole-4-carboxylate

Reaction 7

The aspartate α-amino group is then added:

5′-Phosphoribosyl-5-aminoimidazole-4-N-succinocarboxamide → (Fumarate) → 5′-Phosphoribosyl-5-aminoimidazole-4-carboxamide

Reaction 8

In the next step, the aspartate carbon chain is removed in the form of fumarate (i.e., the succinate moiety is cleaved from the succinocarboxamide):

5′-Phosphoribosyl-5-aminoimidazole-4-N-succinocarboxamide → (Fumarate) → 5′-Phosphoribosyl-5-aminoimidazole-4-carboxamide

Reaction 9

The second formyl group is then added by N^{10}-formyl-FH$_4$:

[Reaction scheme: 5'-Phosphoribosyl-5-aminoimidazole-4-carboxamide + N^{10}-formyl-FH$_4$ → 5'-Phosphoribosyl-5-formamidoimidazole-4-carboxamide + FH$_4$]

Reaction 10

The second ring closure reaction yields IMP:

[Reaction scheme: 5'-Phosphoribosyl-5-formamidoimidazole-4-carboxamide → Inosine 5'-monophosphate (IMP) + H$_2$O]

Formation of AMP and GMP

AMP and GMP are synthesized from IMP by the pathways shown in Figure 11-4. GMP is generated from IMP by the addition of an amino group on C-2, whereas AMP is generated from IMP by replacing the keto on C-6 with an amino.

Regulation of Purine Nucleotide Synthesis

Only three of the enzymes of de novo purine synthesis are regulated (Fig. 11-5). Each is subject to feedback inhibition.

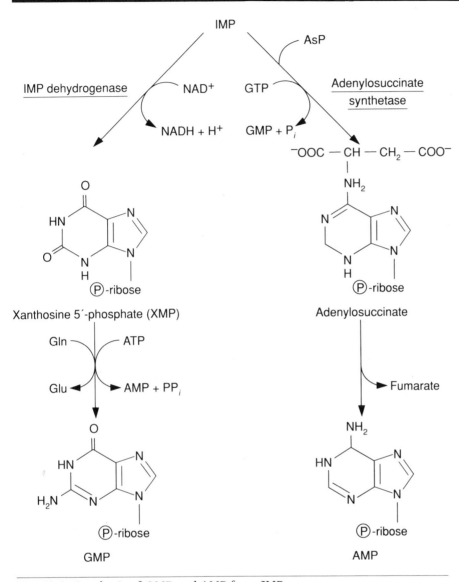

Fig. 11-4. Synthesis of GMP and AMP from IMP.

1. *PRPP amidotransferase* is inhibited by IMP, GMP, and AMP.
2. *IMP dehydrogenase* is inhibited by GMP.
3. *Adenylosuccinate synthetase* is inhibited by AMP.

Figure 11-5 shows the reactions catalyzed by these enzymes. PRPP amidotransferase controls the amount of IMP synthesized. IMP dehydrogenase and adenylosuccinate synthetase control the amounts of IMP converted to GMP and AMP, respectively.

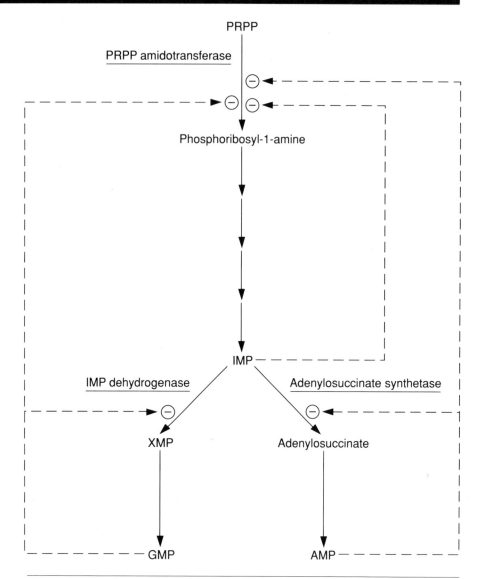

Fig. 11-5. Regulation of purine synthesis. Broken lines indicate negative feedback inhibition of the indicated enzymes.

Interconversion of Purine Nucleotides

The cellular levels of AMP and GMP are kept in balance by enzymes that catalyze their interconversion (Fig. 11-6).

1. The first step in both the conversion of GMP to AMP and the conversion of AMP to GMP is the formation of IMP:
   a. AMP is converted to IMP by *AMP deaminase*.
   b. GMP is converted to IMP by *GMP reductase*.
2. IMP is then converted by the reactions of purine synthesis shown in Figure 11-4 to whichever purine nucleotide is needed.

Nucleotide Metabolism

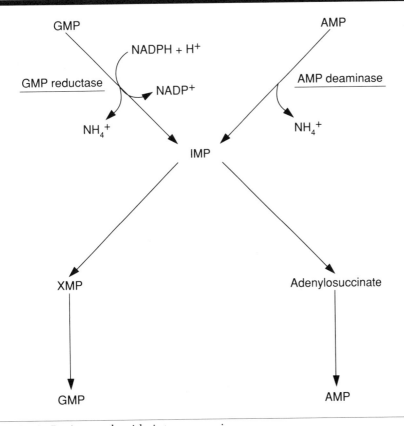

Fig. 11-6. Purine nucleotide interconversion.

AMP deaminase is inhibited by GDP, GTP and P_i, and activated by ATP. GMP reductase is inhibited by xanthine monophosphate (XMP) and activated by GTP.

DE NOVO SYNTHESIS OF PYRIMIDINE NUCLEOTIDES

The first phase in the de novo synthesis of pyrimidine nucleotides is the synthesis of UMP, which occurs in six steps. The pyrimidine ring is constructed and then attached to the ribosyl moiety.

Reaction 1

The first step is the formation of carbamoyl phosphate:

$$\text{Glutamine} + CO_2 + 2\,ATP \xrightarrow[2\,ADP + P_i]{Glu} \underset{\text{Carbamoyl phosphate}}{H_2N-\overset{\overset{O}{\|}}{C}-OPO_3^{2-}}$$

This reaction is catalyzed by the cytosolic enzyme *carbamoyl phosphate synthetase II*, which is the regulated enzyme of pyrimidine nucleotide synthesis. Carbamoyl phosphate synthetase II should not be confused with the mitochondrial enzyme carbamoyl phosphate synthetase I, which participates in the urea cycle.

Reaction 2

In reaction 2, the α-amino group of aspartate displaces the phosphate from carbamoyl phosphate to yield *N*-carbamoylaspartate:

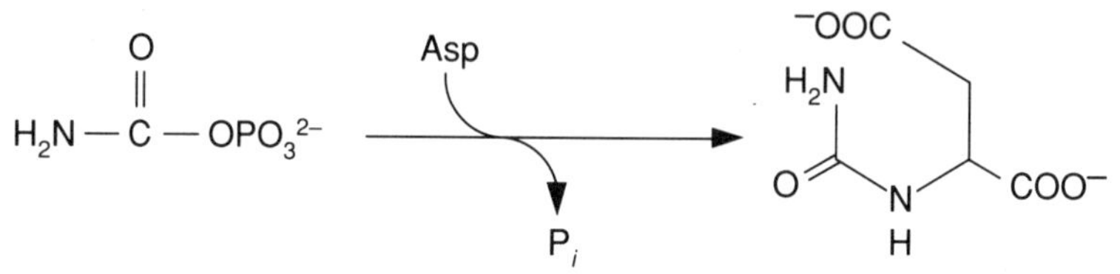

$$H_2N-\overset{O}{\overset{\|}{C}}-OPO_3^{2-} \xrightarrow[P_i]{Asp} \text{N-Carbamoylaspartate}$$

Carbamoyl phosphate → *N*-Carbamoylaspartate

This step is catalyzed by *aspartate carbamoyltransferase*.

Reaction 3

Next, *N*-carbamoylaspartate undergoes ring closure to generate *dihydroorotate*:

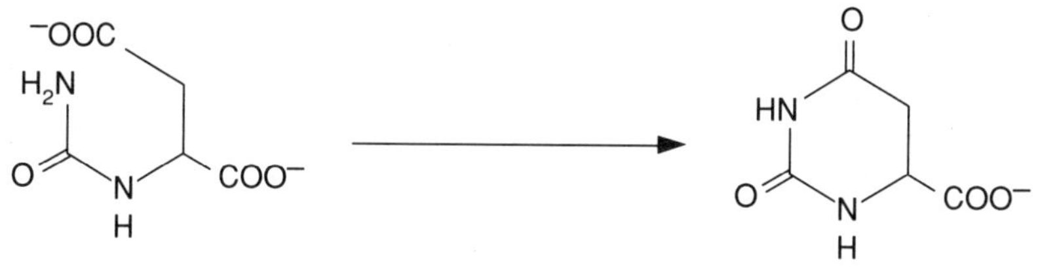

N-Carbamoylaspartate → Dihydroorotate

Reaction 4

Dihydroorotate is oxidized to yield *orotate*:

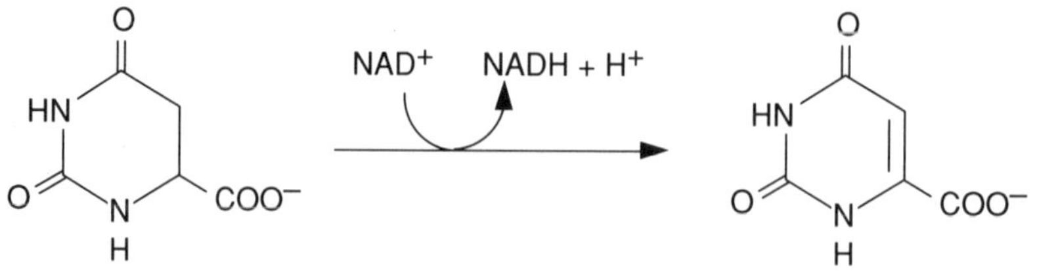

Dihydroorotate $\xrightarrow{NAD^+ \rightarrow NADH + H^+}$ Orotate

Reaction 5

Orotate reacts with PRPP to produce orotidine 5'-monophosphate, the initial nucleotide in de novo pyrimidine synthesis:

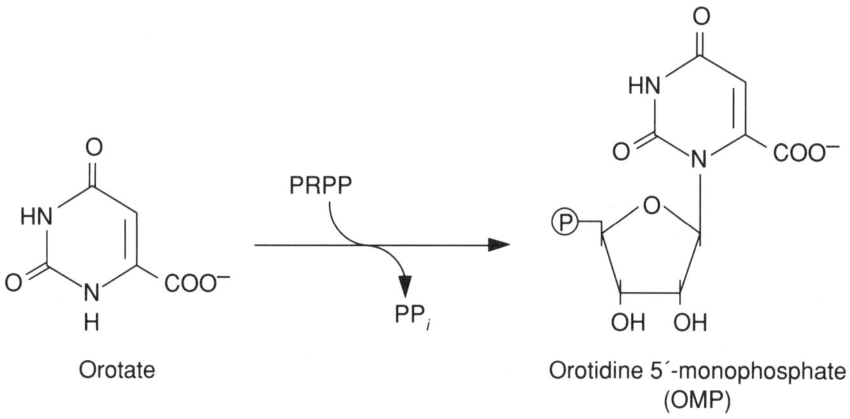

Reaction 6

Finally, the orotate moiety is decarboxylated to uracil, yielding UMP:

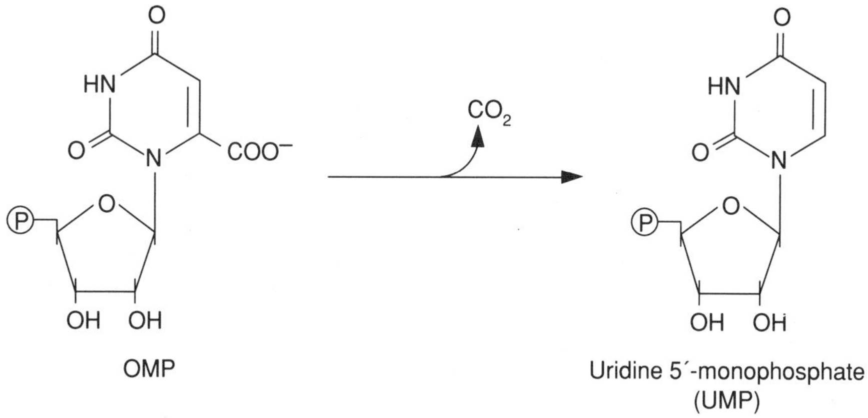

Localization of de Novo Pyrimidine Synthesis

It should be noted that the six enzyme activities required for the de novo synthesis of UMP are located on only three polypeptide chains: the activities that catalyze reactions 1 and 2 are on one chain; the activities that catalyze reactions 3, 5, and 6 on another; and the activity that catalyzes reaction 4 on the third. The first two polypeptide chains are located in the cytosol, whereas the protein that catalyzes reaction 4 is located in the mitochondria. Consequently, intermediates must cross the inner mitochondrial membrane during pyrimidine synthesis.

Synthesis of Cytidine Nucleotides

The cytidine nucleotides are formed from uridine nucleotides. UMP is converted by *UMP kinase* to UDP, which can be converted to UTP by *nucleotide diphosphate kinase*. UTP is then converted to CTP by *CTP synthetase*:

UTP + Gln + ATP → CTP + Glu + ADP + P_i

The synthesis of thymidine nucleotides from uridine nucleotides is discussed later.

Regulation of de Novo Pyrimidine Synthesis

Figure 11-7 shows the regulation of de novo pyrimidine synthesis. Carbamoyl phosphate synthetase II is inhibited by UTP, OMP decarboxylase is inhibited by UMP and CMP, and CTP synthetase is inhibited by CTP.

Perspective: Deoxyribonucleotides

> Deoxyribonucleotides differ from ribonucleotides only in having a hydrogen in place of the hydroxyl on the 2′ carbon. Ribonucleotides are used in the synthesis of RNA, whereas deoxyribonucleotides are used in the synthesis of DNA. Deoxyribonucleotides are synthesized by the reduction of the 2′ group of ribonucleotides from a hydroxymethyl to methylene.

SYNTHESIS OF DEOXYRIBONUCLEOTIDES

The reduction of ribonucleotides to deoxyribonucleotides is catalyzed by *nucleotide diphosphate reductase* (Fig. 11-8). The essential features of this reaction are as follows:

1. Nucleotide diphosphate reductase recognizes as substrates only the ribonucleoside diphosphates ADP, GDP, UDP, and CDP. The products of the reaction are the deoxyribonucleoside diphosphates dADP, dGDP, dUDP, and dCDP.

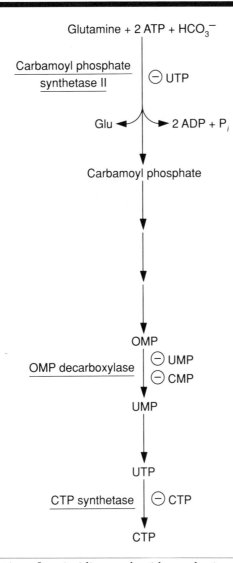

Fig. 11-7. Regulation of pyrimidine nucleotide synthesis.

2. The reduction of each of the four substrates is subject to allosteric regulation by other nucleoside triphosphate(s).
3. The reduction of the substrate nucleotide is accompanied by the oxidation of a closely associated sulfhydryl-containing protein, *thioredoxin*. Thioredoxin is subsequently re-reduced by NADPH and the enzyme thioredoxin reductase.
4. The reduction of thioredoxin renders the nucleotide diphosphate reductase able to catalyze the reduction of another nucleotide diphosphate.

dADP, dGDP, and dCDP (as well as thymidine monophosphate, dTMP, described in the next section) are converted to the corresponding triphosphates by a series of cellular kinases.

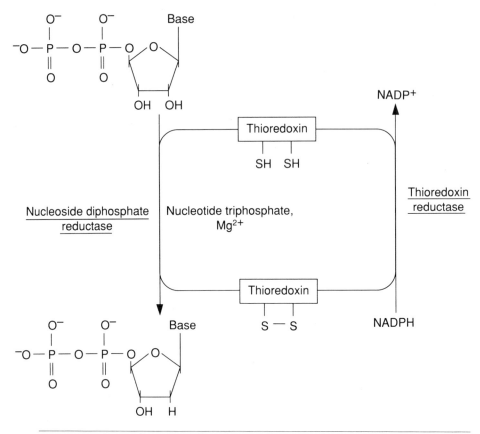

Fig. 11-8. The nucleoside diphosphate reductase reaction.

Regulation of Nucleotide Diphosphate Reductase

Nucleotide diphosphate reductase is solely responsible for supplying the deoxyribonucleotides required for DNA synthesis, and consequently is finely regulated. Table 11-2 gives the positive and negative effectors associated with the reduction of each of the substrates of this enzyme.

SYNTHESIS OF DEOXYTHYMIDINE MONOPHOSPHATE

The nucleotides used in DNA synthesis are dATP, dGTP, dCTP, and dTTP (deoxythymidine triphosphate). The first three are obtained directly from the products of the nucleotide diphosphate reductase reaction, as

Table 11-2. Regulation of Nucleoside Diphosphate Reductase

Substrate	Positive Effector	Negative Effector
CDP	ATP	dATP, dTTP
UDP	ATP	dATP, dTTP
ADP	dGTP	dATP
GDP	dTTP	dATP

described above. dTTP, however, is formed by a more complex pathway from dUDP. First, dUDP is converted to dUMP. The uracil moiety is then converted to thymidine by methylation at C-5, yielding dTMP (deoxythymidine monophosphate) (Fig. 11-9). The methyl donor is N^5,N^{10}-methylene-FH$_4$, which is reduced to dihydrofolate (FH$_2$) during the reaction. dTMP is then converted to dTTP. (It should be noted that the "d" prefix is often omitted from the abbreviations for the thymidine ribonucleotides: TTP and TMP are the same as dTTP and dTMP.)

Some dTMP is also obtained from dCMP, which can be deaminated to dUMP by dCMP deaminase:

$$dCMP \rightarrow dUMP + NH_4^+$$

dUTP can also be hydrolyzed to dUMP and pyrophosphate by dUTPase:

$$dUTP \rightarrow dUMP + PP_i$$

This reaction not only contributes to the supply of dUMP but is believed to prevent the accumulation of dUTP in cells, thus preventing dUTP from being incorporated into DNA.

Interconversion of Pyrimidine Nucleotides

The cell contains an array of enzymes that interconvert the pyrimidine nucleotides. These reactions are important for maintaining the relative levels of the nucleotides (especially dTTP)—if one pyrimidine becomes depleted, it can be replenished from another that may be present in excess.

Under most circumstances, the goal of pyrimidine interconversion is to ensure an adequate supply of dTTP for DNA synthesis. The formation of dTTP depends on the supply of dUMP, which can be formed from either UMP or CMP. Therefore, pyrimidine interconversion consists primarily of

1. Conversion of UMP to dUMP
2. Conversion of CMP to dUMP
3. Conversion of dUMP to dTTP

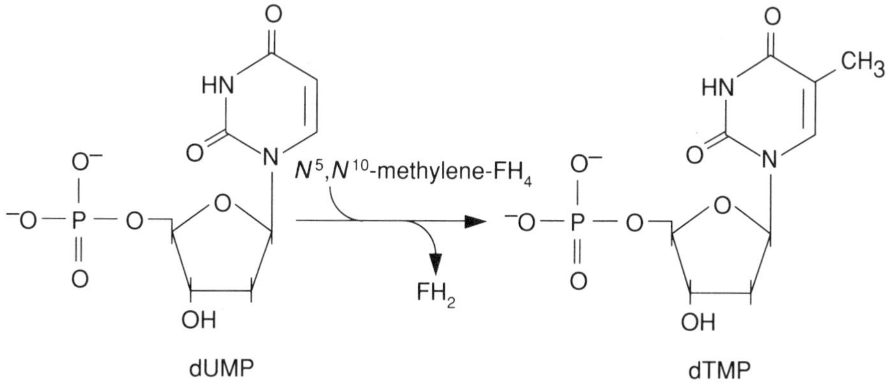

Fig. 11-9. Synthesis of deoxythymidine-5'-monophosphate (dTMP).

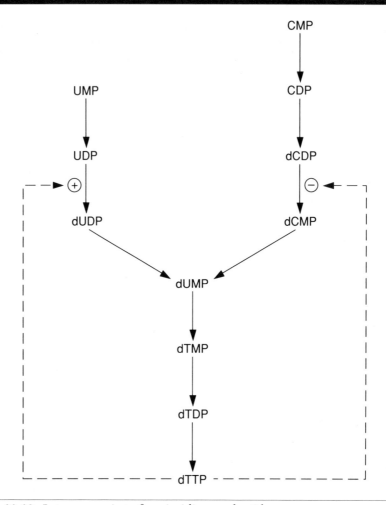

Fig. 11-10. Interconversion of pyrimidine nucleotides.

When the supply of dTTP is adequate, dTTP exerts feedback inhibition over the conversions of UMP and CMP to dUMP (Fig. 11-10).

NUCLEOTIDE DEGRADATION

The degradative process for nucleotides begins with the monophosphate form. The sequence of nucleotide degradation is as follows:

Nucleotides → nucleosides → free bases → final excretion products

Degradation of Purine Nucleotides

AMP and GMP (the predominant purine nucleoside monophosphates) are ultimately converted to uric acid, which is excreted in the urine. The pathways of purine degradation are shown in Figure 11-11.

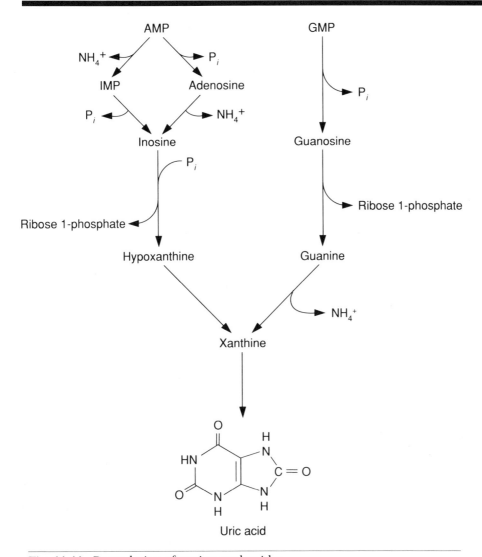

Fig. 11-11. Degradation of purine nucleotides.

AMP Degradation

1. AMP is converted to the nucleoside inosine in one of two ways:
   a. Conversion of AMP to IMP by AMP deaminase, followed by dephosphorylation to yield inosine
   b. Direct dephosphorylation of AMP to yield adenosine, which is converted to inosine by adenosine deaminase
2. Inosine is then converted to the free base hypoxanthine (with loss of the sugar phosphate moiety) This reaction is catalyzed by purine nucleotide phosphorylase.
3. Xanthine oxidase catalyzes the oxidation of hypoxanthine to xanthine and the oxidation of xanthine to uric acid.

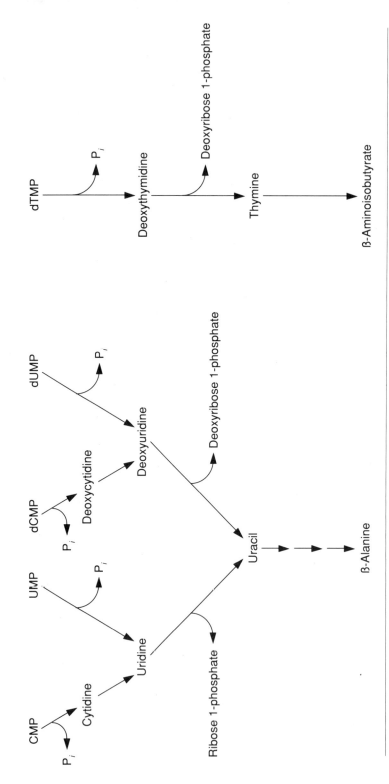

Fig. 11-12. Degradation of pyrimidine nucleotides.

GMP Degradation

1. GMP is converted to guanosine by a nucleotidase.
2. The sugar phosphate is cleaved off by a phosphorylase, yielding the free base guanine.
3. Guanine is deaminated to xanthine, which is metabolized to uric acid as described above for AMP.

Degradation of Pyrimidine Nucleotides

As in the case of purines, pyrimidine nucleotides are degraded first to free bases, which are then converted to the urinary excretion products. Figure 11-12 shows the pyrimidine degradation pathways.

1. CMP, UMP, dCMP, and dUMP all yield the free base uracil. Uracil is degraded to β-alanine, which is excreted in the urine.
2. dTMP yields the free base thymine, which is degraded to the urinary excretory product β-aminoisobutyrate.

SALVAGE PATHWAYS OF PURINE AND PYRIMIDINE SYNTHESIS

A portion of the free bases generated during nucleotide degradation are used to resynthesize nucleotides. As discussed earlier, the use of these salvage pathways saves the cell a considerable amount of energy. In the salvage reactions, free purine and pyrimidine bases react with PRPP to form the monophosphate derivative:

$$\text{Free base} + \text{PRPP} \rightarrow \text{nucleoside 5'-monophosphate} + PP_i$$

Purine Salvage

Two enzymes catalyze the salvage of the purines. Guanine and hypoxanthine are salvaged by the enzyme *hypoxanthine-guanine phosphoribosyltransferase* (HGPRTase):

$$\text{Guanine} + \text{PRPP} \rightarrow \text{GMP} + PP_i$$

$$\text{Hypoxanthine} + \text{PRPP} \rightarrow \text{IMP} + PP_i$$

The salvage of adenine is catalyzed by *adenine phosphoribosyltransferase* (APRTase):

$$\text{Adenine} + \text{PRPP} \rightarrow \text{AMP} + PP_i$$

Both enzymes are inhibited by their products: HGPRTase by GMP and IMP, and APRTase by AMP.

Pyrimidine Salvage

The free pyrimidine bases orotate, uracil, and thymine are converted to monophosphates by the enzyme *pyrimidine phosphoribosyltransferase*:

$$\text{Orotate} + \text{PRPP} \rightarrow \text{OMP} + \text{PP}_i$$
$$\text{Uracil} + \text{PRPP} \rightarrow \text{UMP} + \text{PP}_i$$
$$\text{Thymine} + \text{PRPP} \rightarrow \text{TMP} + \text{PP}_i$$

This enzyme does not catalyze a reaction involving cytosine.

Nucleoside Phosphorylases

The cell contains nucleoside phosphorylases that can catalyze reactions of the type

$$\text{Base} + \text{ribose 1-phosphate} \rightleftharpoons \text{nucleoside} + P_i$$

The resulting nucleoside could be phosphorylated by the appropriate kinase to yield a nucleotide. However, under normal intracellular conditions, nucleoside phosphorylases function primarily to degrade nu-

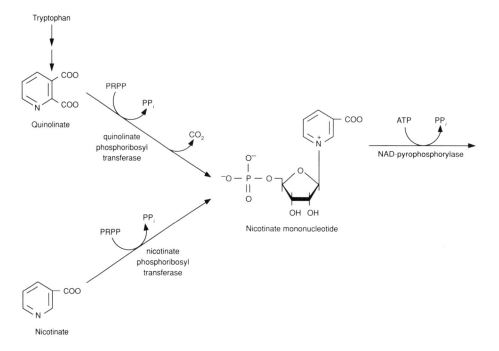

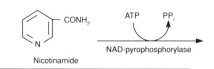

Fig. 11-13. Synthesis of NAD$^+$.

cleotides and do not contribute significantly to purine and pyrimidine salvage.

Nucleotide Kinases

Nucleotides primarily function at the triphosphate level. The cell contains an array of *kinases* that collectively perform the conversions

Nucleoside → monophosphate → diphosphate → triphosphate

Some of these kinases have strict specificities and some do not.

REGULATION OF NUCLEOTIDE METABOLISM: OVERVIEW

Not all the regulatory mechanisms of nucleotide metabolism are clearly understood. However, the levels of some of the enzymes involved in nucleotide synthesis have been observed to increase during the S phase of the cell cycle, when nucleic acid synthesis is rapid and nucleotide degradation is probably at a minimum. Nucleotide synthesis and degradation most likely show coordinate regulation, operating maximally under opposed cellular conditions.

SYNTHESIS OF NUCLEOTIDE COENZYMES

The nucleotide coenzymes NAD^+, $NADP^+$, FAD, and coenzyme A all share two features:

1. The structure of the coenzyme includes an adenine nucleotide.
2. Synthesis of the coenzyme depends on a component from the diet.

The syntheses of these coenzymes are presented to illustrate the contribution of the adenosine nucleotide to their structures.

NAD^+ and $NADP^+$

NAD^+ consists of a nicotinamide ribonucleotide connected via a diphosphate to adenosine. NAD^+ can be synthesized starting from either nicotinate, nicotinamide, or the tryptophan degradation product quinolinate (Fig. 11-13). Quinolinate and nicotinate both react with PRPP to yield nicotinate mononucleotide, whereas nicotinamide reacts with PRPP to yield nicotinamide mononucleotide. Nicotinamide mononucleotide is converted directly to NAD^+ by accepting an AMP from ATP. Nicotinate mononucleotide also accepts an AMP from ATP, yielding nicotinate adenine dinucleotide, which then transaminates with glutamine to produce NAD^+. NAD^+ is converted to $NADP^+$ by phosphorylation at the 2′ carbon of the adenosine moiety, a reaction catalyzed by NAD^+ kinase.

The amount of NAD^+ synthesized from quinolinate depends on the amount of tryptophan that is obtained from the diet and not used for other cellular purposes (recall that tryptophan is a nutritionally essential amino acid). NAD^+ synthesis is inhibited by NAD^+, $NADP^+$, and nicotinamide mononucleotide.

FAD

Figure 11-14 shows the synthesis of FAD. The essential dietary precursor of FAD is riboflavin (vitamin B_2). ATP donates an AMP unit to riboflavin phosphate to generate FAD.

Coenzyme A

The synthesis of coenzyme A requires the vitamin pantothenate (Fig. 11-15). Panthothenate is first phosphorylated to 4-phosphopantothenate. In the next two reactions, cysteine is added and decarboxylated to yield 4-phosphopantotheine. ATP donates an AMP unit to 4-phosphopantotheine to yield dephosphocoenzyme A, which is phosphorylated at the 3′ position of adenosine to produce coenzyme A.

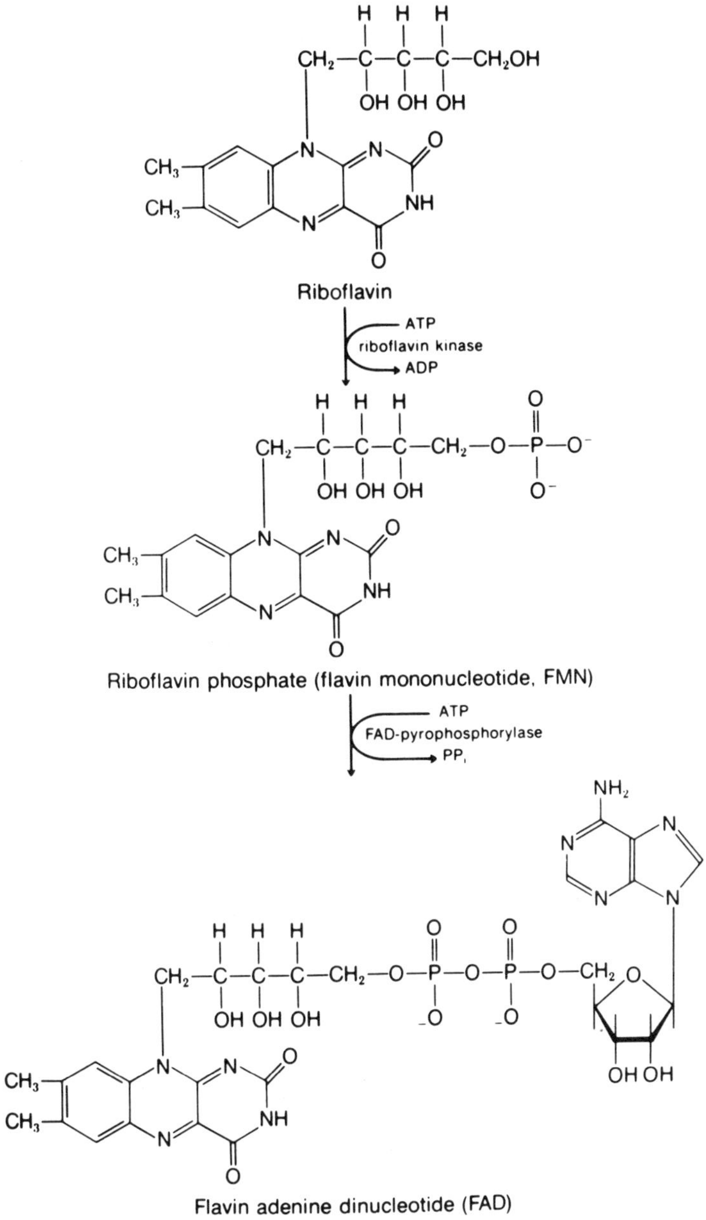

Fig. 11-14. Synthesis of FAD. (From Devlin TM: Textbook of Biochemistry with Clinical Correlations. Wiley, New York, 1986, with permission.)

NUCLEOTIDE METABOLISM AND CANCER CHEMOTHERAPY

Because of their unregulated growth, many tumor cells have a greater demand for nucleotides than normal cells, and should be susceptible to agents that block nucleotide synthesis. A number of anticancer agents

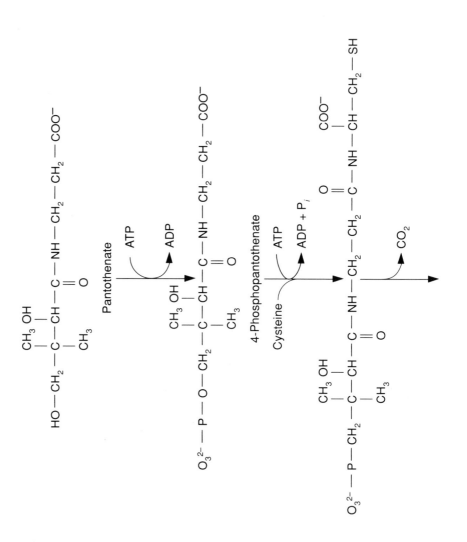

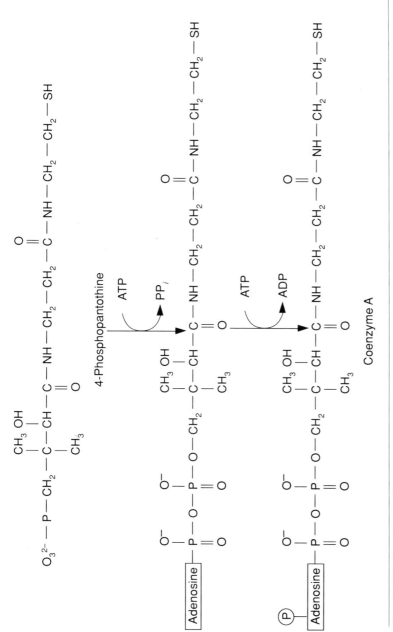

Fig. 11-15. Synthesis of coenzyme A.

work on this principle; the action of methotrexate was discussed in the Clinical Correlation to Chapter 2.

CLINICAL CORRELATION: GOUT

The end product of purine nucleotide metabolism is uric acid. Uric acid is only slightly soluble in aqueous media, and is normally present in body fluids at levels near saturation. If de novo purine synthesis is enhanced or if purine salvage is interrupted, the production of uric acid increases, and the concentration of uric acid in the body fluids and urine rises (hyperuricemia). Uric acid crystals may then precipitate in certain joints of the body, causing gout. The joints most often affected are those of the big toe, heel, ankle, and knee. The deposition of uric acid crystals causes acute episodes of gouty arthritis characterized by severe joint pain along with chills, shivering, and fever. These episodes may last from a few days to a few weeks, and increase in frequency with age. Gout is characterized in most patients by the combination of hyperuricemia and acute arthritic episodes.

Three enyzme abnormalities have been identified as causing hyperuricemia: glucose 6-phosphatase deficiency, HGPRTase deficiency, and the presence of genetic variants of PRPP synthetase that have increased activity. Each of these abnormalities increases the de novo synthesis of purine nucleotides. Glucose 6-phosphatase deficiency causes a rise in glucose 6-phosphate in cells. Excess glucose 6-phosphate is believed to be shunted into the pentose phosphate pathway, causing increased production of ribose 5-phosphate, and thus, ultimately, increased nucleotide synthesis. PRPP synthetase produces PRPP for nucleotide synthesis. An increase in the activity of this enzyme enhances nucleotide synthesis. HPGRTase is an enzyme of the purine salvage pathway. A deficiency of this enzyme causes an increase both in the de novo synthesis of purines and in the degradation of purines to uric acid. Complete HGPRTase deficiency causes the very severe neurologic abnormalities known as Lesch-Nyhan syndrome.

Treatment

Gout treatment is aimed at alleviating the arthritic episodes and controlling serum uric acid. The arthritic episodes represent an inflammatory response to the deposition of uric acid crystals in the joints, and are effectively managed with nonsteroidal antiinflammatory drugs. Serum uric acid may be controlled by administering *uricosuric* drugs, which increase the excretion of uric acid. Probenecid is a uricosuric agent widely used in the treatment of gout. Alternatively, serum uric acid may be reduced by administering *allopurinol*, a drug that inhibits uric acid production. Allopurinol is a competitive inhibitor of xanthine oxidase, the enzyme that catalyzes the conversion of xanthine to uric acid. In some cases gout is treated by the combined administration of allopurinol and a uricosuric. Gouty patients who receive appropriate treatment can expect to lead normal, symptom-free lives.

SUGGESTED READING

1. Henderson JF, Paterson ARP: Nucleotide Metabolism. Academic Press, Orlando, FL, 1973
2. Hoffee PA, Jones ME: Purine and pyrimidine metabolism. Methods Enzymol, Vol 51, 1978
3. Wyngaarden JB: Regulation of purine biosynthesis and turnover. Adv Enzyme Regul 14:25, 1976
4. Jones ME: Pyrimidine nucleotide biosynthesis in animals: Genes, enzymes, and regulation of UMP biosynthesis. Annu Rev Biochem 49:253, 1980
5. Stanbury JB, Wyngaarden JB, Fredrickson DS, et al: The Metabolic Basis of Inherited Disease. 5th Ed. McGraw-Hill, New York, 1983
6. Bondy PK, Rosenberg LE: Metabolic Control and Disease. 8th Ed. WB Saunders, Philadelphia, 1980

STUDY QUESTIONS

Directions: For each of the following multiple choice questions, choose the most appropriate answer.

1. Which of the following groups consists of a pyrimidine base, a purine nucleoside, and a pyrimidine nucleotide?
   A. Uracil, adenosine, GMP
   B. Thymine, adenosine, dTMP
   C. Adenine, cytidine, AMP
   D. Thymidine, guanosine, CMP

2. The synthesis of both purine and pyrimidine nucleotides requires the involvement of PRPP and
   A. glutamine, aspartate, and CO_2
   B. formate, glutamine, and aspartate
   C. glycine, aspartate, and CO_2
   D. glutamine, glycine, and CO_2

3. The initial product of purine nucleotide synthesis is
   A. OMP
   B. AMP
   C. XMP
   D. IMP

4. PRPP amidotransferase is inhibited by
   A. IMP, XMP, GMP
   B. IMP, GMP, AMP
   C. GMP, XMP, AMP
   D. AMP, IMP, XMP

5. Which compound is the immediate precursor to dTMP?
   A. dUMP
   B. UMP
   C. OMP
   D. dUDP

6. Which one of the following reactions is catalyzed in mammalian cells by a phosphoribosyltransferase?
   A. Adenine + GMP → guanine + AMP
   B. Adenine + ribose 1-phosphate → adenosine + P_i
   C. Adenine + PRPP → AMP + PP_i
   D. Adenine + PRPP → ADP + P_i

Directions: Answer the following questions using the key outlined below:
(A) if 1, 2, and 3 are correct
(B) if 1 and 3 are correct
(C) if 2 and 4 are correct
(D) if only 4 is correct
(E) if all four are correct

7. The degradation of dTMP ultimately results in the formation of
   1. β-aminobutyrate
   2. β-alanine
   3. β-isoalanine
   4. β-aminoisobutyrate

8. Nucleotide diphosphate reductase catalyzes the reduction of the substrate(s)
   1. CDP
   2. UDP
   3. GDP
   4. ADP

9. The adenosine structure is present in
   1. NAD and NADP
   2. FAD^+
   3. Coenzyme A
   4. FMN

10. The biochemical abnormalities associated with gout result in
    1. increased salvaging of pyrimidine nucleotides
    2. increased salvaging of purine nucleotides
    3. increased de novo synthesis of pyrimidine nucleotides
    4. increased de novo synthesis of purine nucleotides

12

STRUCTURE OF NUCLEIC ACIDS

Structure of DNA/Structure of RNA

The student should be able to:

1. Diagram the phosphodiester bond of nucleic acids, list the nucleotide components of DNA and RNA, and define and illustrate base complementarity.
2. Identify the distal and proximal nuclease cleavage sites.
3. Distinguish the three types of right-handed DNA helix and describe key features of the left-handed helix.
4. Describe the mechanism of heat denaturation of DNA and define DNA hyperchromism.
5. Describe the process of DNA renaturation, and define and calculate $C_0t_{1/2}$.
6. Define DNA hybridization.
7. Describe DNA superhelices, nucleosomes, and the role of topoisomerases.
8. Define and distinguish palindromes, moderately repetitive DNA, highly repetitive DNA, and single-copy DNA.
9. Describe the basic structural features of the three kinds of RNA involved in protein synthesis, and distinguish the secondary and tertiary structures of tRNA.

Learning Objectives

THE NUCLEIC ACIDS

Nucleic acids are linear polynucleotides in which the nucleotides are joined by phosphodiester bonds between adjacent sugar moieties. *Deoxyribonucleic acid (DNA)* is a deoxyribonucleotide polymer, whereas *ribonucleic acid (RNA)* is a ribonucleotide polymer. The nucleotide sequence of DNA contains the genetic information of the cell, which is expressed ultimately in the synthesis of proteins. The first step in the expression of genetic information is the DNA-directed synthesis of RNA. There are three major classes of RNA: *messenger RNA (mRNA)*, which carries the genetic information that directs the synthesis of proteins; *transfer RNA (tRNA)*, which translates the genetic code into the correct amino acid sequence; and the *ribosomal RNAs* (*rRNAs*), which are integral components of the ribosomes. Eukaryotic cells also contain a class of very small RNA molecules that are not involved in protein synthesis and which are found in the nucleus as protein-RNA conjugates. This chapter explores the structural features of DNA and the three major RNA classes.

Perspective: DNA Structure

1. DNA contains the purines adenine and guanine and the pyrimidines cytosine and thymine. The number of purine nucleotides present exactly equals the number of pyrimidine nucleotides.
2. The nucleotides are joined by phosphodiester bonds between the 3' hydroxyl of one nucleotide and the 5' phosphate of the next (Fig. 12-1).
3. Mammalian DNA usually exists as a double helix, in which two DNA polymers are wound around a central helical axis. The strands are held together by hydrogen bonding between complementary nucleotides.
4. By convention, the nucleotides in a nucleic acid are read from the end with a free 5' phosphate to the end with a free 3' hydroxyl (i.e., in the 5' → 3' direction).

The deoxyribonucleotides in DNA may be abbreviated dA, dG, dC, and dT, or, more simply, A, G, C, T, with an initial d (usually omitted) to indicate that the strand is a deoxyribonucleic acid. Thus, the sequence in Figure 12-1 may be written dAdTdG, d-ATG, or simply ATG. The phosphates may also be indicated by writing the sequence as ApTpGp.

DNA STRUCTURE

Nucleases

Nucleases—enzymes that hydrolyze nucleic acid phosphodiester bonds—have been instrumental in elucidating the structure of DNA. Nucleases exhibit a variety of specificities. For example, some can hydrolyze both DNA and RNA, whereas others hydrolyze only DNA or RNA; some hydrolyze single- and double-stranded nucleic acids, whereas others hydrolyze only one or the other. All nucleases are either *exonucleases* or *endonucleases*. Exonucleases hydrolyze either 3' or 5' terminal phosphodiester bonds and release single nucleotides, whereas endonucleases cleave phosphodiester bonds within the chain. Endonucleases usually recognize and cleave a bond associated with a specific base sequence. A phosphodiester bond can be cleaved at two sites (Fig. 12-2)—either distal to (d) or proximal to (p) the phosphorus. Cleavage at the d site creates a free 5' phosphate group, whereas cleavage at the p site creates a free 3' phosphate group.

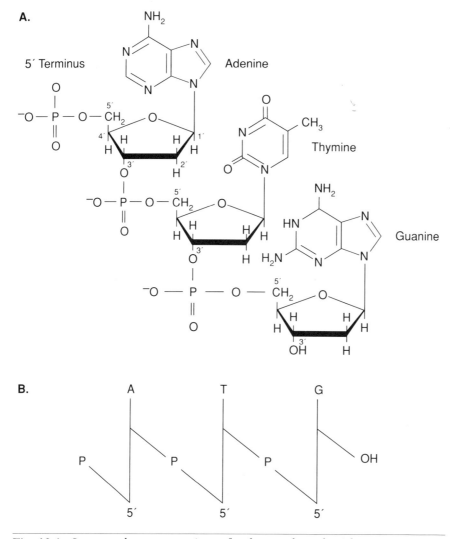

Fig. 12-1. Structural representations of a deoxypolynucleotide segment.

The DNA Double Helix

The two strands of a double helix are held together by complementary hydrogen bonds between the strands. The complementarity is due to the specific base-pairing of purines on each strand with pyrimidines on the other strand. As shown in Figure 12-3, adenine complements thymine, with the formation of two hydrogen bonds between the bases, and guanine complements cytosine, with the formation of three hydrogen bonds between the bases. Thus, in DNA, adenine always pairs with thymine and guanine always pairs with cytosine. The optimal hydrogen bonding between base-paired nucleotides is achieved when the nucleo-

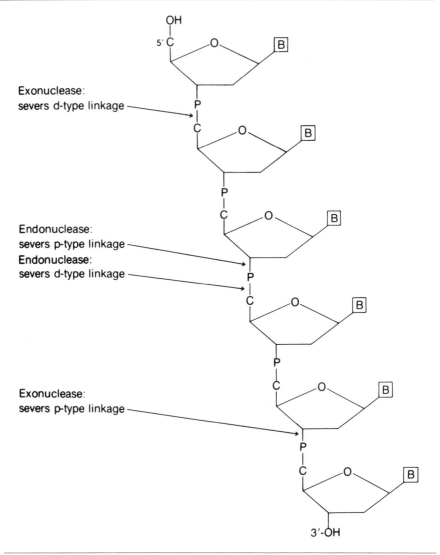

Fig. 12-2. The distal (*d*) and proximal (*p*) nuclease cleavage sites. (From Devlin TM: Textbook of Biochemistry with Clinical Correlations. Wiley, New York, 1986, with permission.)

tides are oriented so that the bases are coplanar and face each other. This is the orientation found in DNA double helices. As a result, the two strands are *antiparallel* (Fig. 12-4)—one strand is oriented in a 5' → 3' direction, and the complementary strand in a 3' → 5' direction. The base pairs form the core of the helix, with the sugar-phosphate backbones coiling around the outside. The sugar-phosphate backbones are hydrophilic, whereas the base pairs in the interior of the helix are hydrophobic. The bases are stacked so that they are not exposed to water, but only to each other via hydrophobic and hydrogen bond interactions.

Four types of DNA double helix have been identified. Three are *right-*

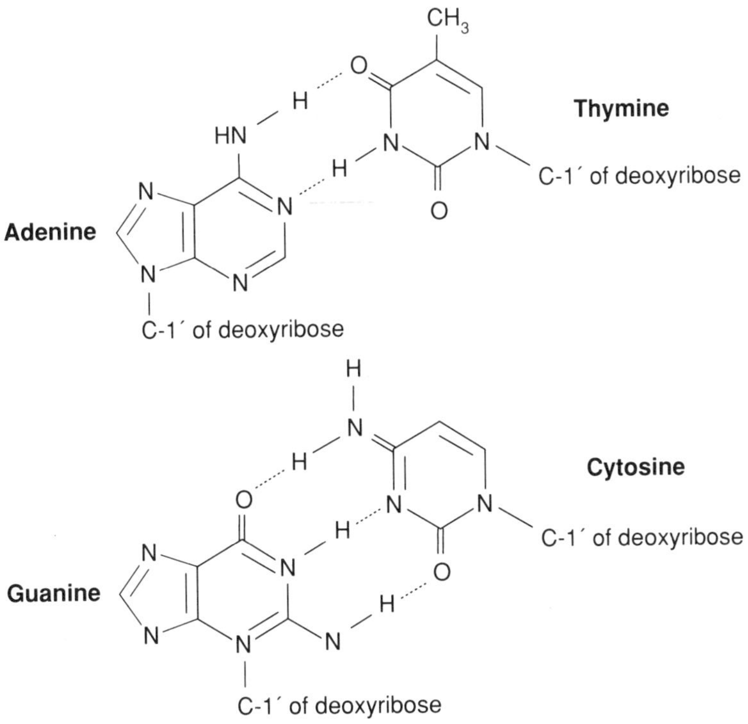

Structure of Nucleic Acids 325

Fig. 12-3. Hydrogen-bonded A-T and G-C base pairs.

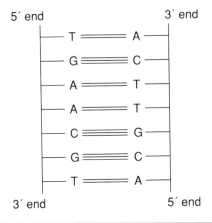

Fig. 12-4. The antiparallel nature of the DNA double helix.

handed (wind to the right), and one is *left-handed*. Double helices are also characterized by the rise per turn (also called the *pitch*), measured in angstroms (Å), and by the number of base pairs per turn.

Right-Handed Helices

A-DNA, B-DNA, and C-DNA are right-handed DNA double helices. B-DNA (which corresponds to the structure originally deduced by Watson and Crick) is the form most abundant in cells.

1. *A-DNA* has a pitch of 28 Å and 11 base pairs per turn.
2. *B-DNA* has a pitch of 34 Å and 10 base pairs per turn.
3. *C-DNA* has a pitch of 33 Å and 9 base pairs per turn.

Clearly, A-DNA is the most condensed form. As can be seen in Figure 12-5, the base pairs in B-DNA and C-DNA are perpendicular to the axis of the helix, whereas the base pairs in A-DNA are tilted relative to the axis (Fig. 12-5). The tilted arrangement of base pairs in A-DNA results in denser packing and shortens the rise per turn.

The sugar-phosphate backbones of the right-handed helices form a pair of winding ridges on the surface of the helix, which are separated by two depressions or grooves that contain the paired bases. These two grooves differ in width. The wider one is called the *major groove* and the narrower one the *minor groove*.

Left-Handed Helices

Interspersed between the right-handed helical regions of mammalian DNA are short segments of left-handed helix. This form is called *Z-DNA* because the sugar-phosphate backbones have a noticeable zig-zag appearance. Z-DNA has a pitch of 37 Å and about 12 base pairs per turn; it is thus more extended than the right-handed helices. Only the equivalent of the minor groove is present in Z-DNA. Z-DNA may play a role in the regulation of gene expression.

DNA Denaturation

If a DNA sample is heated, the purine-pyrimidine hydrogen bonds between the strands break, allowing the strands to separate. This process is called *heat denaturation* of DNA. DNA can also be denatured by exposure to high pH, low ionic strength, or substances that disrupt hydrogen bonds. During heat denaturation, A-T base pairs break at a lower temperature than G-C base pairs because A-T base pairs are held together by two hydrogen bonds whereas G-C base pairs are held together by three.

DNA denaturation, or "melting," can be monitored by the increase in ultraviolet absorbance of the sample. Because of the ring structure of the bases, nucleic acids show significant absorption in the ultraviolet near 260 nm. When the strands separate and the bases are exposed to

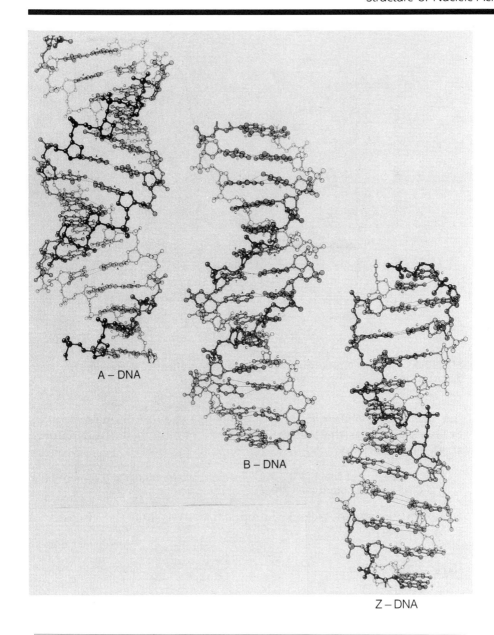

Fig. 12-5. Structures of A-, B-, and Z-DNA. (From Dickerson RE: The DNA and how to read it. Sci Am 249:100, 1983, with permission. Illustration © Irving Geiss.)

solvent, this absorbance increases dramatically, a phenomenon called the *hyperchromic effect* (Fig. 12-6). The hyperchromic effect can be used to monitor the progress of DNA denaturation, which occurs over a relatively narrow temperature range. The temperature that causes a 50 percent rise in absorbance (50 percent denaturation of the sample) is called

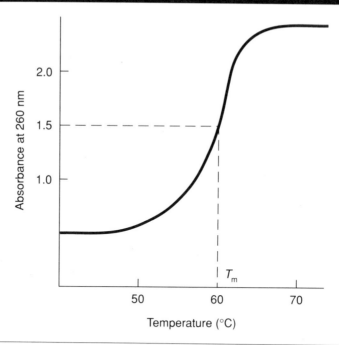

Fig. 12-6. A representation of the hyperchromic effect observed during DNA denaturation.

the *melting temperature* T_m. T_m is also called the *transition temperature* because it reflects the transition from the double helix to denatured single strands. Thus,

T_m is a measure of the AT/GC ratio of a sample: the higher the content of G-C base pairs, the higher the T_m.

Denaturation also decreases the optical activity and the viscosity of DNA samples.

DNA Renaturation

Under appropriate conditions, completely separated DNA strands can renature or *anneal* to form the native double helix. Annealing occurs in two steps:

1. Two complementary strands must first realign properly.
2. The complementary hydrogen bonds between the strands must form.

Realignment is usually slow, the rate depending on the concentration of complementary strands. It occurs by the chance formation of hydrogen bonds between complementary sequences on the two strands. For renaturation to occur, this process, called *nucleation*, must result in correct alignment of the whole strands. Complementary hydrogen bonds then form rapidly between the nucleation sites. Partially denatured DNA samples whose strands have not been completely dissociated renature rapidly by a process corresponding to step 2 above.

The rate of renaturation is equal to the change in the concentration C of single-stranded DNA with time t, and is proportional to the square of the concentration of single-stranded DNA:

$$\frac{dC}{dt} = kC^2$$

where k is a constant. Integration yields

$$\frac{C}{C_0} = \frac{1}{1 + kC_0 t}$$

where C_0 is the concentration of single-stranded DNA at time $t = 0$. C/C_0, the fraction of DNA that has been renatured, is proportional to the product $C_0 t$ of the starting concentration of denatured DNA times the time of renaturation. C_0 and t are usually known, and C/C_0 may be calculated from the change in ultraviolet absorption (hyperchromic effect).

A plot of C/C_0 versus $C_0 t$ gives the rate of renaturation (Fig. 12-7). This plot also yields $C_0 t_{1/2}$, which is the value of $C_0 t$ when $C/C_0 = 0.5$. $C_0 t_{1/2}$ is a measure of DNA complexity. Large DNAs with unique nucleotide sequences are very complex, and have high values of $C_0 t_{1/2}$ (i.e., they renature slowly). DNA complexity and $C_0 t_{1/2}$ decrease with an increase in the number of repetitive base sequences; consequently, $C_0 t_{1/2}$ is a useful index of the amount of repeated sequences in a DNA sample.

DNA Hybridization

Hybridization occurs when DNA strands from different sources are allowed to align and form complementary base pairs, yielding a *hybrid*. The complementary association of RNA and DNA strands is also called hybridization. Since hybridization requires a degree of similarity between the strands, it has been used as a measure of the evolutionary relatedness of different species. mRNA/DNA hybrids are used to identify the DNA sequences that code for particular proteins.

DNA TOPOLOGY

Eukaryotes have DNA in both the nucleus and the mitochondria. Nuclear DNA is linear, whereas mitochondrial DNA is circular. In both cases, the DNA is highly folded and/or twisted into condensed superstructures. This section describes the DNA superstructures found in cells.

Superhelices

Most DNA is twisted into supercoils (superhelices) (Fig. 12-8). Supercoils form when the DNA helix is strained by being either overwound or underwound; supercoiling relieves the strain. This type of strain can be

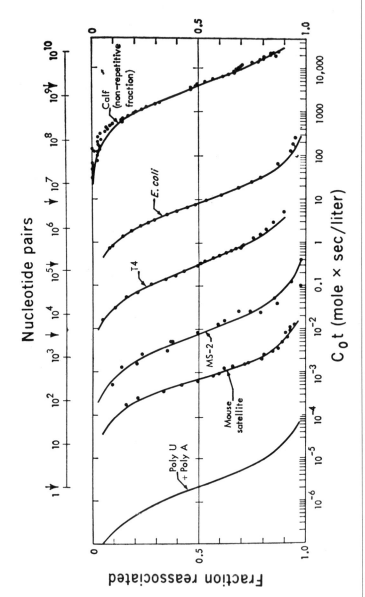

Fig. 12-7. C_0t plot of DNA from different sources (mouse satellite DNA, MS-2 viral DNA, T$_4$ viral DNA, and *Escherichia coli* DNA). Also shown is the C_0t plot of synthetic poly-U and poly-A reassociation. (From Britten RJ, Kohne DE: Repeated sequences in DNA. Science 161:529, 1968, with permission.)

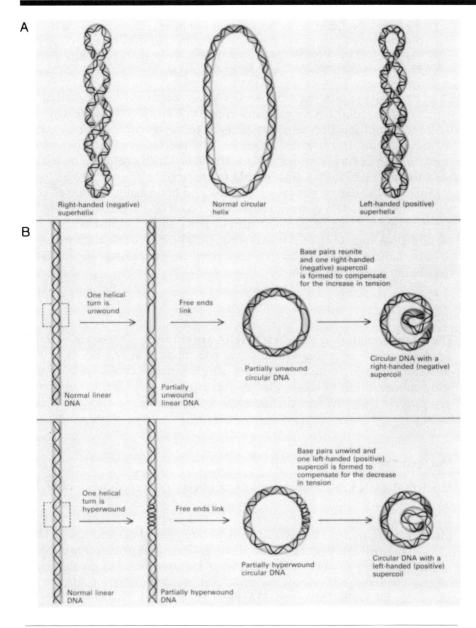

Fig. 12-8. The formation of positive and negative supercoils in DNA. (From Darnell J, Lodish H, Baltimore D: Molecular Cell Biology. Scientific American Books, New York, 1986, with permission.)

introduced into a circular DNA molecule by rotating one end of the double helix relative to the other before ring closure. Alternatively, the ring can be temporarily opened, the ends rotated relative to each other, and the ring resealed. As shown in Figure 12-8, supercoiling occurs in the direction opposite to the overwinding or underwinding that introduced the strain:

1. *Underwinding* (rotation in the direction opposite to the direction of normal double-helical turns) produces a *negative superhelix*.
2. *Overwinding* (rotation in the same direction as the normal double helix) produces a *positive superhelix*.

Most naturally occurring circular DNAs are negatively supercoiled.

Strain in circular DNA molecules can also be relieved by allowing the hydrogen bonding between the strands to be disrupted over a short segment so that the strands separate. This mechanism is called *bubble formation*. In natural circular DNAs, strain is usually relieved by a combination of supercoiling and bubble formation.

Linear DNAs can be supercoiled if one end is transiently rotated relative to the other. The two ends must otherwise be fixed to maintain the supercoiling. In fact, the interaction of DNA with proteins tends to fix certain DNA segments relative to others, making possible the formation of superhelices. Supercoiling is important in increasing the compactness of chromosomal DNA.

Topoisomerases

The relaxation and supercoiling of DNA are catalyzed by enzymes called *topoisomerases* (enzymes that change the topography of DNA). *Type I topoisomerases* introduce temporary single-strand breaks in DNA and allow supercoiling to relax. *Type II topoisomerases* generate supercoils in a process that involves the formation and resealing of double-strand breaks.

Nucleosomes

Much of the DNA in the nucleus is associated with a specific set of highly basic proteins called *histones*. There are five groups of histones: H1, H2A, H2B, H3, and H4. Because of their high content of positive charges, histone proteins interact with the negative phosphate groups of DNA to form neutral DNA-protein complexes called *nucleosomes* (Fig. 12-9A).

The core of a nucleosome consists of $1\frac{3}{4}$ turns of DNA double helix (about 140 base pairs) wound around a protein aggregate that consists of two copies each of histones H2A, H2B, H3, and H4. This core structure is the unit that is isolated when native DNA is degraded by micrococcal nuclease. The native nucleosome (also called the *chromatosome*) includes an additional 60-base-pair length of double helix (*linker DNA*) that is complexed with a molecule of histone H1. *Chromatin* (chromosomal DNA) consists largely of nucleosomes linked in tandem; nucleosomes therefore constitute the structural repeat unit of chromatin. Under the electron microscope, relaxed chromatin has the appearance of beads on a string (Fig. 12-9B). A variety of nonhistone proteins are also associated with chromatin. These proteins are usually present in small amounts, and are believed to play a role in the regulation of gene expression.

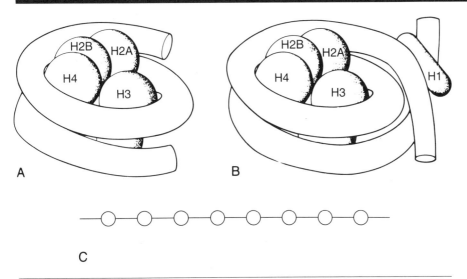

Fig. 12-9. (A) Structure of the nucleosome core. (B) Structure of the intact nucleosome. (C) The bead-on-a-string structure of dispersed chromatin. (Figs A and B from Devlin TM: Textbook of Biochemistry with Clinical Correlations. Wiley, New York, 1986, with permission.)

Structural Levels of Chromosomal DNA

The nuclear DNA in eukaryotes exists in the form of chromosomes, each containing a single linear DNA molecule. Chromatin represents the simplest level of structure in chromosomes. Other structures result from the sequential folding of chromatin into more and more condensed configurations. Located in the center of the chromosome is a very protein-rich region called the *protein scaffold*. Specific regions of chromatin are anchored into the protein scaffold, and the segments between these regions form large chromatin loops. These loops fold independently, and illustrate several levels of increasingly condensed folding. Chromatin is dispersed in resting (S-phase) cells, and becomes condensed after replication. The sequence of folding in the packing of chromatin is believed to be as follows:

1. The chromatin folds into a *nucleofilament* structure, in which the nucleosomes form a regular array and are much closer together than in dispersed chromatin (Fig. 12-10A).
2. The nucleofilaments continue to fold and twist to form dense, fiber-like structures called *helical solenoids* (Fig. 12-10A).
3. Each helical solenoid then folds once more to form a knoblike structure (Fig. 12-10B).

NUCLEOTIDE SEQUENCE PATTERNS IN EUKARYOTIC DNA

Genomic DNA has a number of distinct sequence patterns: palindromes, single-copy DNA, moderately repetitive DNA, and highly repetitive DNA.

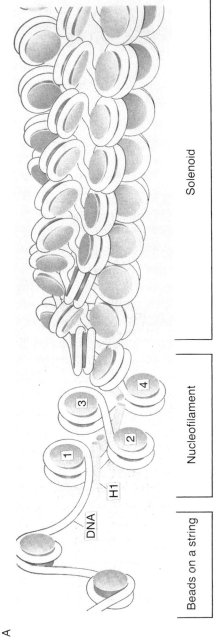
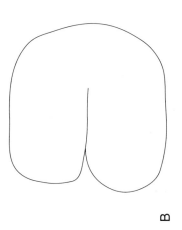

Fig. 12-10. The structural levels of chromatin. (Fig. A From Kornberg RD, Klug A: The nucleosome. Sci Am, Feb: 64, 1981, with permission.)

Palindromes

Palindromes are symmetrical DNA segments, four to six base pairs long, that have the same sequence on the two strands when read in opposite directions. The following sequence is a typical palindrome:

$$-\text{GAATTC}-$$
$$-\text{CTTAAG}-$$

Palindromes can serve as cleavage sites for *restriction endonucleases*, which are nucleases that recognize specific palindromic sequences and cleave both DNA strands. Many palindromic sequences are also recognized by *DNA methylases*, which methylate certain bases and thereby protect the palindrome against cleavage by restriction endonucleases.

Single-Copy DNA

DNA sequences that are present in only one copy are referred to as *unique* or *single-copy* DNA. Most single-copy DNA is believed to code for proteins. In addition, some single-copy DNA represents *pseudogenes*—DNA segments that have significant sequence homology with functional genes but are not expressed, presumably because of mutations that block expression. In addition, *introns* within functional genes and *spacers* between genes contribute to the content of single-copy DNA (see Chapter 14).

Moderately Repetitive DNA

Moderately repetitive DNA sequences are repeated from a hundred to a thousand times. The repeated sequence is several hundred to several thousand base pairs long. Moderately repetitive DNA sequences are sometimes organized in tandem arrays separated by nontranscribed spacer segments (Fig. 12-11A), but usually they are found interspersed among transcribed single-copy DNA segments (Fig. 12-11B).

Highly Repetitive DNA

Highly repetitive DNA usually consists of short segments (about 20 nucleotides or less) that are repeated many thousand to several million times. This DNA is mostly located in the *centromere*, the central region of the chromosome. Highly repetitive DNA has a base composition and density distinct from that of the total genome, so when chromosomal DNA is sheared into small pieces and subjected to density gradient centrifugation, the fragments of highly repetitive DNA appear as a small peak or satellite band distinct from the major peak of chromosomal DNA. For this reason, highly repetitive DNA is also called *satellite DNA*.

Renaturation of DNA

When eukaryotic DNA is fragmented, denatured, and renatured, the renaturation curve is observed to be triphasic (Fig. 12-12), indicating that denaturation occurs at three different rates. Highly repetitive DNA

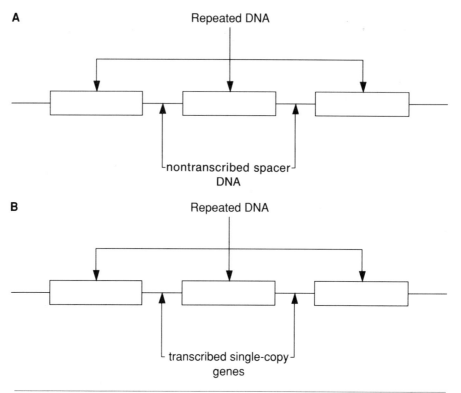

Fig. 12-11. (**A**) Tandem array of repeated DNA segments separated by nontranscribed spacer DNA. (**B**) Repeated DNA interspersed among transcribed DNA segments.

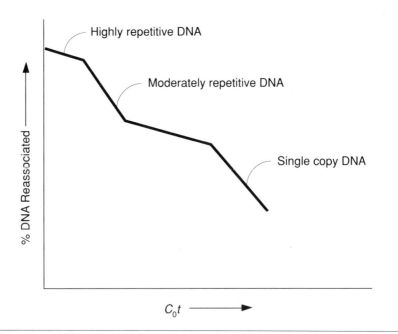

Fig. 12-12. Triphasic curve for reassociation of total DNA from a eukaryotic cell. The three zones represent the reassociation of highly repetitive, moderately repetitive, and single-copy DNA.

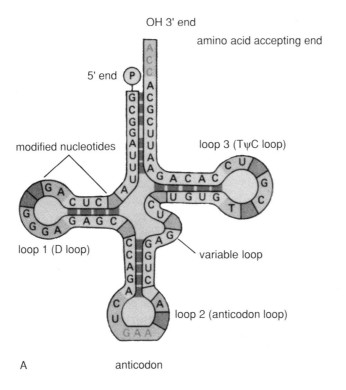
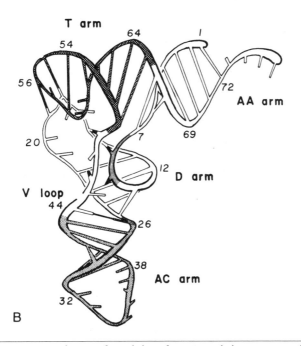

Fig. 12-13. Primary and secondary (**A**) and tertiary (**B**) structure of yeast phenylalanine tRNA. (Fig. A from Alberts B, Bray D, Lewis L et al: Molecular Biology of the Cell. Garland Publishing, New York, 1983, with permission. Fig. B from Sussman JL, Kim SH: Three-dimensional structure of a transfer RNA in two crystal forms. Science 192:853, 1976, with permission.)

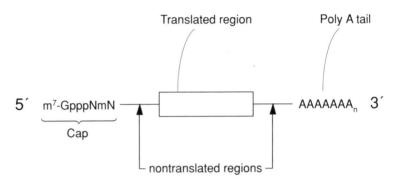

Fig. 12-14. **(A)** The general structure of eukaryotic mRNA. (*Figure continues.*)

sequences renature very rapidly, moderately repetitive sequences renature at an intermediate rate, and single-copy DNA is slowest. The rate of renaturation of a given sequence is directly proportional to the probability of its encountering a complementary sequence.

THE STRUCTURE OF RNA

RNA molecules are linear polyribonucleotides. The purine bases incorporated into RNA are adenine and guanine and the pyrimidine bases are cytosine and uracil. In RNAs, some of the bases are modified after incorporation. RNA is single stranded, and is synthesized from a DNA strand that serves as a template. (Only one strand of the DNA double helix serves as the template.)

As mentioned earlier, there are three major kinds of RNA—mRNA, tRNA, and rRNA. Certain structural properties are common to to all types of RNA. RNA molecules can have *secondary* and *tertiary* levels of structure that are analogous to the secondary and tertiary structure of proteins. The secondary structure of RNA molecules consists of base stacking and intramolecular complementary base pairing. RNA tertiary structure is mediated by hydrogen bonding and hydrophobic interactions between different regions of the RNA molecule. This section introduces the structure of the three kinds of RNA. Their function in protein synthesis is covered in Chapter 14.

tRNA

The tRNAs bind amino acids and translate the nucleotide sequence of mRNAs. There is one or more specific tRNA for each amino acid. tRNA molecules are the smallest of the major classes of RNA, containing about 75 to 90 nucleotides. tRNAs routinely have a high content of modified bases.

All tRNAs have a consistent *cloverleaf* secondary structure (Fig. 12-13A), which is the arrangement that gives maximum intramolecular complementary base-pairing. This structure consists of three double-

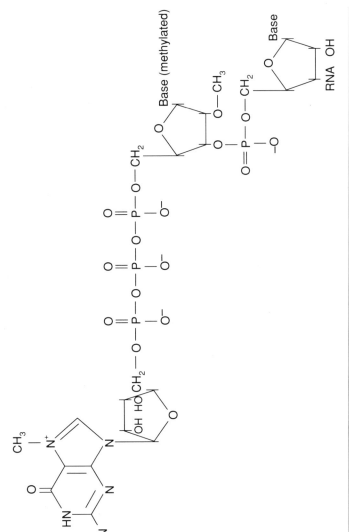

Fig. 12-14 (*Continued*). **(B)** The chemical structure of the 5′ mRNA cap.

stranded arms terminating in loops, a double-stranded stem, and a variable extra loop. Proceeding clockwise from the 3' end, one encounters

1. *The stem* (also known as the *acceptor stem*), which is the region that interacts with amino acids. The actual amino acid binding site is the 3' end of the tRNA molecule, which has the invariant sequence CCA.
2. The TψC loop, so called because it always contains the sequence TψC, where ψ is the modified base pseudouracil.
3. The *variable* or *extra loop*, which varies greatly in size among different tRNAs.
4. The *anticodon loop*, which includes the triplet anticodon that interacts with mRNA.
5. The *D loop*, so called because of its high content of modified dihydrouracil residues.

tRNA is further folded into a tertiary structure that resembles an upside-down capital L (Fig. 12-13B).

rRNA

The rRNAs are the nucleic acid components of the ribosomes. Eukaryotic ribosomes contain four rRNAs of different size, which are usually identified by their sedimentation coefficients as 28S, 18S, 5.8S, and 5S. As will be described in Chapter 14, the intact (80S) eukaryotic ribosome, which contains about 75 different proteins, consists of a larger 60S subunit and a smaller 40S subunit. The 18S rRNA is a component of the 40S subunit, and the other three rRNAs are components of the 60S subunit. rRNAs have elaborate secondary structures involving numerous areas of intrachain base pairing. The commonest form of base modification in rRNAs is methylation at the 2'-hydroxyl of the ribose moiety.

mRNA

mRNAs serve as the templates that direct protein synthesis. Unlike rRNAs and tRNAs, mRNAs have a relatively short half-life. Two striking features of mRNA structure are the presence of a "cap" at the 5' end and a poly A "tail" at the 3' end (Fig. 12-14). The cap structure protects the 5' terminus from exonucleases. The poly A tail is believed to have a role in transporting mRNA out of the nucleus. However, not all mRNA has a poly A tail, and the role of this structure remains unclear.

NUCLEIC ACID STRUCTURE: KEY POINTS

Like proteins, nucleic acids have primary, secondary, and tertiary structure:

1. The *primary structure* is the nucleotide sequence.
2. In the case of DNA, the *secondary structure* is the double helix. In the case of RNA, secondary structure is formed by intramolecular base pairing and base stacking.
3. The *tertiary structure* is the folded conformation adopted by the secondary structure. In the case of DNA, the tertiary structure consists of

the superhelices, nucleosomes, and other condensed configurations. An example of RNA tertiary structure is the folding of tRNA molecules into an L shape.

SUGGESTED READING

1. Cantor CR, Schimmel PR: Biophysical Chemistry. Part 1. The Conformation of Biological Macromolecules. Freeman, San Francisco, 1980
2. Saenger W: Principles of Nucleic Acid Structure. Springer-Verlag, New York, 1983

STUDY QUESTIONS

Directions: Answer the following questions using the key outlined below:
- **(A)** if 1, 2, and 3 are correct
- **(B)** if 1 and 3 are correct
- **(C)** if 2 and 4 are correct
- **(D)** if only 4 is correct
- **(E)** if all four are correct

1. Nucleases cleave
   1. RNA
   2. DNA
   3. the distal site of phosphodiester bonds
   4. the proximal site of phosphodiester bonds

2. G-C base pairs
   1. have three hydrogen bonds
   2. do not show hypochromicity
   3. are more difficult to disrupt than A-T base pairs
   4. have two hydrogen bonds

3. Mammalian cells contain
   1. right-handed DNA double helices
   2. left-handed DNA double helices
   3. three forms of right-handed helix
   4. two forms of left-handed helix

4. Palindromic sequences in DNA
   1. may be recognized by topoisomerases
   2. may be recognized by restriction endonucleases
   3. are usually methylated in order to be recognized and acted on by restriction endonucleases
   4. may be recognized by DNA methylases

5. RNA molecules from mammalian cells
   1. are single stranded
   2. contain regions of intramolecular base pairing
   3. can have tertiary conformations
   4. contain interstrand base-paired regions

6. When one strand of DNA from human liver is complementarily base paired with a strand of DNA from rat liver, the process involves or is referred to as
   1. hybridization
   2. chain aligment
   3. nucleation
   4. hydrogen bond formation

7. DNA superhelices
   1. are only present in circular DNA
   2. are exemplified by bubble formation
   3. generate strain in DNA molecules
   4. may be positive or negative

Directions: For each of the following multiple choice questions, choose the most appropriate answer.

8. The histone core of nucleosomes consists of two copies of each of the following histone proteins
   A. H2A, H2B, H3, H4
   B. H2A, H1, H2B, H3
   C. H2A, H1, H3, H4
   D. H1, H2A, H2B, H4

9. DNA samples A through D are found to exhibit the following melting temperatures. Which sample contains the greatest G+C content?
   A. 57°C
   B. 53°C
   C. 55°C
   D. 59°C

10. DNA samples A through D exhibit the following values for $C_0t_{1/2}$. Which sample probably consists of highly repeated sequences?
    A. 10^{-2} mol·sec·liter^{-1}
    B. 10^{-3} mol·sec·liter^{-1}
    C. 10^{-4} mol·sec·liter^{-1}
    D. 10^{-1} mol·sec·liter^{-1}

13

DNA REPLICATION, MUTATION, AND RECOMBINATION

DNA Replication/Mutation and Repair/ Recombination/Clinical Correlation

The student should be able to:

1. Illustrate the semiconservative nature of DNA replication.
2. Describe the roles of the different DNA polymerases of *E. coli* in the replication and repair of DNA.
3. Describe the process of DNA replication in prokaryotes: list the events required to initiate replication; understand the role of the RNA primer and of topoisomerases and helicases; diagram the replication fork; and distinguish the events on the leading and lagging strands.
4. Contrast DNA replication in eukaryotes and in prokaryotes.
5. Define "mutation" and list the different types of mutation.
6. Describe at least two mechanisms for repairing mutations.
7. Define recombination and recombinant DNA, and distinguish general from site-specific recombination.
8. Diagram plausible steps for a gene cloning procedure.

Learning Objectives

Perspective: DNA Replication

The following section covers the replication of cellular DNA. The replication system of *Escherichia coli* is described in detail, and the distinctive features of replication in eukaryotes are then considered. The essential points of DNA replication are as follows:

1. The complementary strands of the DNA double helix separate, and each strand serves as a template for the synthesis of a new complementary strand.
2. Because both parent strands serve as templates, DNA replication

(continued)

is *semiconservative*—each daughter duplex contains one strand from the parent duplex and one newly synthesized daughter strand.
3. DNA synthesis is performed by *DNA polymerases*. Free deoxyribonucleotide triphospates (dATP, dGTP, dCTP, and dTTP) base pair with the template strand, and the polymerase joins the deoxyribonucleotides by phosphodiester bonds. Therefore, the nucleotide sequence of the parent strand dictates the nucleotide sequence of the daughter strand.
4. The new DNA strands are always synthesized in the 5′ → 3′ direction, which corresponds to the 3′ → 5′ direction of the parent strand (i.e., the two strands are antiparallel; Chapter 12.)

DNA REPLICATION IN *E. COLI*

E. coli contains three different DNA polymerases: *DNA polymerases I, II, and III*. DNA polymerase III is the enzyme primarily responsible for the synthesis of long daughter-strand segments during DNA replication. DNA polymerase I also plays a role in DNA replication, as discussed below. The role of DNA polymerase II is not known. All three enzymes are *DNA-dependent DNA polymerases*, that is, they use one DNA strand as a template for synthesizing another DNA strand. The key features of DNA replication in *E. coli* may be summarized as follows:

1. *Bidirectional mode of replication.* Replication begins at a specific site on the *E. coli* genome, and proceeds in both directions around the circular DNA molecule.
2. *Unzipping of the helix.* The two strands of the double helix must be separated for replication to occur. Strand separation creates two *replication forks* (Fig. 13-1), which are the sites of DNA synthesis. These replication forks migrate in opposite directions along the DNA duplex as replication proceeds.

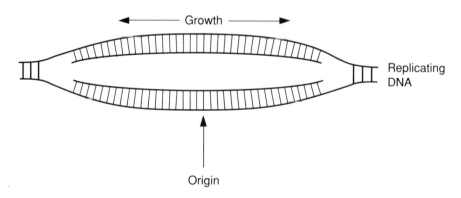

Fig. 13-1. DNA replication shortly after initiation, showing the two replication forks proceeding in opposite directions (bidirectional replication).

3. *Priming.* DNA polymerases require a free 3' hydroxyl, onto which the first deoxyribonucleotide is attached. This requirement is satisfied in replication initiation by the synthesis of a short *RNA primer* that is later excised.
4. *Elongation.* DNA polymerase III occupies the replication fork and replicates both strands. One strand, called the *leading strand*, is synthesized continuously in the 5' → 3' direction. The other strand, the *lagging strand*, is synthesized in short 5' → 3' segments called *Okazaki fragments*. An RNA primer is required for each Okazaki fragment.
5. *Joining of the Okazaki fragments.* DNA polymerase I excises the RNA primers from the Okazaki fragments and fills the resulting gaps with deoxyribonucleotides. The fragments are then ligated by *DNA ligase* to form a continuous strand.
6. *Termination.* The details of replication termination are not understood. In circular chromosomes, replication is believed to continue until the two replication forks meet (Fig. 13-2). In linear chromosomes, replication proceeds until a replication fork meets either another replication fork or the end of the chromosome.

Initiation

DNA replication starts at a site called the *origin of replication*. There is only one origin in the bacterial DNA molecule. The mechanism of initial strand separation at the origin is not clearly understood. Once the strands have been separated, the first RNA primer is laid down in a process that is believed to involve both RNA polymerase and the enzyme *primase*. Primase is part of a complex, the *primasome*, that also includes six or seven other proteins. Primase is believed to recognize specific single-stranded nucleotide sequences and to synthesize an RNA primer fragment 3 to 5 nucleotides long. During the initiation of replication, an RNA primer is synthesized on only the leading strand of each replication fork, and replication begins first on this strand.

DNA Unwinding

As discussed in Chapter 12, DNA exists as a negatively supercoiled helix. Therefore, if the replication fork is to migrate, the superhelix must be unwound and the double helix separated into single strands. This is achieved by the cooperative action of ATP-dependent *helicases* and topoisomerases. *E. coli* has two different helicases, called helicase II and the rep protein. The helicases unwind the superhelix and destabilize the hydrogen bonds between the paired bases, causing the strands to separate. As the double helix is unwound, significant tension (overwinding) is created in the remaining superhelical segment. This strain is alleviated by topoisomerases, which introduce temporary cuts or nicks in the DNA, allowing the strands to relax. Left to themselves, the single strands released by helicase would tend to reassociate. That is prevented by specific proteins called *single-stranded binding proteins*, which bind to and stabilize single-stranded segments of DNA. Overall, therefore, the helicases, topoisomerases, and binding proteins collaborate to generate and maintain the single-stranded template for DNA synthesis.

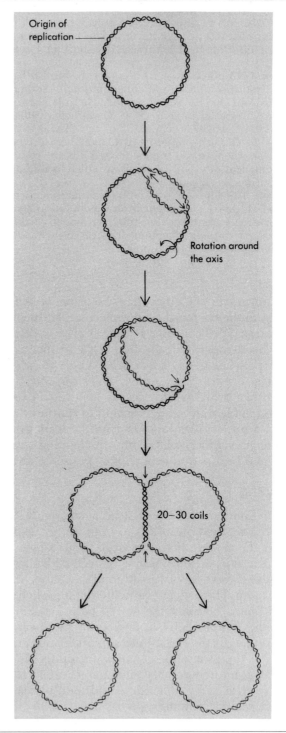

Fig. 13-2. An illustration of the bidirectional replication of circular DNA. (From Watson JD, Hopkins NH, Roberts JW et al: Molecular Biology of the Gene. Benjamin-Cummings, Menlo Park, CA, 1987, with permission.)

The Replication Fork

The actual events of DNA replication occur behind the helicase in the replication fork (Fig. 13-3). DNA polymerase III occurs as a dimer, with one subunit replicating the leading strand and the other replicating the lagging strand. It is not immediately obvious how the two subunits of the dimer manage to elongate the two strands in opposite directions—one toward the replication fork and the other away from it. The answer is that the lagging strand is flipped over or looped so that it can be read "backward" (in the 3' → 5' direction) while the polymerase as a whole moves "forward" (toward the replication fork). A loop of free lagging strand accumulates in front of the polymerase during the synthesis of each Okazaki fragment. The lagging-strand subunit dissociates from the lagging strand at the end of each Okazaki fragment, and then reassociates at the next primer to copy the accumulated loop of free lagging strand.

The events on the two strands are quite different, and are best considered separately.

The Leading Strand

1. A single RNA primer is synthesized at the origin of replication by the primasome complex.

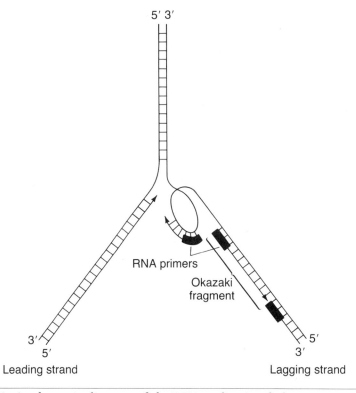

Fig. 13-3. A schematic diagram of the DNA replication fork.

2. DNA polymerase III starts the new complementary DNA strand at the 3' end of the primer. The new strand is elongated continually in the 5' → 3' direction, following the progress of the replication fork.
3. The new double helix assumes its native conformation.

The Lagging Strand

1. The primasome moves along the single-stranded region of the replication fork. It recognizes certain nucleotide sequences that are spaced along the lagging strand, and synthesizes RNA primers complementary to them.
2. Starting from a primer, DNA polymerase III extends the strand in the 5' → 3' direction (Fig. 13-4).

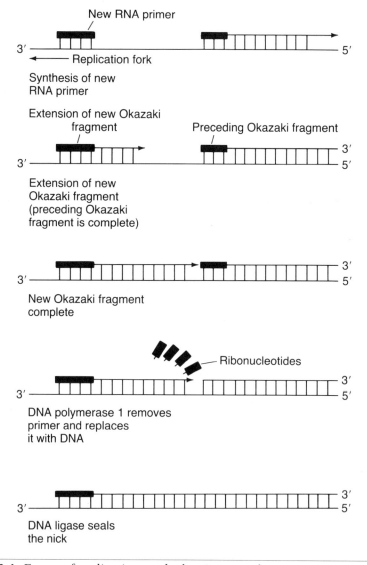

Fig. 13-4. Events of replication on the lagging strand.

3. DNA polymerase III eventually encounters the 5' end of the preceding RNA primer, and dissociates from the lagging strand. The DNA-RNA adduct formed constitutes an Okazaki fragment. When DNA polymerase III finds the next primer, it starts a new Okazaki fragment.
4. The RNA primers are removed from the Okazaki fragments by DNA polymerase I, which simultaneously synthesizes DNA to fill in the gap. The RNA nucleotides are removed and replaced with deoxyribonucleotides starting at the 5' end of the primer.
5. When the RNA primer has been removed and replaced, the last phosphodiester bond is synthesized by DNA ligase—i.e., the final nick is closed.
6. As in the case of the leading strand, the new double helix assumes its native conformation.

These events are repeated on the lagging strand until its replication is complete. Okazaki fragments vary in length from 100 to 200 nucleotides in eukaryotes to from 1000 to 2000 nucleotides in prokaryotes.

DNA REPLICATION IN EUKARYOTES

Replication in eukaryotes differs from replication in prokaryotes in three major ways. First, eukaryotic cells contain much more DNA than prokaryotic cells. Second, the actual movement of DNA polymerase is much slower in eukaryotes than in prokaryotes. Third, eukaryotic DNA has multiple initiation sites, so many replication forks operate at once on each chromosome (Fig. 13-5). The main reason the replication fork migrates more slowly in eukaryotes than in prokaryotes is that the

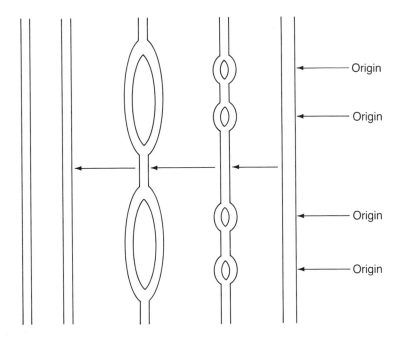

Fig. 13-5. Eukaryotic DNA has multiple origin of replication sites.

highly organized structure of eukaryotic chromosomal DNA must be systematically dismantled before replication. For example, nucleosomal DNA (Chapter 12) must be uncoiled and separated from the histone proteins before replication. The resulting slow rate of replication is compensated for by the presence of many replication forks. During replication, new histones and other chromosomal proteins are synthesized to accommodate the newly synthesized DNA.

Perspective: Mutation and Repair of DNA

Because the base sequence of DNA is the sole repository of genetic information, it must be protected against alteration. Any change in the nucleotide sequence is called a *mutation*; changes that affect only a single base pair are called *point mutations*. Although DNA is quite stable and mutations are uncommon, cells cannot survive without mechanisms to repair damaged DNA

MUTATION AND REPAIR OF DNA

Types of Mutations

Mutations that involve the *substitution* of one base pair for another are called *point mutations*. These mutations do not change the DNA reading frame. The *insertion* or *deletion* of one or more base pairs, on the other hand, does change the DNA reading frame, and these mutations are therefore called *frameshift mutations*. Base substitutions can be subdivided into *transitions* and *transversions*:

1. In a *transition*, the purine on one strand is replaced by another purine, and the pyrimidine on the other strand is replaced by another pyrimidine. A-T → G-C and C-G → T-A are examples of transitions.
2. In a *transversion*, the purine on one strand is replaced by a pyrimidine, and the pyrimidine on the other strand is replaced by a purine. A-T → C-G and T-A → G-C are examples of transversions.

Figure 13-6 diagrams the occurrence of a transition. A frameshift mutation causes a shift in the reading frame of the nucleotide sequence (Chapter 14), and therefore radically changes the information carried by the downstream DNA. In summary, the classes of mutation are:

1. Base substitutions
   a. Transitions
   b. Transversions
2. Frameshift mutations
   a. Insertions
   b. Deletions

Mutagenesis

Mutations are caused primarily by three factors: exposure to chemicals, irradiation, and base-pairing mishaps that occur during replication. Examples of these types of mutagenesis are given below.

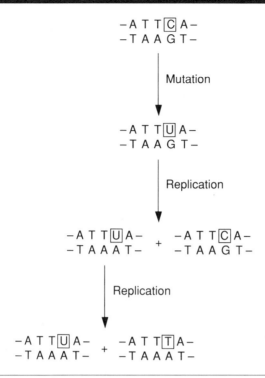

Fig. 13-6. Formation of a transition-type mutation. The net change is from C-G to T-A.

Chemical Mutagenesis

Chemicals usually cause mutations by modifying a base in such a way as to change its base-pairing properties. If the change is not repaired, a different complementary nucleotide is selected during replication, and a base substitution results. Figure 13-7 illustrates how nitrous acid changes the base-pairing properties of susceptible nucleotides.

Another kind of chemical mutagenesis results in frameshift mutations. Flat polycyclic molecules, such as acridines, can intercalate between base pairs in the double helix, forcing the base pairs abnormally far apart. An extra base may then be inserted in the daughter strand during replication.

Irradiation

Both ultaviolet and x rays can cause mutations. One type of irradiation damage that results in base substitution occurs as follows. The ring amino and keto groups of the nucleotide bases exist in equilibrium with their imino and enol forms, respectively. Normally, only a minute fraction of bases are in imino/enol forms that have base pairing properties different from those of the amino/keto forms. Irradiation increases the energy level of the bases to the point where the imino/enol forms are

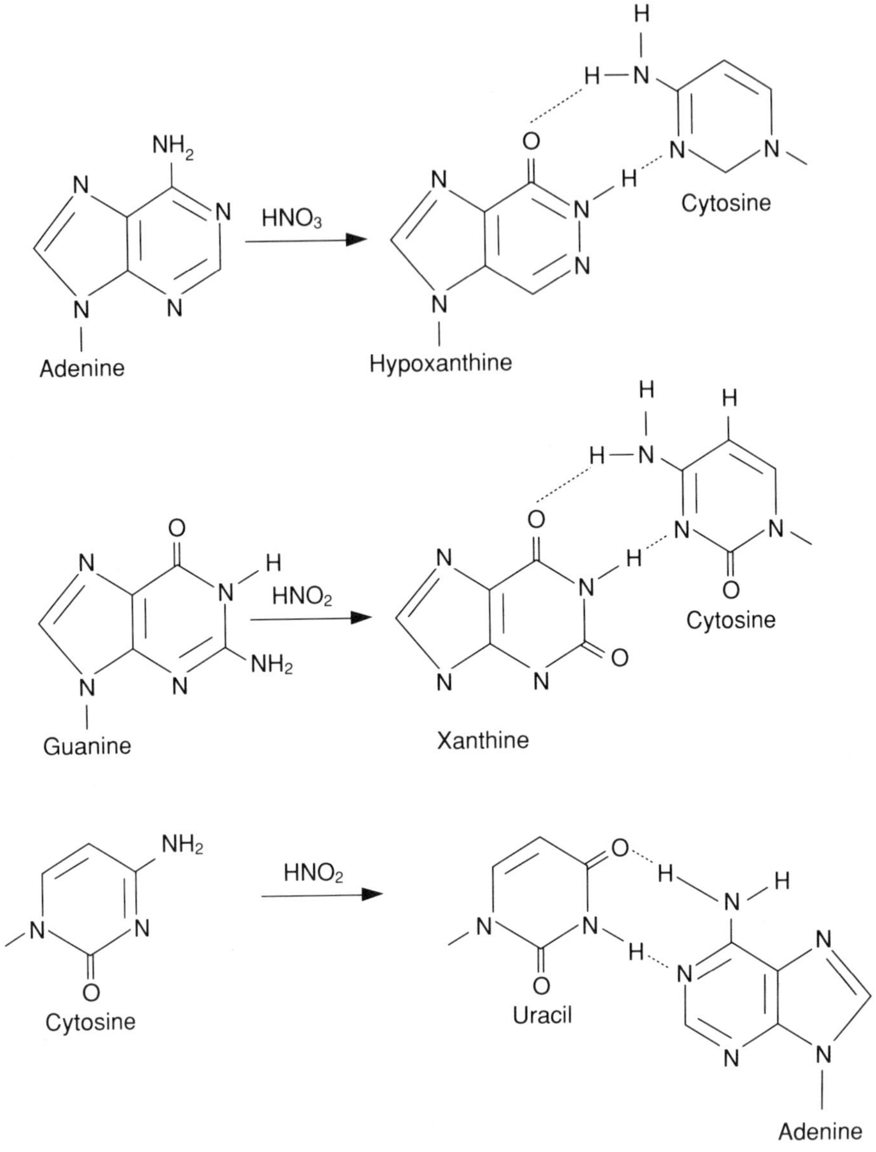

Fig. 13-7. Chemical modification of DNA bases by nitrous acid. Adenine and cytosine are altered in such a way as to change their base-pairing specificities. Guanine is also modified, but its base pairing is not altered.

present in substantial amounts. If these forms are present during replication, base substitutions can result.

Ultraviolet irradiation can also cause structural changes in the ring systems of certain bases. For example, when adjacent pyrimidines (especially thymines) are irradiated, their pyrimidine rings can become joined to form a cyclobutane ring, as shown in Fig. 13-8. The resulting

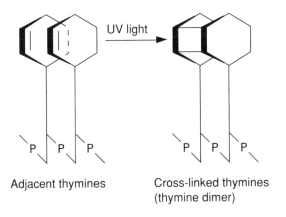

Fig. 13-8. Formation of thymine dimers in DNA.

structure, called a *thymine dimer*, distorts the double helix in such a way as to block replication.

Replication Errors

During replication, a base on either the template or the daughter strand may become looped out to form a one-base hairpin loop. As shown in Figure 13-9, these loops result in deletions (if formed on the parent strand) or insertions (if formed on the daughter strand), and thus cause frameshift mutations. Changes of this kind are called *slippage errors*, and most often occur at the branch point of the replication fork when it is moving through areas that have a high content of identical bases.

Under normal conditions, DNA polymerase is very accurate, and almost never incorporates a mispaired nucleotide. However, errors can be generated intentionally by incorporating into the replication mixture a base analog that resembles one of the usual nucleotides but has different base-pairing properties. This technique results in base substitutions.

Mutagenesis: Key Points

To summarize, the methods of generating mutations and the kinds of mutations they cause are as follows:

1. Exposure to chemicals:
    Base substitutions
    Frameshifts
2. Ultraviolet irradiation:
    Base substitutions
    Structural modification of bases
3. Slippage errors:
    Frameshifts
4. Intentional exposure of the replication mixture to base analogs:
    Base substitutions

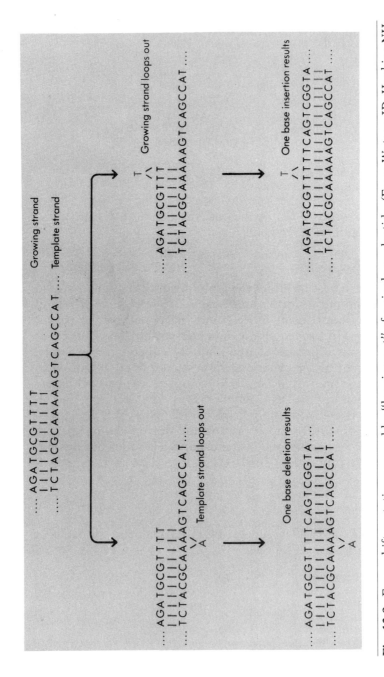

Fig. 13-9. Frameshift mutations caused by "looping out" of a single nucleotide. (From Watson JD, Hopkins NH, Roberts JW et al: Molecular Biology of the Gene. Benjamin-Cummings, Menlo Park, CA, 1987, with permission.)

DNA Replication, Mutation, and Recombination

REPAIR OF DNA

Damaged DNA is repaired by a variety of mechanisms. The specific mechanism used depends largely on the nature of the damage. In many cases, the complementary strand supplies the information needed to repair the damaged strand correctly. Some examples of DNA repair are presented below.

Photoreactivation Repair of Thymine Dimers

One mechanism that repairs thymine dimers involves the enzyme *photolyase*, which is present in most cells. Photolyase binds to the DNA segment containing the dimer. The enzyme is then activated by absorption of visible or near-UV light. Activated photolyase catalyzes the

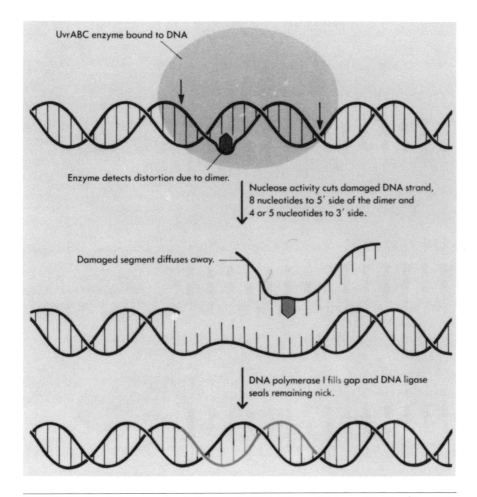

Fig. 13-10. Excision repair of damaged DNA. (From Watson JD, Hopkins NH, Roberts JW et al: Molecular Biology of the Gene. Benjamin-Cummings, Menlo Park, CA, 1987, with permission.)

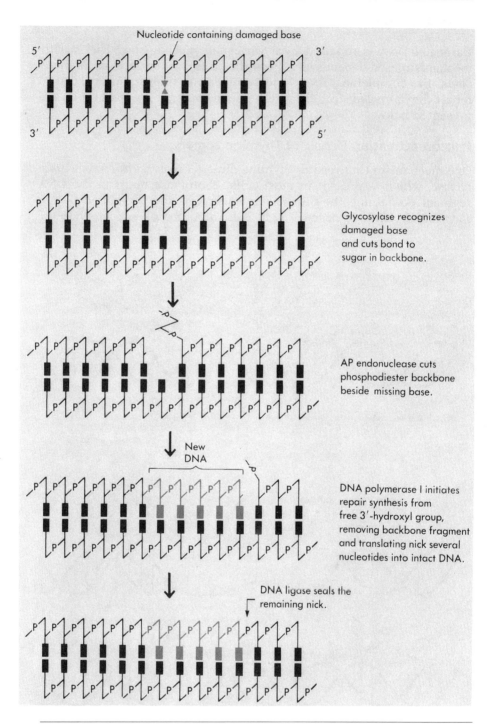

Fig. 13-11. Excision repair of damaged DNA via the formation of an apurinic or apyrimidinic site. (From Watson JD, Hopkins NH, Roberts JW et al: Molecular Biology of the Gene. Benjamin-Cummings, Menlo Park, CA, 1987, with permission.)

cleavage of the thymine dimer cyclobutane ring, restoring the original thymine residues. Defects in the repair of thymine dimers cause the disease xeroderma pigmentosum, which is discussed in the clinical correlation at the end of this chapter.

Excision and Replacement

In repair by excision and replacement, a segment of a DNA strand containing a site of damage is removed and then replaced with bases complementary to the undamaged strand. The damaged segment is detected, removed, and replaced by the combined action of UvrABC endonuclease, DNA polymerase I, and DNA ligase. In the case of thymine dimer repair, for example (Fig. 13-10), the UvrABC endonuclease recognizes the damage and makes the initial cuts on either side of the thymine dimer. The damaged segment then dissociates from the double helix, the gap is filled in by DNA polymerase I, and DNA ligase seals the final nick.

Another type of excision and repair involves removal and replacement of only the damaged base (Fig. 13-11). The *N*-glycosidic bond linking the base moiety to the sugar-phosphate backbone is cleaved, and the damaged base dissociates, leaving behind an intact sugar-phosphate backbone. The empty base site is called an *apurinic* or an *apyrimidinic* site, depending on whether a purine or a pyrimidine was removed; both types are also called *AP sites*. AP sites can be generated by *N*-glycosylases and may also form spontaneously. Another group of enzymes, the AP endonucleases, hydrolyze the phosphodiester bond at AP sites. DNA polymerase I removes the resulting free sugar phosphate and fills in the gap, and DNA ligase closes the nick.

Reversion

Substitution mutations sometimes spontaneously repair themselves by reverting to the original base.

Perspective: DNA Recombination and Recombinant DNA

DNA recombination is a naturally occurring process by which nucleotide segments from different DNA molecules are joined or "combined" to yield new DNA molecules containing portions of the nucleotide sequences of both starting DNA molecules. *Recombinant DNA* is the product of DNA recombination. Restriction endonucleases can also be used to achieve DNA recombination with isolated DNA sequences in vitro. This section first describes DNA recombination as it occurs inside cells. The procedures involved in DNA recombination in vitro are then reviewed.

DNA RECOMBINATION

Three types of recombination occur in cells: *general recombination*, *site-specific recombination*, and *transposition*. General recombination involves the exchange of material between homologous double helices.

Site-specific recombination occurs between short homologous sequences, and usually results in the insertion of one DNA segment into another. In transposition, the recombination is accomplished by *transposons*, highly mobile DNA segments that do not require sequence homology in order to insert themselves into DNA.

General Recombination

In general recombination, the actual site of recombination can be anywhere in the region of homologous DNA. The enzymes involved are mostly not known; in *E. coli*, the RecA and RecBC enzymes have been shown to play a role. Figure 13-12 shows the process of general recombination, which may be described as follows:

1. The homologous sections of two double helices align.
2. Adjacent strands of the two double helices are nicked at matching locations.
3. The strands are ligated so that the broken strand of one helix is joined to the broken strand of the other. The resulting four-stranded arrangement is called a *Holliday structure*.
4. The point where the two double helices are joined is called the *branch point*. The branch point can migrate linearly. As it does so, more and more of the strands are exchanged, leading to the formation of more and more extensively recombined species.
5. As shown in Figure 13-12, the Holliday structure may isomerize by rotation of the double helices around vertical and/or horizontal axes through the branch point.
6. The Holliday structure may be cleaved by an endonuclease and the resulting strands ligated to yield recombinant double helices (Fig. 13-12).

Site-Specific Recombination

Site-specific recombination occurs between relatively short homologous DNA sequences. These sequences bind enzymes that catalyze the strand cleavage and rejoining reactions that result in recombination. Site-specific recombination results in the insertion of one DNA molecule into another. An excellent example of site-specific recombination is the incorporation of bacteriophage DNA into the *E. coli* genome, as shown in Figure 13-13. The homologous insertion sites on the two DNA molecules align, and recombination is catalyzed by the protein integrase. Integrase requires an additional protein called integration host factor.

Transposons

Certain stretches of DNA, called *transposons* or *insertion sequences*, are capable of becoming incorporated into other DNA molecules at sites where the two DNA molecules have no sequence homology. Transposons range in size from about 1 kilobase (1,000 bases) to about 40 kilobases. They code for all the enzymes required to insert them into DNA sequences, and may contain other genes as well. Transposons are capable

DNA Replication, Mutation, and Recombination   361

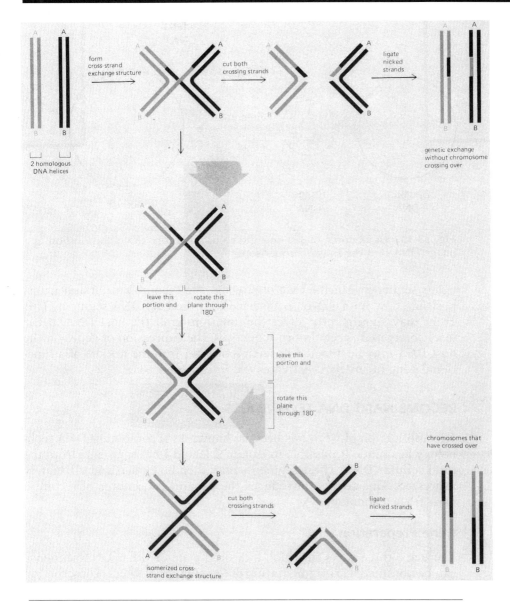

Fig. 13-12. An illustration of general recombination. (From Alberts B, Bray D, Lewis J et al: Molecular Biology of the Cell. Garland Publishing, New York, 1983, with permission.)

of moving about within a single chromosome or from one chromosome to another.

Effects of DNA Recombination

The overall effect of recombination is to introduce new combinations of genetic information. General recombination involving strand isomerization results in the exchange of DNA segments between homol-

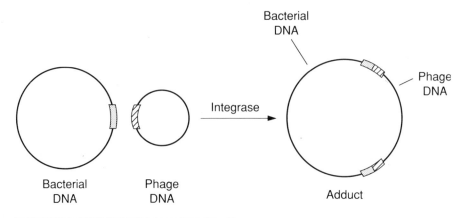

Fig. 13-13. An example of site-specific recombination: the incorporation of λ phage DNA into the E. coli chromosome.

ogous sequences. In the case of site-specific recombination and transposition, new information is inserted into the host DNA segment. This event may significantly affect the functioning of the host DNA. If the insertion contains regulatory sequences, the expression of genes on the host DNA may be affected. Insertion of DNA into the middle of a functional gene normally inactivates the interrupted gene.

RECOMBINANT DNA TECHNIQUES

The application of what has become known as recombinant DNA technology has made it possible to isolate selected DNA segments from the total cellular DNA. These segments may then be cloned and ultimately expressed. This section introduces the general procedures of recombinant DNA technology.

Gene Preparation

There are three ways in which a desired gene or other DNA sequence can be obtained: (1) fragmentation of the cellular DNA, (2) construction of a complementary DNA from an appropriate mRNA, and (3) direct chemical synthesis.

Fragmentation of Cellular DNA

The objective of fragmentation is to generate gene-sized DNA fragments. DNA can be fragmented mechanically by subjecting it to shearing forces. The breaks produced by this technique are random, and the fragmentation pattern usually cannot be reproduced. The fragments also tend to be very large. Instead, fragmentation is usually performed using *restriction endonucleases* (Chapter 12), which cleave DNA at specific sites. A combination of restriction endonucleases can be used to cleave the cellular DNA into gene-sized fragments. The fragmentation pattern is very reproducible.

Construction of Complementary DNA

If the desired gene is one that codes for a protein, it may be possible to prepare *complementary DNA (cDNA)* from the mRNA for that protein. This process is shown in Figure 13-14. First, the mRNA is used as a template on which the enzyme *reverse transcriptase* (an RNA-dependent DNA synthetase) synthesizes a complementary DNA strand, yielding an RNA/DNA hybrid. The RNA strand is then removed by treatment with alkali, and the single-stranded cDNA is converted to double-stranded cDNA by DNA polymerase I. As shown in Figure 13-14, a synthetic poly T oligonucleotide, which complexes with the poly A tail of the mRNA (chapter 12), serves as the primer for reverse transcriptase. The primer for DNA polymerase I is a small hairpin loop that forms naturally at the 3' end of the single-stranded cDNA. This hairpin loop is then removed by a nuclease, yielding the mature double-stranded cDNA gene.

Direct Chemical Synthesis

Moderately long stretches of DNA can be synthesized by chemical methods in vitro. Therefore, if the base sequence of the desired DNA segment is known, it may be possible to synthesize it directly.

Cloning

The next step is to clone the desired gene—i.e., to create many copies of it. Cloning is accomplished by incorporating the gene into a recipient DNA molecule, which is introduced into a host cell in which it is replicated. Either a *bacteriophage* DNA molecule or a *plasmid* is used as the recipient. Bacteriophages are DNA viruses that reproduce in bacterial cells. Plasmids are small, accessory DNA molecules of bacterial cells. They occur in varying numbers in the cell, reproduce independently of the main DNA ring, and are readily exchanged between bacterial cells.

To clone a gene, a recombinant DNA adduct is made that consists of the bacteriophage or plasmid DNA plus the desired gene. The bacteriophage or plasmid DNA is called the *cloning vector* and the desired gene is called *passenger DNA*. For the following reasons, vector-passenger adducts are usually constructed using restriction endonucleases:

1. Restriction endonucleases cleave the double helix at specific sequences. Therefore, if both vector and passenger DNA are cleaved by the same restriction endonuclease, the cleaved regions in the two molecules will be homologous.
2. Many restriction endonucleases make cuts that are staggered slightly on the two strands. Therefore, each cut end has a small projection of single-stranded DNA. These complementary single-stranded projections tend to anneal if they encounter each other.

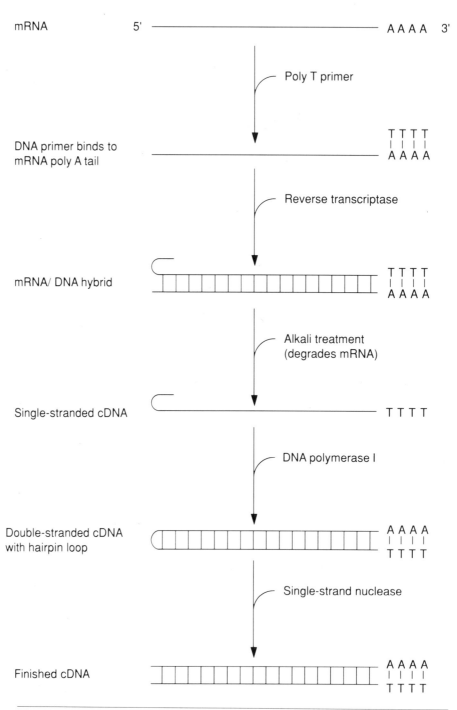

Fig. 13-14. Preparation of double-stranded cDNA from an mRNA template.

3. For these two reasons, if a restriction endonuclease that makes staggered cuts is used to cleave both vector and passenger DNA, the products can be associated by complementary base pairing and subsequently joined by DNA ligase to yield vector-passenger adducts. This process is diagrammed in Figure 13-15.

The recombinant vector-passenger adduct is then allowed to enter appropriate host cells via infection. Once in the host, the vector-passenger adduct is reproduced (cloned) during replication of the host cell genome (Fig. 13-16).

Detection of Clones

The detection of clones containing passenger DNA usually relies on the choice of vector. Vectors are generally selected that confer on the host some easily detected property. For example, an *E. coli* plasmid that

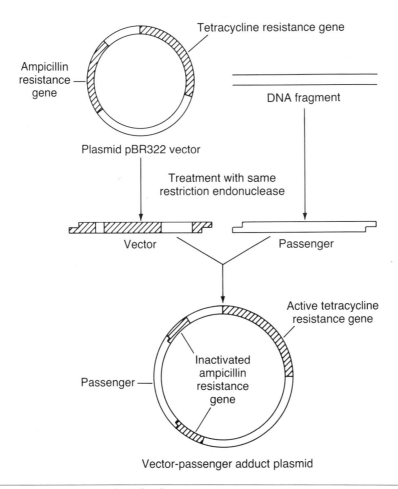

Fig. 13-15. General method for forming vector-passenger DNA adducts, showing the commonly used cloning vector plasmid pBR322.

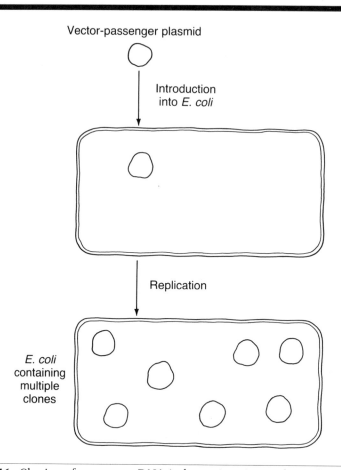

Fig. 13-16. Cloning of passenger DNA in bacteria using a plasmid vector.

carries genes for resistance to the antibiotics tetracycline and ampicillin is a common vector. The appropriate restriction endonuclease can be used to insert the passenger DNA into one of the drug resistance genes, which is thereby inactivated. This process is called *insertional inactivation*. Host cells containing vector-passenger adducts can then be selected because they are resistant to one of the antibiotics but susceptible to the other. In contrast, host cells that received no vector are susceptible to both antibiotics, and host cells that received only nonrecombinant vectors are susceptible to neither.

Detection of Clones Carrying the Desired Passenger DNA

When the DNA to be used in recombination is obtained by fragmenting the total cellular DNA, all the fragments (i.e., the fragment mixture) are used in making vector-passenger adducts and in cloning. The resulting unsorted mixture of clones is called a *genomic library*. Only a very few of the recombinant vectors carry the desired gene. These are usually selected on the basis of hybridization techniques in which a radioactive

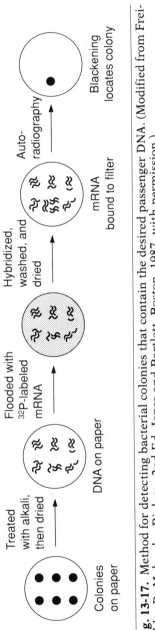

Fig. 13-17. Method for detecting bacterial colonies that contain the desired passenger DNA. (Modified from Freifelder D: Molecular Biology. 2nd Ed. Jones and Bartlett, Boston, 1987, with permission.)

cDNA or mRNA is used as a hybridization probe. Figure 13-17 illustrates this procedure, which is carried out as follows.

1. Cells are replica-plated onto paper (i.e., colonies grown on agar from single cells are blotted onto paper so that the source colony for each replica spot can be identified).
2. The plated cells are lysed with alkali, releasing their DNA onto the paper.
3. The DNA on the paper is dried.
4. The paper spots are hybridized with radioactive mRNA or cDNA that corresponds to the desired gene. After hybridization, the paper is washed to remove nonhybridized radioactivity.
5. Autoradiography is performed to detect hybrids.

Since only colonies containing the passenger DNA of interest will hybridize with the radioactive probe, only those spots on the replica will be visible on the autoradiograph. When the passenger gene is cloned and expressed by the host, immunologic detection of the gene product may be an alternative means of identifying colonies containing the desired passenger DNA.

CLINICAL CORRELATION: XERODERMA PIGMENTOSUM

Clinical Symptoms

Xeroderma pigmentosum is a recessive inherited disorder caused by a defect in the ability to repair DNA damaged by ultraviolet light. The clinical symptoms are variable and tend to be progressive. They may appear in the first few years of life as an abnormal reddening of the skin, edema, and blistering that occur after exposure to sunlight. Excessive freckling, keratosis, and skin cancer may develop later. The most common manifestations of the disease are recurrent skin cancer and pigmentary changes on exposed areas of skin. A second form of xeroderma pigmentosum is characterized by neurologic abnormalities in addition to skin problems. The neurologic symptoms include progressive mental retardation, microcephaly, and areflexia. These patients also exhibit a variety of ocular abnormalities, including atrophy of the lower lid.

Biochemical Defect

Both forms of xeroderma pigmentosum are caused by a deficiency in the repair of pyrimidine dimers induced by ultraviolet light. The exact identity of the deficient enzyme or enzymes is not known. The cell contains more than one system capable of repairing pyrimidine dimers; a defect in any of these could potentially lead to xeroderma pigmentosum.

Treatment

The most common approach to managing xeroderma pigmentosum consists of reducing the exposure to sunlight, treating skin lesions by surgical removal and chemotherapy as they appear, and genetic counseling. Patients are advised to wear dark glasses and protective clothing and to use appropriate sunscreen lotions and thick makeup. Genetic counseling is given to inform the patient and his or her family about the inherited nature of the disease.

SUGGESTED READING

1. Boyer PD (ed): Nucleic Acids. In: The Enzymes. Vol. 14(A). Academic Press, Orlando, FL, 1981
2. Kornberg A: DNA Replication. Freeman, San Francisco, 1980
3. Kornberg A: DNA Replication. Supplement. Freeman, San Francisco, 1984
4. Recombination at the DNA Level. Cold Spring Harbor Symp Quant Biol, Vol. 49, 1986
5. Friedberg EC: DNA Repair. Freeman, San Francisco, 1980
6. Watson JD, Hopkins NH, Roberts JW et al: Molecular Biology of the Gene. 4th Ed. Vol. 1. Benjamin-Cummings, Menlo Park, CA, 1987

STUDY QUESTIONS

Directions: Answer the following questions using the key outlined below:
- **(A)** if 1, 2 and 3 are correct
- **(B)** if 1 and 3 are correct
- **(C)** if 2 and 4 are correct
- **(D)** if only 4 is correct
- **(E)** if all four are correct

1. Which of the following statements about DNA replication is or are not true?
    1. Each of the DNA strands serves as a template, but not at the same time.
    2. DNA replication is semiconservative.
    3. Eucaryotic chromosomes have only one replication initiation site.
    4. Helicases are important in the process of initiation.

2. The primasome
    1. contains a primase
    2. activates transcription
    3. contain primase and several other proteins that cooperate to synthesize the RNA primers required in replication
    4. contains several proteins that function to stabilize the replication fork

3. Elements associated with the replication fork include
    1. DNA polymerase I
    2. primase
    3. DNA polymerase III
    4. Okazaki fragments

4. General recombination
    1. occurs in long regions of homologous sequences
    2. does not require homologous DNA
    3. involves the formation of a Holliday structure
    4. does not involve rotation of the participating double helices

5. Site-specific recombination
   1. involves rotation of the double helices
   2. requires long homologous sequences but also utilizes transposons
   3. only requires short palindromic sequences in order for recombination to occur
   4. requires short homologous sequences

6. True statements concerning xeroderma pigmentosum include
   1. Xeroderma pigmentosum is associated with unusual skin sensitivity to sunlight
   2. Xeroderma pigmentosum is due to defective ability to repair pyrimidine dimers
   3. Xeroderma pigmentosum patients are advised to wear protective clothing and use sunscreen lotions.
   4. There is no increased risk of skin cancer among xeroderma pigmentosum patients.

Directions: For each of the following multiple choice questions, choose the most appropriate answer.

7. In the repair of thymine dimers, which of the following enzymes would most likely not be involved?
   A. UvrABC endonuclease
   B. DNA ligase
   C. DNA polymerase I
   D. Primase

8. When one purine-pyrimidine base pair in the parental DNA is replaced by a different purine-pyrimidine base pair in the daughter DNA, the mutation is called a
   A. transversion
   B. reversion
   C. transition
   D. site-specific mutagenesis

Items 9–11 Consider the following mutations:

A. CCTC → CCAC   B. CCTC → CCCTC   C. CCTC → CCCC

9. Which mutation is most likely to result in an altered reading frame?

10. Which mutation corresponds to a transversion?

11. Which mutation corresponds to a transition?

Items 12–15 Consider the following terms:
   A. Reverse transcriptase
   B. Genomic library
   C. Passenger DNA
   D. Restriction endonuclease

Choose the term that is most closely associated with each of the following statements.

12. Double-stranded cDNA that is to be cloned

13. An enzyme used to prepare the vector-passenger DNA adduct

14. An enzyme used in the synthesis of cDNA

15. An example of recombinant DNA

14

GENE EXPRESSION

Transcription/Translation/Gene Expression in Mitochondria/
Clinical Correlation

The student should be able to:

Learning Objectives

1. Define gene expression.
2. Distinguish the roles of the different eukaryotic RNA polymerases.
3. Describe the events in the initiation, elongation, and termination of transcription, and define the roles of promoter and terminator sequences.
4. Define the posttranscriptional processing of RNA, and give examples for each kind of RNA.
5. Distinguish introns and exons.
6. Describe at least two examples of transcriptional control of gene expression.
7. Name four properties of the genetic code.
8. Describe the events in initiation, elongation, and termination of protein synthesis.

Perspective: Transcription

Genetic information is *expressed* when the information content of DNA is used to determine cellular properties and to direct cellular activities. The properties and activities of the cell are ultimately determined by its proteins and enzymes, so genetic information is expressed when DNA directs the synthesis of proteins. However, gene expression is mediated by RNA. mRNA codes for the amino acid sequence of proteins, and rRNA and tRNA also participate directly in protein synthesis. Therefore, the initial event in gene expression is *transcription*, the DNA-directed synthesis of RNA.

TRANSCRIPTION

RNA synthesis is catalyzed by *RNA polymerases*. These enzymes use a DNA strand as a template to synthesize a complementary RNA strand. Only one strand of the double-stranded helix is used as a template; the product is complementary single-stranded RNA. In prokaryotes, a single

RNA polymerase synthesizes all types of RNA. In eukaryotes, the transcription of nuclear DNA is divided among three RNA polymerases: rRNA, mRNA, and tRNA are synthesized by RNA polymerases I, II, and III, respectively, except that RNA polymerase III also synthesizes 5S rRNA. RNA polymerases do not transcribe DNA indiscriminately. Instead, they recognize and interact initially with specific control sequences in DNA that determine where transcription begins. These interactions are described below.

Promoters and Terminators

RNA polymerases recognize and bind to certain sequences, called *promoters*, on the DNA template. Binding of RNA polymerase to a promoter initiates RNA synthesis. *Terminator* sequences at the end of the gene cause the RNA polymerase to stop transcribing and dissociate from the DNA.

Prokaryotic Promoters

Prokaryotic promoters are about 55 base pairs long, and include the site of transcription initiation. The promoter contains two regions, each about six base pairs long, that are highly conserved. Although the sequence of these regions varies slightly from promoter to promoter, it is possible to derive for each of them an idealized or optimal sequence called a *consensus sequence*. One of the consensus sequences in the promoter is centered at about the -10 position relative to the first transcribed base of the template (designated position $+1$) and the other is located at about -35. The -10 region, also called the *Pribnow box*, has the consensus sequence

$$\text{TATAAT}$$

and the -35 region has the consensus sequence

$$\text{TTGACA}$$

RNA polymerase interacts with DNA in the areas of the consensus sequences. This interaction is important in the process of initiating transcription.

Eukaryotic Promoters

Each of the eukaryotic RNA polymerases recognizes a particular type of promoter region. The promoter for RNA polymerase I (the polymerase that transcribes the rRNA genes) is estimated to extend about 140 base pairs upstream (i.e., to the 5' side) of the transcription start site. Within this region, the key sequences for initiation are a highly conserved TTT at about -30 and a G at -16. rRNA genes occur in tandem arrays separated by spacer sequences. These spacers contain repeated promoters that stimulate the transcription of downstream genes.

mRNA is synthesized by RNA polymerase II. The promoters recognized by this polymerase usually have two very highly conserved consensus sequences (Fig. 14-1). The first, located at −25, has the sequence

$$\text{TAT}^{T}_{A}\text{ATA}$$

and is called the *TATA* or *Hogness box*. The second consensus sequence, called the *CAAT box*, is located further upstream and has the sequence

$$\text{GGCC}^{T}_{A}\text{ATCT}$$

In addition, multiple copies of a GC-rich sequence are often found upstream from the TATA box. The CAAT box and the GC-rich sequences are less highly conserved than the TATA box.

The action of RNA polymerase II can also be stimulated by regions called *enhancers*, which may contain several hundred base pairs and may be separated from the nearest promoter by several thousand base pairs. Not only can enhancers exert their influence at a distance, but they can stimulate promoters both upstream and downstream. The activity of enhancers is tissue-specific. They may be involved in controlling the differential gene expression of different tissues.

RNA polymerase III, which transcribes the genes for tRNA and 5S rRNA, recognizes an internal promoter—i.e., a promoter located within the transcribed region of the gene. The internal promoter required by RNA polymerase III consists of conserved sequences called the *A* and *B blocks*. The distance between these blocks varies in different genes. In the case of the 5S rRNA gene, the promoters bind additional protein factors which enhance transcription. Transcription by RNA polymerase III is also affected by sequences that flank the genes on the upstream and downstream sides.

Terminators

Both eukaryotic and prokaryotic RNA polymerases recognize termination sequences (*terminators*) on the template strand, which signal the end of transcription. Termination in prokaryotes is of two types. In ρ-*independent termination*, the terminator sequence alone causes chain termination. In ρ-*dependent termination*, a protein factor called rho protein (ρ) is required. In ρ-independent genes, the terminator has a GC-rich palindrome that precedes a sequence of adenine residues on the template strand (Fig. 14-2). When the GC-rich palindome is transcribed, it undergoes intrachain base pairing to form a stem-and-loop structure that is believed to cause the RNA polymerase to dissociate from the DNA template. ρ-Dependent terminators code for weaker stem-and-loop structures. The exact role of the ρ factor is not clear, but it is believed to cause the RNA polymerase to dissociate from the DNA template by binding at the stem structure.

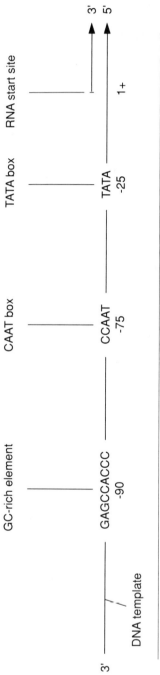

Fig. 14-1. A diagram of a typical eukaryotic promoter region recognized by RNA polymerase II.

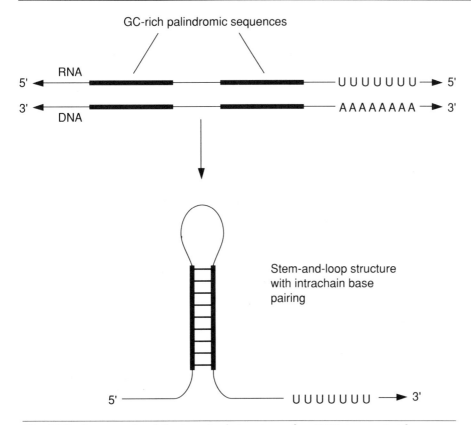

Fig. 14-2. An example of the terminal structure of an RNA transcript for a gene that shows ρ-independent termination.

Promotion and Termination: Key Points

1. Both prokaryotic and eukaryotic RNA polymerases recognize promoter regions, which bind the polymerase and induce it to start transcription.
2. Most kinds of promoters lie upstream of their associated gene, but some are internal.
3. All RNA polymerases recognize terminator sequences on the template strand. Some bacterial terminators require the action of the protein factor ρ, whereas others do not.

RNA Polymerases: Properties

Prokaryotic RNA Polymerases

Prokaryotic RNA polymerases are multisubunit enzymes. The RNA polymerase of *E. coli* has been studied extensively. It consists of five different subunits: α, β, β', σ, and ω. The intact holoenzyme comprises two copies of the α subunit and one of each of the others ($\alpha_2\beta\beta'\sigma\omega$). The entire enzyme has a molecular weight of about 450 kilodaltons and covers a DNA segment approximately 60 base pairs long. The σ subunit is only loosely bound to the holoenzyme, and can dissociate to leave a core enzyme. The σ subunit and the core enzyme have distinct functions. The holoenzyme is capable of binding loosely to the double helix and moving along it in search of a promoter. The σ subunit recognizes the promoter, and probably interacts with both the -35 and the -10 consensus sequences. After transcription has been initiated, the σ subunit is released, and the remaining core enzyme elongates the RNA chain. As described in the Clinical Correlation at the end of this chapter, the antibiotic rifampin acts by inhibiting bacterial RNA polymerases.

Eukaryotic RNA Polymerases

The three nuclear RNA polymerases of eukaryotes all have many subunits (at least 10), but the details of their quaternary structure are not known. Eukaryotic RNA polymerases require the assistance of auxiliary proteins called *transcriptional factors* in order to initiate transcription at the correct sites. α-Amanitin, one of the toxins in the poisonous mushroom *Amanita phalloides*, differentially inhibits eukaryotic RNA polymerases. The three polymerases can be distinguished by their response to this toxin: RNA polymerase II is extremely sensitive to it, RNA polymerase III is less sensitive, and RNA polymerase I is insensitive.

The Process of Transcription

For transcription to occur, RNA polymerase must progressively unwind the DNA double helix. Ribonucleotides form complementary base pairs with the template strand, and the RNA polymerase catalyzes the formation of $5' \rightarrow 3'$ phosphodiester bonds. The immediate product of transcription is called the *primary transcript*. A primary transcript represents one *transcriptional unit* of DNA—the segment of template strand between an initiation and a termination site. Transcriptional units vary in size, and may contain more than one gene. Transcription may be divided into three phases: initiation, elongation, and termination. The process is much better understood in prokaryotes (especially *E. coli*) than in eukaryotes.

Transcription in Prokaryotes

Initiation. Chain initiation involves three events:

1. Binding of RNA polymerase to the promoter region
2. Local separation of the double helix to produce a single-stranded template
3. Synthesis of the first phosphodiester bond of the new RNA molecule

The complex formed between the DNA promoter and the RNA polymerase holoenzyme is called the *closed complex* because the DNA is double helical. RNA polymerase in the closed complex form is believed to interact primarily with the -35 consensus sequence. Starting within the −10 consensus sequence and proceeding downstream, the RNA polymerase unwinds 10 to 20 base pairs of the double helix. The strands separate, forming the *open complex*. RNA synthesis begins at a specific start site on the template strand (no primer is needed). Usually the first RNA nucleotide is A or G; RNA polymerases prefer to start the chain with a purine and may scan the template until a suitable purine start site is found. Synthesis of the first phosphodiester bond completes chain initiation.

Chain Elongation. Chain elongation involves two processes:

1. The DNA helix is progressively unwound and separated at the front of the polymerase, and is allowed to anneal and rewind at its rear.
2. The catalytic part of the RNA polymerase rides the crest of the template single strand, synthesizing the growing RNA chain.

The σ subunit dissociates from the RNA polymerase holoenzyme after the RNA chain is initiated. The dissociation of the σ subunit from the open complex makes the interaction between RNA polymerase and the template strand more secure. An additional factor, called *nus A*, probably binds reversibly to actively transcribing RNA polymerases and influences both the rate of RNA synthesis and chain termination.

The RNA polymerase, DNA, and nascent RNA chain constitute the *elongation complex* (Fig. 14-3). The single-stranded DNA region is usually about 17 base pairs long. A short stretch of nascent RNA (about 12 bases long) is base-paired with the template strand. This RNA-DNA hybrid is progressively disrupted by the reforming double helix as the polymerase moves along the DNA, and the nascent RNA chain extends away from the DNA.

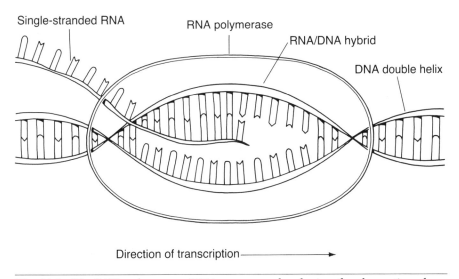

Fig. 14-3. The RNA polymerase/DNA open complex during the elongation phase of transcription.

Termination. When the polymerase encounters a terminator sequence, chain elongation ceases, the primary transcript is released, and the polymerase dissociates from the DNA. As described above, termination may be either ρ-dependent or ρ-independent.

Transcription in Eukaryotes

Transcription is more complicated in eukaryotes than in prokaryotes. As described above, eukaryotic transcription is influenced by enhancer sequences as well as by promoters and terminators. In addition,

1. All three eukaryotic RNA polymerases require the presence of additional protein *transcriptional factors* for optimal transcription.
2. Because of the nucleosome structure of chromatin, the unwinding of eukaryotic DNA to produce the single-stranded template is more complicated than in prokaryotes.

Available evidence indicates that transcriptionally active eukaryotic genes have nucleosomes that are structurally altered in a way consistent with unwinding of the DNA from around the nucleosome core. The DNA appears to be less condensed and more disorganized than in nontranscribing DNA. Such changes in the structure of chromatin would certainly enhance transcription by RNA polymerase. However, RNA polymerase alone is not believed to cause these dramatic structural changes.

RNA Processing

RNA processing consists of the conversion of the primary transcript into mature, functional RNA. In many cases, flanking sequences at either end of the transcript and spacer sequences between coding regions are removed by endonucleases. Eukaryotic primary transcripts—particularly nascent mRNAs—often contain noncoding sequences (*introns*) within genes, which are spliced out during processing.

rRNA

rRNA Processing in Bacterial Cells. Bacterial ribosomes contain 16S, 23S, and 5S RNAs. All three are transcribed as part of a single 30S primary transcript (Fig. 14-4). The genes in this transcript are separated by noncoding *spacers*. In addition, a *leader sequence* separates the 16S rRNA gene from the promoter, and a *trailer sequence* follows the last gene. As shown in Figure 14-4, the spacer between the 16S and 23S genes contains one or two tRNA genes, and the trailer after the 5S gene may also contain tRNAs. Ribosomal proteins begin to associate with the 30S precursor while transcription is still under way. As soon as the transcript is complete, the enzyme RNase III cleaves it to yield pre-16S, pre-23S, and pre-5S rRNAs (as well as pre-tRNAs). The mature rRNAs and tRNAs are produced by further nuclease activity. The mature rRNAs then associate with the rest of the ribosomal proteins to form the small and large ribosomal subunits.

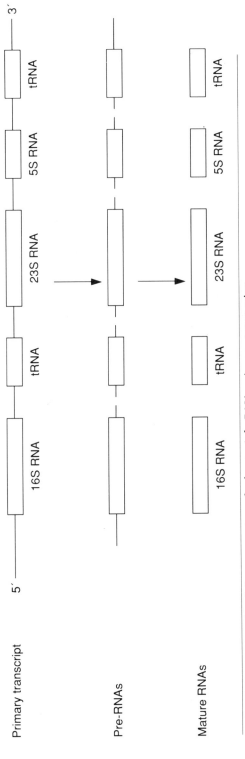

Fig. 14-4. An example of the processing of a bacterial rRNA primary transcript.

rRNA Processing in Eukaryotic Cells. Eukaryotic ribosomes have four RNA components: 18S, 28S, 5.8S, and 5S. The genes for all but the 5S component are transcribed as part of a single 45S precursor. Figure 14-5 shows the processing of this precursor. The transcript includes a 5' leader sequence and usually a 3' trailer. Ribosomal proteins begin to associate with the precursor while it is still being transcribed. The methylation of specific nucleotides also begins during transcription. 5S rRNA undergoes no processing: the primary transcript is mature 5S rRNA. As shown in Figure 14-5, the mature 18S rRNA becomes part of the small 40S ribosomal subunit, and the other rRNAs are incorporated into the large 60S subunit.

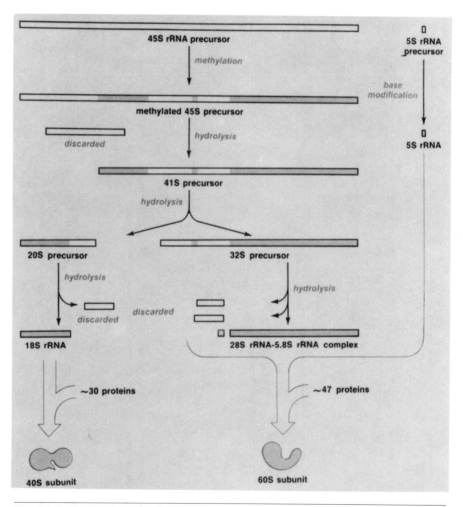

Fig. 14-5. Processing of eukaryotic rRNA. (From McGilvery RW: Biochemistry: A Functional Approach. 3rd Ed. WB Saunders, Philadelphia, 1983, with permission.)

tRNA

tRNA Processing in Bacterial Cells. Bacterial tRNA transcripts have 3' and 5' flanking sequences that are later removed, and some transcripts include several tandem tRNAs separated by spacers. As shown in Figure 14-6, tRNA transcripts are processed by specific RNases to yield mature tRNAs. The best characterized of these RNases, *RNase P*, removes the 5' flanking sequence. It consists of a protein and a small RNA fragment. Under certain conditions, the RNA fragment alone can catalyze the reaction. Another enzyme, *RNase D*, removes the 3' flanking region, exposing the CCA amino acid acceptor sequence. Various enzymes catalyze the specific base modifications that result in mature tRNA. Finally, a specific nucleotidyl transferase enzyme replaces the essential CCA terminal sequence if it becomes cleaved off.

tRNA Processing in Eukaryotes. In eukaryotes, as in bacteria, tRNA genes are transcribed as part of larger precursors that undergo processing. Precursors containing several tRNAs are known in yeast, but have not been well documented in other eukarotes.

Splicing in Immature tRNAs and rRNAs

In some eukaryotes, certain tRNA and rRNA genes are interrupted by sequences that are spliced out during processing. This splicing occurs in addition to the processing of the primary transcript already described.

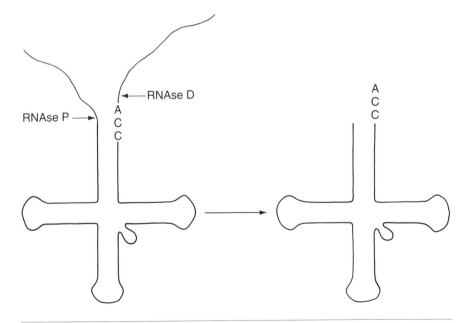

Fig. 14-6. The processing of precursor tRNA.

mRNA

Mature mRNAs code for the amino acid sequences of proteins. Non-translated flanking regions occur on either side of each coding region, and contain the codons that initiate and terminate protein synthesis (translation).

mRNA Processing in Prokaryotes. Translation of a prokaryotic mRNA often begins before transcription is complete. Simultaneous transcription and translation is possible in bacteria for two reasons:

1. No mRNA processing occurs.
2. There is no nucleus, so both the transcription and the translation machinery have simultaneous access to the DNA.

The bacterial genome is organized so that genes for proteins that have related functions or participate in a common pathway are arranged in tandem on a segment of DNA called an *operon*. Each operon has a single promoter and is transcribed as a unit. The mRNA from an operon therefore codes for several proteins, and is called *polygenic* (*polycistronic*). Each protein-coding region on the mRNA has its own start and stop codons.

mRNA Processing in Eukaryotes. Mature eukaryotic mRNAs are *monocistronic*: they code for only one protein. The processing of nuclear mRNA consists of the following steps:

1. Attachment of the 5' 7-methylguanylate cap
2. Attachment of the 3' poly A tail
3. Splicing out of noncoding introns

The *5' 7-methylguanylate cap* (see Fig. 12-14B) is added only to mRNAs. First, a guanylate is added by an unusual 5' → 5' linkage to the first transcribed nucleotide of the mRNA. The guanylate is then methylated at N-7. This cap structure is believed to aid in translation.

A *poly A tail*—a string of about 30 to 150 adenine residues—is added to the 3' end of many mRNAs immediately after transcription. First, a specific endonuclease recognizes the sequence AUAAA in the 3' region of the precursor mRNA, and cleaves off the 3' end of the transcript downstream from this sequence. The enzyme *poly A polymerase* then constructs the poly A tail on the new 3' end. The function of the poly A tail is not completely understood; some mRNAs have no poly A tail but are competently exported to the cytoplasm and translated. Poly A tails are shorter when observed in the cytoplasm than when constructed in the nucleus.

Because of the presence of intragenic noncoding sequences (introns), mRNA primary transcripts are much larger than mature mRNAs. These introns interrupt the coding sequence and must be removed. The segments of coding sequence are called *exons*. Figure 14-7 shows the processing of the ovalbumin mRNA primary transcript, which has several

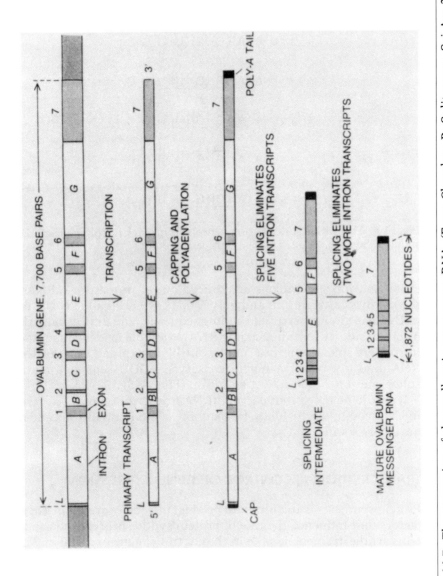

Fig. 14-7. The processing of the ovalbumin precursor mRNA. (From Chambon P: Split genes. Sci Am 244:60, 1980, with permission.)

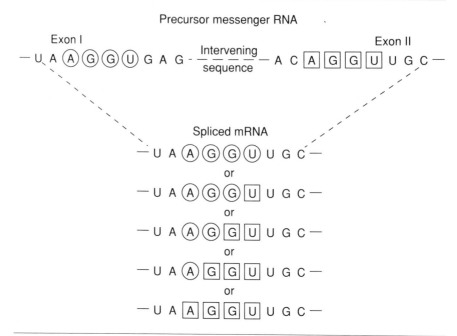

Fig. 14-8. The mRNA splice site sequence. (From Crick F: Split genes and mRNA splicing. Science, 204:265, 1969, with permission.)

exons and introns. Introns are spliced out of immature mRNAs after the methylguanylate cap and poly A tail have been added. The mature mRNA has an uninterrupted coding sequence, which is usually flanked on either side by some noncoding RNA. Available evidence suggests that introns are spliced out by a complex involving a small nuclear RNA (U1 RNA) and a protein. A highly conserved AGGU sequence marks the splice sites on either side of an intron (Fig. 14-8).

It should be noted that not all mRNA transcripts contain introns. The mRNAs coding for histones, for example, contain no introns and do not have poly A tails.

TRANSCRIPTIONAL CONTROL OF GENE EXPRESSION

Protein synthesis is the ultimate product of gene expression. Therefore, factors that influence the rate of protein synthesis or the choice of proteins synthesized participate in the control of gene expression. Much of this control is exerted at the level of transcription.

Transcriptional Control in Prokaryotes

Bacterial proteins are generally of three types: *constitutive, inducible,* and *repressible.*

Constitutive proteins are present in the cell in more or less constant amounts; the cell is "programmed" to produce these proteins at a constant rate.

Inducible proteins are synthesized in response to specific stimuli, and are otherwise absent or present in low amounts.

Repressible proteins are synthesized under normal conditions, but their synthesis is diminished by specific stimuli.

Specific Regulatory Proteins: Repressors and Activators

Control of transcription in prokaryotes is mediated largely by *repressors* and *activators*—regulatory proteins that bind to the DNA and either repress or enhance transcription of the downstream gene or operon.

Repressors. Repressors bind to a region of DNA called the *operator*, which is located immediately downstream from the promoter and may overlap it. When the repressor is bound to the operator, RNA polymerase cannot bind to the promoter and the gene is not transcribed. The activity of most repressors is controlled by a small molecule, which is usually chemically related to the product or substrate of the affected enzyme(s).

1. The repressors of repressible genes are controlled by a metabolite called a *corepressor*. The repressor-corepressor complex binds to the operator and prevents transcription.
2. The repressors of inducible genes are controlled by a small molecule called an *inducer*. These repressors are normally bound to the operator, preventing transcription. Binding of the inducer to the repressor causes the repressor to dissociate from the DNA, which allows transcription to occur. Inducers are usually metabolically related to the substrate of the induced enzyme(s).

Activators. Activators are regulatory proteins that bind to a site near the promoter but different from the operator. Activators are usually positive-acting: they enhance the binding of RNA polymerase to the promoter, and thus stimulate transcription. Like repressors, activators may be controlled by small-molecule effectors. A given activator may bind to more than one site in the genome and influence a number of separate operons. Proteins of this type can also inhibit transcription of some operons; in this case, the protein is a negative-acting regulatory protein. Operons that are not associated with upstream operators are expressed constitutively.

Transcriptional Control in Prokaryotes: Key Points

Repressors
1. Repressor proteins bind to operator sites, and prevent transcription by preventing the binding of RNA polymerase.
2. Many repressors are controlled by an effector molecule, which is usually a metabolite of the affected pathway. These cofactors are of two types: corepressors and inducers.
   a. Corepressors cause their repressor proteins to bind to the operator. Repressor proteins of this type regulate repressible operons.
   b. Inducers cause their repressor proteins to dissociate from the operator. Repressor proteins of this type regulate inducible operons.

Activators

3. Activators are regulatory proteins that bind to a site other than the operator. Activators usually stimulate transcription but in some cases inhibit it.

The E. coli lac Operon

A classic example of transcriptional control in prokaryotes is the *lac* operon (lactose operon) of *E. coli* (Fig. 14-9). The *lac* operon has three structural genes, which code for proteins involved in the utilization of β-galactosides. In addition, the operon contains an operator site and a *regulatory gene* that codes for the operon's repressor protein. The *lac* operon is inducible: the repressor protein remains bound to the operator and prevents transcription unless inducer molecules (various β-galactosides) are present. Binding of an inducer inactivates the repressor and causes it to dissociate from the operator. The three structural genes of the *lac* operon are transcribed as a single polycistronic mRNA.

Transcriptional Control in Eukaryotes

The control of transcription in eukaryotes is both more complicated and less well understood than in prokaryotes. It is generally agreed that proteins called *transcriptional factors*, which interact directly with DNA, are involved.

Steroid hormones have long been known to stimulate the transcription of specific genes in their respective target tissues. These lipid-soluble hormones enter the cell and associate with a cytosolic receptor protein. The resulting hormone-receptor complex enters the nucleus, binds to specific DNA sites, and directly activates the transcription of certain genes (see Chapter 15). Steroid hormones also appear to complex with other receptor proteins, which then interact with the steroid-induced mRNAs so as to increase their stability.

Perspective: Protein Synthesis

The process of protein synthesis is called *translation* because it involves the translation of mRNA coding sequences into the correct amino acid sequences of proteins. Translation is mediated by tRNA:

1. Each amino acid is bound by one or more tRNAs that are specific for that amino acid.
2. The three-nucleotide *anticodon* on each tRNA base-pairs with the complementary three-nucleotide *codon* segments on mRNA.

As a result, each tRNA anticodon recognizes only certain mRNA codons. When a tRNA anticodon recognizes and binds an mRNA codon during protein synthesis, the amino acid carried by the tRNA is added to the polypeptide chain—i.e., the nucleotide sequence of the codon is translated as the amino acid carried by the tRNA. Thus, codons are said to code for specific amino acids.

(continued)

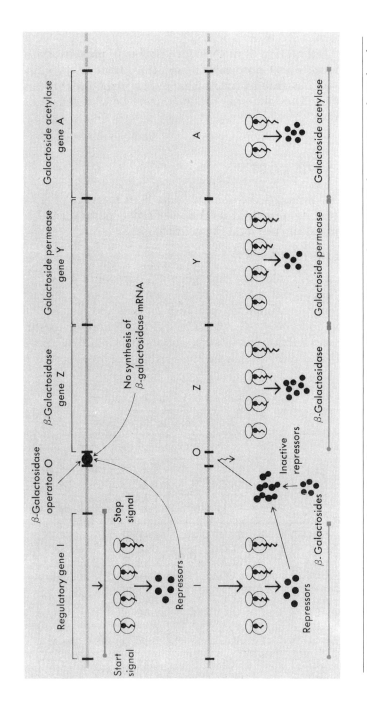

Fig. 14-9. The regulation of the *E. coli lac* operon. (From Watson JD, Hopkins NH, Roberts JW et al: Molecular Biology of the Gene. Benjamin-Cummings, Menlo Park, CA, 1987, with permission.)

> Codons are read from left to right (5' to 3'). The genetic code consists of the codons that code for each of the amino acids incorporated into proteins, plus start and stop codons that initiate and terminate translation.
>
> The interaction of mRNA with tRNAs in protein synthesis is not a random event occurring in solution. Instead, protein synthesis is an orchestrated process that takes place on the surface of ribosomes. The following sections describe the elements of protein synthesis.

THE GENETIC CODE

Table 14-1 presents the genetic code. Four features of the code require elaboration: (1) punctuation codons; (2) degeneracy; (3) the wobble concept; and (4) the universality of the code.

Punctuation Codons

As shown in Table 14-1, the genetic code contains *initiation* or *start codons* that signal the start of the coding sequence and *termination* or *stop codons* that signal its end. In both prokaryotes and eukaryotes the initiation codon is AUG, which is also the codon for methionine. When this codon functions as a start codon, it is read by special initiation tRNAs that are distinct from the methionine tRNAs that read internal

Table 14-1. The Genetic Code

5' Base	Middle Base				3' Base
	U	C	A	G	
U	UUU ⎤ Phe UUC ⎦ UUA ⎤ Leu UUG ⎦	UCU ⎤ UCC ⎥ Ser UCA ⎥ UCG ⎦	UAU ⎤ Tyr UAC ⎦ UAA[b] Stop UAG[b] Stop	UGU ⎤ Cys UGC ⎦ UGA[b] Stop UGG Trp	U C A G
C	CUU ⎤ CUC ⎥ Leu CUA ⎥ CUG ⎦	CCU ⎤ CCC ⎥ Pro CCA ⎥ CCG ⎦	CAU ⎤ His CAC ⎦ CAA ⎤ Gln CAG ⎦	CGU ⎤ CGC ⎥ Arg CGA ⎥ CGG ⎦	U C A G
A	AUU ⎤ AUC ⎥ Ile AUA ⎦ AUG[a] Met	ACU ⎤ ACC ⎥ Thr ACA ⎥ ACG ⎦	AAU ⎤ Asn AAC ⎦ AAA ⎤ Lys AAG ⎦	AGU ⎤ Ser AGC ⎦ AGA ⎤ Arg AGG ⎦	U C A G
G	GUU ⎤ GUC ⎥ Val GUA ⎥ GUG[a] ⎦	GCU ⎤ GCC ⎥ Ala GCA ⎥ GCG ⎦	GAU ⎤ Asp GAC ⎦ GAA ⎤ Glu GAG ⎦	GGU ⎤ GGC ⎥ Gly GGA ⎥ GGG ⎦	U C A G

[a] Initiation (start) codons.
[b] Termination (stop) codons.

methionine codons. In prokaryotes the initiation tRNA carries the modified amino acid *N*-formylmethionine, whereas in eukaryotes it carries methionine. The initial amino terminal residue of all proteins is therefore either methionine or *N*-formylmethionine. In most proteins this residue is removed immediately after chain initiation by a special aminopeptidase. The codon GUG has been observed to function as a start codon on rare occasions in prokaryotes.

The termination codons are UAA, UAG, and UGA. These codons have no complementary tRNAs. Instead, they are recognized by proteins called *release factors*, which catalyze the release of the polypeptide chain from the ribosome complex. Prokaryotes have two release factors, one recognizing UAA and UAG and the other recognizing UAA and UGA. In eukaryotic cells, one release factor recognizes all three stop codons.

Degeneracy

Most amino acids are coded for by more than one codon. This property of the genetic code is referred to as *degeneracy*. Degeneracy results from the fact that most amino acids bind to more than one specific tRNA. Serine, for example, binds to multiple specific tRNAs, each with a different anticodon.

The Wobble Concept

Although 61 codons specify amino acids, both prokaryotes and eukaryotes have fewer than 61 tRNAs. As it turns out, many tRNAs are capable of translating more than one codon. It was observed that the codons read by a single tRNA always differ only in the third position—i.e., in the position complementary to the 5′ nucleotide of the tRNA anticodon. This observation led to the formulation of the *wobble concept* and the wobble rules. Because of the geometry of the tRNA molecule, the base in the 5′ position of the anticodon (the *wobble position*) can pair with a base other than its normal complement, provided the resulting ribose-ribose distance is close to the ribose-ribose distance for the standard base pair. Therefore, U in the 5′ anticodon position can base pair with either A or G, and inosine (I) in the 5′ anticodon position can base pair with A, U, or C.

Degeneracy and Wobble. It is important to remember that degeneracy and wobble, although they appear similar, are not the same:

> *Degeneracy* occurs when multiple tRNAs that carry the same amino acid read different codons.
> *Wobble* occurs when one tRNA reads more than one codon.

In both cases, more than one codon is translated as a single amino acid.

The Universality of the Genetic Code

The nuclear DNA of all organisms expresses the same genetic code. However, the mitochondrial DNA of eukaryotes employs a slightly different genetic code (Table 14-2). Eukaryotes therefore have two functioning genetic codes.

Table 14-2. Differences between the Nuclear and Mitochondrial Genetic Codes

	Translation	
Nuclear Codon	Nuclear Code	Mitochondrial Code
AUA	Ile	Met
UGA	STOP	Trp
AGA	Arg	STOP
AGG	Arg	STOP

SYNTHESIS OF AMINOACYL tRNAs

Before a tRNA can participate in protein synthesis, it must be charged with the appropriate amino acid. The amino acid is carried in acyl linkage by the 3' hydroxyl of the ribose moiety on the 3' adenosine residue of the tRNA (see Fig. 12-13). tRNAs are charged with their appropriate amino acid by a series of specific *aminoacyl-tRNA synthetases*. These enzymes catalyze a two-step process involving initial phosphorylation of the amino acid by ATP. The reaction is driven by the subsequent hydrolysis of PP_i. Each amino acid has a specific aminoacyl-tRNA synthetase. The aminoacyl-tRNA synthetases are among the most discriminating enzymes found in cells.

PROTEIN SYNTHESIS

Protein synthesis (translation) consists of three phases: initiation, elongation, and termination. The mRNA is read in the 3' to 5' direction, which corresponds to the amino terminal to carboxy terminal direction of the polypeptide chain. Protein synthesis in prokaryotes and eukaryotes will be considered separately.

Protein Synthesis in Prokaryotes

Initiation

Translation in bacteria is initiated in three steps. In addition to the two ribosomal subunits and mRNA, the process requires three protein *initiation factors* (IF1, IF2, and IF3), GTP, and the tRNA that carries the modified *N*-formylmethionine starter amino acid (see above under Genetic Code). The complex of fMet with its tRNA is abbreviated fMet-$tRNA_f$. The process of initiation is as follows (Fig. 14-10):

Step 1 The three initiation factors bind to the 30S (small) ribosome subunit. GTP participates in the binding process, and is believed to be attached to IF2.

Step 2 Both fMet-$tRNA_f$ and the mRNA associate with the complex generated in step 1, and IF3 is released. The order of association of fMet-$tRNA_f$ and mRNA does not matter. The aggregate formed in this step is called the *30S initiation complex*.

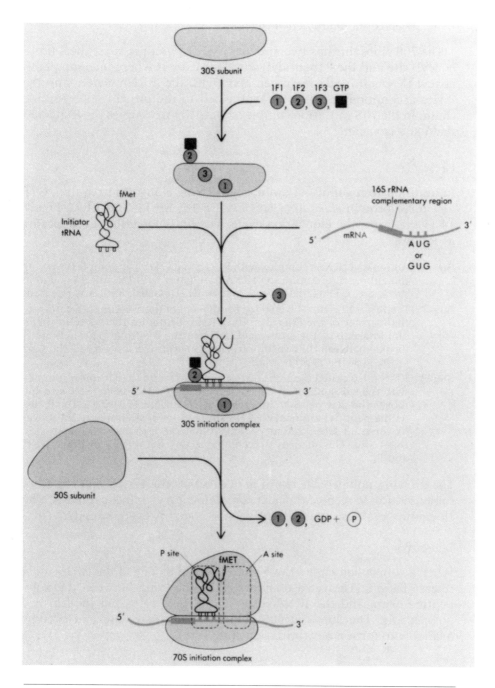

Fig. 14-10. Initiation of translation in prokaryotes. (From Watson JD, Hopkins NH, Roberts JW et al: Molecular Biology of the Gene. Benjamin-Cummings, Menlo Park, CA, 1987, with permission.)

Step 3 The 50S subunit joins the 30S initiation complex. The bound GTP is hydrolyzed, and IF1, IF2, GDP, and P_i are released. The product of this step is the *70S initiation complex*.

The 70S ribosome has two aminoacyl-tRNA binding sites: the A (aminoacyl) site and the P (peptidyl) site. The A site is where incoming aminoacyl tRNAs initially hydrogen bond with the mRNA codon, and the P site accommodates the tRNA attached to the growing polypeptide chain. In the 70S initiation complex, fMet-tRNA$_f$ occupies the P site and the A site is vacant.

Elongation

In addition to a supply of aminoacyl-tRNAs, elongation requires GTP and three protein *elongation factors*—EF-Tu, EF-Ts, and EF-G. Figure 14-11 diagrams the elongation cycle. The first elongation cycle occurs as follows:

Step 1 Aminoacyl tRNAs complex with EF-Tu and GTP. The complex carrying the appropriate aminoacyl tRNA base-pairs with the mRNA codon in the A site. GTP is hydrolyzed, and an EF-Tu-GDP complex is released.

Step 2 *Peptidyl transferase*, an enzyme component of the 50S subunit, catalyzes the transfer of the fMet from the tRNA located in the P site to the α-amino group of the aminoacyl tRNA in the A site. A peptide bond is formed between the two amino acids, generating a dipeptide—the start of the growing peptide chain.

Step 3 EF-G (also called the *translocase*) binds GTP and then forms a complex with the ribosome. This event ejects the free tRNA from the P site and moves the new peptidyl aminoacyl tRNA from the A site to the P site. As the peptidyl aminoacyl tRNA shifts to the P site, the mRNA moves by three nucleotides (one codon) so that the next codon is exposed in the A site. This process is accompanied by the release of EF-G, GDP, and P_i.

The growing polypeptide chain now occupies the P site, and the ribosome is ready to receive the next complementary aminoacyl tRNA. The elongation cycle is repeated until a stop codon is encountered.

Termination

When a stop codon enters the A site, it is read by one of the two protein release factors. The release factors cause the bond between the polypeptide chain and the tRNA to be hydrolyzed, releasing the finished polypeptide. The 30S and 50S subunits then dissociate and become available to form a new initiation complex.

Protein Synthesis in Eukaryotes

Initiation

Initiation of protein synthesis is more complicated in eukaryotes than in prokaryotes. In particular, it requires the participation of many more

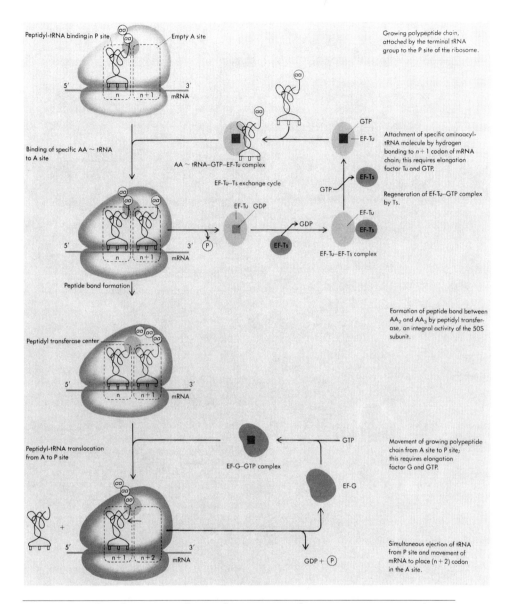

Fig. 14-11. The elongation phase of protein synthesis in procaryotes. (From Watson JD, Hopkins NH, Roberts JW et al: Molecular Biology of the Gene. Benjamin-Cummings, Menlo Park, CA, 1987, with permission.)

initiation factors, which are called *eukaryotic initiation factors* (*eIFs*). The process is diagrammed in Figure 14-12.

Step 1   The 40S ribosomal subunit binds eIF3 and eIF4C to form the 43S ribosomal complex.

Step 2   eIF2, GTP, and Met-tRNA$_i$ (the initiating tRNA) form a complex, which binds to the 43S ribosomal complex.

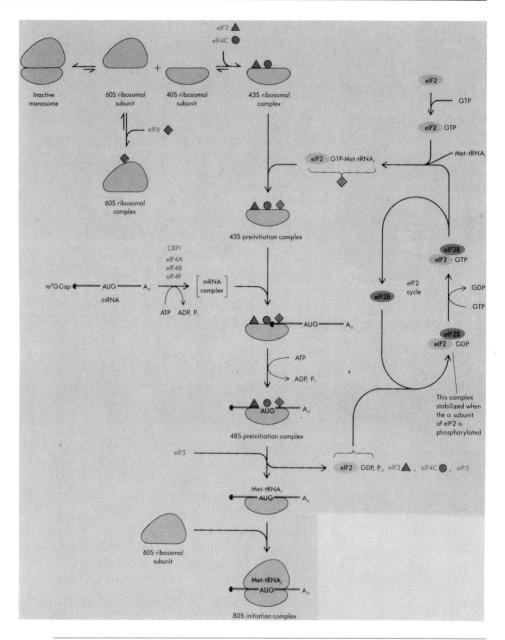

Fig. 14-12. Initiation of protein synthesis in eukaryotes. (Courtesy of Dr. Joan A. Steitz.)

Step 3   mRNA binds to the 43S complex in a process that involves eIF4E (the cap binding protein), eIF4A, eIF4B, and eIF1. The binding of mRNA is accompanied by hydrolysis of ATP to ADP and P_i. This step forms the *48S preinitiation complex.*

Step 4   eIF5 then mediates the binding of the 60S ribosomal subunit to form the *80S initiation complex*. GDP, P_i, and the bound initiation factors are released in this step.

As in the prokaryotic initiation complex, the initiator Met-tRNA$_i$ is located in the P site and the A site is ready for an incoming aminoacyl tRNA.

Elongation

Eukaryotes employ three elongation factors: EF1α, EF1βα, and EF2. Eukaryotic elongation is similar to the corresponding process in prokaryotes.

Step 1   EF1α forms a complex with GTP and an aminoacyl tRNA. This complex transfers the aminoacyl to the A site of the ribosome, with the concurrent hydrolysis of GTP.
Step 2   Peptidyl transferase catalyzes the formation of the peptide bond that transfers the initiating methionine or the growing polypeptide chain to the A site tRNA.
Step 3   EF2 (the eukaryotic translocase) mediates the release of the uncharged tRNA, the movement of the peptide-bearing tRNA to the P site, and the movement of the mRNA to the next codon. GTP is hydrolyzed in this step.
Step 4   EF1βα catalyzes a GTP/GDP exchange reaction with EF1α, which prepares EF1α to deliver another aminoacyl tRNA to the ribosome.

Termination

Eukaryotes have only one release factor, which reads all three stop codons. As in prokaryotes, the release factor hydrolyzes the bond between the peptide and the tRNA, releasing the peptide. GTP is concurrently hydrolyzed. The factors eIF3 and eIF6 then cause the 80S ribosome to dissociate into 40S and 60S units, which are ready for a new round of protein synthesis.

Posttranslational Modification of Proteins

Many proteins are modified after translation. Proteins may also undergo proteolytic cleavage, covalent modification of amino acid side chains, aggregation into quaternary structures, and the attachment of prosthetic groups.

Polyribosomes

As soon as the ribosome has moved beyond the initiation site on the mRNA, another ribosome can bind and begin translation. Assemblies of multiple ribosomes on a single mRNA strand are called *polyribosomes* or *polysomes*.

SYNTHESIS OF PROTEINS FOR EXPORT

Proteins destined for an extracellular location are synthesized on membrane-bound ribosomes. These proteins all have a highly hydrophobic amino segment called the *signal peptide*. The signal peptide becomes inserted in the membrane and guides the rest of the polypeptide chain through the membrane. At some point during chain elongation, the signal peptide is cleaved off by a specific *signal peptidase*. In bacterial cells

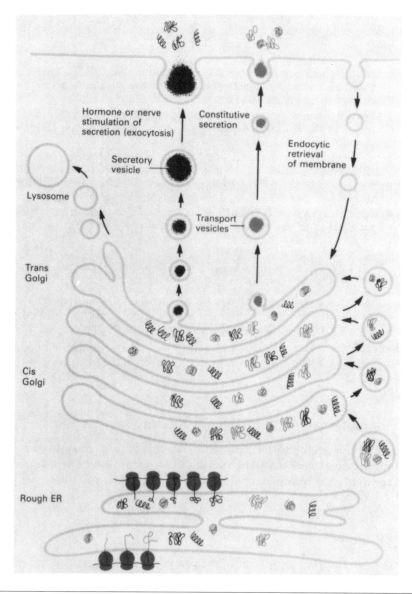

Fig. 14-13. The synthesis and secretion of secretory proteins. (From Darnell J, Lodish H, Baltimore D: Molecular Cell Biology. Scientific American Books, New York, 1986, with permission.)

such ribosomes are attached to the cell membrane, whereas in eukaryotes they are attached to the endoplasmic reticulum (ER). The proteins are secreted into the lumen of the ER, and are eventually exported from the cell via the golgi apparatus (Fig. 14-13). This process is called *co-translational excretion*. It must be noted that for some exported proteins (such as ovalbumin), no signal peptide has been detected.

In the case of proteins that are destined not for export but for incorporation into the membrane, the signal sequence is not cleaved off, and the protein remains inserted in the membrane. This process is called *co-translational insertion*. In eukaryotes, such proteins are transported from the ER to the plasma membrane by mechanisms that are not well understood.

TRANSLATIONAL CONTROL OF GENE EXPRESSION

Evidence for the control of gene expression at the level of translation is not as abundant as evidence for transcriptional control. However, the association of ribosomes with mRNA initiation sites may sometimes be under control. In mammalian reticulocytes, a mechanism involving the phosphorylation of eIF2 has been shown to inhibit the translation of hemoglobin mRNA.

GENE EXPRESSION IN MITOCHONDRIA

Mitochondria contain their own tRNAs and rRNAs, which are coded for by the mitochondrial genome. The mitochondrial DNA also codes for about 19 other proteins. Mitochondria have a completely independent protein synthesizing system, including their own RNA polymerase and ribosomes that are smaller than their cytosolic counterparts and resemble bacterial ribosomes. Mitochondria have fewer tRNAs than either bacterial cells or the eukaryotic cytoplasm, and because of the altered genetic code of mitochondria, some of these tRNAs recognize different codons.

The mitochondrial genome is a circular double helix. During transcription, both DNA strands are transcribed into very long RNA molecules that undergo processing to yield mature RNAs. In many instances, tRNAs appear as spacers between mRNA segments. Initiation codons are usually very close to the 5' end of mRNAs resulting in very short leader sequences.

Although the mitochondria contain their own genome, most mitochondrial proteins are coded for by the nuclear genome.

CLINICAL CORRELATION: ANTIBIOTICS THAT INTERFERE WITH GENE EXPRESSION

A number of antibiotics, some of which are described below, act by inhibiting gene expression in bacteria. These agents do not affect gene expression in the host, although they may have other side effects.

Tetracyclines. Tetracylines bind to the A site of prokaryotic ribosomes, preventing aminoacyl tRNAs from binding.

Erythromycin. Erythromycin binds to the 30S ribosomal subunit and disturbs the normal mRNA-tRNA interaction. Distorted ribosomes are produced which misread the genetic code. Streptomycin also inhibits the initiation of translation.

Rifampin. Rifampin inhibits the initiation of transciption by binding to the β subunit of bacterial RNA polymerase.

SUGGESTED READING

1. McClure W: Mechanism and control of transcription initiation in prokarytotes. Annu Rev Biochem 54:171, 1985
2. Platt T: Transcription termination and the regulation of gene expression. Annu Rev Biochem 55:339, 1986
3. Kozak M: Comparison of initiation of protein synthesis in prokaryotes, eukaryotes, and organelles. Microbiol Rev 47:1, 1983
4. Hunt T: The initiation of protein synthesis. Trends Biochem Sci 5:178, 1980
5. Clark B: The elongation step of protein synthesis. Trends Biochem Sci 5:207, 1980
6. Laskey CT: Peptide chain termination. Trends Biochem Sci 5:234, 1980
7. Leader DP: Protein biosynthesis on membrane-bound ribosomes. Trends Biochem Sci 4:205, 1979
8. Watson JD, Hopkins NH, Roberts JW et al: Molecular Biology of the Gene. 4th Ed. Vol. 1. Benjamin-Cummings, Menlo Park, CA, 1987

STUDY QUESTIONS

Directions: For the following multiple choice questions, choose the most appropriate answer.

1. Which of the following enzymes catalyzes the synthesis of mRNA in eukaryotic cells?
   A. RNA polymerase I
   B. RNA polymerase II
   C. RNA polymerase III
   D. RNA polymerase P

2. Most prokaryotic promoters are characterized by
   A. two highly conserved consensus sequences
   B. one highly conserved consensus sequence
   C. the absence of conserved consensus sequences
   D. a Pribnow box at -35

3. The termination of translation in both prokaryotic and eukaryotic cells requires
   A. release factor proteins, one of which is ρ
   B. only ρ
   C. no additional protein factors
   D. only release factor proteins

4. Polyribosomes are formed because
   A. mRNAs have multiple ribosome binding sites
   B. one ribosome can participate in the translation of multiple mRNAs
   C. after translation is initiated, the ribosome binding site on mRNA becomes accessible for the binding of another ribosome
   D. the rate of translation is slow

Directions: Answer the following questions using the following key:

(A) if 1, 2, and 3 are correct
(B) if 1 and 3 are correct
(C) if 2 and 4 are correct
(D) if only 4 is correct
(E) if all four are correct

5. Sequences are characteristic of eukaryotic promoters include:
   1. TGCT
   2. TATA
   3. GTGT
   4. CAAT

6. Bacterial and eukaryotic RNA polymerases
   1. recognize terminator sites
   2. may require a protein called ρ for termination
   3. recognize promoter sites
   4. always require ρ for termination

7. RNA polymerases
   1. require a single-stranded DNA as template
   2. may use double-stranded DNA as template
   3. synthesize 5' to 3' phosphodiester bonds between ribonucleotides
   4. synthesize 3' to 5' phosphodiester bonds between ribonucleotides

8. RNA processing can includes the following steps:
   1. removal of flanking sequences
   2. removal of introns
   3. removal of intergene spacer sequences
   4. removal of exons

9. Features of the genetic code include
   1. punctuation codons
   2. degeneracy
   3. universality
   4. one codon per amino acid

10. Which of the following compound(s) would most likely NOT be used to treat a bacterial infection?
    1. Erythromycin
    2. Streptomycin
    3. Rifampin
    4. α-Amanitin

11. Poly A tails are added to
    1. prokaryotic mRNA
    2. eukaryotic rRNA
    3. prokaryotic rRNA
    4. eukaryotic mRNA

Items 12 and 13. Match the following terms with the most appropriate statements:

(A) Repressor
(B) Corepressor
(C) Inducer
(D) Operator

12. A protein that binds the operator region of DNA transcriptional units and contributes to the regulation of gene expression.

13. A segment of DNA that contributes to the regulation of gene transcription.

15

BIOCHEMICAL ENDOCRINOLOGY

Mechanisms of Hormone Action/Steroid Hormones/
Hormones of Calcium Metabolism/Thyroid Hormones/
Pancreatic Hormones/Catecholamines/
Pituitary and Hypothalamic Hormones/
Clinical Correlation

The student should be able to:

Learning Objectives

1. Define "hormone."
2. Describe the two general mechanisms of hormone action; identify the classes of hormones associated with each mechanism; and identify two second messengers.
3. List the classes of steroid hormones and their physiologic roles; identify the precursor of the steroid hormones and the rate-limiting enzyme of steroid hormone biosynthesis; describe the regulation of synthesis and secretion for each group of steroid hormones; and identify the general reactions involved in the inactivation of steroid hormones.
4. List the hormones that regulate Ca^{2+} metabolism and describe their roles.
5. Describe the synthesis, biologic actions, and regulation of the thyroid hormones.
4. List the pancreatic hormones and their physiologic functions, and describe the processing of preproinsulin.
5. Identify the catecholamine hormones and their mechanism of action via the β- and α-adrenergic receptors.
6. Describe the general function of the hypothalamic hormones.
7. List the major hormones of the anterior pituitary, their functions, and their regulation.
8. List the hormones of the posterior pituitary, the mechanism by which they are synthesized and secreted, their functions, and their regulation.

**Perspective:
Biochemical
Endocrinology**

An organism's ability to respond to stimuli depends on adquate internal communication. The *endocrine* or *hormone system* constitutes a communication network among the cells and tissues of the body. Information in the form of specific chemicals is released into the circulation and modulates the metabolic processes of specific target tissues. The blood-borne chemical messengers of the endocrine system are *hormones,* and the tissues that release them constitute the *endocrine glands. Endocrinology* is the study of hormones in action. *Biochemical endocrinology* is the study of hormones as vital chemicals, and explores the chemical transformations involved in their synthesis and release, action, inactivation, and elimination. This chapter first covers the two main mechanisms of hormone action and then considers the biochemistry of the various hormone groups.

GENERAL MECHANISMS OF HORMONE ACTION

All hormones exert their effects by binding to specific hormone receptor proteins located either on or in target cells.

1. *The peptide hormones and catecholamines* bind to receptors located on the external surface of the plasma membrane of target cells, and do not enter the cell.
2. *The lipid-soluble hormones* (*the steroid and thyroid hormones*) enter target cells and bind to intracellular receptors.

In both cases, the effects of the hormone are mediated by the *hormone-receptor complex*. However, cell-surface and internal receptors mediate hormone action by very different mechanisms. Complexing of a hormone with a cell-surface receptor stimulates the production of a *second messenger* substance within the cell, which modifies the activity of specific cellular enzymes. Intracellular hormone-receptor complexes, by contrast, act directly on the DNA and influence the transcription of specific genes.

Mechanism of Action of Hormones with Membrane-Bound Receptors: Second Messengers

The action of hormones that bind to membrane-bound receptors is mediated by second messengers as follows:

1. The hormone binds to receptors on the surface of a target cell.
2. Binding of the hormone to the receptor triggers the formation of one or more second messenger substances within the cell.
3. The second messenger or messengers modify the activities of specific cellular enzymes, often via an enzyme cascade.
4. The affected enzymes mediate changes in the affected cellular processes.

Because catecholamines and peptide hormones (the "first messengers") do not enter cells, second messengers are used to transmit the hormonal message into the cell.

Cyclic AMP as a Second Messenger

Cyclic AMP is a very common second messenger. Most peptide hormones act by increasing the intracellular concentration of cAMP. Cyclic AMP is synthesized by the enzyme *adenylate cyclase* in the following reaction:

$$ATP \xrightarrow{Mg^{2+}} cAMP + P_i$$

How does binding of the hormone to the receptor activate adenylate cyclase?

Activation of Adenylate Cyclase

Adenylate cyclase is a membrane-bound enzyme that is found in the plasma membranes of most mammalian cells. The active form of adenylate cyclase consists of a complex of a catalytic protein (C) with a *guanine-nucleotide-binding regulatory protein* (referred to as a G protein or N protein) that carries bound GTP. The G protein constitutes the communication link between hormone receptor proteins and the catalytic subunit. G protein consists of three subunits: α, β, and γ. The G protein can bind either GDP or GTP, and also has a sluggish GTPase activity. The system works as follows (Fig. 15-1):

1. Hormone (H) binds to the membrane receptor protein (R), forming a hormone-receptor (H-R) complex. This event triggers a conformational change in the receptor protein.
2. The conformational change in the receptor protein causes it to complex with a G protein carrying bound GDP, forming a ternary H-R-G(GDP) complex.
3. The formation of this ternary complex causes GDP to dissociate from the G protein and to be replaced by GTP. This event causes both the hormone and the G protein to dissociate from the receptor protein to yield H, R, and G(GTP). The receptor experiences a brief refractory period before it can again be activated by hormone.
4. G(GTP) associates with the adenylate cyclase catalytic protein (C), forming a C-G(GTP) complex. This active complex catalyzes the formation of cAMP.
5. Eventually, the GTPase activity of the G protein hydrolyzes its bound GTP to GDP. G(GDP) dissociates from the catalytic protein, inactivating adenylate cyclase. G(GDP) is now ready to bind to another H-R complex.

The overall activity of adenylate cyclase at any given moment thus depends on the proportion of G proteins that carry GTP rather than GDP. This ratio, in turn, depends on how often hormone molecules bind to receptor proteins. Therefore, the concentration of the appropriate hormones in the circulation determines the concentration of cAMP in the cell. Adenylate cyclase may be associated with hormone receptors spe-

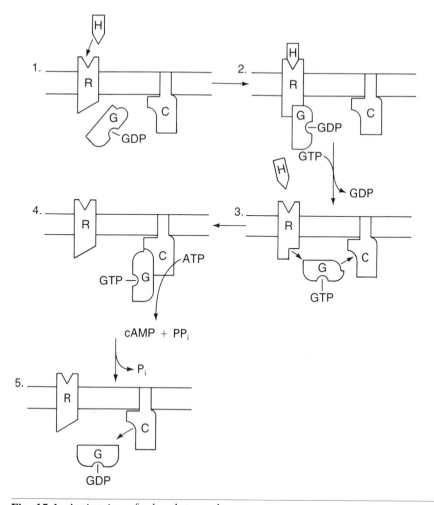

Fig. 15-1. Activation of adenylate cyclase.

cific for different hormones, allowing target cells to be sensitive to different hormonal stimuli.

Action of cAMP in Cells

Cyclic AMP acts by stimulating *protein kinases* (Chapter 4). Protein kinases consist of catalytic and regulatory subunits. Cyclic AMP binds to the regulatory subunit and causes it to dissociate, which activates the catalytic subunit. Activated protein kinases catalyze the phosphorylation of specific enzymes by reactions of the following type:

$$\text{Enzyme} + \text{ATP} \xrightarrow{\text{Protein kinase}} \text{Enzyme} - P_i + \text{ADP}$$

The enzymes may be either activated or inhibited by phosphorylation. The action of protein kinases is opposed by the action of *phosphoprotein phosphatases*, which dephosphorylate specific enzymes:

$$\text{Enzyme} - \text{P}_i \xrightarrow{\text{Phosphoprotein phosphatase}} \text{Enzyme} + \text{P}_i$$

Cyclic AMP is inactivated by *phosphodiesterases*, which catalyze the conversion of cAMP to AMP. The activity of phosphodiesterases may also be regulated by hormones. Different phosphodiesterases display different enzymic properties.

Calcium Ion as a Second Messenger

Calcium ion is another hormonal second messenger. Some relevant facts about Ca^{2+} are as follows:

1. The extracellular $[Ca^{2+}]$ is much higher than the cytosolic $[Ca^{2+}]$. The cell membane has mechanisms to move Ca^{2+} either into or out of the cell.
2. Tissues contain intracellular Ca^{2+} pools, for example in the mitochondria, smooth endoplasmic reticulum, and secretory granules.
3. The activity of several enzymes is sensitive to $[Ca^{2+}]$.
4. Certain hormones alter $[Ca^{2+}]$ in the cytosol either by promoting an influx of Ca^{2+} or by stimulating release of stored intracellular Ca^{2+}.

Binding of certain hormones to their cell surface receptors results in an increase in intracellular $[Ca^{2+}]$, which in turn influences cellular processes. As described in Chapter 4, the binding of epinephrine to α-adrenergic receptors leads to the release of Ca^{2+} from the endoplasmic reticulum via the second messengers inositol triphosphate and diacylglycerol. In this case, Ca^{2+} acts as a "third messenger," and may more appropriately be referred to simply as an intracellular messenger. Changes in intracellular $[Ca^{2+}]$ affect cellular metabolism largely by means of the calcium binding protein *calmodulin*.

Calmodulin

Figure 15-2 shows the structure of calmodulin (also called Ca^{2+}-dependent regulatory protein). Calmodulin is the major Ca^{2+} binding protein in the cell. Each calmodulin has four Ca^{2+} binding sites. Binding of Ca^{2+} to all four sites triggers a conformational change that activates the protein. Active calmodulin can complex with a number of cellular enzymes, stimulating their activity (Fig. 15-3).

In summary, the steps by which calmodulin mediates the effects of a hormone-triggered rise in intracellular $[Ca^{2+}]$ are as follows:

1. The hormone binds to the receptor.
2. The hormone-receptor complex causes a rise in intracellular $[Ca^{2+}]$, usually by way of an influx of extracellular Ca^{2+}.
3. Ca^{2+} binds to calmodulin and activates it.
4. Activated calmodulin binds to certain enzymes and stimulates their activity.

Other Second Messengers

There is evidence that cyclic GMP may act as a second messenger with effects opposite to those of cAMP. The physiologic importance of cGMP in cells remains to be established, however. One or more second mes-

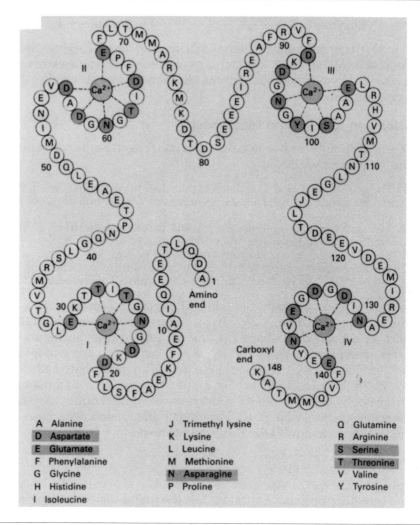

Fig. 15-2. Calmodulin primary structure, showing the Ca^{2+} binding sites. (From Cheung WY: Calmodulin. Sci Am 246(6):62, 1982, with permission.)

sengers are also believed to be involved in the action of insulin, but their identity is not yet known.

Mechanism of Action of Lipid-Soluble Hormones: Cytosolic and Nuclear Receptors

Steroid and thyroid hormones are lipid soluble and enter cells by passive diffusion. They complex with intracellular receptors, and the hormone-receptor complex exerts its effects by binding to to the chromatin and stimulating the expression of certain genes. The receptors for steroid hormones are located in the cytosol, whereas the thyroid hormone receptors are in the nucleus.

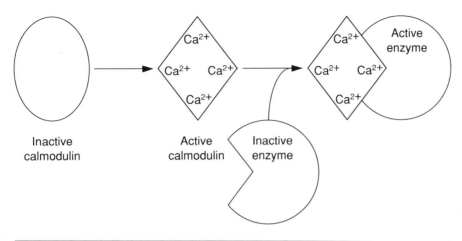

Fig. 15-3. Activation of a cellular proteins by calmodulin.

Action of Steroid Hormones. The steps involved in the action of steroid hormones are as follows (Fig. 15-4):

1. The hormone enters the cell by passive diffusion, and complexes with its cytosolic receptor.
2. Binding of the hormone activates the receptor via a conformational change. The complex is then translocated to the nucleus.
3. In the nucleus, the hormone-receptor complex binds to the chromatin and stimulates the expression of certain genes, thus ultimately altering the protein composition of the cell and influencing cellular metabolism.

Action of Thyroid Hormones. The mechanism of action of thyroid hormones is similar to that of steroid hormones, except that the receptors for the thyroid hormones are located in the nucleus.

THE STEROID HORMONES

The steroid hormones are all derived from cholesterol. They are divided into two functional classes: the *sex hormones* and the *corticosteroids*. The sex steroids are synthesized mainly in the gonads and regulate sexual differentiation and function, whereas the corticosteroids are synthesized mainly in the adrenal cortex and control a variety of physiologic functions. Table 15-1 lists the major steroid hormones.

The Sex Hormones

1. The sex hormones fall into three groups: *androgens*, *estrogens*, and *progestins* (essentially progesterone).
2. These hormones regulate the development and maintenance of the reproductive organs and of secondary sexual characteristics.
   a. *Androgens and estrogens* are responsible for the development, maintenance, and function of the reproductive system and secondary sexual characteristics in males and females, respectively.

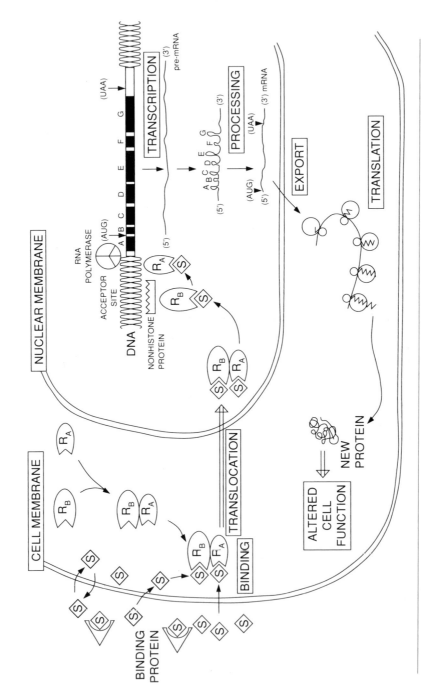

Fig. 15-4. Mechanism of action of steroid hormones. (From Chan L, O'Malley B: Steroid hormone action: recent advances. Ann Intern Med 89:694, 1978, with permission.)

Table 15-1. The Major Steroid Hormones

Hormone	Main Site of Synthesis
Sex steroids	
Androgens	Leydig cells of testis
Testosterone	
Dihydroxytestosterone	
Estrogens	Follicle cells of ovaries
Estradiol	
Estriol	
Estrone	
Progestins	Corpus luteum
Progesterone	
Corticosteroids	
Glucocorticoids	Adrenal cortex
Cortisol	
Cortisone	
Corticosterone	
Mineralocorticoids	Adrenal cortex
Aldosterone	

  b. *Progesterone* plays a role in the menstrual cycle and also functions to maintain pregnancy and to stimulate related phenomena such as breast enlargement.

The Corticosteroids

3. The corticosteroids fall into two functional groups: the *mineralocorticoids* and the *glucocorticoids*.
   a. The *mineralocorticoids* regulate the balance of sodium and potassium in the extracellular fluid.
   b. The *glucocorticoids* promote protein degradation, gluconeogenesis, and glycogen synthesis. They also have antiinflammatory properties.

Biosynthesis of Steroid Hormones

The primary sites of steroid hormone synthesis are listed in Table 15-1. *Androgens* are produced mainly in the Leydig cells of the testes, and *estrogens* in the follicle cells of the ovary. However, the testes produce small amounts of estrogens, and both the ovaries and the adrenal cortex produce small amounts of androgens. *Progesterone* is produced mainly by corpus luteum cells in the ovaries. In the adrenal cortex, the *mineralocorticoids* are produced by the glomerulosa cells and the *glucocorticoids* by the fasciculata and reticularis cells. It should be noted that several other tissues in mammals synthesize steroid hormones to a minor degree.

The following points are essential to an understanding of steroid hormone biosynthesis.

1. All the steroid hormones are synthesized through the common intermediate *pregnenolone*.

2. The overall pathway of steroid hormone biosynthesis is the same in all tissues. The assortment of steroid hormones actually produced by a given tissue or cell depends on the enzymes of the steroid biosynthetic pathway that are expressed in that tissue or cell. For example, Leydig cells in the testis express mainly the enzymes that lead to androgens, and to a lesser extent the enzymes that convert androgens to estrogens.
3. The synthesis and secretion of steroid hormones is controlled by the action of other hormones.
4. The reactions involved in steroid hormone biosynthesis are mostly hydroxylations, reductions, and oxidations. Some of these reactions take place in the mitochondria and endoplasmic reticulum.

Figure 15-5 shows the branching pathway of steroid hormone biosynthesis. The overall pathway may be viewed as consisting of five segments.

Segment 1: *Cholesterol to pregnenolone.* This step consists of three reactions carried out by an enzyme complex called *cholesterol desmolase*, which is the rate-limiting enzyme of steroid hormone biosynthesis.

Segment 2: *Pregnenolone to testosterone.* 17-α-Hydroxypregnenolone is an important intermediate in this pathway. Testosterone, the most abundant male hormone, is converted to estradiol and to its very potent derivative *dihydrotestosterone* via different pathways.

Segment 3: *Testosterone to estradiol.* This sequence of reactions is carried out by an enzyme complex called *aromatase* that is associated with the endoplasmic reticulum. These reactions create an aromatic ring in the steroid hormone structure. Estradiol, the most important of the estrogens, is subsequently converted to the less active estrogens estrone and estriol. Estrone may also be obtained from 4-androstene-3,17-dione by the action of aromatase.

Segment 4: *Pregnenolone to aldosterone.* The sex hormone progesterone and the glucocorticoid corticosterone are synthesized in the course of this segment.

Segment 5: *17-α-Hydroxypregnenolone to cortisol.* Under normal conditions, cortisol is the only accumulated product of this segment.

Transport of Steroid Hormones in the Blood

Steroid hormones are carried in the blood bound to a specific binding protein called *transcortin*. Steroid hormones may also bind loosely to albumin. In both cases, the bound steroids are in equilibrium with free steroids in solution.

Regulation of Steroid Hormone Synthesis

Steroid hormones are not stored, but are secreted as soon as they are synthesized. The synthesis of steroid hormones is controlled by various other hormones.

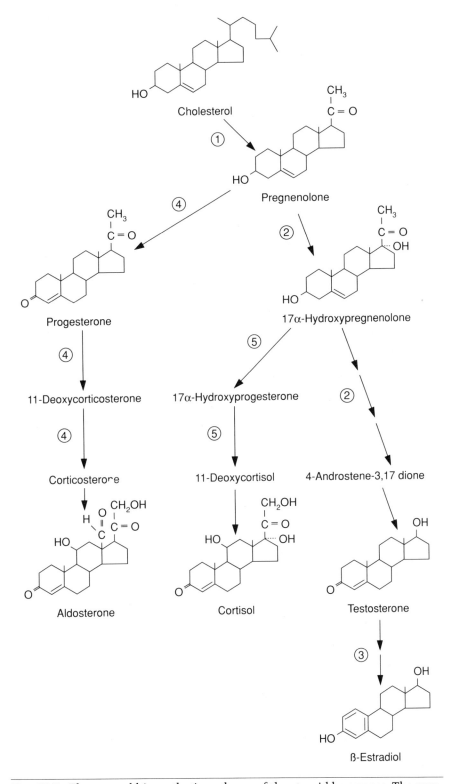

Fig. 15-5. The general biosynthetic pathway of the steroid hormones. The numbers correspond to the pathway segments discussed in the text.

Glucocorticoids

The blood level of glucocorticoids is controlled by a feedback system involving the hypothalamus, the pituitary, and the adrenals (Fig. 15-6). When the blood level of glucocorticoids drops, the hypothalamus secretes *corticotropin releasing factor* (CRF), which in turn stimulates the release of *adrenocorticotropic hormone* (ACTH) from the pituitary gland. ACTH then stimulates the adrenal cortex to synthesize the glucocorticoids cortisol and corticosterone. In humans, cortisol is the major glucocorticoid. As the blood level of cortisol rises, it interacts with receptors in both the hypothalamus and the pituitary, inhibiting the production, respectively, of CRF and ACTH. ACTH also exerts feedback inhibition on the release of CRF from the hypothalamus.

Aldosterone

The synthesis of the mineralocorticoid aldosterone responds to the balance of Na^+ and K^+ in the extracellular fluid via the *renin-angiotensin system*. A decrease in the Na^+/K^+ ratio stimulates the kidney to secrete the proteolytic enzyme *renin*. Renin cleaves the inactive plasma globular protein *angiotensinogen*, releasing the decapeptide *angiotensin I*, which is ultimately converted to the active ocatapeptide *angiotensin II* by an enzyme called *converting factor*. Angiotensin II binds to receptors on the glomerulosa cells of the adrenal cortex, and enhances the syn-

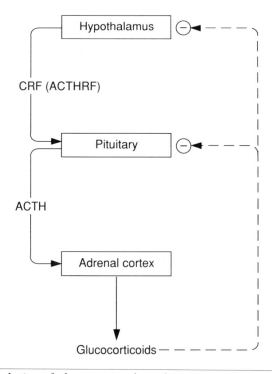

Fig. 15-6. Regulation of glucocorticoid synthesis.

thesis of aldosterone by stimulating the conversion of cholesterol to pregnenolone and of corticosterone to aldosterone, which functions to restore the Na^+/K^+ levels.

The Sex Hormones

Secretion of the sex steroids is controlled by the hypothalamus and pituitary (Fig. 15-7). When blood levels of the sex hormones decrease, the hypothalamus secretes *luteinizing hormone releasing factor* (LHRF) and/or *follicle stimulating hormone releasing factor* (FSHRF). These hormones stimulate the pituitary to secrete *luteinizing hormone* (LH) and *follicle stimulating hormone* (FSH), respectively. In women, LH promotes the secretion of progesterone by the corpus luteum, whereas in men it promotes the secretion of testosterone by the testicular Leydig cells. FSH stimulates the production and secretion of estradiol by the ovaries. High blood levels of sex hormones exert feedback inhibition on the relevant pituitary and hypothalamic hormones.

Metabolism and Excretion of Steroid Hormones

The metabolism of steroid hormones ultimately results in their inactivation and excretion in the urine as methyl, sulfate, and/or glucuronide conjugates. The liver is the major site of steroid hormone inactivation.

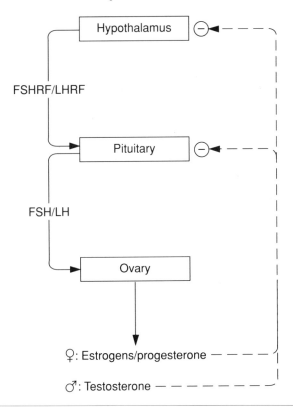

Fig. 15-7. Regulation of sex hormone synthesis.

Testosterone

Testosterone is metabolized as follows:

1. Inactivation by one of two alternative pathways:
   a. Reduction of the 3-keto group to a hydroxyl and reduction of the 4,5 double bond
   b. Oxidation of the 17-hydroxyl group, followed by reduction of the 4,5 double bond and reduction of the keto group to a hydroxyl
2. The resulting metabolites conjugate with sulfate and/or glucuronate, and are excreted in the urine.

Estrogens

Estradiol is converted to estrone and estriol by oxidation and hydroxylation, respectively. Both estrone and estriol are conjugated with sulfate and/or glucuronate and excreted in the urine. Some estrogens may also be excreted as methylated derivatives.

Progesterone

A major pathway of progesterone metabolism is reduction to pregnanediol. Progesterone is also metabolized by hydroxylation reactions and by a combination of hydroxylations and reductions. The resulting products are conjugated with sulfate and/or glucuronide and excreted in the urine.

Corticosteroids

For the most part, the corticosteroids undergo only reduction before being conjugated. The extent of reduction varies for different hormones. The resulting metabolites are conjugated mostly with glucuronide (although some sulfate conjugates are formed), and excreted in the urine.

CALCIUM-REGULATING HORMONES: PARATHYROID HORMONE, CALCITONIN, AND 1,25-DIHYDROXYCHOLECALCIFEROL

Almost all the calcium in the body is bound in the bones. However, the small percentage that is in solution plays important physiologic roles, and is tightly regulated. Calcium ion can serve as a hormone intracellular messenger. It also participates in some enzyme reactions, in blood coagulation, in the conduction of nerve impulses, and in muscle contraction. To maintain serum $[Ca^{2+}]$ in the proper range, Ca^{2+} is deposited to or withdrawn from the skeletal and tissue stores. These processes are controlled by three hormones: *parathyroid hormone* (PTH), *calcitonin*, and *1,25-dihydroxycholecalciferol*. The major counterion to Ca^{2+} is phosphate; therefore, these three hormones also affect the concentration and distribution of phosphate.

PTH and calcitonin are polypeptides; PTH is synthesized in the parathyroid glands and calcitonin in the parathyroids and the thyroid.

1,25-Dihydroxycholecalciferol, on the other hand, is a derivative of vitamin D. The synthesis of this compound from vitamin D is initiated in the liver and completed in the kidney.

Parathyroid Hormone

Physiologic Effects. PTH is secreted by the parathyroids in response to low serum $[Ca^{2+}]$. It acts to increase serum $[Ca^{2+}]$ via effects on the kidney, bone, and intestine, as listed in Table 15-2. Note that PTH acts directly on kidney and bone, whereas its effect on the intestine is mediated by 1,25-dihydroxycholecalciferol.

Biosynthesis. PTH is a 84 amino acid polypeptide. It is synthesized on ribosomes associated with the endoplasmic reticulum as a larger precursor of 115 amino acids. Successive removal of two oligopeptides from the amino terminal end converts the precursor to mature PTH, which is ultimately secreted via the golgi apparatus (see the section on the synthesis of extracellular proteins in Chapter 14).

Calcitonin

Physiologic Effects. Calcitonin acts to lower serum $[Ca^{2+}]$ by exerting effects opposed to those of PTH (Table 15-2). Calcitonin secretion is stimulated by high serum $[Ca^{2+}]$ and inhibited by low serum $[Ca^{2+}]$.

Biosynthesis. The 32 amino acid polypeptide calcitonin is synthesized in the parathyroid and thyroid glands. Like PTH, it is initially synthe-

Table 15-2. Actions of PTH, Calcitonin, and 1,25-Dihydroxycholecalciferol

Hormone	Site	Effects
PTH	Kidney	1. Increases renal tubular resorption of Ca^{2+} and P_i
		2. Stimulates the production of 1,25-dihydroxycholecalciferol, which in turn stimulates mobilization of bone Ca^{2+}
	Bone	3. Increase the resorption of bone Ca^{2+} (i.e., decreases Ca^{2+} binding and increases bone erosion)
	Intestine	4. Exerts an indirect effect: the 1,25-dihydroxycholecalciferol synthesized by the kidney in response to PTH stimulates intestinal absorption of Ca^{2+}
Calcitonin	Kidney	1. Decreases renal tubular resorption of Ca^{2+} and P_i
	Bone	2. Decreases resorption of bone Ca^{2+} and P_i
1-25-Dihydroxycholecalciferol	Bone	1. Stimulates Ca^{2+} resorption in a manner that requires PTH
	Intestine	2. Stimulates absorption of Ca^{2+} and P_i by intestinal epithelial cells

sized as a larger precursor; peptides are cleaved from both the amino and carboxyl ends to yield the mature hormone.

Mechanism of Action of PTH and Calcitonin

As mentioned, PTH and calcitonin have largely opposite effects. However, they both interact with their target cells via specific membrane receptors that use cAMP as a second messenger.

1,25-Dihydroxycholecalciferol (Vitamin D)

1,25-Dihydroxycholecalciferol is synthesized from vitamin D. 1,25-Dihydroxycholecalciferol functions to maintain serum $[Ca^{2+}]$ at the levels required for bone mineralization. When vitamin D is administered, a lag of several hours usually precedes any observed physiologic changes, presumably due to the conversion of vitamin D to 1,25-dihydroxycholecalciferol.

Physiologic Effects. 1,25-Dihydroxycholecalciferol exerts its effects primarily on the bone and intestine, as shown in Table 15-2. Like other steroid derivatives, 1,25-dihydroxycholecalciferol is lipid soluble. It enters its target cells by diffusion, and binds to a cytoplasmic receptor. The hormone-receptor complex enters the nucleus and stimulates the expression of certain genes.

Biosynthesis. Figure 15-8 shows the biosynthesis of 1,25-dihydroxycholecalciferol. Vitamin D (cholecalciferol) is initially hydroxylated at the 25 position by cholecalciferol 25-hydroxylase. This reaction occurs in the liver. The 25-hydroxy derivative binds to a specific binding protein and is transported in the blood to the kidney, where it is hydroxylated at the 1 position to yield 1,25-dihydroxycholecalciferol. 1,25-Dihydroxycholecalciferol is synthesized in response to low serum levels of Ca^{2+} and P_i, and exerts feedback inhibition over its own synthesis. PTH also stimulates the synthesis of 1,25-dihydroxycholecalciferol.

Metabolism and Excretion. Most 1,25-dihydroxycholecalciferol is inactivated by oxidation of the side chain. The major route of excretion for metabolites is via the bile; very little is eliminated in the urine.

Integration of Calcium Metabolism

Figure 15-9 shows how the combined effects of PTH, calcitonin, and 1,25-dihydroxycholecalciferol regulate serum $[Ca^{2+}]$.

HORMONES OF THE THYROID GLAND

The thyroid gland produces the thyroid hormones *thyroxine* (T_4) and *triiodothyronine* (T_3), in addition to some calcitonin (Fig. 15-10). T_3 and T_4 are secreted as a mixture by the thyroid gland.

Cholecalciferol

25-Hydroxycholecalciferol

1, 25-Dihydroxycholecalciferol

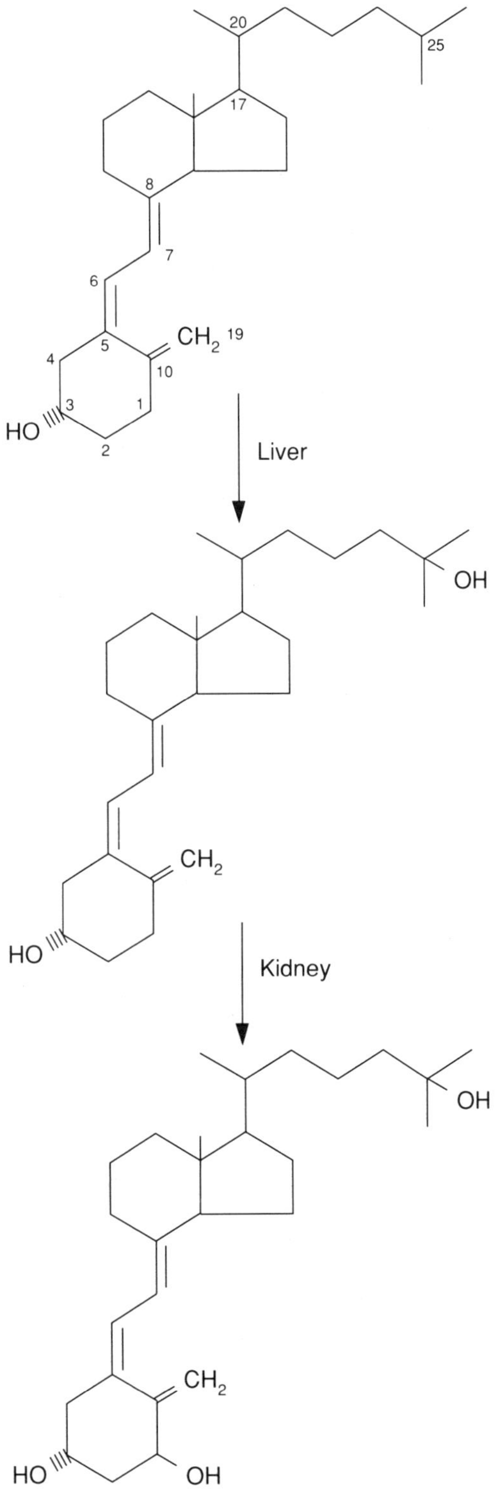

Fig. 15-8. Synthesis of 1,25-dihydroxycholecalciferol from vitamin D.

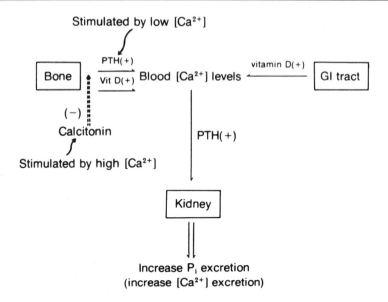

Fig. 15-9. The combined effects of PTH, calcitonin, and vitamin D on Ca^{2+} metabolism. (From Devlin TM: Textbook of Biochemistry with Clinical Correlations. Wiley, New York, 1986, with permission.)

Actions of T_3 and T_4

The thyroid hormones increase the basal metabolic rate, stimulate oxygen utilization, and increase the rates of carbohydrate, lipid, and protein metabolism. These hormones are also essential for growth and development. Thyroid hormones bind to receptors located in nucleus of

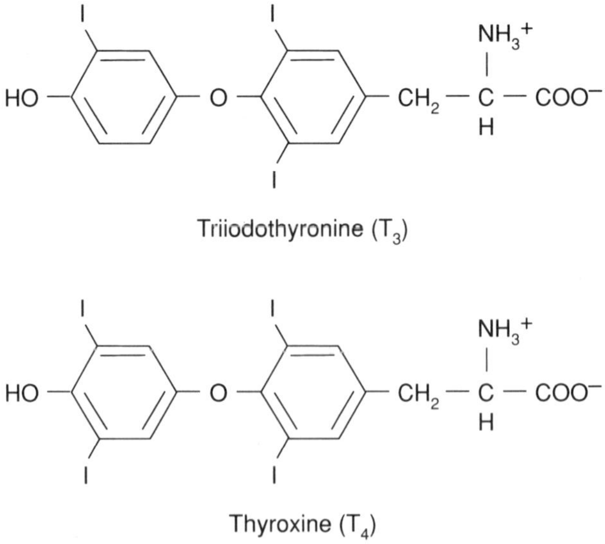

Fig. 15-10. Structures of thyroxine (T_4) and triiodothyronine (T_3).

target cells, where the resulting hormone-receptor complexes stimulate the expression of specific genes.

Biosynthesis of Thyroid Hormones

The follicular cells of the thyroid gland perform two activities that are essential for the synthesis of thyroid hormones: (1) they accumulate iodide, and (2) they synthesize thyroglobulin. *Thyroglobulin* is a large glycoprotein (MW 660,000). Certain tyrosine residues of thyroglubulin are iodinated by *thyroid peroxidase* to mono- and diiodotyrosine derivatives. The iodinated residues are then covalently coupled: in some cases two diiodotyrosine residues are joined, whereas in other cases a diiodotyrosine is joined to a monoiodotyrosine. The iodinated thyroglobulin is then degraded by intracellular proteases to yield the hormones T_4 and T_3. The ratio of T_4 to T_3 produced is approximately 4:1. However, T_4 can be converted to T_3 by deiodination. In fact, most of the physiologic activity of the thyroid hormones can be attributed to T_3.

Regulation of Thyroid Hormone Synthesis

The levels of the thyroid hormones are controlled by the hypothalamus and pituitary, as shown in Figure 15-11. When the blood levels of the thyroid hormones fall too low, the hypothalamus secretes *thyroid re-*

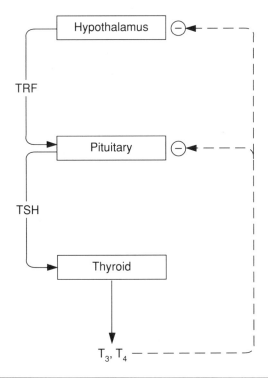

Fig. 15-11. Regulation of thyroid hormone synthesis.

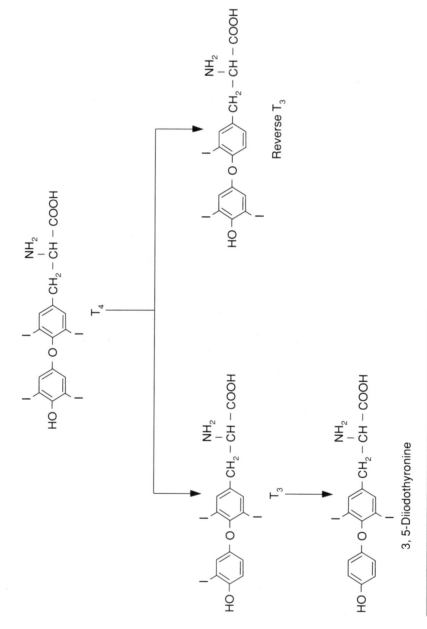

Fig. 15-12. Thyroid hormone inactivation.

leasing factor (TRF), which stimulates the anterior pituitary to secrete *thyroid stimulating hormone* (TSH). TSH then stimulates the thyroid to synthesize and release T_4 and T_3. High blood levels of T_3 and T_4 inhibit the secretion of TRF and TSH, which inhibits thyroid secretion. The hypothalamic hormone *somatostatin* also inhibits the secretion of TSH.

Metabolism and Excretion of Thyroid Hormones

The metabolic deactivation of the thyroid hormones is shown in Figure 15-12. Not only can T_4 be deiodinated to the more active T_3, but it can be deiodinated to the less active form *reverse T_3*. In addition, T_3 can be deiodinated to a variety of inactive or much less active compounds.

Deiodination of thyroid hormones is carried out by two deiodinases, which are present in liver, muscle, kidney, anterior pituitary, thyroid, and brain. The side chain group of the thyroid hormones usually is deaminated and decarboxylated to yield an acetic acid moiety. The free hydroxyl group may be conjugated with sulfate or glucuronic acid and excreted in the urine. Small quantities of mostly conjugated derivatives are excreted in the bile.

THE PANCREATIC HORMONES

The pancreatic hormones are *glucagon*, *insulin*, and *somatostatin*, secreted respectively by the A, B, and D cells of the pancreatic islets of Langerhans. All three are peptides. Their major functions are as follows (Table 15-3):

1. Insulin and glucagon regulate the availability of fuel molecules by controlling the blood levels of glucose, fatty acids, and amino acids. They interact to control the storage of fuel molecules and their degradation and mobilization. Many of the effects of these hormones were discussed in Chapters 4 through 7.
2. *Insulin* has mainly *anabolic* effects: it functions to decrease the blood

Table 15-3. The Major Effects of Insulin, Glucagon, and Somatostatin on Fuel Molecule Availability

Hormone	Effects
Insulin	Promotes fuel storage: 1. Increases the synthesis of proteins, glycogen, and triacylglycerols 2. Stimulates uptake and utilization of glucose 3. Inhibits gluconeogenesis
Glucagon	Promotes fuel utilization: 1. Raises serum glucose by increasing glycogen degradation and gluconeogenesis 2. Increases protein degradation, and mildly increases fatty acid degradation
Somatostatin (pancreatic)	Inhibits the secretion of insulin and glucagon

levels of glucose, amino acids, and fatty acids by promoting their uptake by cells and their utilization in the synthesis of glycogen, proteins, and triacylglycerols.
3. *Glucagon* has mainly *catabolic* effects, opposing the effects of insulin: it promotes the degradation of glycogen, triacylglycerols, and proteins, and the release of glucose, fatty acids, and amino acids into the blood.
4. *Somatostatin* inhibits secretion of both glucagon and insulin.

Insulin

Actions of Insulin

Insulin affects the metabolism of carbohydrates, lipids, and proteins, and its primary target tissues are muscle, liver, and adipose tissue. Insulin has three basic types of effect:

1. It stimulates the cellular uptake of glucose, amino acids, and K^+ in specific target tissues.
2. It modulates the activity of key metabolic enzymes.
3. It promotes growth.

Effects on Carbohydrate Metabolism
1. Insulin stimulates the entry of glucose into cells. Glucose is rapidly phosphorylated within the cell, and thereby prevented from leaving. Thus, insulin tends to lower the blood glucose.
2. Insulin modulates enzymes of carbohydrate metabolism so as to stimulate glycolysis and glycogen synthesis and inhibit glycogen breakdown.

Effects on Lipid Metabolism.
Many of the effects of insulin on fatty acid metabolism are described in detail in Chapter 7.

1. Insulin stimulates the synthesis of fatty acids from carbohydrate precursors.
2. Insulin inhibits the mobilization of fatty acids in adipose tissue.
3. Insulin enhances both the synthesis of LDLs in the liver and the activity of the membrane-bound lipoprotein lipase, resulting in increased availability of fatty acids to the tissues.

Effects on Protein Metabolism
1. Insulin enhances cellular uptake of amino acids.
2. Insulin stimulates protein synthesis (the increased synthesis of some transport proteins leads to increased amino acid uptake).

Mechanisms of Action of Insulin

Insulin binds to membrane receptors on its target cells. Some of the effects of insulin—the stimulation of amino acid and glucose uptake—occur at the plasma membrane, whereas other effects are intracellular. The two types of insulin effects probably occur by different mechanisms. Many details concerning insulin action remain obscure. It is known that stimulation of glucose uptake by adipose cells is accompanied by an increase in the number of glucose permease transport systems on the cell surface; the binding of insulin to its receptor causes intracellular vesicles containing the permeases to fuse with the cell membrane (Fig. 15-13).

No molecule has been unequivocally identified as an insulin second messenger. However, the binding of insulin to its receptor causes the release of small peptides (MW ~2,000) from the receptor into the cytoplasm. These peptides may be second messengers.

Biosynthesis and Secretion of Insulin

Mature insulin is synthesized by way of two precursors:

1. *Mature insulin* (MW ~6,000) consists of an A peptide and a B peptide joined by two disulfide bridges.
2. *Proinsulin*, the precursor of insulin, is a single polypeptide chain comprising the A and B peptides plus a C peptide between them.
3. *Preproinsulin*, the precursor of proinsulin, has an additional 23 amino acids on the amino terminal end of proinsulin.

Preproinsulin is synthesized by ribosomes on the endoplasmic reticulum. It is inserted into the lumen of the endoplasmic reticulum during synthesis. Figure 15-14 shows the processing of the insulin precursor. Preproinsulin is cleaved in the endoplasmic reticulum to proinsulin, which is transported to the golgi apparatus and stored. Proinsulin in the golgi apparatus is gradually cleaved to yield mature insulin and the C peptide, both of which are ultimately secreted into the circulation.

Insulin secretion is regulated by the serum levels of glucose, amino acids, fatty acids, and ketone bodies. When the serum levels of these fuel molecules increase, insulin secretion is increased. The level of glu-

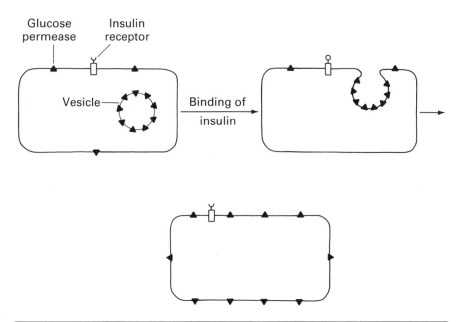

Fig. 15-13. Effect of insulin on the number of cell surface glucose permease transport systems.

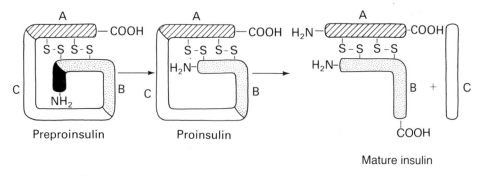

Fig. 15-14. Processing of preproinsulin.

cose is the single most potent stimulant of insulin secretion; its effect is enhanced by Ca^{2+}. Somatostatin inhibits insulin secretion.

Glucagon

Actions of Glucagon

The actions of glucagon generally oppose those of insulin. Glucagon raises the blood level of glucose, promotes the breakdown of glycogen, and mildly stimulates triacylglycerol degradation. In addition:

1. Gluconeogenesis is stimulated.
2. Liver protein synthesis is inhibited, thus increasing urea production and providing amino acid carbon chains to meet the demands of increased gluconeogenesis.
3. Fatty acid synthesis is mildly inhibited.
4. Ketogenesis (the synthesis of ketone bodies) is stimulated.
5. Mobilization of fatty acids from adipose tissue is mildly stimulated.

Mechanism of Action of Glucagon

Glucagon acts via the second messenger cAMP. The control of glycogen metabolism by glucagon is explored in detail in Chapter 4 (see Figs. 4-12 and 4-13).

Biosynthesis and Secretion of Glucagon

The 29 amino acid glucagon peptide (MW ~3,500) is synthesized as a much larger precursor (MW ~8,000) by cleavage of peptide segments from both the amino and carboxyl ends. Glucagon secretion by pancreatic B cells is stimulated by low blood levels of glucose and fatty acids. High concentrations of serum amino acids also stimulate glucagon secretion. Glucagon secretion is enhanced by Ca^{2+} and is inhibited by somatostatin.

Somatostatin

Somatostatin, the third pancreatic hormone, inhibits the secretion of insulin and glucagon by the pancreas and also the production of renin by the kidneys. Somatostatin is believed to inhibit insulin and glucagon release by blocking the entry of Ca^{2+} into pancreatic A and B cells; Ca^{2+} uptake by these cells is necessary for secretion of insulin and glucagon. Somatostatin is also produced in small amounts by the hypothalamus. Hypothalamic somatostatin inhibits secretion of the pituitary hormones ACTH, TSH, and GH.

Like insulin and glucagon, somatostatin is a peptide hormone. It is synthesized by cleavage of a much larger prosomatostatin peptide. Secretion of mature somatostatin from the pancreatic D cells is stimulated by serum glucose, amino acids, and pancreozymin.

THE CATECHOLAMINE HORMONES

The two catecholamine hormones—epinephrine and norepinephrine—are synthesized by the adrenal medulla. Figure 15-15 shows the structures of these hormones. Epinephrine and norepinephrine also function as neurotransmitters, as does the catecholamine dopamine. The catecholamine hormones tend to promote the utilization of fuel molecules, and therefore support the actions of glucagon and oppose those of insulin; the cooperative action of epinephrine and glucagon on glycogen metabolism was described in Chapter 4. However, unlike the pancreatic hormones, the catecholamines are secreted only during stressful or emergency situations such as exercise, exposure to cold, fright, and hypoglycemia. They act very rapidly, and their effects wear off within minutes.

Mechanisms of Action of the Catecholamine Hormones

Catecholamines bind to membrane receptors on target cells. There are four catecholamine receptors (called *adrenergic receptors*), which mediate different biologic responses to epinephrine and norepinephrine. Different target cells carry different combinations of the receptors. Both epinephrine and norepinephrine bind to all four types of adrenergic receptors, but the two hormones exhibit different potencies at different receptors. The adrenergic receptor classes and the relative potencies of epinephine and norepinephrine bound to them are shown in Table 15-4.

β-Adrenergic Receptors. Binding of epinephrine or norepinephrine to β_1- and β_2-adrenergic receptors results in the stimulation of adenylate cyclase, with cAMP as intracellular hormonal messenger.

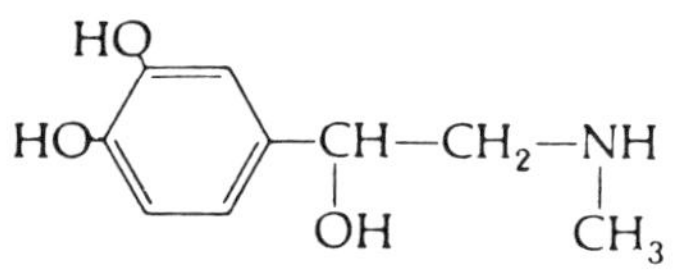

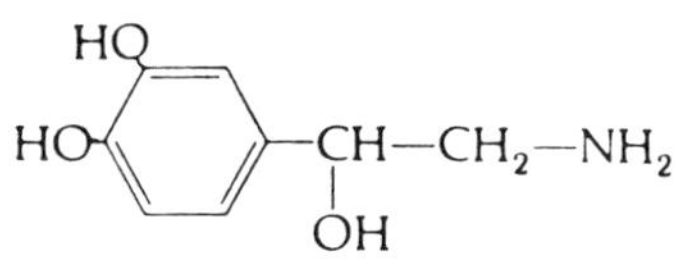

Fig. 15-15. Structures of epinephrine and norepinephrine.

Table 15-4. Relative Potencies of Catecholamine Hormones Bound to Adrenergic Receptors

Receptor	Relative Potency
α-Adrenergic receptors	
α_1-receptors	Epinephrine $\geq$ norepinephrine
α_2-receptors	Epinephrine $\approx$ norepinephrine (response varies widely among tissues)
β-Adrenergic receptors	
β_1-receptors	Epinephrine $\geq$ norepinephrine
β_2-receptors	Epinephrine $\gg$ norepinephrine

α-Adrenergic Receptors. Binding of norepinephrine or epinephrine to the α_1- and α_2-adrenergic receptors causes different responses:

1. Binding to α_1-receptors activates an enzyme with phospholipase C-like activity. This enzyme hydrolyzes membrane-associated phosphatidylinositol-4,5-bisphosphate to inositol triphosphate and diacylglycerol, both of which act as second messengers. As described in Chapter 4, inositol triphosphate stimulates the release of Ca^{2+} from intracellular stores. This Ca^{2+} participates in mediating the hormone response.
2. Binding of catecholamine to α_2-adrenergic receptors *inhibits* adenylate cyclase and lowers intracellular cAMP. This decrease in cAMP mediates the hormone responses associated with binding to the α_2-adrenergic receptors.

Biologic Effects of the Catecholamine Hormones

The catecholamines function to make glucose and fatty acids available by promoting rapid degradation of glycogen and triacylglycerols. Table 15-5 lists the major biologic effects of catecholamine binding to each of the four receptors. The variety of biologic responses triggered by cate-

Table 15-5. Physiologic Effects of Catecholamine Binding to the Adrenergic Receptors

Catecholamine Receptor	Effects
α_1-receptor	Contraction of smooth muscle
α_2-receptor	Relaxation of smooth muscle
	Inhibition of lipolysis in adipose tissues
	Inhibition of insulin release
β_1-receptor	Contraction of cardiac muscle
	Stimulation of lipolysis
β_2-receptor	Relaxation of smooth muscle
	Stimulation of glycogen degradation and gluconeogenesis
	Stimulation of insulin and glucagon release

cholamines is due both to the variety of catecholamine receptors and to individual responses of different tissues.

Storage and Secretion of the Catecholamine Hormones

The catecholamine hormones are stored in adrenal medulla cells in granules that contain a high concentration of ATP. Secretion of catecholamines is stimulated by the nervous system as follows:

1. The neurotransmitter acetylcholine binds to receptors on adrenal medulla cells
2. Depolarization of the cell membrane is accompanied by an influx of Ca^{2+}.
3. Catecholamines are secreted by exocytosis.

The secretory process is believed also to involve the contractile proteins actin, myosin, and troponin C.

Metabolism and Excretion of Catecholamines

The synthesis, metabolism, and excretion of the catecholamines are discussed in detail in Chapter 10 (see Figs. 10-3 and 10-4). In brief, catecholamines are metabolized (mainly in the liver) by *O*-methylation and/or oxidative deamination, and the products mostly form sulfate or glucuronide conjugates. Catecholamine metabolites are excreted in the urine.

THE HYPOTHALAMIC AND PITUITARY HORMONES

The hypothalamus and pituitary are located adjacent to each other at the base of the brain. The pituitary consists of posterior and anterior regions, which have different functions. The *posterior pituitary* (neurohypophysis) is connected by neuronal axons to the hypothalamus. It secretes only two hormones, vasopressin and oxytocin. These hormones are actually synthesized in the hypothalamus and travel in nerve axons to the posterior pituitary. The *anterior pituitary* (adenohypophysis), which is connected to the hypothalamus by a special portal circulation, secretes hormones that influence a wide variety of body functions. All the pituitary hormones are peptides and act via second messengers. In many cases the second messenger is cAMP; in other cases the specific second messenger has not been identified.

The secretion of anterior pituitary hormones is controlled by special *releasing* and *inhibiting factors* produced by the hypothalamus. The hypothalamus secretes these factors into the hypothalamic-pituitary portal circulation, which carries them directly and exclusively to the anterior pituitary. The secretion of hypothalamic hormones is controlled by both neural and hormonal signals. Therefore, the hypothalamic-pituitary system integrates neural and endocrine control of physiology.

Table 15-6 lists the major hormones of the anterior pituitary and their respective hypothalamic releasing and inhibiting factors.

Table 15-6. The Major Anterior Pituitary Hormones, Their Hypothalamic Releasing and Inhibiting Factors, and Their Major Effects

Anterior Pituitary Hormone	Hypothalamic Releasing Factors[a]	Hypothalamic Inhibiting Factors[a]	Major Biologic Effects
ACTH Adrenocorticotropic hormone (corticotropin)	CRF (ACTHRF)	Somatostatin[b]	Stimulates synthesis and secretion of adrenocortical steroids
TSH Thyroid stimulating hormone (thyrotropin)	TRF	Somatostatin	Stimulates synthesis and secretion of thyroid hormones
GH Growth hormone (somatotropin)	GHRF	Somatostatin	Stimulates skeletal and soft tissue growth Stimulates mobilization of fatty acids from adipose tissue Inhibits the response to insulin
FSH Follicle stimulating hormone	FSHRF		In the female, stimulates growth of graafian follicles and estrogen production In the male, stimulates sperm development
LH Luteinizing hormone	LHRF		In the female, promotes ovulation and stimulates the synthesis of progesterone In the male, stimulates testosterone secretion
Prolactin	PRF	PIF	Stimulates and maintains lactation

[a] RF indicates releasing factor; IF indicates inhibiting factor.
[b] Somatostatin is also secreted by the pancreas. It inhibits the production of ACTH, TSH, and GH by the pituitary, of insulin and glucagon by the pancreas, and of renin by the kidney.

Hypothalamic Hormones

The hypothalamus secretes specific releasing factors for all the anterior pituitary hormones. In addition, it secretes two inhibitory hormones: somatostatin, which inhibits the secretion of TSH, GH, and ACTH, and prolactin inhibiting factor (PIF). Recall that somatostatin is also secreted by the pancreas. The secretion of hypothalamic hormones is controlled by both neuronal and hormonal signals. Each of the hypothalamic releasing factors is under feedback inhibition by the blood levels of its respective pituitary hormone; rising serum levels of a pituitary hormone inhibit production of its releasing factor.

Anterior Pituitary Hormones

ACTH

Physiologic Actions. The primary function of ACTH is to stimulate the synthesis and secretion of the corticosteroids, and to participate in maintaining the normal functioning of the adrenal cortex. Specifically, ACTH

1. enhances corticosteroid synthesis by stimulating the conversion of cholesterol to pregnenolone
2. stimulates the proliferation of adrenal cortex cells

ACTH acts via cAMP as a second messenger.

Synthesis and Secretion. ACTH is produced by cleavage of a 31 kilodalton glycoprotein, which is also the precursor for the peptide hormone β-lipotropin. Figure 15-16 illustrates the processing of this precursor. Secretion of ACTH is stimulated by the hypothalamic factor CRF and is under feedback inhibition by adrenocortical steroids (see Fig. 15-6). Somatostatin also inhibits ACTH secretion.

Thyrotropin

Physiologic Actions. TSH stimulates the synthesis and secretion of the thyroid hormones T_3 and T_4. Specifically, TSH stimulates:

1. The uptake of iodide by the thyroid gland
2. The fixation of iodide in tyrosine residues of thyroglobulin
3. Degradation of thyroglobulin to T_3 and T_4
5. Secretion of T_3 and T_4

TSH acts via cAMP as a second messenger.

Synthesis and Secretion. Human TSH is a 28 kilodalton glycoprotein consisting of α and β subunits. TSH, FSH, and LH all have identical α subunits, and their β subunits are distinct but similar. The synthesis and secretion of TSH are stimulated by the hypothalamic factor TRF

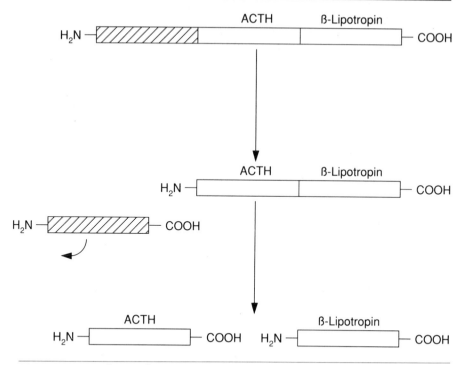

Fig. 15-16. Synthesis of ACTH.

and inhibited by somatostatin. TSH secretion is also under feedback inhibition by circulating T_3 and T_4 (see Fig. 15-11).

Growth Hormone (Somatotropin, GH)

Physiologic Actions. Growth hormone:

1. Stimulates skeletal and soft tissue growth (on the molecular level, GH stimulates RNA and protein synthesis)
2. Stimulates the mobilization of fatty acids from adipose tissue
3. Inhibits the responses of target tissues to insulin

GH binds to a surface receptor on its target cells. A second messenger for GH has not been identified, and its intracellular mechanism of action is not known. Some of the effects of GH are due to the action of a group of peptides called *somatomedins*, which are released from the liver upon stimulation by GH. These peptides, which usually have a molecular weight less than 10,000, have growth-promoting effects on various tissues. Somatomedins bind to target cell surface receptors.

Synthesis and Secretion. Human GH is a polypeptide with molecular weight 21,125, and is derived from a larger precursor. The release of GH is stimulated by hypothalamic GHRF and inhibited by somatostatin. GH secretion is also influenced by a variety of environmental and physiologic stimuli, some of which are listed in Table 15-7.

Table 15-7. Nonhormonal Factors that Influence the Secretion of Growth Hormone

Factor	Stimulatory	Inhibitory
Serotonin	+	
Obesity		+
Exercise	+	
Stress	+	
Sleep	+	
High blood glucose		+

FSH and LH

The gonadotropins FSH and LH both function to stimulate the production of sex hormones, thus ultimately promoting the development and maintenance of the gonads. The specific actions of these hormones are as follows:

FSH
1. In females, FSH stimulates the growth of graafian follicles and the production of estrogen.
2. In males, FSH stimulates the development of sperm cells.

LH
1. In females, LH promotes ovulation and stimulates the secretion of progesterone.
2. In males, LH stimulates the synthesis of testosterone.

Both FSH and LH act via cAMP as a second messenger.

Synthesis and Secretion. FSH and LH are glycoprotein α,β dimers with molecular weights of 34,000 and 25,520, respectively. As mentioned above, the α subunits of FSH, LH, and TSH are identical, and their β subunits are similar. Both the α and β subunits carry N-linked carbohydrate moieties. The secretion of LH and FSH is controlled by the respective hypothalamic releasing factors, which are produced in response to both hormonal and neural stimuli. The sex hormones exert feedback repression over their respective hypothalamic and pituitary hormones (see Fig. 15-7). FSH target cells synthesize a peptide called *inhibin* which directly inhibits pituitary secretion of FSH.

Prolactin

Physiologic Actions. Prolactin stimulates and maintains lactogenesis (milk production) in women after childbirth. The hormone acts via a specific membrane receptor. Increased synthesis of mRNA and protein occurs within hours of a rise in prolactin. Prolactin receptors have also been detected on cells of the ovary and testis, but gonadotropic effects of prolactin on these tissues have not been demonstrated.

Synthesis and Secretion. Prolactin is similar in structure to growth hormone and shows some weak growth hormone activity. Human prolactin has a molecular weight of 23,000, and is probably synthesized as a larger precursor. Secretion of prolactin is stimulated by pregnancy, stress, sleep, and nursing. In addition, the hypothalamus secretes both a prolactin releasing factor and a prolactin inhibiting factor.

Posterior Pituitary Hormones: Oxytocin and Vasopressin

Physiologic Actions

Vasopressin. Vasopressin (antidiuretic hormone, ADH) helps to control the osmotic pressure of the extracellular fluid by stimulating water resorption by the kidney tubules. Cyclic AMP is the second messenger for vasopressin.

Oxytocin. Oxytocin induces uterine contractions during pregnancy and childbirth, and during nursing it causes breast contractions that expel the milk. Oxytocin binds to surface receptors, but its second messenger has not been identified.

Synthesis and Secretion

Both vasopressin and oxytocin are octapeptides. They are synthesized within hypothalamic neurons as precursor polypeptides, which yield not only vasopressin and oxytocin but also hormone carrier proteins called *neurophysins*. The hormone-neurophysin complex is packaged in membrane vesicles which travel down the axons of the hypothalamic neurons to the posterior pituitary, where they are stored until secretion. Vasopressin is secreted in response to a rise in osmotic pressure of the extracellular fluid. Secretion of oxytocin is induced by breast-feeding (suckling) and by the stimuli that initiate labor.

CLINICAL CORRELATION: THYROTOXICOSIS

Thyrotoxicosis (hyperthyroidism) is due to excessive production of triiodothyronine and thyroxine by an overstimulated thyroid gland. Clinical features are weakness, nervousness, weight loss despite an increased food intake, ocular discomfort, warm skin, tachycardia (accelerated heartbeat), and goiter. The serum level of thyroid hormones may be increased tenfold.

The cause of thyroid overstimulation has not been identified for all patients. Plasma levels of TSH usually are not elevated and often are depressed. However, a thyroid stimulating immunoglobulin (TSI) is detectable in many thyrotoxicosis patients. TSI binds to TSH receptors on thyroid cells and stimulates the synthesis of thyroid hormones. Since TSI, unlike TSH, is not under feedback control by the thyroid hormones, it stimulates the thyroid gland continuously and results in hyperthyroidism. The mechanism by which TSI is generated is not understood.

The treatment of thyrotoxicosis usually involves administering thyrotoxic drugs (agents that interfere with thyroid function) to relieve the hyperthyroidism. Different antithyroid drugs interfere with thyroid metabolism at various points, including

1. The uptake of iodine by thyroid cells
2. The iodination of thyroglobulin
3. The conversion of T_4 to the more active T_3
4. The secretion of thyroid hormones

The ophthalmic complications associated with thyrotoxicosis—puffy eyelids and exophthalmia (protruding eyeballs)—may require separate treatment, which is aimed at preventing corneal damage and preserving vision. The ophthalmic changes apparently represent an inflammatory response, and have been observed to regress spontaneously.

SUGGESTED READING

1. Lefkowitz RJ (ed): Receptor Regulation. Chapman and Hall, London, 1981
2. DeGrout LJ, Stanbury JB: The Thyroid and Its Diseases. 4th Ed. McGraw-Hill, New York, 1976
3. Porter R, Fitzsimmons DW (eds): Polypeptide Hormones: Molecular and Cellular Aspects. Cib Found Symp 41, 1976
4. Bajaj JS (ed): Insulin and Metabolism. Elsevier, Amsterdam, 1977
5. Schulster D, Burstein S, Cooke BA: Molecular Endocrinology of the Steroid Hormones. Wiley, New York, 1976
6. James VHT (ed): The Adrenal Gland. Raven Press, New York, 1979
7. Krieger D (ed): Neuroendocrinology. Sinauer, Sunderland, MA, 1980
8. Body PK, Rosenberg LE (eds): Metabolic Control and Disease. WB Saunders, Philadelphia, 1980
9. Schwartz V: A Clinical Companion to Biochemical Studies. WH Freeman, San Francisco, 1978
10. Wilson JD, Foster DW: Williams Textbook of Endocrinology. 7th Ed. WB Saunders, Philadelphia, 1985

STUDY QUESTIONS

Directions: For each of the following multiple choice questions, choose the most appropriate answer.

1. Which of the following hormones binds to a membrane receptor located on the surface of its target cell?
   A. Testosterone
   B. Progesterone
   C. T_3
   D. Norepinephrine

2. Hormone-stimulated increases in intracellular cAMP usually result in
   A. phosphorylation of certain proteins
   B. synthesis of inositol triphosphate and diacylglycerol
   C. activation of phospholipase C
   D. release of intracellular Ca^{2+}

3. An imbalance in the Na^+/K^+ ratio of the extracellular fluid would likely stimulate the secretion of
   A. cortisone
   B. aldosterone
   C. corticosterone
   D. all corticosteroids

4. All of the following statements are true EXCEPT that
   A. androgens are synthesized only in the Leydig cells of the testes.
   B. estrogens are synthesized primarily in follicle cells of the ovary.
   C. estrogens are synthesized from androgens.
   D. all steroid horones are synthesized from cholesterol.

5. Decreased blood levels of which of the following hormones is likely to stimulate the secretion of CRF?
   A. cortisol
   B. aldosterone
   C. estrone
   D. progesterone

6. Renal absorption of Ca^{2+} is stimulated by
   A. calcitonin
   B. 1,25-dihydroxycholecalciferol
   C. vitamin D
   D. parathyroid hormone

7. All of the following hormones are involved in regulating the release of T_3 and T_4 EXCEPT
   A. TRF
   B. TSH
   C. Somatostatin
   D. Inhibin

8. All of the following statements are true EXCEPT that
   A. hypothalamic hormones are all peptide hormones.
   B. most hypothalamic hormones stimulate the secretion of hormones from the anterior pituitary.
   C. the secretion of hypothalamic hormones is controlled only by neuronal stimuli.
   D. some hormones synthesized in the hypothalamus reach the pituitary in nerve axons.

9. All of the following hormones are secreted by the anterior pituitary EXCEPT
   A. FSH
   B. Prolactin
   C. Growth hormone
   D. Oxytocin

Directions: Answer the following questions using the key outlined below:
(A) if 1, 2, and 3 are correct
(B) if 1 and 3 are correct
(C) if 2 and 4 are correct
(D) if only 4 is correct
(E) if all four are correct

---

10. Cholesterol desmolase
    1. converts cholesterol to 17-α-hydroxypregnenolone.
    2. is the rate-limiting enzyme of steroid hormone biosynthesis.
    3. only participates in the synthesis of progesterone.
    4. catalyzes the conversion of cholesterol to pregnenolone.

11. Which of the following statements is (are) true?
    1. The liver is the major site of steroid hormone inactivation.
    2. Steroid hormone inactivation may include oxidation and/or reduction of the active species.
    3. The final metabolic products are mostly excreted in the urine as sulfate or glucuronide conjugates.
    4. The kidney and intestinal mucosal cells are the primary site of steroid hormone inactivation.

12. Thyroxine and triiodothyronine
    1. contain respectively one and two iodide atoms per molecule
    2. stimulate the basal metabolic rate
    3. are synthesized via the iodination of free tyrosine molecules.
    4. bind to receptor proteins that are present in the nucleus of target cells.

13. Glucagon
    1. lowers serum glucose
    2. increases the synthesis of triacylglycerols
    3. increases protein synthesis
    4. stimulates glycogen degradation

14. The catecholamines
    1. can oppose the action of insulin
    2. are usually synthesized as needed and secreted immediately
    3. are secreted in response to emergency situations
    4. bind to a single type of membrane-bound receptor

15. Which of the following hormones are involved in the stimulation of the synthesis of the sex steroids?
    1. FSH
    2. LHRF
    3. LH
    4. FSHRF

16

BLOOD COMPONENTS

Plasma Proteins/Hemoglobin/Blood Clotting/
Fibrinolysis/Anticoagulants/Clinical Correlation

Learning Objectives:

The student should be able to:

1. Define blood plasma and blood serum, and know the major protein fractions of plasma.
2. Describe the cooperative nature of hemoglobin oxygen binding.
3. Describe the factors involved in the loading of oxygen by hemoglobin in the lungs and in oxygen unloading in the tissues.
4. Describe the contribution of hemoglobin to the transport of carbon dioxide.
5. Describe the events involved in hemostasis.
6. Describe the intrinsic and extrinsic pathways of blood clotting and the role of vitamin K in clotting.
7. Describe the process of fibrinolysis by which clots are dissolved.

Perspective: Blood

Blood consists of water, suspended cells, and a complex mixture of dissolved solutes. *Blood plasma* is the water–solute fraction that remains after the suspended cells are removed. *Blood serum* is obtained when both the cells and the clotting factors are removed. The major solutes of plasma by weight are proteins, which account for about 70 percent of the total; small organic molecules account for about 20 percent and inorganic salts for about 10 percent. The principle organic small molecules are amino acids, carbohydrates, lipids, cellular metabolites, and excretion products. The major inorganic ions are bicarbonate, chloride, phosphates, sulfate, iodide, calcium, sodium, potassium, and magnesium. The precise concentrations of these solutes reflect the physiologic activities of the tissues.

PLASMA PROTEINS

When blood plasma is subjected to electrophoresis, five major protein fractions are detected. In order of decreasing electrophoretic mobility, these fractions are albumin and the α_1-, α_2-, β-, and γ-globulins. The electrophoretic pattern of plasma proteins is very reproducible, and is the basis for assigning an individual protein to a given fraction. Albumin is the most abundant plasma protein.

Albumin

Albumin is a single polypeptide chain of approximately 580 amino acids which is folded into three structural domains. It is synthesized in the liver and secreted into the plasma, and is one of the few plasma proteins that has no carbohydrate moieties.

As mentioned in previous chapters, albumin binds a variety of substances and serves as a protein carrier in transporting them between tissues. For example, albumin is responsible for transporting the fatty acids released by adipose tissue and for transporting bilirubin from extrahepatic tissues to the liver. Albumin also plays an important role in regulating plasma osmotic pressure. It has a significant effect on osmotic pressure for two reasons:

1. It is a relatively small protein, and is therefore abundant not only in terms of weight but also in molar concentration.
2. At the pH of plasma, albumin carries several negative charges. These charges cause positive ions to move from cells into the plasma, increasing the plasma osmotic pressure.

The Globulins

The globulin proteins are a heterogeneous mixture. Many globulins have been purified and studied, but the biologic function is not known for all of them, and some globulins have received only cursory study. As mentioned above, the globulins are divided into α-, β-, and γ-globulins on the basis of electrophoretic mobility. Some representative members of each class are identified below.

α-Globulins

1. *Retinol-binding protein:* Binds the fat-soluble vitamin retinol.
2. *Ceruloplasmin:* Functions in the transport of copper and in the oxidation of ferrous iron. Iron must be oxidized to the ferric state before it can be incorporated into transferrin.
3. *Protease inhibitors:* Protect the body from random intravascular proteolysis. An important member of this group is α_2-macroglobulin.

β-Globulins

1. *Lipoproteins:* Function in transporting triacylglycerols, cholesterol, and phospholipids, as discussed in Chapter 7.

2. *Transferrin:* Transports Fe^{3+} to the tissues.
3. *Hemopexin:* Binds heme and thereby prevents it from being excreted by the kidneys.

γ-Globulins

The primary γ-globulins are the immunoglobulins (Chapter 23).

Plasma Enzymes as Diagnostic Tools

Normally, the only enzymes present in the plasma in substantial amounts are the enzymes involved in clotting. High concentrations of other enzymes in the plasma usually represent the release of cellular enzymes from damaged tissues. Abnormal plasma enzymes are therefore diagnostically useful. For example, an increase in the level of serum amylase is associated with pancreatitis, and the plasma concentrations of glutamate-oxaloacetate transaminase and lactate dehydrogenase increase after myocardial infarction. Liver damage may elevate plasma levels of several characteristic enzymes.

HEMOGLOBIN

Hemoglobin is important for the transport of oxygen, the transport of carbon dioxide, and the maintenance of blood pH. Hemoglobin is contained in the red cells (erythrocytes). Normal adult hemoglobin is an $\alpha_2\beta_2$ tetramer. The subunits are held together by noncovalent interactions, which are especially extensive between the α and β subunits. Each of the four subunits contains a heme. The heme iron is the binding site for oxygen, so each hemoglobin can carry four molecules of oxygen. X-ray studies show that the α and β subunits, which are similar in structure, are composed mainly of several segments of α-helix. Some of these segments wrap around the heme, enclosing it in a very hydrophobic microenvironment called the *heme pocket*. Oxygenated hemoglobin is called *oxyhemoglobin*, and deoxygenated hemoglobin is called *deoxyhemoglobin*. It should be noted that fetal hemoglobins have a higher oxygen affinity than adult hemoglobin, which enables them to absorb oxygen efficiently from the maternal blood in the placenta.

Perspective: Hemoglobin Function

Hemoglobin binds oxygen and transports it to the tissues. Oxygen binding or *loading* occurs in the alveolar capillaries of the lung, whereas unloading of oxygen occurs in the tissues. The loading and unloading of oxygen by hemoglobin are strongly influenced by the opposed conditions of carbon dioxide partial pressure (PCO_2) and pH found in the lungs and tissues: PCO_2 is low and pH is high in the lungs, whereas PCO_2 is high and pH low in the tissues. CO_2 diffuses freely across cell membranes. The events in the binding and delivery of oxygen are as follows:

(continued)

1. Oxygen from the inspired air diffuses into red blood cells in alveolar capillaries.
2. Oxygen binds to hemoglobin in the red cells.
3. Oxyhemoglobin is transported to the tissues, where the oxygen is released and diffuses from the red cells to the tissue cells.

HEMOGLOBIN FUNCTION

Cooperativity

The binding of oxygen to hemoglobin is *cooperative* (Chapter 2)—the binding of each successive oxygen increases the binding affinity of hemoglobin for the next oxygen, so that the fourth oxygen binds much more readily than the first. This property is also called *heme-heme interaction*. The cooperative binding of hemoglobin is described by the *Hill equation*:

$$Y = 100 \frac{(P/P_{1/2})^n}{1 + (P/P_{1/2})^n}$$

where P is the partial pressure of oxygen, Y is the percent of heme sites that are oxygenated, $P_{1/2}$ is the partial pressure of oxygen at which half of the heme sites are oxygenated, and n is the Hill coefficient. The Hill coefficient is an index of the degree of cooperativity exhibited: the higher this coefficient, the more cooperative is the binding and the more strongly sigmoidal will be a plot of Y versus P (a Hill plot). A value of $n = 1$ indicates zero cooperativity and yields a nonsigmoidal Hill plot. Figure 16-1 shows a Hill plot for normal adult hemoglobin. The value of n for adult hemoglobin is approximately 2.8.

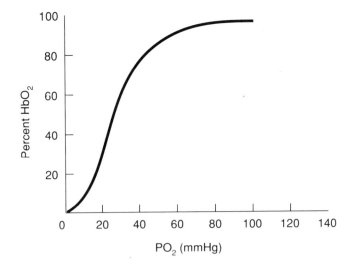

Fig. 16-1. The sigmoidal hemoglobin oxygen-binding curve.

Mechanism of Cooperative Oxygen Binding

The most widely accepted model of hemoglobin function assumes that hemoglobin has two extreme conformations (Fig. 16-2): a *tense* or T state and a *relaxed* or R state. The T state has lower oxygen affinity and the R state has higher oxygen affinity; deoxyhemoglobin is usually in the T state and oxyhemoglobin in the R state. When an oxygen molecule binds to a subunit of deoxyhemoglobin, it causes the subunit to change toward the conformation found in the R state. This change is transmitted via subunit interactions to an adjacent subunit, which in turn adopts the R configuration and has a higher oxygen affinity. Binding of oxygen to this subunit similarly affects binding to the third subunit, and so forth. Similarly, the *loss* of oxygen from one subunit lowers the oxygen affinity of adjacent subunits.

Transport and Delivery of Oxygen and Carbon Dioxide

The partial pressure of oxygen in the lungs (approximately 100 mmHg) is such that hemoglobin in red cells leaving the lungs is almost 100 percent saturated—that is, the reaction

$$Hb + 4O_2 \rightleftharpoons Hb \cdot 4O_2$$

proceeds almost to completion. The unloading of oxygen from hemoglobin in the tissue capillaries is due to two factors:

1. The much lower partial pressure of oxygen in the tissues (approximately 35 mmHg)
2. The presence in the tissues of substances that lower the oxygen binding

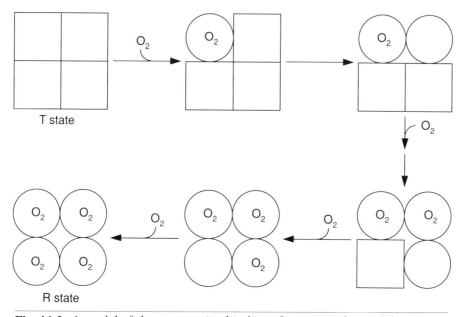

Fig. 16-2. A model of the cooperative binding of oxygen to hemoglobin.

affinity of hemoglobin—specifically, H^+, CO_2, and 2,3-bisphosphoglycerate.

At the low PO_2 found in the tissues, the oxygen binding reaction tends to proceed to the left.

The Bohr Effect

Hemoglobin is able to bind hydrogen ions. Binding of hydrogen ion lowers the oxygen affinity of hemoglobin, so that the following reaction is shifted to the right:

$$HbO_2 + H^+ \rightleftharpoons HbH^+ + O_2$$

An increase in PCO_2 also causes oxygen to dissociate from hemoglobin. This effect is due primarily to the fact that increases in PCO_2 also increase $[H^+]$. Carbon dioxide in the plasma diffuses into the red cells, where the enzyme *carbonic anhydrase* converts it into carbonic acid. Carbonic acid largely dissociates into bicarbonate and hydrogen ion:

$$CO_2 + H_2O \xrightleftharpoons{\text{Carbonic anhydrase}} H_2CO_3 \rightleftharpoons HCO_3^- + H^+$$

The resulting hydrogen ions may then function to displace more O_2 from hemoglobin. The dependence of oxygen affinity on $[CO_2]$ and/or $[H^+]$ is known as the *Bohr effect*. The Bohr effect is important in the delivery of oxygen in the tissues, the loading of oxygen in the lungs, and the excretion of carbon dioxide in the lungs.

Oxygen Delivery

The Bohr effect is important in the delivery of oxygen to the tissues. When blood arrives in the tissues, carbon dioxide enters the red cells and is converted to carbonic acid, bicarbonate, and H^+ as described above. The resulting fall in pH enhances the release of oxygen from hemoglobin.

The release of oxygen from hemoglobin is also stimulated by the compound *2,3-bisphosphoglycerate* (BPG). Red cells contain a high concentration of BPG, which is synthesized from the glycolytic intermediate 1,3-bisphosphoglycerate. BPG binds to deoxyhemoglobin in a 1:1 molar ratio. Binding of BPG stabilizes the T conformation of deoxyhemoglobin, and therefore promotes the release of oxygen from hemoglobin:

$$HbO_2 + BPG \rightleftharpoons Hb\text{-}BPG + O_2$$

In the lungs, where PO_2 is high, hemoglobin is almost completely oxygenated despite the presence of BPG. In the tissues, where PO_2 is much lower, BPG increases the efficiency of oxygen dissociation from hemoglobin. The effect of BPG is very important in oxygen-depleted tissues. In general, BPG ensures maximal release of oxygen from hemoglobin in tissues. Figure 16-3 shows the effect of $[H^+]$ and BPG on hemoglobin binding curves.

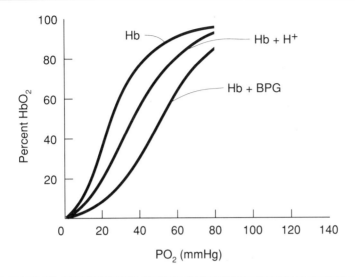

Fig. 16-3. The effect of H^+ and BPG on hemoglobin oxygen-binding curves.

In summary, the factors that contribute to the delivery of oxygen to the tissues are as follows:

1. The low tissue PO_2
2. The Bohr effect caused by the decrease in pH in the tissues
3. The effect of BPG

Transport of Carbon Dioxide

Carbon dioxide is transported in the blood from the tissues to the lungs, where it diffuses into the alveolar air and is exhaled with the breath. Carbon dioxide is transported in the blood in two forms:

1. As carbamoyl groups attached to the N-terminal amino groups of the hemoglobin subunits
2. In solution in the blood in the form of the components of the HCO_2^-/H_2CO_3 buffer system.

Hemoglobin participates in both types of transport.

Formation of Carbamino Groups

A small fraction of the CO_2 that diffuses into red cells does not form carbonic acid, but instead reacts directly with the unprotonated amino groups of hemoglobin to form carbamino groups:

$$Hb\text{-}NH_2 + CO_2 \rightleftharpoons Hb\text{-}NHCOO^- + H^+$$

Carbamino groups are formed primarily on the N-terminal amino groups of the hemoglobin subunits. Deoxyhemoglobin is much more active in the formation of carbamino groups than oxyhemoglobin.

Therefore, the formation of carbamino groups promotes oxygen release, whereas oxygen binding promotes CO_2 release. The sequence of events associated with the transport of CO_2 in the form of carbamino groups on hemoglobin therefore is as follows:

1. In the tissue capillaries, CO_2 diffuses into red cells.
2. Some CO_2 binds to hemoglobin to form carbamino groups.
3. In the lungs, the carbamino-Hb binds oxygen, which causes CO_2 to be released. CO_2 diffuses into the alveolar air and is exhaled.

Because only the uncharged N-terminal α-amino groups participate in the carbamino reactions, only a small fraction of the total CO_2 carried in the blood travels in the form of carbamino groups.

Plasma Bicarbonate and Blood pH

Most of the carbon dioxide that diffuses into the red cells is converted to carbonic acid. Dissociation of carbonic acid in the red cells produces H^+ as well as HCO_3^-. HCO_3^- and H_2CO_3 reenter the plasma and are carried to the lungs as components of the HCO_3^-/H_2CO_3 buffer system. If the resulting hydrogen ion were to diffuse into the plasma, it would significantly alter plasma pH. Instead, as described above in connection with the Bohr effect, much of this H^+ binds to hemoglobin. The uptake by hemoglobin of the protons produced by the conversion of CO_2 to HCO_3^- is called the *isohydric shift*. It allows CO_2 produced in the tissues to be transported to the lungs as HCO_3^-.

Hemoglobin does not bind all of the hydrogen ion produced by carbonic acid dissociation; only about 0.8 mole of protons are bound per mole of oxygen released in the tissues. Some of the remaining hydrogen ion is buffered by plasma proteins. A slight difference in pH remains between arterial and venous blood. As discussed in Chapter 17, the HCO_3^-/H_2CO_3 system is the primary buffer system involved in the regulation of extracellular pH. The ratio of HCO_3^- to H_2CO_3 is kept constant at 20:1, but the absolute concentrations of the buffer components vary with the quantity of carbon dioxide being transported in the blood.

The bicarbonate produced in the red cells tends to diffuse down its concentration gradient into the plasma. Because H^+ binds to hemoglobin, this flux of bicarbonate out of the red cell is not accompanied by an equal flux of hydrogen ion. The positive charge that would tend to build up in the red cells is neutralized by an influx of chloride ions from the plasma. In the lungs, the process is reversed: HCO_3^- moves from the plasma back into the red cells, and chloride moves out into the plasma. This exchange is referred to as the *chloride shift*.

Hemoglobin Function: Overview

The events involved in the transport of oxygen and carbon dioxide between the lungs and tissues may be summarized as follows:

In the Tissues. PCO_2 is high and PO_2 is low.

1. Carbon dioxide diffuses into the red cells, and is converted into carbonic acid by carbonic anhydrase. Carbonic acid largely dissociates into HCO_3^- and H^+:

$$CO_2 + H_2O \rightleftharpoons H_2CO_3 \rightleftharpoons H^+ + HCO_3^-$$

2. The low PO_2, high PCO_2, and rise in $[H^+]$ all combine to drive oxygen off of hemoglobin. The presence of BPG maximizes the efficiency of oxygen unloading.
3. Much of the H^+ produced by dissociation of carbonic acid is bound by hemoglobin. Some carbon dioxide is also bound by hemoglobin in the form of carbamate:

$$Hb\text{-}NH_2 + CO_2 \rightarrow Hb\text{-}HNCOO^-$$

4. The bicarbonate produced by carbonic acid dissociation mostly diffuses into the plasma, where it is carried to the lungs in the form of the HCO_3^-/H_2CO_3 plasma buffer couple.

In the Lungs. PO_2 is high and PCO_2 is low.

1. The oxygenation of hemoglobin is accompanied by the release of H^+ and CO_2 (carbamate) from hemoglobin:

$$H^+\text{-}Hb\text{-}NHCOO^- + O_2 \rightarrow NH_2\text{-}Hb\text{-}O_2 + H^+ + CO_2$$

2. The hydrogen ions released from hemoglobin protonate bicarbonate to produce carbonic acid, which in turn dissociates to CO_2 and H_2O:

$$CO_2 + H_2O \rightleftharpoons H_2CO_3 \rightleftharpoons H^+ + HCO_3^-$$

These reactions lower the concentration of bicarbonate in the red cells. Plasma bicarbonate then diffuses into the red cells, where it is converted to CO_2 and H_2O. The flux of bicarbonate into the red cells is accompanied by an opposing flux of chloride.

3. Dissolved CO_2 is in equilibrium with CO_2 in the alveolar air. The CO_2 produced from carbonic acid therefore is exhaled with the breath.

Perspective: Hemostasis

When a vessel is injured, the first necessity is to stop the bleeding—i.e., to achieve hemostasis. Three processes participate in hemostasis:

1. Platelet cells in the blood become sticky and adhere to the site of injury, clinging both to exposed endothelial tissue and to each other.
2. The stimulated platelets release vasoconstrictors, which induce the blood vessel to constrict.
3. Blood clotting occurs at the site of the injury, resulting in the formation of a fibrin hard clot or *thrombus*.

The combined actions of vessel constriction and platelet accumulation result in the formation of a *platelet plug*, which in some cases may stop the bleeding. However, long-term hemostasis depends on the formation of a fibrin hard clot. Wound healing then proceeds, and after the damaged vessel is healed, the clot is removed by fibrinolysis.

BLOOD CLOTTING

A blood clot consists of cross-linked monomers of a protein called *fibrin*. Fibrin monomers are normally present in the circulation in the inactive form *fibrinogen*. Fibrinogen is converted to fibrin by two *clotting mechanisms* involving enzyme cascades. Fibrin monomers associate spontaneously to form a *fibrin soft clot*, and are then enzymatically cross-linked to form the insoluble *fibrin hard clot*.

As shown in Figure 16-4, blood clotting (coagulation) can be initiated by two mechanisms: the *intrinsic pathway* and the *extrinsic pathway*. The intrinsic pathway is triggered by the presence of a damaged vessel wall or other abnormal surface, whereas the extrinsic pathway is triggered by the release of special substances from traumatized tissue. The clotting system consists of a number of enzymes called *clotting factors* plus some noncatalytic proteins. The enzymes are all present in the circulation in the form of inactive zymogens. The cascade arrangement of these enzymes allows the signal for clotting to be amplified.

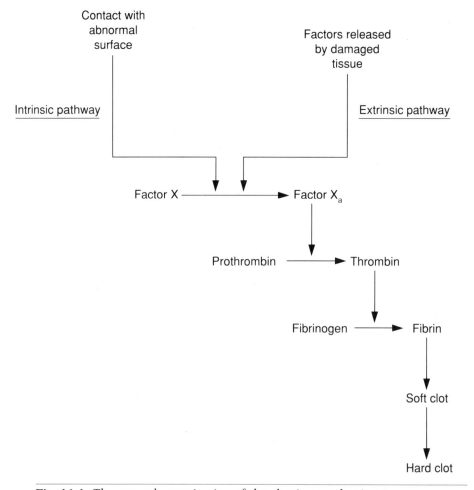

Fig. 16-4. The general organization of the clotting mechanism.

Blood Components

As can be seen from Figure 16-4, the segment of the clotting pathway from prothrombin to hard clot formation is common to both the extrinsic and intrinsic pathways. This common segment will be described first.

The Common Segment: Prothrombin to the Hard Clot

The common segment of the clotting pathway starts with Factor X_a (a subscript *a* indicates an activated clotting factor). This segment has four reactions.

Reaction 1: Activation of Prothrombin to Thrombin

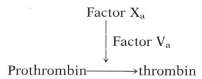

Two proteins, Factors X_a and V_a, participate in the conversion of prothrombin to thrombin. As shown in Figure 16-5, prothrombin, Factor X_a, and Factor V_a all bind to a phospholipid surface (usually the surface of a platelet). Calcium is necessary for the binding of prothrombin and Factor X_a. The resulting complex is called the *catalytic complex*. Factor X_a makes two proteolytic cleavages, and thrombin and an amino terminal peptide are released. The function of Factor V_a is not known exactly; it probably helps to orient prothrombin and Factor X_a relative to each other. Thrombin consists of two peptides joined together by a disulfide bridge.

Reaction 2: Activation of Fibrinogen to Fibrin

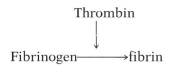

Thrombin catalyzes the activation of fibrinogen to fibrin. Fibrinogen consists of three pairs of subunits forming an $\alpha_2\beta_2\gamma_2$ complex. Thrombin cleaves an arginine-glycine peptide bond in each of the α and β chains, releasing two each of the A and B fibrinopeptides and thus generating fibrin. Fibrin monomers then associate by noncovalent interactions to form the *fibrin soft clot*.

Reaction 3: Activation of FSF

The *fibrin stabilizing factor* FSF (also called Factor XIII) is responsible for cross-linking the fibrin monomers of the soft clot to yield the hard clot. Inactive FSF is present in both platelets and plasma; platelet FSF is an a_2 dimer, whereas plasma FSF is an a_2b_2 tetramer. Thrombin makes a cut in each *a* chain, releasing two peptides and FSF a'_2 or a'_2b_2. The dimer FSF a'_2 is active, but the tetramer a'_2b_2 is inactive until the *b* subunits dissociate, a process that requires Ca^{2+}. The two versions of

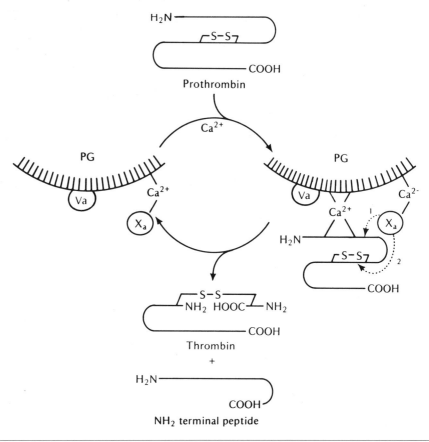

Fig. 16-5. Model of the catalytic complex for the activation of prothrombin. (From Smith EL, Hill RL, Lehman IR et al: Principles of Biochemistry. Mammalian Biochemistry. 7th Ed. McGraw-Hill, New York, 1983, with permission.)

FSF activation are thus as follows:

Platelet FSF: a_2 (zymogen) ⟶ a'_2 (active)
                                    ↑
                                Thrombin
                                    ↓
Plasma FSF: a_2b_2 (zymogen) ⟶ a'_2b_2 (inactive) ⟶
                                                               a'_2 (active) + b_2

Reaction 4: Cross-Linking of Fibrin

$$\text{Fibrin soft clot} \xrightarrow{\text{FSF}(a'_2)} \text{fibrin hard clot}$$

Active FSF a'_2 is a transglutaminase. It catalyzes the formation of covalent end-to-end cross-links between glutamine γ-amide groups on one fibrin monomer and lysine ε-amino groups on an adjacent monomer (Fig. 16-6). The resulting hard clot is insoluble and remains in place until degraded during the process of fibrinolysis.

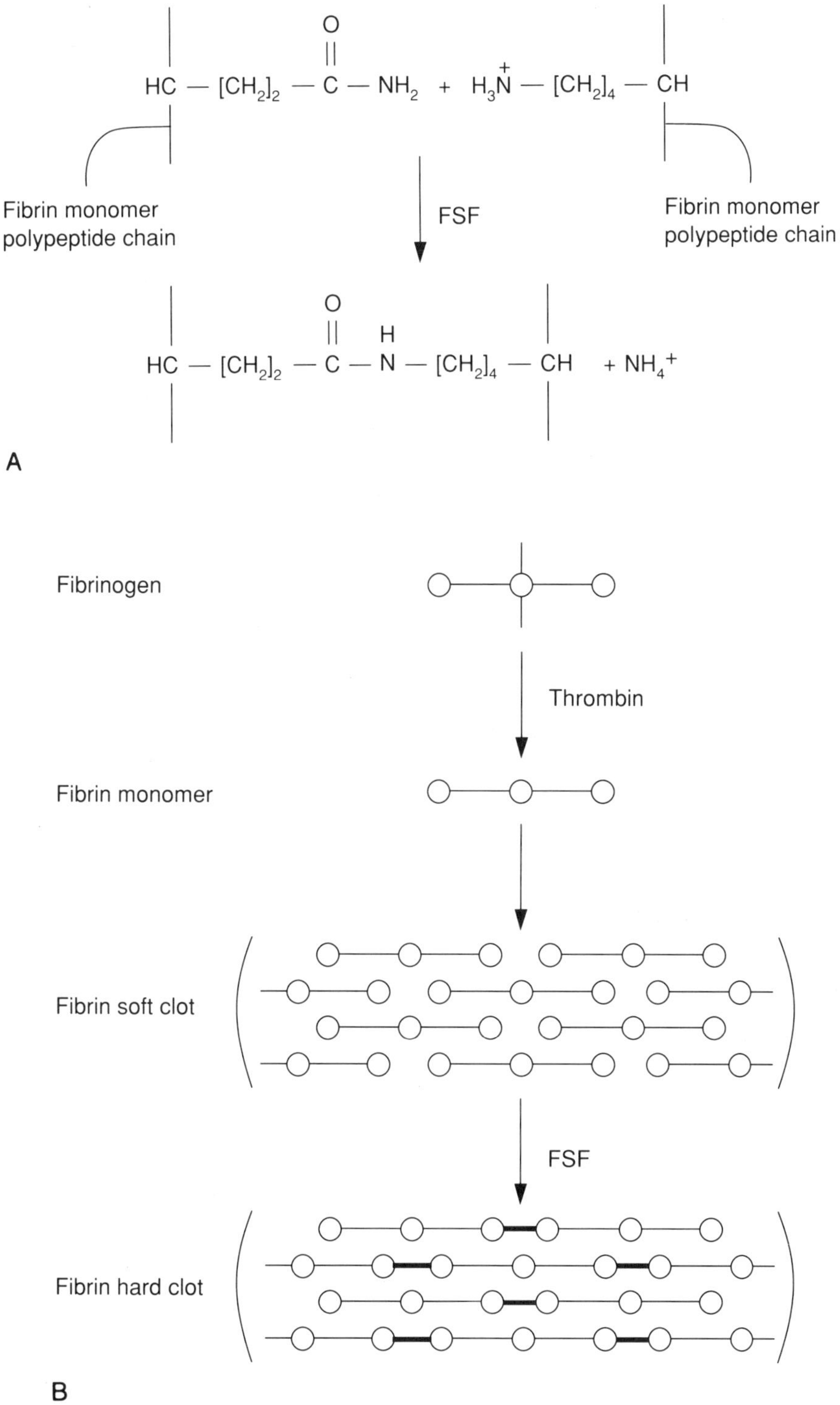

Fig. 16-6. **(A)** The formation of covalent cross-links between fibrin monomers by activated FSF. **(B)** Formation of the fibrin hard clot.

The Extrinsic Pathway

A clot is formed much more rapidly by the extrinsic pathway (Fig. 16-7) than by the intrinsic pathway—clotting by the extrinsic pathway may occur in 10 to 12 seconds, whereas the intrinsic pathway requires several minutes. The steps of the extrinsic pathway are as follows:

1. A protein called *tissue factor* is released by damaged tissue.
2. In the presence of phospholipid, tissue factor stimulates a very limited amount of activity in the zymogen form of Factor VII.
3. This semiactive Factor VII generates a small amount of Factor X_a and/or thrombin.
4. This small amount of thrombin then activates Factor VII to the fully active Factor VII_a, which in turn activates the normal quantity of Factor X to Factor X_a.

The mechanism by which Factor X is activated is similar to the mechanism described above for prothrombin. Factor VII_a, Factor X, and tissue factor form a complex on a phospholipid surface. The binding of Factor VII_a and Factor X to the surface requires Ca^{2+}. Factor VII_a converts Factor X to Factor X_a by limited proteolysis.

The Intrinsic Pathway

Figure 16-8 shows the steps of the intrinsic pathway. The intrinsic pathway is initiated by a series of events called the *contact stage*, in which clotting factors are activated by making contact with certain surfaces. The proteins involved in the contact stage are Factor XII, prekallikrein,

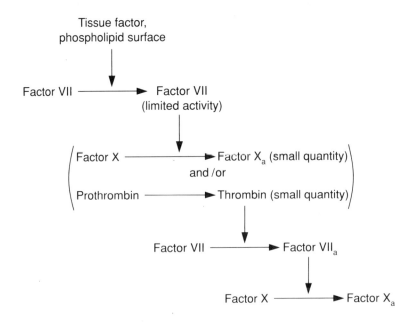

Fig. 16-7. The extrinsic pathway of clot formation.

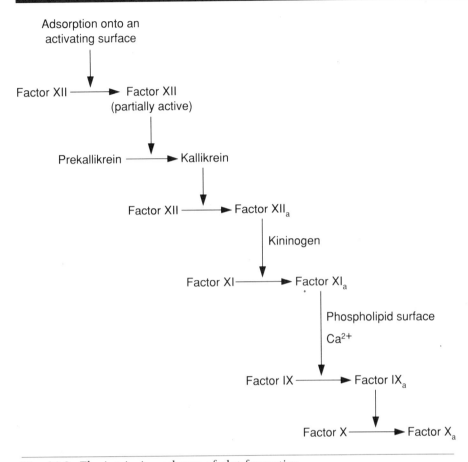

Fig. 16-8. The intrinsic pathway of clot formation.

Factor XI, and high molecular weight kininogen. The steps of the contact stage are as follows:

1. Factor XII zymogen is partially activated by adsorption onto platelets or collagen.
2. The adsorbed Factor XII zymogen activates prekallikrein to kallikrein.
3. Kallikrein in turn converts Factor XII zymogen to Factor XII$_a$.
4. Factor XII$_a$ activates Factor XI to Factor XI$_a$. High molecular weight kininogen is required in this reaction as a nonenzymatic activator.

The steps of the contact stage do not require Ca^{2+}. The rest of the intrinsic pathway proceeds as follows.

5. Factor XI$_a$, the product of the contact stage, activates Factor IX to Factor IX$_a$ in a process that requires phospholipid and Ca^{2+}.
6. Factor IX$_a$ then activates Factor X to Factor X$_a$ in a process that—again—is similar to the activation of prothrombin. Factor IX$_a$, Factor X, and Factor VIII$_a$ (a nonenzymic activator) bind to a phospholipid surface, where Factor IX$_a$ activates Factor X by proteolytic cleavage. Calcium is required in this process.

General Comments: Extrinsic and Intrinsic Pathways

The product of both the extrinsic and the intrinsic pathways is Factor X_a, which initiates the common segment of the coagulation pathway. Thrombin is known to activate both Factor V in the common segment and Factor VIII in the intrinsic pathway. However, Factor VIII is used prior to, and Factor V concurrent with, the activation of thrombin by Factor X_a. It is likely that the small amount of thrombin generated at the initiation of the extrinsic pathway also participates in the activation of Factors V and VIII.

The Role of Vitamin K in Blood Clotting

Vitamin K influences the performance of Factors VII, IX, X, and prothrombin. All these proteins are synthesized in the liver and contain Ca^{2+} binding sites that are essential in their activation. The Ca^{2+} binding sites consist of a cluster of γ-carboxyglutamyl residues in the amino terminal region of the protein. These γ-carboxyglutamyl residues are formed by posttranslational modification as shown in Figure 16-9. The liver carboxylase that catalyzes this reaction requires vitamin K as a cofactor. Because the γ-carboxyglutamyl cluster is essential for normal binding to the phospholipid surface, vitamin K deficiency results in the synthesis of clotting factors that are not activated or are only slightly activated. The result is impaired clotting and a tendency toward hemorrhage. Some anticoagulants, such as warfarin, act as vitamin K antagonists.

FIBRINOLYSIS

After a damaged vessel has healed sufficiently, the clot is no longer needed. *Fibrinolysis,* the enzymatic dissolution of the fibrin clot, is achieved by the enzyme *plasmin,* which normally exists in zymogen form as *plasminogen.* Fibrinolysis occurs in two steps:

1. *Activation of plasminogen to plasmin.* Plasminogen is activated by urokinase, a protease synthesized in the kidney. Similar proteases synthesized in other tissues also activate plasminogen.
2. *Proteolysis of the fibrin hard clot.* Plasmin is a protease that specifically degrades fibrin to soluble cleavage products.

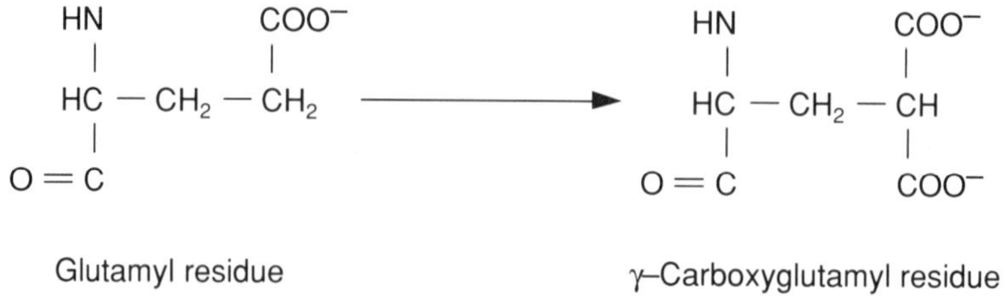

$$\begin{array}{c}\text{HN} \\ | \\ \text{HC} - \text{CH}_2 - \text{CH}_2 \\ | \\ \text{O} = \text{C}\end{array} \quad \begin{array}{c} \text{COO}^- \end{array} \quad \longrightarrow \quad \begin{array}{c}\text{HN} \\ | \\ \text{HC} - \text{CH}_2 - \text{CH} \\ | \\ \text{O} = \text{C}\end{array} \quad \begin{array}{c} \text{COO}^- \\ | \\ \\ | \\ \text{COO}^- \end{array}$$

Glutamyl residue   γ–Carboxyglutamyl residue

Fig. 16-9. The vitamin K-dependent γ-carboxylation of glutamate residues.

Plasmin is inactivated by a specific inhibitor, α_2-antiplasmin, which rapidly inactivates plasmin molecules that are not associated with a fibrin clot.

INHIBITION OF CLOTTING: ANTICOAGULANTS

The most widely used strategies to prevent clotting in drawn blood take advantage of the Ca^{2+} requirement of clotting. The blood is protected from clotting by adding substances that bind Ca^{2+}, such as oxalate, fluoride, and citrate. The anticoagulant must be added immediately in order to be effective.

Heparin, a natural anticoagulant, can be used to prevent clot formation in vivo. Heparin administered as a drug or released by the tissues forms a complex with a plasma protein, antithrombin III. This complex is an effective inhibitor of several coagulation factors.

CLINICAL CORRELATION: CLASSIC HEMOPHILIA

The hemophilias are genetic diseases characterized by impaired clotting. Affected individuals may bleed severely from minor injuries or from surgical and dental procedures. In some cases, spontaneous bleeding into tissues may occur. In a test tube the blood clots very slowly. Different hemophilias are due to genetic deficiencies in different clotting factors, and may involve either decreased synthesis of the factor or synthesis of a defective factor.

Classic hemophilia involves a deficiency of Factor VIII, which is synthesized in normal amounts but is functionally defective. The clinical severity of the disease varies widely due to variation in the extent of the genetic defect. Classic hemophilia is inherited as an X-linked recessive trait, and therefore occurs mainly in males, who inherit it from mothers who are usually heterozygous carriers. Classic hemophilia has also been observed in a few females homozygous for the trait.

Diagnosis and Treatment

The most reliable way to differentiate classic hemophilia from other hemophilias is to demonstrate that the defect involves Factor VIII. This may be achieved by testing whether the patient's plasma is able to improve the clotting time of blood from a known classic hemophiliac; if the clotting time is improved, then the patient's disorder does not involve Factor VIII. Alternatively, the amount of Factor VIII in the plasma can be measured by reacting the patient's blood with antibodies to normal Factor VIII.

The major treatment for serious bleeding in classic hemophiliacs is transfusion with Factor VIII-rich plasma concentrate. For minor wounds, pressure over the site of injury and/or local application of powdered bovine Factor VIII is often sufficient.

SUGGESTED READING

1. Putnam FW (ed): The Plasma Proteins. 2nd Ed. Vols. 1–3. Academic Press, Orlando, FL, 1975, 1979
2. Ratnoff OD: Hereditary disorders of hemostasis. In Stanbury JB, Wyngaarden JB, Frederickson DS et al (eds): The Metabolic Basis of Inherited Disease. 5th Ed. McGraw-Hill, New York, 1983
3. Jackson CM, Nemerson Y: Blood coagulation. Annu Rev Biochem 49:765, 1980

STUDY QUESTIONS

Directions: Answer the following questions using the key outlined below.

- (A) if 1, 2, and 3 are correct
- (B) if 1 and 3 are correct
- (C) if 2 and 4 are correct
- (D) if only 4 is correct
- (E) if all four are correct

1. Albumin plays is important in
   1. transporting a variety of substances in the blood
   2. regulating plasma pH
   3. regulating plasma osmotic pressure
   4. transporting substances across tissue cell membranes

2. Plasma enzymes other than those used in blood clotting
   1. are usually members of the α_1-globulin group
   2. are always members of the α_1-globulin group
   3. are never present
   4. when present in large quantities, have usually been released into the plasma from tissues damaged by injury or disease

3. Substances that enhance the release of oxygen from hemoglobin include
   1. 2-phosphoglycerate
   2. 2,3-bisphosphoglycerate
   3. pyruvate
   4. dilute [H^+]

4. Hemoglobin functions in the transport of CO_2 from the tissues to the lungs by
   1. the formation of carbamino groups
   2. the formation of CO_2-heme iron (ferrous) complexes
   3. enhancing the formation of plasma HCO_3^-, which is subsequently carried to the lungs in the circulation
   4. the formation of CO_2-heme iron (ferric) complexes

Directions: For each of the following multiple choice questions, choose the most appropriate answer.

5. The major protein fraction of blood plasma is
   A. β_1-globulins
   B. α_2-globulins
   C. γ-globulins
   D. albumin

6. All of the following statements concerning oxygen binding to hemoglobin are true EXCEPT that
   A. oxygen binds to plasma hemoglobin.
   B. the oxygen affinity of hemoglobin is reduced by 2,3-BPG.
   C. the $P_{1/2}$ for oxygen is greater in the lungs than at the tissues.
   D. oxygen binding to hemoglobin is cooperative.

7. All of the following reactions are involved in the blood clotting process EXCEPT
   A. the transglutaminase reaction
   B. glutamate γ-carboxylation
   C. proteolysis of prothrombin to thrombin by factor X_a
   D. glutamate β-carboxylation

Items 8 through 10. Consider the following terms relative to blood clotting
   A. Intrinsic pathway
   B. Extrinsic pathway
   C. Common segment
   D. Fibrinolysis

Match the term that is most closely associated with statements below.

8. A blood clotting mechanism that is triggered when certain clotting factors make contact with a surface.

9. The more rapid blood clotting mechanism.

10. The process of clot dissolution.

17

The Extracellular Fluid

Regulation of Volume, Osmotic Pressure, and pH

The student should be able to:

1. Describe the mechanism by which the extracellular fluid volume is regulated.
2. Describe the mechanism by which extracellular fluid pH is regulated.
3. Describe the mechanism by which the lungs compensate for changes in extracellular fluid pH.
4. Describe the three renal mechanisms for regulating extracellular fluid pH.
5. Describe the intracellular buffering mechanism that compensates for changes in extracellular fluid pH.

Learning Objectives

The body water is divided into two compartments: the *intracellular fluid* (ICF) located inside the cells, and the *extracellular fluid* (ECF) located outside the cells. The ICF and ECF have distinctly different solute compositions, which are suited to their differing physiologic roles (Table 17-1). The intracellular fluids provide the aqueous matrix in which cellular reactions take place. Different cells and intracellular compartments may have different solute concentrations and pH values, reflecting the needs of the reactions that occur in them. The composition of the ICF is important to cellular functioning. Some solutes distribute freely between the cell and the ECF. Mammals have highly developed mechanisms for regulating the ECF, which indirectly influence the regulation of the ICF.

This chapter addresses the mechanisms involved in regulating the volume, osmotic pressure, ionic composition, and pH of the ECF. The major subcompartments of the ECF are the blood plasma and the interstitial fluid (the fluid surrounding tissue cells). Minor extracellular fluid compartments include the cerebrospinal fluid, lymph, synovial fluid, and aqeous humor.

Perspective: The Extracellular Fluid

Table 17-1. Electrolyte Composition (mg/L) of Extracellular and Intracellular Fluid

	Extracellular Fluid		
Ion	Plasma	Interstitial	Intracellular Fluid
Na^+	152	143	14
K^+	5	4	157
Ca^{2+}	5	5	—
Mg^{2+}	3	3	25
Cl^-	113	117	—
HCO_3^-	27	27	10
PO_4^{3-}	2	2	113

REGULATION OF ECF VOLUME

The volume of the ECF is controlled by altering the quantity of Na^+, the major cation of the ECF. The extracellular concentration of Na^+ is tightly regulated. Therefore, a change in the total quantity of Na^+ in the ECF is accompanied by a matching change in ECF volume to maintain a constant concentration of Na^+. A net excretion of Na^+ stimulates a loss of water and a reduction of ECF volume, and vice versa. The volume of the ECF is controlled by regulating the renal excretion of Na^+. The following are key points concerning the relationship between the Na^+ content and the volume of the ECF:

1. The Na^+ concentration of the ECF is kept constant.
2. The ECF volume is controlled by regulating the total Na^+ content of the ECF.
3. The Na^+ content of the ECF is regulated by the kidney. If the ECF volume (Na^+ content) is too high, the kidney excretes more Na^+; if the ECF volume or Na^+ content is too low, the kidney excretes less Na^+.

Mechanism of Renal Control of ECF Volume

The renal excretion of Na^+ is controlled mainly by the steroid hormone aldosterone (Chapter 15). Aldosterone stimulates resorption of Na^+ from renal tubules back into the blood, and therefore promotes the excretion of a low-Na^+ urine. A decrease in blood volume stimulates aldosterone secretion via a series of events that is not entirely understood. When the blood volume decreases to a certain level, signals are sent from the brain that result in the release of renin by the kidneys. As described in Chapter 15, renin cleaves circulating angiotensinogen to ultimately yield the active form angiotensin II, which stimulates the secretion of aldosterone from the adrenal cortex.

To a certain extent, the ECF Na^+ content is also regulated by the glomerular filtration rate. An increase in ECF Na^+ content (increase in ECF volume) increases the glomerular filtration rate and leads to increased excretion of Na^+.

REGULATION OF ECF OSMOLARITY

The osmolarity of the ECF is controlled by a combination of water intake and water excretion. Water intake is controlled by two thirst mechanisms, one responding to a decrease in ICF and the other responding to a decrease in ECF. The excretion of water in the urine is regulated by the pituitary hormone vasopressin (Chapter 15). A rise in ECF osmolarity is sensed by special cells in the hypothalamus called *osmoreceptors*. Stimulation of these cells leads to the release of vasopressin (antidiuretic hormone, ADH) by the pituitary. Vasopressin stimulates the resorption of water from the renal collecting ducts, resulting in the excretion of a more concentrated urine and minimizing the loss of water from the ECF.

Control of ECF Volume and Osmotic Pressure: Key Points

1. ECF volume is regulated primarily by aldosterone, which controls the amount of Na^+ excreted by the kidney and thus the total Na^+ content of the ECF. Aldosterone stimulates resorption of Na^+ by the kidneys, resulting in the excretion of a low-Na^+ urine.
2. ECF osmolarity can be controlled by the mechanisms that regulate water intake and excretion:
   a. Water intake is controlled by the thirst mechanisms.
   b. Water excretion is controlled by the hormone vasopressin, which stimulates resorption of water from the renal collecting ducts and thus promotes the excretion of a concentrated urine.
3. The quantity of Na^+ and water eliminated by the kidney, and therefore the regulation ECF volume and osmolarity, is determined mostly by the circulating levels of vasopressin and aldosterone.

REGULATION OF ECF pH

The extracellular pH is tightly regulated, and is kept very close to 7.4. Normal cellular metabolism produces acidic compounds, which enter the ECF. As described in Chapter 16, the main source of acid in the ECF is carbon dioxide, which is converted in the red cells to carbonic acid which dissociates to H^+ and HCO_3^-. It is not surprising, therefore, that the main buffer involved in regulating extracellular pH is the HCO_3^-/H_2CO_3 couple. Both the lungs and the kidneys participate in regulating the ratio of $[HCO_3^-]$ to $[H_2CO_3]$.

Key points to understanding the regulation of ECF pH are as follows:

1. The major buffer system is HCO_3^-/H_2CO_3. The ratio of $[HCO_3^-]$ to $[H_2CO_3]$ is maintained at approximately 20:1.
2. The lungs control $[H_2CO_3]$ by regulating the amount of CO_2 exhaled with the breath. Remember from Chapter 16 that as blood passes through the lungs, carbonic acid is converted to carbon dioxide and water: $H_2CO_3 \rightarrow CO_2 + H_2O$.
3. The kidneys control $[HCO_3^-]$ by regulating the degree to which HCO_3^- is resorbed in the kidney—if resorption is incomplete, some HCO_3^- is eliminated in the urine.
4. The kidneys also have two mechanisms for eliminating H^+ in the urine.

Regulation of the ECF pH by the lungs is called *respiratory compensation*, and regulation by the kidneys is called *renal compensation*.

Respiratory Compensation

First, it must be recognized that:

1. The concentration of H_2CO_3 in the blood is determined by the rate at which CO_2 leaves the plasma and enters the alveolar air. This rate, in turn, depends on the PCO_2 in the alveolar air.
2. The PCO_2 in the alveolar air is determined by the rate and depth of the breathing. The deeper and more rapid the breathing, the more completely the alveolar air is exchanged for atmospheric air (which has a very low PCO_2), and the lower is the alveolar PCO_2.
3. The rate and depth of breathing is controlled by the respiratory center of the brain, which responds to changes in blood pH.

A change in $[HCO_3^-]/[H_2CO_3]$ can be due to a change in either $[HCO_3^-]$ or $[H_2CO_3]$. In either case, the respiratory center responds with an appropriate change in the rate and depth of breathing. A *decrease* in $[HCO_3^-]/[H_2CO_3]$ (caused by a decrease in $[HCO_3^-]$ or an increase in $[H_2CO_3]$) is accompanied by a drop in pH. This change signals the respiratory center to increase the rate and depth of breathing, which leads to increased loss of CO_2 from the lungs and therefore to a drop in $[H_2CO_3]$, tending to restore the correct $[HCO_3^-]/[H_2CO_3]$ ratio. As the pH rises, the respiratory center senses a pH change in the opposite direction, and the stimulation of respiration is shut off.

An *increase* in the $[HCO_3^-]/[H_2CO_3]$ ratio causes the ECF pH to increase. In this case, the pH increase signals the respiratory center to decrease the rate of breathing. PCO_2 increases, ECF $[H_2CO_3]$ rises, and ECF pH decreases. The respiratory center then senses this downward change in pH and stimulates respiration.

The respiratory compensation for changes in ECH pH is very rapid. However, because the respiratory center responds to the direction of change rather than to the absolute value of pH, respiratory compensation never proceeds to completion. Before the correct 20:1 $[HCO_3^-]/[H_2CO_3]$ ratio is achieved, the respiratory center senses that the change in pH has reversed direction, and adjusts the breathing rate to prevent a continued change in pH.

Renal Compensation

Renal correction of alterations in extracellular pH is slower than respiratory compensation, but renal control is able to completely restore normal pH. The kidneys affect the extracellular fluid pH in two ways: through tubular resorption of bicarbonate and through excretion of hydrogen ions. When the plasma pH is in the normal range, the role of tubular bicarbonate resorption is simply to prevent loss of bicarbonate to the urine. If the plasma pH rises, however, tubular resorption of bicarbonate becomes less complete, so that there is a net excretion of bicarbonate. If plasma pH falls too low, on the other hand, the kidneys excrete hydrogen ion by one or both of two mechanisms: excretion of titratable acid and excretion of protons as ammonium salts.

Bicarbonate Resorption

In the kidneys, HCO_3^- is continually filtered by the glomeruli and resorbed from the tubules. The concentration of HCO_3^- in the ECF (i.e., the ECF pH) is influenced by the degree of tubular resorption of $[HCO_3^-]$. Figure 17-1A shows the mechanism of tubular resorption of HCO_3^-, which takes place as follows:

1. Filtered HCO_3^- and Na^+ are both present in the tubule lumen.
2. In the renal tubular cells, H_2CO_3 is synthesized from CO_2 and H_2O by carbonic anhydrase. H_2CO_3 dissociates into HCO_3^- and H^+.
3. An antiport system (see Chapter 21) in the luminal membrane of the tubule cell transports H^+ into the tubule lumen and simultaneously transports Na^+ from the lumen into the tubule cell.
4. HCO_3^- (from the dissociation of carbonic acid) and Na^+ are transported out of the tubule cell and into the plasma.
5. In the tubule lumen, H^+ reacts with the filtered HCO_3^- to form H_2CO_3, which in turn dissociates to CO_2 and H_2O.
6. The CO_2 diffuses back into the tubule cells, where it either combines with H_2O to form more H_2CO_3 (which participates in another cycle of HCO_3^- resorption) or diffuses into the blood as dissolved CO_2.

Therefore, the net result is that HCO_3^- is resorbed from the tubule filtrate back into the blood. If the body is challenged with excess bicarbonate, resorption is not complete, and the excess bicarbonate appears in the urine. Thus, tubular resorption of bicarbonate provides a mechanism for correcting high plasma pH.

Excretion of Titratable Acid

When the extracellular pH falls too low, the excess protons are excreted in the urine in the form of titratable acid ($H_2PO_4^-$). The amount of acid excreted in the urine may be measured by titrating the urine with OH^- back to the normal plasma pH of 7.4.

The mechanism for the excretion of titratable acid makes use of HPO_4^{2-} in the glomerular filtrate. The ECF contains the phosphate buffer system $HPO_4^{2-}/H_2PO_4^-$. HPO_4^{2-} and $H_2PO_4^-$ are present in the plasma at a ratio of 4:1, and are filtered by the glomerulus in the same ratio.

Figure 17-1B shows the mechanism for the excretion of titratable acid. The first steps of the mechanism are the same as for bicarbonate resorption: carbonic acid is synthesized in the tubule cell and dissociates to H^+ and HCO_3^-; the antiport system excretes H^+ into the lumen in exchange for Na^+; and Na^+ and HCO_3^- are returned to the blood. However, when plasma pH is too low, the tubule cells excrete more H^+ into the lumen than can be used by the bicarbonate resorption mechanism. This excess H^+ combines with luminal HPO_4^{2-} to yield $H_2PO_4^-$, which is excreted in the urine. Thus, for every CO_2 that is converted to

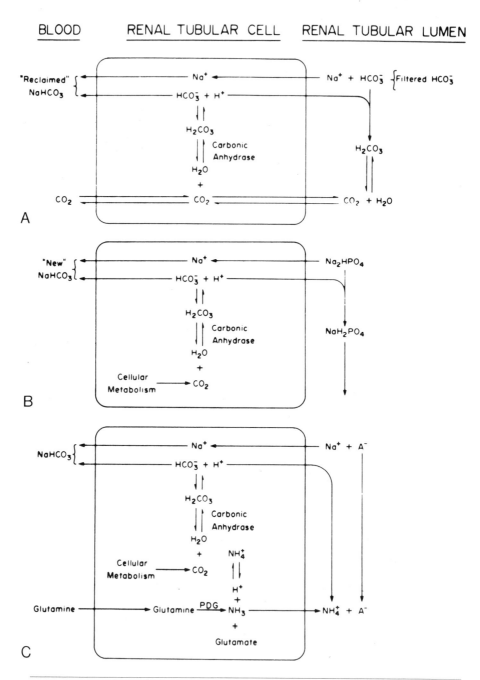

Fig. 17-1. Renal regulation of extracellular fluid pH. (From Smith EL, Hill RL, Lehman IR et al: Principles of Biochemistry. Mammalian Biochemistry. 7th Ed. McGraw-Hill, New York, 1983, with permission.)

HCO_3^- and H^+ in the tubule cell and used in this mechanism, the net result is that

1. one HCO_3^- is returned to the ECF.
2. one H^+ is eliminated in the urine as dihydrogen phosphate.

Both these results help to return low plasma pH to normal. However, the maximum urine acidity that can be achieved is pH 4.5.

Renal Excretion of Ammonium Salts

When the ECF acidity exceeds the compensatory capacities of both HCO_3^- resorption and excretion of titratable acid, the renal excretion of ammonium salts commences. This mechanism is diagrammed in Figure 17-1C. As in the preceding two mechanisms, carbonic acid generated in the tubule cells dissociates into HCO_3^- and H^+. H^+ is transported into the lumen in exchange for Na^+; Na^+ and HCO_3^- enter the blood. Meanwhile, ammonia is generated in the tubule cells by the conversion of glutamine by glutaminase to glutamate and NH_3. NH_3 moves out into the tubule lumen, where it combines with H^+ to form NH_4^+. The charged ammonium cannot re-enter the tubule cells, and is therefore excreted in the urine as an ammonium salt. For each CO_2 converted to HCO_3^- and H^+, the net result of this mechanism is that

1. one HCO_3^- is returned to the ECF.
2. one ammonium ion (i.e., one H^+) is eliminated.

This mechanism makes it possible for the kidney to excrete large quantities of hydrogen ion without further lowering the urine pH, and is therefore capable of correcting severe excesses of acid in the ECF.

Renal Regulation of Extracellular pH: Key Points

The three mechanisms described above allow the kidney to compensate for a very wide range of changes in extracellular pH.

1. *Renal resorption of HCO_3^-* functions to maintain *normal* pH by preventing loss of HCO_3^- in the urine, and also corrects for *increases* in plasma pH by eliminating HCO_3^- in the urine.
2. *Excretion of titratable acid* corrects for mild to moderate acid challenges to the ECF.
3. *Excretion of ammonium salts* corrects for severe acid challenges to the ECF.

CELLULAR REGULATION OF EXTRACELLULAR pH

The intracellular compartment contains proteins and metabolites that have significant buffering capacity. When extracellular pH falls, protons enter cells, and Na^+ and/or K^+ move out. The influx of hydrogen ion is buffered by the intracellular pool of buffer substances. Conversely, a rise in extracellular [HCO_3^-] can be partly compensated by a movement of protons out of cells balanced by an influx of Na^+ and/or K^+.

SUGGESTED READING

1. Pitts RF: Physiology of the Kidney and Body Fluids. 3rd Ed. Year Book, Chicago, 1974
2. Masoro EJ, Siegel PD: Acid-Base Regulation. 2nd Ed. WB Saunders, Philadelphia, 1977
3. Warnock DC, Rector FC Jr: Proton secretion by the kidney. Annu Rev Physiol 41:197, 1979
4. Davenport HW: The ABC of Acid-Base Chemistry. 6th Ed. Chicago University Press, Chicago, 1974

STUDY QUESTIONS

Directions: Answer the following questions using the key outlined below:
- **(A)** if 1, 2, and 3 are correct
- **(B)** if 1 and 3 are correct
- **(C)** if 2 and 4 are correct
- **(D)** if only 4 is correct
- **(E)** if all four are correct

1. Volumetrically major subcompartments of the extracellular fluid include
   1. cerebrospinal fluid
   2. blood plasma
   3. synovial fluid
   4. interstitial fluid

2. Events involved in the respiratory compensation of lowered ECF pH include
   1. a decrease in the PCO_2 of alveolar air
   2. an increase in the breathing rate
   3. exhalation of increased CO_2 in the breath
   4. shallower breathing

3. The kidneys regulate extracellular pH by
   1. resorption of HCO_3^-
   2. excretion of titratible acid
   3. excretion of ammonium salts
   4. excretion of $NaHCO_3$

Directions: For each of the multiple choice questions below, choose the most appropriate answer.

4. The major extracellular cation is
   A. Na$^+$
   B. K$^+$
   C. H$^+$
   D. albumin

5. An increase in the extracellular fluid content of Na$^+$ causes
   A. an increase in extracellular fluid volume
   B. a decrease in extracellular fluid volume
   C. no change in extracellular fluid volume
   D. no change in extracellular fluid volume, but a decrease in intracellular fluid volume

6. Under normal conditions, the ratio of [HCO$_3^-$]/[H$_2$CO$_3$] in the ECF is
   A. 1:20
   B. 2:1
   C. 1:2
   D. 20:1

7. If the extracellular fluid is severely challenged with acid, the change in pH is most likely to be compensated by
   A. renal resorption of HCO$_3^-$
   B. renal resorption of H$_2$PO$_4^-$
   C. renal excretion of ammonium salts
   D. renal excretion of HCO$_3^-$

Items 8 and 9. Consider the following hormones:

   A. Aldosterone
   B. Vasopressin

Match each hormone with the most appropriate statement below.

8. A hormone that stimulates Na$^+$ resorption.

9. A hormone that stimulates water resorption.

18

NUTRITION

Energy-Yielding Nutrients/Water-Soluble Vitamins/
Fat-Soluble Vitamins/Minerals/Clinical Correlation

The student should be able to:

1. Identify the major classes of nutrients used for energy production.
2. Define the biologic value of proteins.
3. List the individual water-soluble and fat-soluble vitamins, and their roles in cellular metabolism.
4. Describe the function and metabolism of the major mineral nutrients.

Learning Objectives

Perspective: Nutrition

Certain substances are required by the body for growth, reproduction, maintenance of body structure and physiologic processes, and repair of damaged tissues. These substances are *nutrients*. Nutrients that can be synthesized by the body from other substances are *nonessential*. Nutrients that cannot be synthesized endogenously in sufficient amounts are *essential*. Nutrition is the study of the overall intake of nutrients, the effects of specific nutrients, and the mechanisms of nutrient utilization. This chapter focuses primarily on the role of nutrients in the maintenance of normal tissue and cell function, and on the mechanisms of nutrient utilization.

There are three general types of nutrients—the energy-yielding nutrients (protein, carbohydrate, and lipid), the vitamins, and the minerals. The *vitamins* are a heterogeneous collection of small organic molecules that usually function in enhancing the utilization of other nutrients and in the maintenance of tissue structures. They are essential nutrients. The vitamins are traditionally divided into the *water-soluble vitamins* and the *fat-soluble vitamins*. Water-soluble vitamins generally function as coenzyme precursors, whereas fat-soluble vitamins play various roles. The *minerals* include the

(continued)

> various inorganic ions that play a role in body chemistry. They can be divided into two broad groups: the *major minerals*, which are present in the body in fairly large quantities, and the *trace minerals*, which are present in minute amounts and mostly serve as prosthetic groups of specific enzymes. The major minerals can serve various functions, including the regulation of water balance and bone development and maintenance.

THE ENERGY-YIELDING NUTRIENTS

Protein

As discussed in previous chapters, cells use amino acids to synthesize proteins and a variety of other nitrogen-containing compounds (Chapters 10 and 14). Amino acids are not stored; any quantities not needed for biosynthesis are degraded and either oxidized to produce energy or converted to carbohydrate or lipid (Chapter 9). As a result, protein amino acids contribute to the energy supply of the body. The amino acids needed to meet cellular demands are obtained both from the degradation of dietary and tissue protein and by endogenous synthesis.

Ten of the 20 protein amino acids cannot be synthesized endogenously in amounts sufficient for growth, and therefore are nutritionally essential (see Table 9-2). The absence of nutritionally essential amino acids from the diet significantly affects both protein synthesis and amino acid metabolism. Suppose, for example, that the diet is deficient in just one essential amino acid, valine. The consequences of this deficiency are extensive:

1. The synthesis of proteins that contain valine (i.e., the deficient amino acid) is inhibited.
2. The other amino acids that would normally be used in the synthesis of these proteins accumulate.
3. Because of this accumulation, more amino acids are degraded and/or used in the synthesis of nitrogenous compounds.
4. Because tissue proteins are in a constant state of turnover—i.e., they are always being degraded and replaced by newly synthesized proteins—a prolonged dietary deficiency in one or more essential amino acids leads to tissue wasting because tissue proteins are not fully replaced.

Biologic Value of Proteins

Different proteins contain different quantities of the various essential amino acids. Some proteins entirely lack one or more essential amino acids. The greater the variety and quantity of essential amino acids in a dietary protein, the more useful it will be to the cell, because its constituent amino acids can more significantly contribute to the pool of essential amino acids required for protein synthesis. The degree of usefulness of a given dietary protein is quantified as its *biologic value*. The biologic value of a protein reflects both its digestibility and its amino acid composition. It is expressed as the percent of absorbed amino acid nitrogen that is actually retained (i.e., used):

$$\text{Biologic value} = \frac{\text{N retained}}{\text{N absorbed}} \times 100$$

Proteins obtained from animal tissues usually contain all the essential amino acids in proportions appropriate for humans, and are also easy to digest. Plant proteins are the opposite—they are usually deficient in one or more essential amino acids, and are more difficult to digest. Therefore, animal proteins have a higher biologic value than plant proteins. Moreover, plant tissues usually contain a lower proportion of protein than animal tissue. The low protein content of plants and the lower biologic value of plant proteins can be compensated by eating a larger quantity of plant protein from a wide variety of sources. Since different plant proteins are likely to be deficient in different essential amino acids, eating proteins from a number of sources should provide the total mix of essential amino acids. An understanding of the biologic value of proteins is important for individuals on vegetarian diets.

Lipids

As described in Chapters 7 and 8, lipids are used primarily as fuel molecules and as components of membranes. Most fatty acids can be synthesized in mammalian cells using precursors derived from amino acid or carbohydrate degradation. However, certain polyunsaturated fatty acids (the linoleate and linolenate series) are nutritionally essential and must be either obtained from the diet or synthesized from essential dietary precursors. Two aspects of lipid nutrition must be considered:

1. What happens when lipid intake exceeds lipid utilization?
2. What happens when the requirement for lipid exceeds lipid intake?

As described in Chapter 8, excess dietary lipid is stored in adipose tissue as triacylglycerols. A long-term excess of dietary lipid therefore leads to obesity. In addition, it is usually associated with elevated serum levels of saturated fatty acids and cholesterol, both of which increase the risk of atherosclerosis ("hardening of the arteries"). This consequence occurs because most dietary fat sources contain more saturated than unsaturated fats. A high intake of saturated fatty acids raises serum cholesterol as well as serum saturated fatty acids. If this condition continues, lipid (primarily cholesterol) may be deposited on the inner walls of blood vessels, forming *plaques*. Continued development of plaques leads to hardening of the arteries, reduced blood flow, and increased blood pressure.

The risk factors and recommendations concerning atherosclerosis include nondietary as well as dietary factors. However, the dietary recommendations include the following:

1. Lowering the intake of saturated fatty acids (which also lowers the serum levels of LDL-cholesterol)
2. Increasing the intake of polyunsaturated fatty acids, which lower serum LDL-cholesterol
3. Reducing the cholesterol intake

Vegetable oils usually contain a greater proportion of polyunsaturated fatty acids than do animal fats.

What happens when the intake of lipid is less than the amount required? When neither stored nor dietary lipid is available to meet the energy needs of cells, more glycogen and amino acids will be degraded, which could deplete the glycogen stores and eventually lead to muscle wasting. Lipids are needed for the synthesis and maintenance of membranes as well as for fuel. Although mammalian cells are capable of synthesizing most membrane lipids, a prolonged very low intake of fatty acids can result in a deficiency of essential fatty acids that will interfere with the maintenance and synthesis of membranes and also with the synthesis of prostaglandins, thromboxanes, and leukotrienes. Essential fatty acid deficiency is very rare because the essential fatty acids are needed in small quantities and are present in a wide variety of foods. The problem is enountered most often in hospitalized patients maintained exclusively on fat-free liquid diets. The primary symptom is a scaly dermatitis.

Carbohydrate

As discussed in detail in Chapters 4 and 5, most carbohydrate is used as fuel. It can either be degraded immediately for energy production or stored as glycogen. When the amount of carbohydrate in the diet exceeds the immediate requirements for energy and for glycogen synthesis, the excess is converted to fatty acids and stored as triacylglycerols. Therefore, a continued excessive intake of carbohydrates also results in obesity.

As discussed at the end of Chapter 7, under conditions of low carbohydrate intake, fatty acids are degraded at a higher rate and ketone bodies are produced. Since ketone bodies are acidic, elevated levels of ketone bodies (ketosis) predisposes the individual to a lowering of blood pH (acidosis). Acidosis caused by ketosis is also called *ketoacidosis*.

FIBER

Fiber in the diet consists mostly of the plant polysaccharides that cannot be digested by human dietary enzymes. Fiber therefore is not considered a nutrient, although some fiber is slightly degraded by the intestinal bacteria. However, fiber is a significant component of the diet. Fiber from different sources has different chemical and physical properties and often slightly different physiologic effects. In general, fiber increases stool bulk and decreases the time that waste material spends in the gastrointestinal tract. Because of its ability to bind water, fiber adds bulk to food in the gastrointestinal tract. The presence of fiber in the diet has been correlated with a decreased incidence of cardiovascular disease and colon cancer. Excess fiber intake, on the other hand, may lower the absorption of mineral nutrients. Fiber therefore should be consumed in moderate amounts.

THE WATER-SOLUBLE VITAMINS

Thiamine (Vitamin B$_1$)

Biologic Functions. Thiamine (Fig. 18-1) reacts with ATP to form thiamine pyrophosphate, the active species in biochemical reactions. Thiamine pyrophosphate is a coenzyme in the decarboxylations of pyruvate and α-ketoglutarate and in transketolase and transaldolase reactions.

Deficiency. Thiamine deficiency is associated with elevated blood levels of pyruvate and lactate and with a reduction in red cell transketolase. Thiamine deficiency causes the clinical disorder known as beriberi. "Dry" beriberi is characterized by weight loss, muscle wasting and weakness, peripheral neuritis, and mental confusion. In "wet" beriberi, edema and an array of cardiac abnormalities may also be present.

Distribution, Requirement, Sources. Thiamine is available in a wide variety of foods, so thiamine deficiency is fairly rare. Thiamine is not stable when heated, so the thiamine content of foods is lowered by cooking. Because thiamine functions in the release of energy from fuel molecules, the thiamine requirement is linked to the caloric intake. The recom-

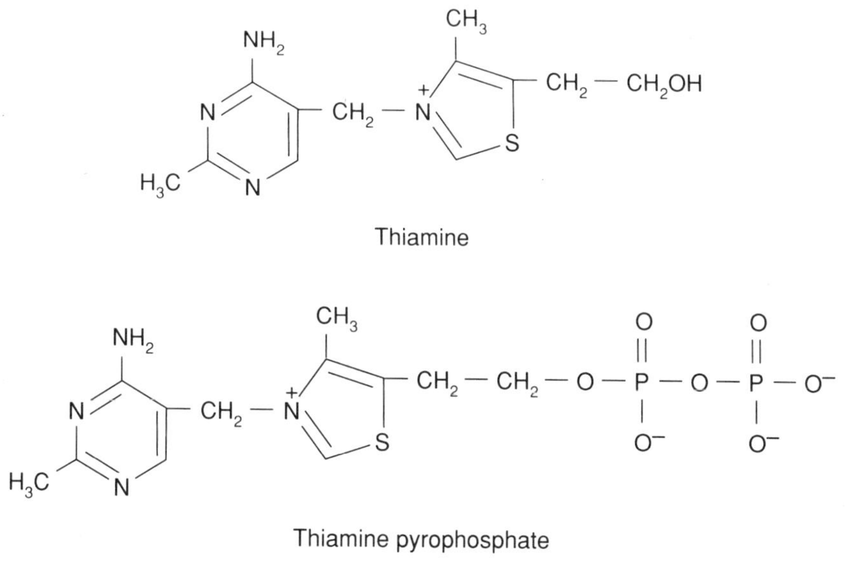

Fig. 18-1. Structures of thiamine (vitamin B$_1$) and thiamine pyrophosphate.

mended daily allowance (RDA) of thiamine is 1.0 to 1.5 mg for most individuals, but the thiamine requirement increases during pregnancy. Important sources of thiamine are buckwheat, bacon, ham, nuts, wheat germ, and brewer's yeast.

Riboflavin (Vitamin B$_2$)

Biologic Functions. The active forms of riboflavin are the electron-carrier coenzymes FAD and FMN. These coenzymes function in a variety of redox reactions catalyzed by oxidases, reductases, and dehydrogenases. Riboflavins are necessary for aerobic respiration and also for tissue maintenance.

Deficiency. Riboflavin deficiency is rare, and, when it occurs, is usually associated with deficiencies in other nutrients. The identification of symptoms due specifically to riboflavin deficiency has therefore been difficult. Symptoms associated with, although not necessarily caused exclusively by, riboflavin deficiency are an oily skin dermatitis, cracked or broken skin lesions on the lips and corners of the mouth, and corneal vascularization. A decline in red cell riboflavin is believed to be the earliest biochemical indicator of riboflavin deficiency.

Distribution, Requirement, Sources. Riboflavin is found in a wide variety of foods, especially liver, kidney, and green leafy vegetables. The RDA for riboflavin is 1.2 mg, but pregnant and lactating women may require more.

Niacin (Vitamin B$_3$)

Biologic Functions. Niacin (also known as nicotinic acid) and its amide derivative nicotinamide are the precursors of the coenzymes NAD^+ and $NADP^+$. The biosynthesis of NAD^+ and $NADP^+$ was described in Chapter 11. NAD^+ and $NADP^+$ are essential coenzymes in numerous cellular redox reactions, for example reactions of energy metabolism, lipid biosynthesis, the pentose phosphate pathway, and amino acid metabolism.

Deficiency. Niacin deficiency causes the disease pellagra. The major symptoms are a dermatitis on skin exposed to the sun, impaired digestion, diarrhea, and mental confusion. Pellagra probably also involves other B-vitamin deficiencies.

Distribution, Requirement, Sources. A very small quantity of niacin is synthesized endogenously from tryptophan (approximately 1 mg/60 mg of ingested tryptophan). Niacin is present mostly in meats, fish, and nuts; lesser quantities are found in several other foods. The RDA for niacin in adults is 13 mg.

Pyridoxine (Vitamin B$_6$)

Biologic Functions. The biologically active forms of pyridoxine are pyridoxamine phosphate and pyridoxal phosphate. These coenzymes are mostly involved in amino acid metabolism. They function in various transaminations and oxidations, in cysteine synthesis, and in certain decarboxylations. A number of pyridoxal phosphate-dependent reactions were described in Chapter 9.

Deficiency. Symptoms of pyridoxine deficiency include irritability, mental confusion, convulsions (in infants), peripheral neuropathy, inflammation of the mouth, and an oily dermatitis. Since many of these symptoms are the same as for deficiency of other B vitamins, the diagnosis is made by observing improvement upon administration of pyridoxine.

Distribution, Requirement, Sources. Since the pyridoxine coenzymes are involved primarily with amino acid metabolism, their daily requirement varies with the daily protein intake. The RDA for adult men and women is respectively 2.2 and 2.0 mg. Most of the body's pyridoxine is concentrated in the liver (the organ most active in amino acid metabolism). Beef liver and wheat germ are especially rich sources of pyridoxine. Various other foods contain smaller amounts.

Panthothenic Acid (Vitamin B$_5$)

Biologic Functions. Pantothenate is a precursor of coenzyme A, described in Chapter 11. Coenzyme A participates in acyl transfer reactions. As described in Chapters 5 through 7, coenzyme A is the carrier for acetyl groups obtained from the degradation of carbohydrate and lipid; coenzyme A also carries the acetyl and malonyl groups used in fatty acid synthesis.

Deficiency. Natural pantothenate deficiency has not been observed in humans. Symptoms of experimental pantothenate deficiency are low energy, restlessness, reduced response to ACTH, and increased sensitivity to insulin.

Distribution, Requirement, Sources. No RDA has been set for pantothenate. Most individuals ingest between 5 and 20 mg of pantothenate per day. Beef liver, peanuts, soybeans, and wheat germ are all rich in pantothenate.

Biotin (Vitamin H)

Biologic Functions. Biotin (Fig. 18-2), which is used by the cell without any modification, is the coenzyme for cellular carboxylation reactions. It participates in carbohydrate and amino acid metabolism. The car-

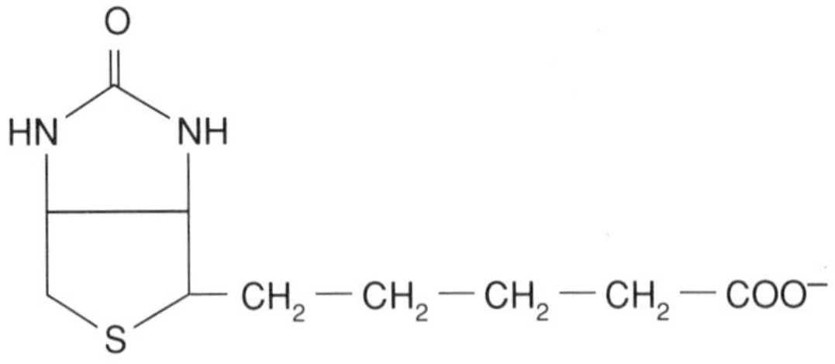

Fig. 18-2. Structure of biotin (vitamin H)

boxylation of acetyl CoA to malonyl CoA in preparation for fatty acid synthesis is an example of a biotin-dependent carboxylation.

Deficiency. Biotin is synthesized by the intestinal flora in amounts believed sufficient to meet the daily needs of humans, and biotin deficiency has not been observed in humans on a normal diet. Raw egg contains a protein called avidin that binds biotin very tightly ($K_a \sim 10^5$). Biotin deficiency has been induced experimentally by eliminating biotin intake and either sterilizing the intestinal tract or feeding raw egg white.

Distribution, Requirement, Sources. No RDA has been set for biotin; the intake of the average healthy adult is 100 to 300 µg/day. Liver, peanuts, and chocolate are rich in biotin.

Folic Acid (Vitamin B$_9$)

Biologic Functions. Folic acid, also called folicin (Fig. 18-3), is converted to tetrahydrofolate (see Fig. 9-16), which is the primary carrier of one-carbon units in the cell. As described in Chapter 9, the main endogenous source of one-carbon units is the conversion of serine to glycine. One-carbon units carried by tetrahydrofolate are used, for example, to methylate dUMP in the synthesis of dTMP (Chapter 11) and in the resynthesis of homocysteine to methionine, which may be converted to S-adenosylmethionine, the general methylating agent in the cell (Chapter 9).

Deficiency. Symptoms associated with folic acid deficiency are megaloblastic and macrocytic anemia, growth failure, and leukopenia. Symptoms remit upon administration of folic acid.

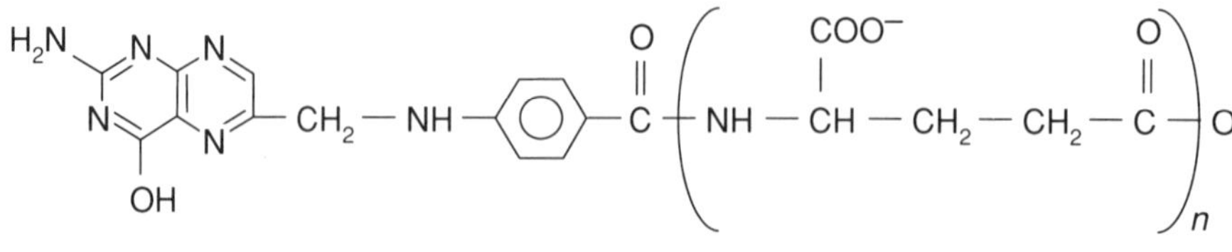

Fig. 18-3. Structure of folic acid (vitamin B$_9$).

Distribution, Requirement, Sources. Folic acid is synthesized by the intestinal bacteria, so the daily dietary requirement is difficult to measure. The RDA for healthy individuals has been set at 400 μg, with higher allowances for pregnant and lactating women. The liver is the major site of folic acid metabolism. Good sources of folic acid are beef and chicken liver, salmon, brewer's yeast, peanuts, and dried lima beans.

Vitamin B_{12}

Biologic Functions. Vitamin B_{12} consists of cobalamin (Fig. 18-4) and its physiologically active derivatives, including cyanocobalamin, methylcobalamin, hydroxocobalamin, and 5-deoxyadenosylcobalamin. Cobalamin is an essential cofactor in the conversion of homocysteine to methionine and in the conversion of methylmalonyl CoA (an intermediate in the degradation of some amino acid carbon chains and propionyl CoA metabolism) to succinyl CoA (Chapters 7 and 9).

Deficiency. Vitamin B_{12} deficiencies have been diagnosed from elevated urine levels of methylmalonate. The clinical symptoms of vitamin B_{12} deficiency are collectively called *pernicious anemia*, and include megaloblastic anemia, changes in the intestinal mucosa, and, in extreme cases, neuropathy. Pernicious anemia is most often due not to an actual lack of B_{12} in the diet but to inadequate absorption resulting from decreased gastric secretion of *intrinsic factor*, a protein that promotes B_{12} absorption. Some vegetarians eating diets inadequate in vitamin B_{12} have also developed pernicious anemia, but without the accompanying gastric and neurologic symptoms.

Distribution, Requirement, Sources. The vitamin B_{12} RDA for healthy individuals has been set at 3 μg; the actual daily intake for most individuals is from 5 to 15 μg/day. Most vitamin B_{12} in the body is concentrated in the liver; some other body parts contain smaller amounts. In general, foods derived from animals are better sources of B_{12} than foods derived from plants.

Vitamin C

Biologic Functions. The active forms of vitamin C are ascorbic and dehydroascorbic acids (Fig. 18-5). These substances participate in reactions and hydroxylations, and also promote the intestinal absorption of iron. They participate in amino acid metabolism, the synthesis of norepinephrine, and the synthesis of collagen.

Deficiency. Vitamin C deficiency causes scurvy. Symptoms of scurvy include fatigue, hemorrhaging, anemia, loose teeth, swollen joints, development of keratotic hair follicles, impaired wound healing, and changes in bone development and growth. Many of these symptoms appear to be the result of the defective collagen synthesized in the ab-

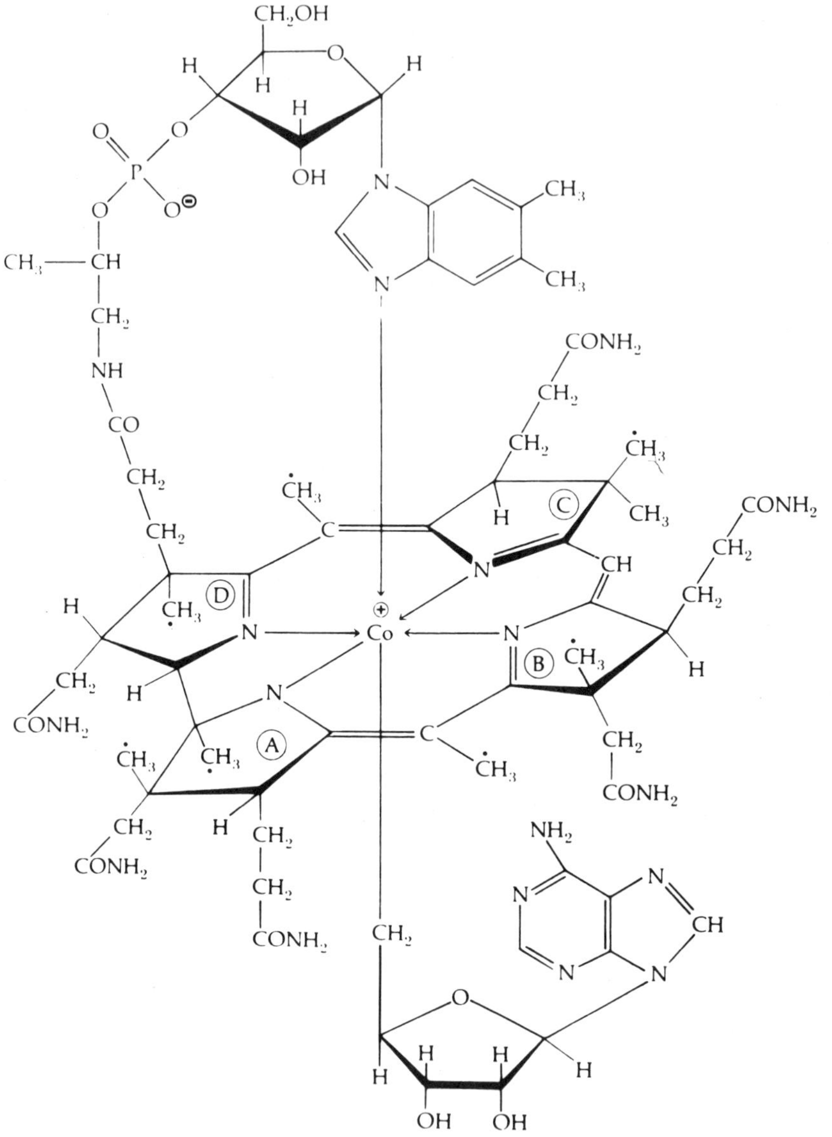

Fig. 18-4. Structure of cobalamin (vitamin B_{12}). (From Smith EL, Hill RL, Lehman IR et al: Principles of Biochemistry: Mammalian Biochemistry. 7th Ed. McGraw-Hill, New York, 1983, with permission.)

sence of vitamin C. Administration of vitamin C cures the disease. Improved eating habits have made scurvy a rare disorder in developed countries. However, chronic diarrhea may contribute to vitamin C deficiency.

Distribution, Requirement, Sources. The **RDA** for vitamin C is 60 mg, based on the amount required to prevent or cure scurvy. More vitamin C may

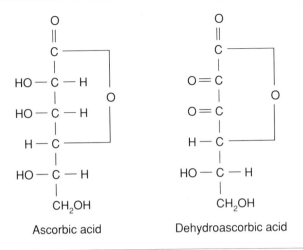

Fig. 18-5. Structures of ascorbic acid and dehydroascorbic acid.

be required during pregnancy and lactation. Good sources of vitamin C are peppers, papaya, persimmon, strawberries, citrus fruits such as oranges and lemons, turnip greens, mustard greens, kale, collard greens, cauliflower, brussels sprouts, and broccoli.

THE FAT-SOLUBLE VITAMINS

Vitamin A

Biologic Functions. Vitamin A is present in foods mainly as carotenoid compounds such as β-carotene. These compounds are converted by the body into the active forms shown in Fig. 18-6: retinol, retinaldehyde, and retinoic acid. Retinol phosphate is used in the synthesis of certain glycoproteins needed for growth and for the secretion of mucus. Retinal and retinoic acid participate in the regulation of certain genes that participate in growth and differentiation. Retinal is needed for normal vision (Chapter 20). Vitamin A also prevents excessive keratinization and is required for the maintenance of healthy epithelial tissue. In this role, vitamin A is thought to protect from infection and cancer.

Deficiency. Vitamin A deficiency causes a wide range of symptoms corresponding to the various biochemical roles of vitamin A derivatives. The earliest symptom is night blindness, due to a deficiency of vitamin A in the cornea. Vitamin A deficiency may also cause xerophthalmia (drying and thickening of the corneal conjunctiva). This condition, which may result in blindness, has been associated with poor dieting and with chronic generalized nutrient deficiency. Epithelial atrophy followed by hyperkeratinization are other symptoms of vitamin A deficiency, as is cessation of bone growth in young children.

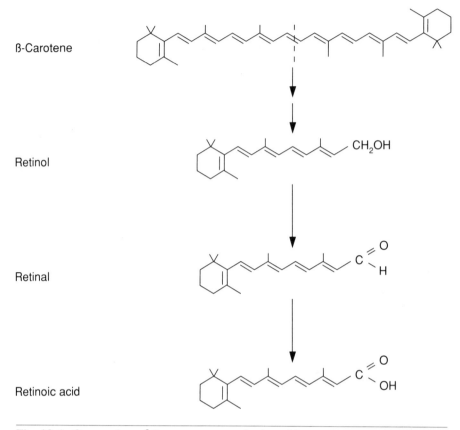

Fig. 18-6. Conversion of β-carotene to retinoic acid.

Vitamin A Toxicity.   Vitamin A is toxic if consumed in excessive amounts. Vitamin A intoxication may result in bone fragility, swellings over the long bones, hypercalcemia, headaches, nosebleeds, nausea, weakness, and loss of appetite.

Distribution, Requirement, Sources.   The RDA for retinol is 1,000 μg for men and 800 μg for women. Additional vitamin A is recommended to rebuild depleted body stores and during pregnancy and lactation. Most of the body's pool of vitamin A is concentrated in the liver in the form of retinol palmitate. Smaller amounts are stored in other tissues. Retinol circulates in the blood bound to a specific transport protein. Foods rich in vitamin A include liver from most food animals, crabmeat, hard cheese, butter, winter squash, papaya, carrots, and dandelion greens.

Vitamin D

Biologic Functions.   The active forms of vitamin D in the body are ergocalciferol and cholecalciferol (Fig. 18-7), which in humans have the same biologic activity. Vitamin D is required in the regulation of cal-

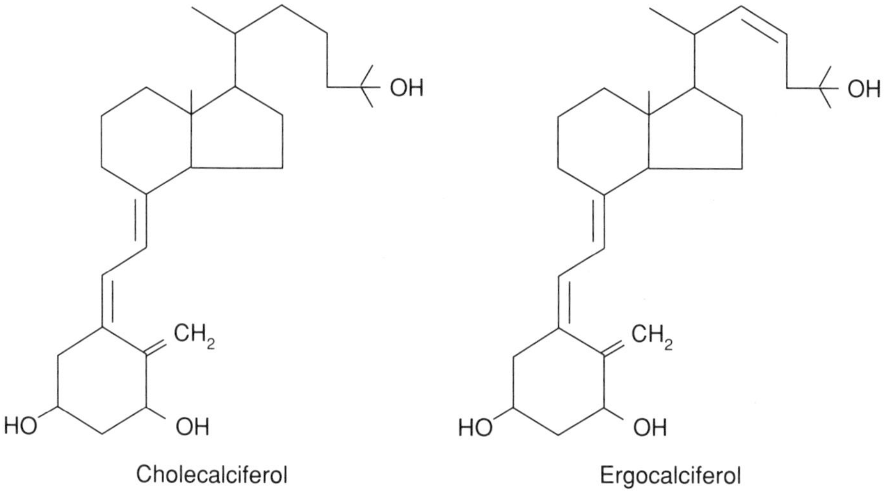

Fig. 18-7. Structure of cholecalciferol and ergocalciferol.

cium metabolism (Chapter 15)—it is involved in the regulation of intestinal calcium absorption and the mobilization of bone calcium.

Deficiency. The major symptom of vitamin D deficiency is generalized weakening of the skeleton, due both to bone demineralization and to failure of bone remineralization. Vitamin D deficiency can be caused by impaired absorption of vitamin D, lack of sunlight, and/or a diet deficient in vitamin D.

Distribution, Requirement, Sources. Because the need for calcium is highest during periods of bone growth, the need for vitamin D is highest in growing children. The RDA for children and adolescents has been set at 10 µg, an amount sufficient to prevent all symptoms of calcium deficiency. For young adults and older adults, the RDAs are 7.5 µg and 5 µg, respectively. Additional vitamin D is recommended during pregnancy and lactation. The compound 7-dehydrocholesterol, present in the skin, can be converted to cholecalciferol on exposure to sunlight. Therefore, some of the requirement for vitamin D can be satisfied endogenously. Ergocalciferol, on the other hand, is synthesized only by plants and therefore must be obtained from the diet. Oily fishes are usually rich sources of vitamin D.

Vitamin E

Biologic Functions. Vitamin E belongs to a family of compounds called the tocopherols, the most abundant and most active species of which is α-tocopherol (Fig. 18-8). The role of vitamin E in human metabolism is not fully understood; it is believed that vitamin E functions as an

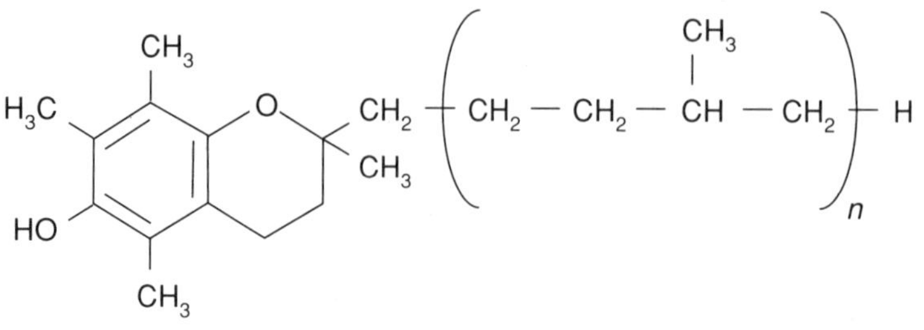

Fig. 18-8. Structure of α-tocopherol.

antioxidant and is involved in protecting membrane lipids and vitamin A from oxidative destruction.

Deficiency. Vitamin E deficiency in humans is associated with the development of ceroid pigments (deposits of oxidized lipids), excretion of creatinine in the urine, sensitivity of red cell membranes to oxidants, and impaired lipid absorption. Adults normally contain large stores of vitamin E, so signs of dietary deficiency are slow to appear.

Distribution, Requirement, Sources. A normal balanced diet provides adequate vitamin E. The RDA for adults is approximately 10 mg of *d*-α-tocopherol. An additional 3 mg is recommended during pregnancy and lactation. It should be noted, however, that increasing the intake of polyunsaturated lipids increases the requirement for vitamin E. Vitamin E is found throughout the body, but is concentrated in the adipose tissue. Cottonseed, peanut, and soybean oils and margarine made with these oils are all good sources, and vitamin E is present in a wide variety of vegetables and fruits.

Vitamin K

Biologic Functions. The structure of vitamin K is shown in Figure 18-9. As described in Chapter 16, vitamin K is required for the correct post-translational γ-carboxylation of several glutamate residues in a number of blood clotting factors.

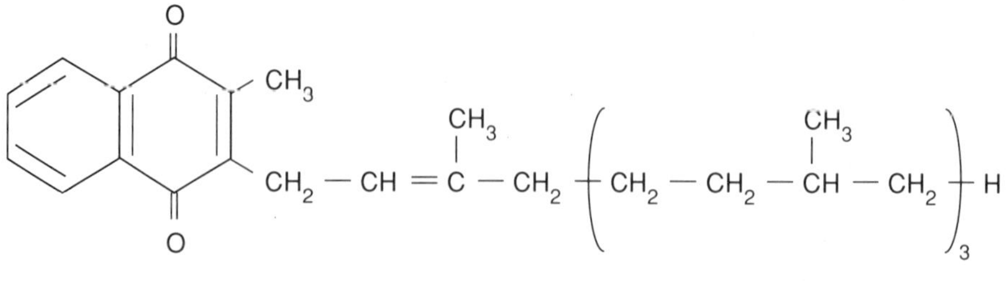

Fig. 18-9. Structure of vitamin K.

Deficiency. The most noticeable effect of vitamin K deficiency is impaired clotting. In newborn infants, vitamin K deficiency appears as hemorrhagic disease of the newborn, a bleeding disorder. This disease is cured as the infant's intestinal flora develops and begins to produce vitamin K.

Distribution, Requirement, Sources. Vitamin K is normally synthesized by the intestinal flora and is absorbed to a varying extent. However, an RDA of 70 to 140 µg has been suggested for adults. Important food sources of vitamin K are brussels sprouts, cabbage, cauliflower, and dark green leafy vegetables.

MINERALS

The Major Minerals

Calcium

Because calcium is the major mineral cation of bone, the body contains a very large calcium pool. Dissolved Ca^{2+} is also essential in a great variety of physiologic and biochemical processes. It is required, for example, in blood clotting, muscle contraction, nerve impulse transmission, the mediation of certain hormonal signals, and as a cofactor in various enzyme reactions. The RDA for calcium has been set at 800 mg, but postmenopausal women may need much higher amounts.

Phosphate

Phosphate is the major mineral anion of bone. Most of the rest of the phosphate in the body is in the form of the nucleotide phosphates and phosphocreatine. In addition, the $HPO_4^{2-}/H_2PO_4^{-}$ couple is an important buffer of the intracellular fluid and the urine (Chapter 17). Most healthy adults ingest between 800 and 1,500 µg/day of phosphorus.

Sodium, Potassium, Chloride

Sodium, potassium, and chloride are all major dissolved ions of body fluids, and are important in regulating the osmolarity and ionic balance of the intracellular and extracellular spaces. Sodium is primarily found in the extracellular fluid and potassium in the intracellular fluid. As described in Chapter 17, the volume of the extracellular compartment is regulated by controlling the amount of sodium in the extracellular fluid.

Chloride functions as a counterion—it moves across cell membranes to compensate for fluxes of other ions in the maintenance of electrical neutrality. Chloride is also excreted in the urine as a counterion to various excreted cations, and thus helps maintain the electrical neutrality of urine.

Magnesium

Mg^{2+} is an important dissolved cation. It participates in nerve transmission, and is required for the activity of a number of enzymes. Magnesium deficiency may result from impaired absorption, metabolic acidosis, and diarrhea. Conditions that result in significant losses of body fluid may lead to magnesium deficiency. Symptoms of magnesium deficiency are tremors, generalized weakness, and neuromuscular irritability. Administration of magnesium relieves these symptoms.

The Trace Minerals

Iron

Most of the iron in the body is bound in hemoglobin; a smaller portion is bound in myoglobin, the heme-containing oxygen-binding protein of tissue cells. Many electron transfer proteins also contain iron, especially the cytochromes.

Iron is efficiently salvaged from degraded heme and reused, so that only small quantities are lost from the body each day. These small losses are counterbalanced by absorption of iron from the intestine. Intestinal absorption of iron is very inefficient; it is estimated that only about 10 percent of the iron present in ingested food is actually absorbed. Iron is better absorbed from some foods than from others. The amount of iron in a normal diet is enough to meet the needs of men and of nonpregnant, nonlactating women, but it is often insufficient for the needs of children and of pregnant or lactating women. These groups usually require iron supplementation. If the diet does not replace the small daily loss of iron, iron-deficiency anemia eventually develops.

Figure 18-10 shows the mechanism by which iron is absorbed, salvaged, and re-used. Intestinal mucosal cells contain a protein, apoferritin, that binds ferric iron and is converted in the process to ferritin. Iron from ferritin is passed to transferrin, the iron-transporting protein of the plasma. Transferrin delivers iron to cells, which store it bound to ferritin and to hemosiderin (a large protein-iron complex). Ferritin iron can be released and used as needed, whereas iron bound to hemosiderin is usually no longer available for tissue use. Accumulation of excess iron leads to the condition hemochromatosis, characterized by high levels of ferritin and hemosiderin iron, especially in the liver and pancreas. This condition may lead to organ dysfunction.

Copper

Copper is an essential cofactor of several enzymes. Copper-containing enzymes are involved in amino acid metabolism, catecholamine metabolism, and oxidation reactions catalyzed by a variety of oxidases, including the enzyme lysyl oxidase, which is responsible for the formation of stable collagen fibers. Copper deficiency inhibits iron absorption and causes an anemia that responds to iron administration.

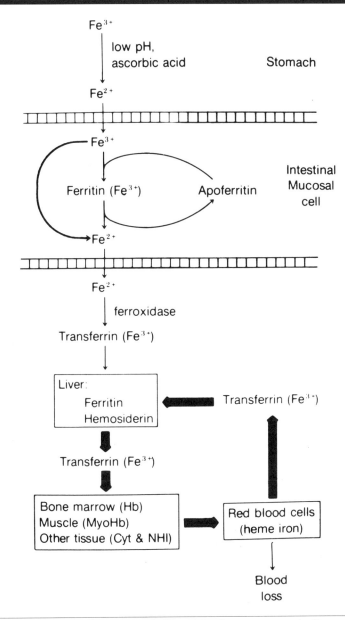

Fig. 18-10. Overview of iron metabolism. The heavy arrows indicate iron reutilization. Hb, hemoglobin; MyoHb, myoglobin; Cyt, cytochrome; NHI, heme iron. (From Devlin TM: Textbook of Biochemistry with Clinical Correlations. 2nd Ed. Wiley, New York, 1986, with permission.)

Copper deficiency also causes bone demineralization and fragile arteries.

Genetic deficiency of the copper binding protein ceruloplasmin causes several symptoms collectively known as Wilson's disease: low levels of serum copper, high levels of copper in the urine, and excessive accumulation of copper in the liver. Impaired intestinal absorption, on the other hand, is observed in a disorder known as Menkes' syndrome. Symptoms of this disorder include growth retardation, white and kinky (steel-wool) hair, bone demineralization, and demyelination of axons.

Zinc

Several enzymes require zinc as an essential cofactor. Zinc is also necessary for the normal development and functioning of the sense of taste, and one of the symptoms of zinc deficiency is decreased acuity of taste and smell. This diminished ability to taste and smell is probably responsible for the loss of appetite and poor growth observed in children with zinc deficiency. In adolescents, zinc deficiency is associated with impaired sexual development. Most of the zinc in the body is bound to the transport/storage protein metalothionein. Cereal diets that are high in phytic acid may cause zinc deficiency because phytic acid impairs zinc absorption. Zinc deficiency has also been associated with alchoholism, chronic renal disease, proteinuria, and severe trauma.

Chromium

Chromium is a cofactor of the protein known as glucose tolerance factor, which facilitates the binding of insulin to its cell receptors and therefore maintains the effectiveness of insulin. Chromium deficiency causes a decrease in the response to insulin and impaired removal of glucose from the circulation. The average daily intake of chromium is approximately 60 μg/day, and the RDA for healthy adults has been set at 50 to 200 μg.

Molybdenum

Molybdenum is an essential cofactor of xanthine and sulfite oxidizing enzymes. Trace quantities of this mineral are required for normal growth.

Selenium

Selenium is an essential component of the glutathione peroxidase system, which catalyzes the reduction of peroxide to water:

$$2\ GSH + H_2O_2 \rightleftharpoons GSSG + 2\ H_2O$$

This reaction is believed to be important in protecting cell components from oxidation by peroxide.

Iodine

As discussed in Chapter 15, iodine is taken up by the thyroid gland and incorporated into the thyroid hormones thyroxine and triiodothyronine. The adult RDA for iodine is 150 μg. Iodine deficiency stimulates the development of goiter (an enlarged thyroid gland due to proliferation of the thyroid epithelial cells) as well as the symptoms of thyroid hormone insufficiency. Goiter used to be common in parts of the world naturally deficient in iodine; the use of iodized salt has greatly decreased the incidence of iodine deficiency.

Fluoride

Small amounts of fluoride can be incorporated into bones and teeth, strengthening them and protecting against decay. A small daily intake of fluoride (1 to 4 mg) reduces the incidence of dental caries.

CLINICAL CORRELATION: OBESITY

Obesity is the accumulation of adipose tissue in the body. It is due primarily to an imbalance between caloric intake and utilization. The causes of obesity are multiple and poorly understood, and different causes or constellations of causes operate in different individuals. Inherited factors, changes in metabolic regulation leading to increased storage of fat, poor dietary practices, inactivity, and psychological and environmental factors can all contribute. Obesity is of concern to health professionals because it predisposes the individual to a number of serious medical problems, including diabetes mellitus, cardiovascular disease, hypertension, and altered lung function.

Treatment

The treatment for obesity is a combination of reduced calorie intake and increased energy utilization (exercise). Both the reduction of calorie intake and the increase in exercise must be tailored to the individual's general health, size, age, sex, and habitual physical activity. In addition, the diet must continue to supply all necessary nutrients in adequate amounts.

SUGGESTED READING

1. Whitney EN, Hamilton EMN: Understanding Nutrition. West Publishing, St. Paul, MN, 1984
2. Tuckerman MM, Turco S: Human Nutrition. Lea & Febiger, Philadelphia, 1983
3. National Academy of Sciences: Recommended Dietary Allowances. 9th Ed. National Academy Press, Washington, D.C., 1980
4. Salans L: Obesity. In Bondy PK, Rosenberg LE (eds): Metabolic Control and Disease. 8th Ed. WB Saunders, Philaelphia, 1980

STUDY QUESTIONS

Directions: For the each of the following multiple choice questions, choose the most appropriate answer.

1. Which one of the following substances is NOT an energy nutrient?
   A. Protein
   B. Carbohydrate
   C. Lipid
   D. Fiber

2. All of the following statements are true EXCEPT that:
   A. animal proteins are more digestible than plant proteins.
   B. the protein concentration inside plant cells is higher than the protein concentration inside animal cells.
   C. animal proteins usually contain all of the essential amino acids.
   D. plant proteins may not always contain all of the essential amino acids.

3. Impaired intestinal absorption of copper causes
   A. Menkes' syndrome
   B. Wilson's disease
   C. loss of the sense of taste
   D. decreased insulin binding to target cells

Items 4 through 7. Consider the following water-soluble vitamins:
   A. Thiamine
   B. Riboflavin
   C. Pyridoxine
   D. Folic acid

Match each statement below with the most approprite vitamin.

4. A vitamin that is converted to a coenzyme which participates in the decarboxylation of α-keto acids.

5. A vitamin that is converted to a coenzyme which is used in a large number of oxidation-reduction reactions.

6. A vitamin that is converted to a coenzyme which participates in one-carbon transfer reactions.

7. A vitamin that is converted to a coenzyme which is used in transmination, oxidation, and decarboxylation of amino acids.

Directions: Answer the following questions using the key outlined below:

(A) if 1, 2, and 3 are correct
(B) if 1 and 3 are correct
(C) if 2 and 4 is correct
(D) if only 4 is correct
(E) if all four are correct

8. Functions of vitamin A include
   1. the synthesis of secretory glycoproteins
   2. regulation of gene expression
   3. light detection by rod cells
   4. prevention of epidermal keratinization

9. Vitamins believed to have an antioxidant function include:
   1. vitamin A
   2. vitamin K
   3. vitamin D
   4. vitamin E

10. Since very little iron is lost from the diet, most individuals do not require iron supplementation. Exceptions are:
    1. adult men
    2. children
    3. nonpregnant women
    4. lactating women

19

MUSCLE CONTRACTION

Types of Muscle/The Contractile Unit/
Mechanism of Contraction/
Energy Sources for Contraction

The student should be able to:

1. Distinguish striated and smooth muscle.
2. Describe the major proteins involved in muscle contraction, and their organization to form the contractile unit of striated muscle.
3. Describe the mechanism of contraction in striated muscle.
4. Understand the roles of phosphocreatine, AMP kinase, and AMP deaminase in providing energy for muscle contraction.

Learning Objectives

> Physical labor of all types requires muscle contraction. Muscles can contract because they contain specific protein assemblies that move relative to each other in a coordinated manner. ATP is the immediate source of energy for muscle contraction. This chapter covers the protein components involved in muscle contraction, their organization into filaments, the organization of filaments into contracting units, and the events of contraction.

Perspective: Muscle Contraction

TYPES OF MUSCLE

Muscles are of three types: *striated, cardiac,* and *smooth*. Striated muscles are called "striated" because they exhibit a regular array of dark and light bands under the light microscope. The contraction of striated muscles is under voluntary control. Cardiac (heart) muscle is similar to striated muscle, but contracts involuntarily. Smooth muscle does not have a striated appearance, and its contraction is also under involuntary control. Skeletal muscle is striated, whereas the uterus, intestinal wall, and arteries contain smooth muscle.

Striated muscle tissue is not all alike. Skeletal muscle consists of two types of fiber: *white* and *red*. Red fibers contain a high concentration of myoglobin, cytochromes, and mitochondria, and are capable of a high rate of oxidative respiration. They readily oxidize either glucose or fatty acids. White fibers contain fewer cytochromes and mitochondria, and derive their energy primarily from the glycolytic degradation of glucose to lactate.

Muscle fibers can also be divided on the basis of their speed of contraction into *fast-twitch fibers*, which contract very rapidly, and *slow-twitch fibers*, which contract more slowly. White muscle fibers are all fast-twitch, whereas red muscle fibers can be either fast-twitch or slow-twitch. Fast-twitch fibers function to generate high muscle tension very quickly, whereas slow-twitch fibers maintain muscle tension over a long period. Therefore, exercise involving continuous activity results in the preferential development of slow-twitch red fibers, whereas exercise consisting of spurts of intense activity tends to develop white fibers.

Smooth muscle cells lack troponin, one of the proteins involved in striated muscle contraction. The mechanism of contraction is slightly different in smooth muscle than in striated muscle. Striated muscle contraction is better understood, and is discussed in this chapter.

THE CONTRACTILE UNIT

This section first introduces the proteins involved in the contraction of striated muscle: actin, tropomyosin, troponin, and myosin. These proteins are organized into the two main components of the contractile unit—the thin and thick filaments—which in turn are organized into sarcomeres, the fundamental contracting units of muscle.

The Proteins of Muscle Contraction

Actin

Actin can exist as a monomer or as a double-helical fibrous polymer. The monomeric unit, called G-actin, is a globular protein of molecular weight 43,000. It normally carries one molecule of bound ATP. In the presence of Mg^{2+}, G-actin monomers associate to form the double helix polymer *F-actin*:

$$n\text{G-actin-ATP} \xrightarrow{Mg^{2+}} \text{F-actin-}n\text{ADP} + n\text{P}_i$$

The ATP associated with G-actin is hydrolyzed during polymerization. However, G-actin that lacks bound ATP is also capable of polymerizing to form F-actin. Each actin monomer has a binding site for myosin.

Tropomyosin

Tropomyosin is a long, rod-shaped protein with a molecular weight of 66,000. It consists largely of two α-helical segments wrapped around each other to form a coiled-coil supersecondary structure (Chapter 1).

Troponin

Troponin is a globular protein (MW 76,000) consisting of three subunits called Tn-T, Tn-I, and Tn-C. The T subunit binds to tropomyosin, the C subunit binds Ca^{2+}, and the I subunit regulates the interaction between actin and myosin.

Myosin

Myosin (MW ~465,000) is by far the largest of the proteins involved in muscle contraction. It is a hexamer consisting of three kinds of subunit: one pair of *heavy chains* (MW 200,000) and two pairs of *light chains*. Both of the light chain proteins vary somewhat among myosins from different sources. The molecule has the structure shown in Fig. 19-1. The long α-helical carboxyl-terminal segments of the two heavy chains are wound around each other to form a coiled-coil rod-shaped tail. The amino-terminal portions of the heavy chains are globular, and associate with the light chains to form a pair of globular heads at the end of the rod-shaped tail. The globular heads of myosin have an ATPase activity and also a binding site for actin. The ATPase activity of myosin is peculiar: the globular heads of isolated myosin can bind and hydrolyze ATP, but they cannot release the resulting ADP and P_i until they also bind to actin.

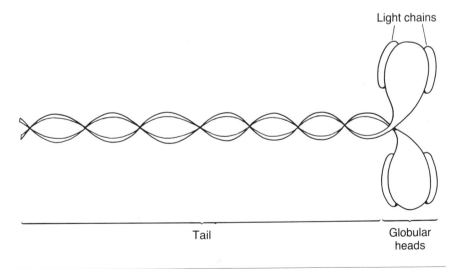

Fig. 19-1. A schematic representation of the structure of myosin.

Myofilaments

The two components of the contractile unit are the *thick* and *thin filaments* (or *myofilaments*), which slide past each other and thereby cause the muscle to contract. The thick filaments consist of myosin, whereas the thin filaments consist of F-actin, tropomyosin, and troponin.

Thin Filaments

Each thin filament consists of an F-actin double helix to which is bound tropomyosin and troponin. Because F-actin is a double helix of globular monomers, it has two helical grooves in its surface. As shown in Figure 19-2, tropomyosin rods are arranged end-to-end along both of these grooves. One troponin (consisting of the three subunits) is associated with each tropomyosin.

Thick Filaments

As shown in Figure 19-3, the thick filaments consist of a staggered array of myosin molecules. The myosins are packed in such a way that the globular heads protrude in a spiral from the cylindrical body of the thick filament. An intact thick filament consists of two complementary myosin bundles butted together stalk-to-stalk so that the globular heads of the bundles are located at the two ends and the center of the filament is bare. A cross-linking protein holds together the stalks of the complementary myosin bundles.

The Sarcomere

The thick and thin filaments combine to form the *sarcomere*, which is the contracting unit of muscle. As shown in Figure 19-4, each sarcomere consists of a central bundle of thick filaments flanked by two bundles of thin filaments. In the relaxed sarcomere, the tips of the thick and thin filaments intermesh slightly. As the sarcomere contracts, the two bundles of thin filaments slide toward each other, intermeshing more extensively with the thick filaments. Sarcomeres are joined end-to-end to form long fibrils within the muscle cell.

Striated muscle is characterized by the light and dark bands of the sarcomere that are visible under the light microscope. The very dark line joining adjacent sarcomeres represents a protein structure into which the thin filaments are anchored. It is called the *Z line*. The light bands consist of thin filaments alone and are called the *I bands*. The thick filaments appear darker than the thin filaments, and make up the *A bands*. A darker area on either side of the A band marks the intermeshing of thin and thick filaments; the width of this darker zone indicates the degree of contraction of the sarcomere. A dark *M line* at the center of the A band represents the cross-linking protein that holds together the complementary halves of the thick filaments.

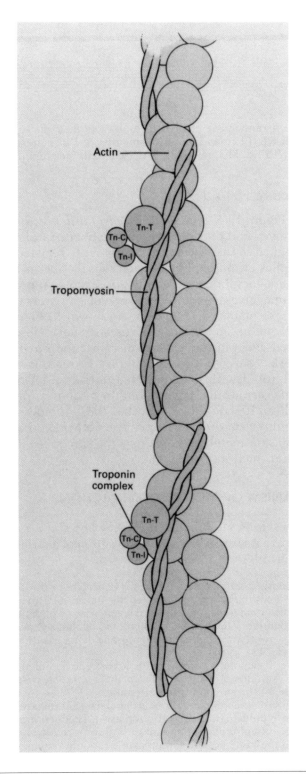

Fig. 19-2. Structure of striated muscle thin filaments. (From Darnell J, Lodish H, Baltimore D: Molecular Cell Biology. Scientific American Books, NY, 1986, with permission.)

Fig. 19-3. The arrangement of myosin molecules in a thick filament. (From Smith EL, Hill RL, Lehman IR et al: Principles of Biochemistry. Mammalian Biochemistry. 7th Ed. McGraw-Hill, NY, 1983, with permission.)

Muscle Structure

As shown in Figure 19-5, *muscle fibrils* (*myofibrils*) consist of many sarcomeres arranged end to end. Each muscle fiber contains many fibrils, which are aligned so that their sarcomeres are in register. Muscle tissue contains bundles of muscle fibers. Each muscle fiber is surrounded by a special plasma membrane called the *sarcolemma*. At regular intervals, the sarcolemma gives rise to a network of transverse invaginations called *transverse tubules* or *T tubules*. As shown in Figure 19-6, the T tubules penetrate among the myofibrils and make close contact with the specialized endoplasmic reticulum called the *sarcoplasmic reticulum*. A network of sarcoplasmic reticulum surrounds each fibril. At the level of the T tubules, the sarcoplasmic reticulum forms large, transverse *terminal cisternae* that closely abut the T tubules, forming the triad of the reticulum. It should be noted that the T tubules and the sarcoplasmic reticulum are distinct membrane networks; the T tubules are part of the plasma membrane, whereas the sarcoplasmic reticulum is an intracellular membrane system.

THE MECHANISM OF MUSCLE CONTRACTION

A striated muscle is stimulated to contract by the arrival of a nerve impulse. The events involved in the contraction of a striated muscle are as follows (Fig. 19-7):

1. A nerve impulse carried by a motor nerve arrives at the *neuromuscular junction*. The impulse is transmitted to the muscle fiber by the neurotransmitter *acetylcholine*, which triggers a wave of depolarization in the sarcolemma. The resting sarcolemma is polarized, with the inside more negative than the outside. (Membrane depolarization is discussed in Chapter 24.)
2. The wave of depolarization spreads across the sarcolemma and also along the systems of T tubules. From the T tubules, it spreads to the membrane of the sarcoplasmic reticulum.
3. When the sarcoplasmic reticulum is depolarized it releases stored Ca^{2+} into the sarcoplasm (muscle cell cytoplasm). The membrane of the sarcoplasmic reticulum has abundant Ca^{2+} pumps that continually pump Ca^{2+} into the lumen of the sarcoplasmic reticulum, where it is stored. Therefore, the sarcoplasmic $[Ca^{2+}]$ is very low in the resting muscle fiber. It is the sudden increase of $[Ca^{2+}]$ caused by membrane depolarization that induces the myofibril to contract.
4. In the resting myofibril, tropomyosin has a conformation that prevents contact between the globular heads of myosin and the myosin binding

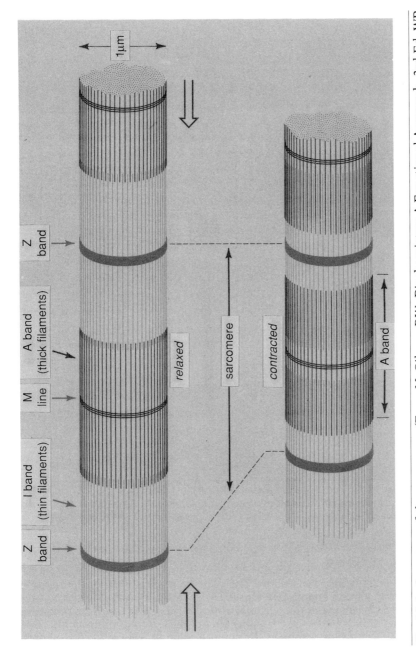

Fig. 19-4. Structure of the sarcomere. (From McGilvery RW: Biochemistry: A Functional Approach. 3rd Ed. WB Saunders, Philadelphia, 1983, with permission.)

498 Biochemistry

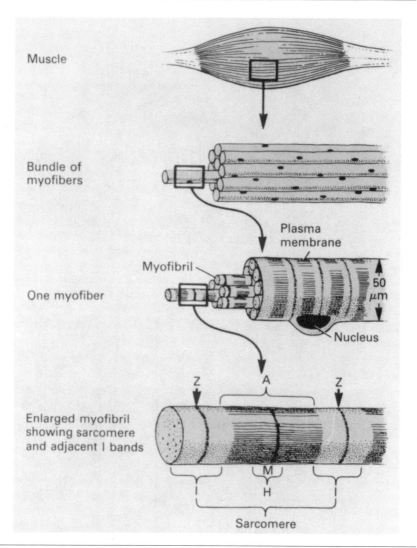

Fig. 19-5. The structural levels of muscle. (From Darnell J, Lodish H, Baltimore D: Molecular Cell Biology. Scientific American Books, NY, 1986, with permission.)

sites of actin. Remember that the globular heads of myosin are able to bind and hydrolyze ATP, but are unable to release ADP and P_i until they also bind actin. Thus, in the resting muscle, the myosin heads contain bound ADP and P_i.

5. When membrane depolarization releases Ca^{2+} into the myofibril, Ca^{2+} binds to the Ca^{2+} binding sites of the C subunit of troponin (Tn-C). The resulting change in the conformation of troponin causes tropomyosin to change position on the actin filament. This movement exposes the myosin binding sites on the actin.

6. The actual contraction of the sarcomere now begins. Several steps are involved. (1) The globular heads of myosin bind to actin, forming *actomyosin*. At this point the myosin globular heads project at an ap-

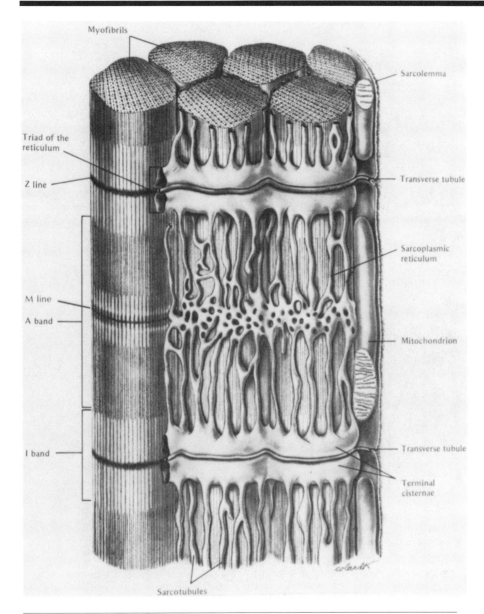

Fig. 19-6. The organization of muscle fibers. (From Bloom WD, Fawcett DW: Textbook of Histology. WB Saunders, Philadelphia, 1965, with permission.)

proximate right angle from the myosin stalks. (2) P_i is released from the globular head. (3) Myosin undergoes a conformational change in which the globular heads fold back toward the stalk. This change pulls the attached thin filament in the direction of the M line. (4) ADP is believed to be released during or immediately after the conformational change in myosin.

7. The myosin head is free to bind another ATP. When it does so, it dissociates from actin. Hydrolysis of the ATP to ADP and P_i prepares the globular head to bind again to actin and repeat the contraction cycle.

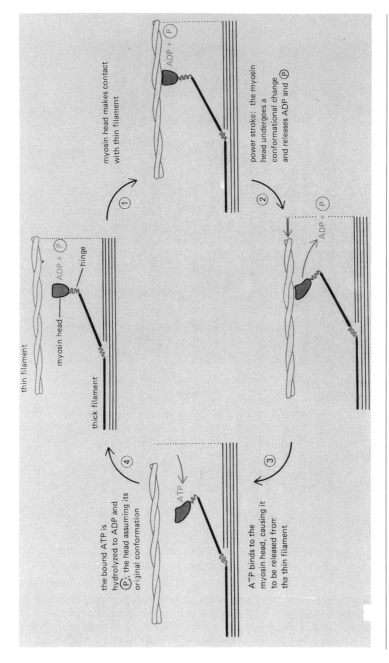

Fig. 19-7. The interactions of actin and myosin during muscle contraction. (From Alberts B, Bray D, Lewis J et al: Molecular Biology of the Cell. Garland, NY, 1983, with permission.)

8. As long as Ca^{2+} and ATP are available, the sarcomere shortens as a result of repeated contraction cycles. The thick and thin filaments intermesh so that each thick filament is surrounded by six thin filaments. Because the globular heads form a spiral projecting from the thick filament, each thick filament interacts with all six surrounding thin filaments.
9. When the nerve impulse stops, Ca^{2+} is pumped back into the sarcoplasmic reticulum, where it is stored. As the level of Ca^{2+} in the myofibril drops, Ca^{2+} is released from Tn-C, allowing tropomyosin once again to cover the myosin binding sites on actin. Muscle contraction comes to a halt, and the sarcomere is returned to its resting length.

One ATP is hydrolyzed for each actin-myosin interaction. Both globular heads of each myosin molecule can bind to the same actin filament. The exact nature of the actin-myosin interaction is not understood.

Muscle Contraction: Key Points

1. In order for the sarcomere to contract, the actin of the thin filaments and the myosin of the thick filaments must interact to form actomyosin.
2. Actin has a high affinity for myosin that contains bound ADP + P_i, but a low affinity for myosin that contains bound ATP. Free myosin heads rapidly hydrolyze bound ATP to bound ADP + P_i.
3. In resting muscle, actin and myosin are prevented from binding by tropomyosin, which covers the myosin binding sites on actin.
4. Binding of ATP or of ADP + P_i to a myosin head causes the head to assume an extended position in which it projects from the stalk at an approximate right angle. Release of the nucleotide and/or P_i causes the head to flex back toward the stalk to an approximate 45 degree angle, causing contraction.

Energy Sources for Contraction

The ATP used by muscle cells is ultimately supplied by glycolysis and oxidative phosphorylation. However, the amount of ATP present in muscle cells is often not sufficient to meet the immediate needs of muscle contraction. Muscle cells also contain high levels of phosphocreatine, which is the indirect energy source for muscle contraction. The ATP used during muscle contraction can be replenished by the following reactions:

1. The phosphorylation of ADP by phosphocreatine
2. The AMP kinase reaction
3. The AMP deaminase reaction

Phosphocreatine

The synthesis and metabolism of phosphocreatine were described in Chapter 10 (see Fig. 10-7), and the role of phosphocreatine in replenishing ATP was mentioned in Chapter 3. Creatine is reversibly phosphorylated to phosphocreatine by ATP when ATP is abundant. This process builds up a reservoir of high-energy phosphate bonds in the form of phosphocreatine. When the concentration of ATP drops, the reaction reverses, and ADP is phosphorylated by phosphocreatine:

$$ATP + creatine \rightleftharpoons ADP + creatine\ phosphate$$

AMP Kinase

The enzyme AMP kinase catalyzes the following reaction:

$$2\ ADP \xrightleftharpoons{\text{AMP kinase}} ATP + AMP$$

This reaction helps to replenish ATP in two ways:

1. Since AMP is a stimulator of phosphofructokinase (the major regulatory enzyme of glycolysis), the AMP kinase reaction stimulates the glycolytic phosphorylation of ADP to ATP.
2. By producing one ATP for every two ADPs, the reaction stretches the supply of available ATP. This reaction becomes important if phosphocreatine is exhausted.

AMP Deaminase

During extreme exertion, when the supply of phosphocreatine has been exhausted, the efficiency of the AMP kinase reaction for replenishing ATP is enhanced by the action of another enzyme, AMP deaminase. AMP deaminase catalyzes a reaction that consumes AMP and therefore "pulls" the AMP kinase reaction toward ATP formation:

$$AMP + H_2O \xrightarrow{\text{AMP deaminase}} IMP + NH_3$$

This reaction becomes important only in extreme cases of muscle exhaustion.

SUGGESTED READING

1. Taylor EW: Chemistry of muscle contraction. Annu Rev Biochem 41:577, 1972
2. Huxley HE: The mechanism of muscle contraction. Sci Am 213(6):18, 1965
3. Carlson FD, Wilkie DR: Muscle Physiology. Prentice-Hall, Englewood Cliffs, NJ, 1974
4. Weber A, Murray JM: Molecular control mechanisms in muscle contraction. Biophys J 15:709, 1973
5. Pollack GH, Sugi H (eds): Contractile Mechanisms in Muscle. Plenum Press, New York, 1984

STUDY QUESTIONS

Directions: Answer the following questions using the key outlined below.

- **(A)** if 1, 2 and 3 are correct
- **(B)** if 1 and 3 are correct
- **(C)** if 2 and 4 are correct
- **(D)** if only 4 is correct
- **(E)** if all four are correct

1. A striated muscle sarcomere contains the proteins
   1. actin
   2. tropomyosin
   3. myosin
   4. troponin

2. Thin filaments consist of
   1. F-actin
   2. tropomyosin
   3. troponin
   4. G-actin

3. All of the following statements are correct EXCEPT that:
   1. muscle cells lack ATP, and instead use phosphocreatine as the energy source for muscle contraction.
   2. muscle cells contain ATP.
   3. muscle cells contain phosphocreatine but not creatine.
   4. muscle cells contain ATP, phosphocreatine, and creatine.

4. During times of extreme physical activity, which of the following reactions may function to replenish depleted cellular supplies of ATP?
   1. $2\ ADP \rightarrow ATP + AMP$
   2. $Creatine + ADP \rightarrow phosphocreatine$
   3. $AMP + H_2O \rightarrow IMP + NH_3$
   4. $2\ Phosphocreatine + AMP \rightarrow ATP$

Directions: For each of the following multiple choice questions, choose the most appropriate answer.

5. All of the following terms are associated with the sarcomere EXCEPT
   A. I band
   B. A band
   C. M line
   D. S line

6. Which of the following proteins or protein complexes is NOT present or is present in only very small amounts in resting (relaxed) muscle?
   A. Tropomyosin
   B. Actomyosin
   C. Actin
   D. Troponin

7. All of the following statements concerning the events of muscle contraction are true EXCEPT that
   A. the sarcoplasmic reticulum releases Ca^{2+} into the sarcoplasm.
   B. one myosin head interacts with one actin filament.
   C. the sarcomere becomes shortened.
   D. the myosin heads and the actin filaments move in the direction of the M-line.

20

BIOCHEMISTRY OF VISION

Structure and Metabolism of the Eye/
Mechanism of Vision/Clinical Correlation

Learning Objectives

The student should be able to:

1. Describe the general structural features of the cornea, lens, and retina.
2. Describe the mechanism of rod cell vision.
3. Distinguish essential features of vision in rod and cone cells.

STRUCTURE AND METABOLISM OF THE EYE

This section presents some points concerning the structure and metabolism of the cornea, lens, and retina.

The Cornea

The cornea is the transparent anterior region of the covering of the eyeball. The cornea is a multilayered structure. Most of its thickness is made up by the collagenous *stroma*, which consists primarily of collagen fibrils that are extremely fine and uniform in size and arrangement. It is the fineness and uniformity of these collagen fibrils that make the cornea transparent. The cornea is avascular. It receives nutrients from the underlying aqueous humor, and it obtains much of its oxygen directly from the air by diffusion across the epithelial membrane. The most characteristic feature of corneal metabolism is the high activity of the pentose phosphate pathway: about 50 percent of the glucose used by the cornea is metabolized by this pathway to yield NADPH. The rest is metabolized by glycolysis and the TCA cycle.

The Lens

Like the cornea, the lens is transparent and avascular, and receives most of its nutrients from the aqueous humor. The anterior surface of the lens consists of a layer of intact cells. The remaining cells of the lens (called the fiber cells) show a progressive decrease in mitochondria and nuclei from the periphery to the nuclear region of the lens. The lens contains three proteins, the α, β, and γ *crystallins*, which constitute about 90 percent of the protein of the lens. The relative absence of mitochondria and nuclei in fiber cells and the precise arrangement of crystallins contribute to the transparency of the lens. Most of the glucose used by the lens is metabolized by the glycolytic pathway; little is metabolized by either the pentose phosphate pathway or the TCA cycle.

The Retina

The retina contains the light-sensitive photoreceptor cells of the eye, as well as nerves and blood vessels. There are two kinds of photoreceptor cells: *rods* and *cones*. The rods are responsible for vision in dim light, whereas the cones are responsible for color vision and for vision in bright light. The rod and cone cells have a very high rate of oxygen consumption, and the glycolytic pathway, TCA cycle, and pentose phosphate pathway are all very active in these cells. Some of the NADPH generated by the pentose phosphate pathway is used in the regeneration of the chromophore 11-*cis*-retinal (see below).

Perspective: Mechanism of Vision

> For vision to occur, an appropriate nerve impulse must be sent via the optic nerve from the photoreceptor cells of the retina to the brain. Before such an impulse can be sent the appropriate signal must be generated. The events that lead to the release of a nerve impulse from the retina occur in the photoreceptor (rod and cone) cells. Vision in both rod and cone cells depends on light-absorbing *visual pigments*.

Light Detection by Rod Cells

The mechanism of vision is better understood in rods than in cones. As shown in Figure 20-1, the rod cell consists of an inner segment and an outer segment. The inner segment contains the nucleus, numerous mitochondria, and other organelles, and also has a synaptic body that makes contact with a bipolar neuron. The outer segment is packed with flat membranous vesicles. The membranes of these vesicles contain *rhodopsin*, the visual pigment of rods. Rhodopsin is a complex of the integral membrane protein *opsin* and the vitamin A derivative 11-*cis*-retinal (Fig. 20-2). The alternating double bonds of 11-*cis*-retinal make it a very strong chromophore. Vision in rod cells occurs as follows:

1. In the nonexcited rod cell, Na^+ ions are pumped out of the inner segment and into the outer segment. There is thus an extracellular flow

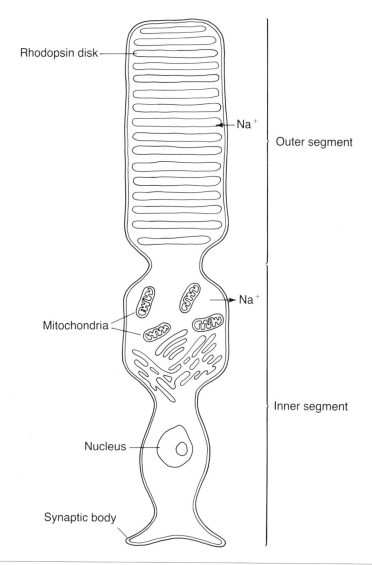

Fig. 20-1. Structure of the rod cell. In the unstimulated cell, Na^+ is pumped out by the membrane of the inner segment and pumped in by the membrane of the outer segment, so that extracellular Na^+ flows from the inner to the outer segment.

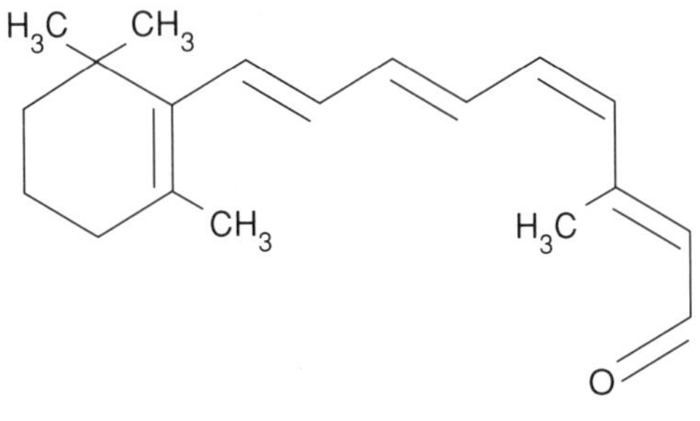

Fig. 20-2. Structure of 11-*cis*-retinal. Compare the structure of all-*trans*-retinal shown in Figure 18-6.

of Na$^+$ from the inner to the outer segment (Fig. 20-1), and the rod cell membrane is not polarized.

2. When 11-*cis*-retinal absorbs light, it is converted to all-*trans*-retinal. The complex of opsin with all-*trans*-retinal is called *bathorhodopsin* (*prelumirhodopsin*). Bathorhodopsin rapidly dissociates to opsin and all-*trans*-retinal (Fig. 20-3).
3. The formation of bathorhodopsin activates the opsin moiety via a conformational change that ultimately causes the plasma membrane of the rod outer segment to become impermeable to Na$^+$. As a result, the rod cell becomes hyperpolarized.
4. The hyperpolarization of the rod cell membrane travels to the synaptic body, where it initiates an impulse in the adjoining nerve. The nerve impulse travels to the brain, and vision occurs.

The process by which opsin activation leads to hyperpolarization probably involves Ca^{2+} and/or cGMP. Two observations support this conclusion:

1. Illumination of rod cells causes a reduction in the cellular content of cGMP via the activation of phosphodiesterase (phosphodiesterase converts cGMP to 5'-GMP).
2. Illumination of the retina causes Ca^{2+} to be released from the rod outer segments into the extracellular space. Ca^{2+} is known to interfere with the transport of Na$^+$.

Regeneration of Rhodopsin

For the visual process to continue, dissociated rhodopsin molecules must be reconstituted. All-*trans*-retinal is converted by the rod cell enzyme retinal isomerase back to 11-*cis*-retinal, which associates with opsin to form rhodopsin (Fig. 20-4). The supply of 11-*cis*-retinal in the

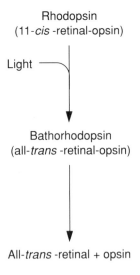

Fig. 20-3. Light-induced dissociation of rhodopsin into all-*trans*-retinal and opsin.

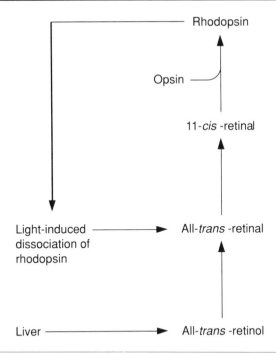

Fig. 20-4 Regeneration and maintenance of 11-*cis*-retinal in rod cells.

retinal cells is also maintained by the transport of all-*trans*-retinol from the liver, bound to retinol binding protein. All-*trans*-retinol is converted in rod and cone cells to all-*trans*-retinal, which is then isomerized to 11-*cis*-retinal.

Light Detection by Cone Cells

Cone cells are responsible for color vision and for vision in bright light. Color vision is possible because cone cells contain three different visual pigments that have absorption maxima respectively in the red, green, and blue regions of the spectrum. These pigments consist of complexes of 11-*cis*-retinal with three different opsins. Color blindness results from the absence or deficiency of any one of the three opsins of cone cells.

CLINICAL CORRELATION: CATARACTS

A cataract is an opacity of the lens caused by the formation of altered protein aggregates. In general, the exact cause of cataract formation is not known. However, cataracts have been observed in diabetic individuals and in individuals with increased plasma levels of galactose 1-phosphate. In both cases, the elevated blood level of the monosaccharide causes the monosaccharide or its derivative(s) to accumulate in the lens, eventually leading to alteration of the protein metabolism and/or structure of the lens. The result is lens opacity. In the case of diabetic individuals, uncontrolled diabetes causes glucose to be underutilized by the lens cells. ATP production declines, resulting in a decrease in the synthesis of lens proteins.

SUGGESTED READING

1. Adler FH: Adler's Physiology of the Eye. 7th Ed. CV Mosby, St. Louis, 1981
2. Graymore CN (ed): Biochemistry of the Eye. Academic Press. Orlando, FL, 1970
3. Hubbell WL: Bownds MD: Visual transduction in vertebrate photoreceptors. Annu Rev Neurosci 2:17, 1974
4. Wald G, Brower PK: Human color vision and color blindness. Cold Spring Harbor Symp Quant Biol 30:345, 1965
5. Schnapf JL, Baylor DA: How photoreceptor cells respond to light. Sci Am 256(4):40, 1987

STUDY QUESTIONS

Directions: Answer the following question using the key outlined below.
- **(A)** if 1, 2, and 3 are correct
- **(B)** if 1 and 3 are correct
- **(C)** if 2 and 4 are correct
- **(D)** if only 4 is correct
- **(E)** if all four are correct

1. Transparent regions of the eye are the
   1. retina
   2. cornea
   3. sclera
   4. lens

Directions: For the following multiple choice questions, choose the most appropriate answer.

2. The retina contains
   A. cone cells but no rod cells
   B. rod cells but no cone cells
   C. photoreceptor cells that are neither cone or rod cells
   D. both cone cells and rod cells

3. Which portion of the eye is exposed to the atmosphere?
   A. The cornea
   B. The lens
   C. The retina
   D. The cone cells

4. The cornea is transparent because of the
   A. fine structure, uniform size, and uniform distribution of its collagen fibrils
   B. random distribution of collagen fibrils
   C. α, β, and γ-crystallins that it contains
   D. high number of mitochondria present

5. All of the following statements are true EXCEPT that:
   A. rods cells are responsible for vision in dark or dim light.
   B. cone cells are responsible for vision in bright light.
   C. cone cells are responsible for color vision.
   D. rod cells are responsible for color vision and vision in bright light.

6. All of the following statements concerning the vision mechanism in rod cells are true EXCEPT that:
   A. light is absorbed by rhodopsin.
   B. the plasma membrane of the rod cell outer segment becomes less permeable to Na^+ when the cell is stimulated by light.
   C. rhodopsin dissociates to opsin and 11-*trans*-retinal.
   D. all-*trans*-retinal is converted back to 11-*cis*-retinal by retinal isomerase.

21

BIOLOGIC MEMBRANES

Membrane Composition/Interaction of Lipids with Water/
Membrane Dynamics/Membrane Asymmetry/
Membrane Transport

The student should be able to:

Learning Objectives

1. List the major functions and major chemical components of biologic membranes and illustrate the bilayer structure of membranes.
2. Distinguish integral and peripheral membrane proteins.
3. Define critical micelle concentration.
4. Explain the effects of temperature, fatty acid chain length, degree of fatty acid unsaturation, and cholesterol content on the fluidity of membranes, and describe the major types of lipid movement in membranes.
5. Explain what is meant by membrane asymmetry.
6. Distinguish nonmediated and mediated transport; facilitated diffusion and active transport; primary and secondary active transport; and symport and antiport transport systems.
7. Define *ionophore* and distinguish the two classes of ionophore.

Perspective: Biologic Membranes

Eukaryotic cells are enclosed by a plasma membrane and also have a variety of internal membranes. This chapter concentrates mainly on the properties of the plasma membrane, but much of the discussion applies equally to intracellular membranes. The plasma membrane performs a variety of functions essential to the cell:

1. It determines the boundaries of the cell, and thus influences their size and shape.
2. It controls the movement of nutrients, metabolites, and other solutes into and out of the cell.
3. It contains receptors for signal molecules (such as hormones).

(continued)

4. It carries cell recognition sites (cell surface glycoproteins) that are involved in cell-cell recognition.
5. It provides a surface for some enzyme reactions.

As mentioned in previous chapters, the fabric of biologic membranes consists of a bilayer of amphipathic lipids. Some of the chemical and physical properties of membranes are as follows:

1. Biologic membranes are composed of lipid, protein, and carbohydrate. Lipid and protein are the major components. Carbohydrate occurs in the form of carbohydrate moieties on glycoproteins and glycolipids, and is present in much smaller quantities. The relative proportions of protein and lipid vary widely.
2. Functional biologic membranes are fluid, and most membrane components are capable of some mobility in the plane of the membrane.
3. Most biologic membranes are highly asymmetric—the inside and the outside of the membrane differ in protein, lipid, and carbohydrate composition.

COMPOSITION OF BIOLOGIC MEMBRANES

Lipids

The lipids usually found in biologic membranes include a variety of phospholipids, cholesterol, and some glycolipids. The most abundant membrane phospholipids are phosphatidylcholine, phosphatidylethanolamine, phosphatidylserine, and sphingomyelin. Cell membranes from different tissues contain different proportions of the individual lipids, and specialized membranes may contain very high concentrations of one or a few lipids. Lipids make up about 50 percent of most membranes, but the proportion of lipid may be much higher or lower in specialized membranes.

Proteins

Proteins perform most of the enzymic, transport, and recognition functions of membranes. Many kinds of proteins are usually associated with a given membrane. Most are present in only a few copies. Membrane proteins are grouped into two classes—*integral* (intrinsic) and *peripheral* (extrinsic)—depending on how easily they can be removed from the membrane. Figure 21-1 shows the relation of these two types to the lipid bilayer.

1. *Integral proteins* are embedded in the lipid bilayer. The protein may be exposed on only one side of the membrane, or it may span the membrane and be exposed on both sides. Integral proteins can be removed from the membrane only by drastic measures (treatment with detergent or organic solvents), which usually disrupt the membrane. Isolated

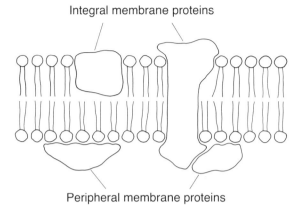

Fig. 21-1. The association of peripheral and integral membrane proteins with the lipid bilayer.

   integral proteins are insoluble in water and are inactive unless allowed to reassociate with lipid.
2. *Peripheral proteins* associate only with the surface of the membrane, either by interacting with an integral protein or by interacting with the polar head groups of the lipid bilayer itself. Peripheral proteins can be removed easily by treating the membrane with aqueous buffers of appropriate ionic strength or pH. These proteins are water soluble when removed from the membrane, and often retain their biologic activity in solution.

Carbohydrate

The carbohydrate components of membranes consist of the oligosaccharide components of glycolipids and glycoproteins, principally glycoproteins. Membranes contain no monosaccharide components.

INTERACTION OF MEMBRANE LIPIDS WITH WATER

As discussed in Chapter 8, membrane lipids are all amphipathic: they have a polar, hydrophilic end and a nonpolar, hydrophobic end. When an amphipathic molecule is placed in water, it orients itself so as to maximize the interaction of the hydrophilic group with water and minimize the interaction of the hydrophobic group with water. Depending on the conditions under which the lipid is added, one of the following three types of structure will be formed (Fig. 21-2). Very few lipid molecules exist free in solution.

1. If a small quantity of amphipathic lipid is added to a container of water, it forms a film or *monolayer* at the air-water interface, with the polar groups facing the water and the nonpolar tails facing the air.
2. After a specific concentration of lipid has been reached, amphipathic lipids spontaneously form spherical structures called *micelles*, in which the polar groups face the aqueous medium and the nonpolar tails are hidden in the center of the sphere, away from water. The concentration

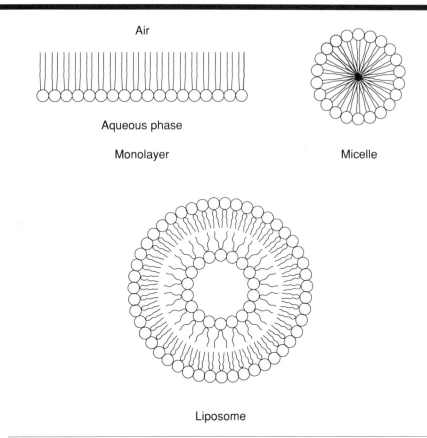

Fig. 21-2. The structures of lipid monolayers, micelles, and liposomes.

at which a given amphipathic lipid begins to form micelles is called the *critical micelle concentration* (CMC).
3. Under appropriate conditions (an adequate concentration of lipid and agitation of the medium), amphipathic lipids form bilayers. A bilayer consists of two monolayers arranged so that the polar head groups face the medium and the nonpolar tails face each other. Bilayers spontaneously form spherical structures called *liposomes*.

Biologic membranes consist of lipid bilayers. The two layers are held together by the strong attractive forces among the hydrophobic tails. Integral membrane proteins are also anchored into the membrane by hydrophobic interactions.

MEMBRANE DYNAMICS

Figure 21-3 shows the *fluid mosaic model*, which approximates the organization of biologic membranes. According to this model, the proteins float in a fluid bilayer, and both the proteins and the lipids are free to move in the plane of the membrane. This model accounts not only for the properties of integral and peripheral membrane proteins but also for the observed mobility of both protein and lipid components.

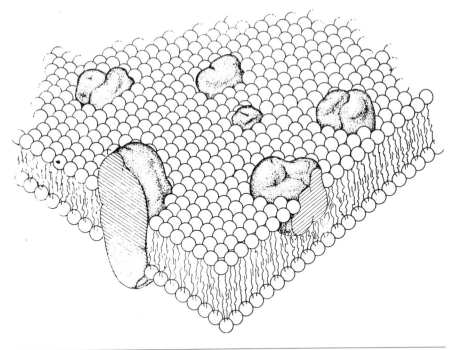

Fig. 21-3. The fluid mosaic model of biologic membranes. (From Singer SJ, Nicolson GL: The fluid mosaic model of the structure of cell membranes. Science 175:723, 1972, with permission.)

The *fluidity* of membranes is influenced by temperature, fatty acid chain length, the degree of unsaturation of fatty acid chains, and the cholesterol content. The factors that affect membrane fluidity are as follows.

1. *Temperature.* Fluidity increases with increasing temperature.
2. *Fatty acid chain length.* The shorter the fatty acid chains of the constituent phospholipids, the more fluid the membrane.
3. *Degree of unsaturation of fatty acids.* The greater the degree of unsaturation of the fatty acid chains, the more fluid the membrane.

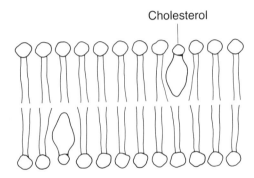

Fig. 21-4. The association of cholesterol with lipid bilayers.

4. *Content of cholesterol.* Fluidity decreases with increasing cholesterol content. The planar cholesterol molecule is both shorter and more rigid than the fatty acid chains of phospholipids. It intercalates lengthwise between phospholipids, with its hydroxyl group facing the aqueous phase (Fig. 21-4). Because cholesterol is more rigid than phospholipids, it tends to decrease the fluidity of membranes. The cholesterol molecule is not as long as membrane fatty acids so its effect on the center of the bilayer is limited.

Movement of Lipids in the Bilayer

In general, membrane lipids can move within the plane of their own monolayer, but cannot easily "flip-flop" from one monolayer to the other. The following sorts of movement are observed (Fig. 21-5):

1. Lipid molecules rapidly exchange places with neighboring molecules in the same monolayer—i.e., they are capable of lateral diffusion.
2. A pool of lipid molecules that interact strongly with each other may migrate as a group across the membrane surface. The lipids within the pool also move relative to each other, but apparently at a different rate than the pool as a whole.
3. The fatty acid tails of membrane phospholipids are very flexible, especially near the methyl end. Therefore, the membrane is most fluid toward the center of the bilayer.
4. Lipids undergo rapid rotation in the monolayer.
5. Flip-flop of membrane lipids from one monolayer to the other occurs, but at a very low rate.

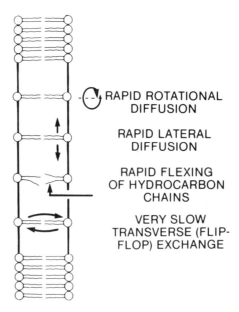

Fig. 21-5. Types of movement exhibited by membrane lipids. (From Devlin TM: Textbook of Biochemistry with Clinical Correlations. Wiley, New York, 1986, with permission.)

Movement of Proteins in the Bilayer

Like lipids, proteins are mostly free to migrate laterally in the membrane. They do not flip-flop. Proteins generally diffuse less rapidly in the plane of the membrane than lipids, and some proteins are more localized to a particular area of membrane than others. In some cases the movement of a membrane protein is determined by its association with cytoplasmic microfilaments or microtubules. Some proteins are strongly associated with specific membrane lipids. In these cases, the mobility of the protein is influenced by the mobility of the associated lipids.

Interaction of Integral Proteins with the Lipid Bilayer

The region of integral proteins that is embedded in the lipid bilayer is composed of amino acids with hydrophobic side chains. The strong interaction of the hydrophobic side chains with the nonpolar tails of the membrane lipids determines the depth and orientation at which the protein is anchored in the membrane. Very often, the embedded part of an integral protein consists of one or several α-helices. For example,

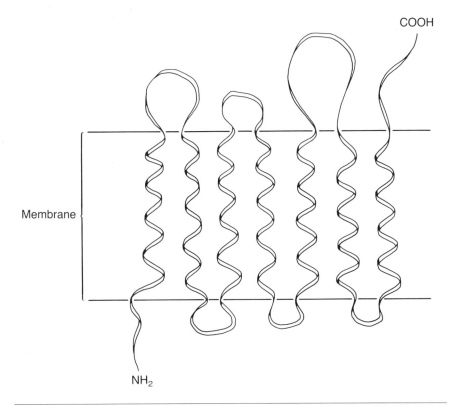

Fig. 21-6. Bacteriorhodopsin has seven helical segments inserted into the bilayer.

the embedded portion of bacteriorhodopsin (a bacterial light-sensitive protein) consists of seven parallel segments of α-helix (Fig. 21-6).

MEMBRANE ASYMMETRY

In most membranes, the protein, carbohydrate, and lipid compositions of the two monolayers are distinctly different.

Protein Asymmetry

1. Individual membrane proteins either span the bilayer or are associated with the inner or outer monolayer. In either case, the orientation of the protein in the membrane is specific.
2. All copies of a membrane protein display the same orientation.

Integral membrane proteins are inserted into the membrane of the endoplasmic reticulum while they are being synthesized by the ribosome (Chapter 14). These proteins are eventually incorporated into the plasma membrane via vesicle carriers.

Carbohydrate Asymmetry

Plasma membrane carbohydrates are associated with the external monolayer, in accordance with their role in cell recognition. When carbohydrate is associated with intracellular membranes, the carbohydrate moieties are located on the luminal rather than the cytoplasmic surface.

Lipid Asymmetry

The two monolayers of a bilayer usually have different lipid compositions. The types of lipid found in each may be the same, but they are generally present in different proportions.

MEMBRANE TRANSPORT

The transport of substances across membranes is either *nonmediated* or *mediated*. Nonmediated transport consists simply of the diffusion of a membrane-permeant substance through the membrane down its molecular gradient. A protein transporter or transport mechanism is not involved. Properties of nonmediated transport are as follows:

1. It is not specific. Any molecule capable of permeating across the membrane can engage in nonmediated transport. In general, substances must be lipid-soluble to diffuse across the membrane.
2. In the case of uncharged solutes, nonmediated transport is driven by the concentration gradient of the substance across the membrane. In the case of charged solutes, both the concentration gradient and the electrical gradient (membrane potential) contribute. The substance moves down its gradient, and net transport stops when the gradient has been dissipated.

3. The rate of diffusion is proportional to the steepness of the gradient, the surface area of the membrane, and the permeability of the substance through the membrane.

Mediated transport involves specific membrane proteins (cotransporters, translocases, permeases, and transport systems) which facilitate the movement of the substance across the membrane. The properties of mediated transport systems are as follows:

1. They are specific—i.e., each transporter will transport only certain substances.
2. Mediated transport displays saturation kinetics (Fig. 21-7).
3. Mediated transport can be inhibited by appropriate inhibitors.
4. Mediated transport can be either *facilitated diffusion* or *active transport*.
   a. In *facilitated diffusion*, the transport system merely helps transported substances to permeate through the membrane so that they can flow down their gradients. Energy is not expended.
   b. In *active transport*, the transport system expends energy to transport substances against their gradients.

Active transport systems are classified into *primary* and *secondary* types depending on the source of the energy for transport.

1. In *primary active transport*, the energy for transport is supplied by the simultaneous hydrolysis of ATP (Fig. 21-8).
2. In *secondary active transport* (Fig. 21-9), the transport of two substances is coupled, and the energy released as one substance flows down its gradient is used to drive the transport of the other substance against its gradient. Thus, the gradient that drives secondary active transport is maintained by primary active transport of a different substance.

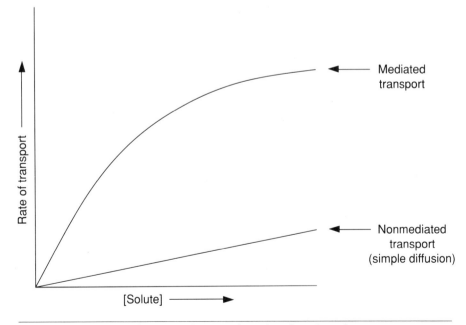

Fig. 21-7. Initial rate kinetics of nonmediated and mediated transport systems.

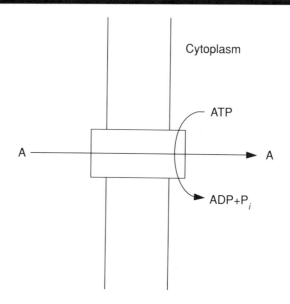

Fig. 21-8. Primary active transport.

Transport systems (facilitated diffusion or active transport) in which the transport of two compounds is coupled can be either *symport* or *antiport*. In symport systems, both molecules are transported across the membrane in the same direction, whereas in antiport systems, they are transported across the membrane in opposite directions. Table 21-1 lists a few examples of the various types of transport systems found in mammals.

Ion Transport Systems

Some of the most important active transport systems are the ones that maintain the ionic gradients across cell membranes. The Na^+ and K^+ gradients are maintained by a *Na^+-K^+ ATPase antiport pump* that transports Na^+ out of the cell and K^+ into the cell. Ion transport is driven by the direct hydrolysis of ATP, and therefore is primary active transport. The inward gradient of Na^+ created by the Na^+-K^+ pump is used to power a number of secondary symport systems (for example, the

Table 21-1. Examples of Solutes Transported in Mammalian Cells

Substance Transported	Type of Transport Mechanism
Monosaccharides	Facilitated diffusion or Active transport (Na^+ symport)
Amino acids	Active transport (Na^+ symport)
Ca^{2+}	Active transport
Na^+-K^+	Active transport (antiport system)

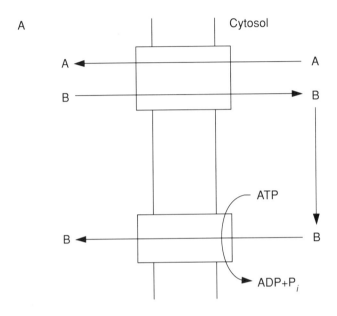

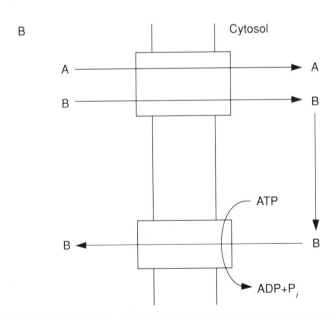

Fig. 21-9. Secondary transport: **(A)** antiport; **(B)** symport. Energy is supplied for the transport of A against its concentration gradient by the co-transport of B down its concentration gradient. B is then transported back across the membrane with the expenditure of ATP.

systems that transport amino acids and monosaccarides into cells). It is also important in the generation and propagation of nerve impulses (Chapter 23).

Another active primary transport system is the *Ca^{2+} pump* that transports Ca^{2+} out of the cytosol. This pump is very abundant in the membrane of the sarcoplasmic reticulum, and keeps the Ca^{2+} content of the myofibrils low in the resting muscle (Chapter 19). Like the Na^+-K^+ pump, the Ca^{2+} is a primary active transport system.

IONOPHORES

Ionophores are antibiotics that facilitate the transport of monovalent and divalent cations across membranes. Ionophores are grouped into two types—*ion carriers* and *ion channels*—according to their mechanism of action (Fig. 21-10).

Ion Carriers. The ion-carrier ionophores either have a cyclic structure or are capable of forming a cyclic structure. The inside of the ring contains several oxygens that can chelate a cation. The outside of the ring is hydrophobic, so a complex of a cation with the ionophore is capable of diffusing across the membrane. The antibiotic valinomycin (Fig. 21-11) carries K^+ across membranes.

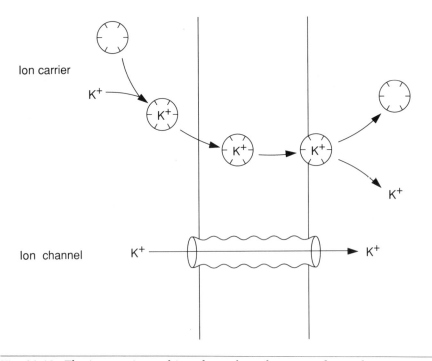

Fig. 21-10. The ion carrier and ion channel mechanisms of ionophore action.

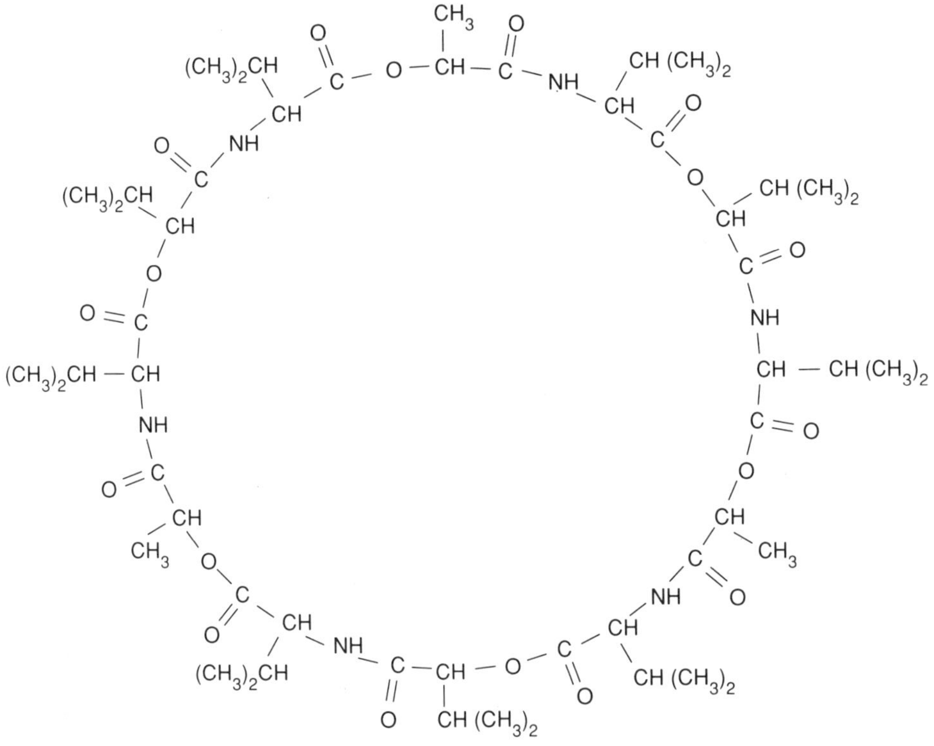

Fig. 21-11. Structure of valinomycin.

Ion Channels. Ion-channel ionophores insert themselves into the membrane in such a way as to create a pore or channel through which ions can flow. For example, the antibiotic gramicidin A (Fig. 21-12) forms a helical structure in the membrane that has a central hydrophilic pore large enough to accommodate ions. Two molecules of gramicidin A associate to span the full width of the membrane. Ions diffuse through the membrane via the hydrophilic pore.

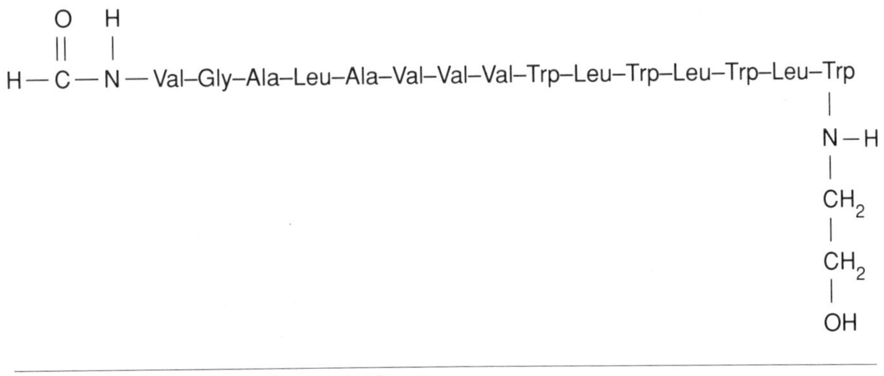

$$\text{H-C(=O)-N(H)-Val-Gly-Ala-Leu-Ala-Val-Val-Val-Trp-Leu-Trp-Leu-Trp-Leu-Trp-NH-CH}_2\text{-CH}_2\text{-OH}$$

Fig. 21-12. Structure of gramicidin A.

SUGGESTED READING

1. Finean JB, Mitchell RH (eds): Membrane Structure. Elsevier, New York, 1981
2. Housley MD, Stanley KK: Dynamics of Biological Membranes. Wiley, New York, 1982
3. Psote G, Crooke ST (eds): New Insights into Cell and Membrane Transport Processes. Plenum Press, New York, 1986
4. Semenza G, Kinne R (eds): Membrane transport driven by ion gradients. Ann NY Acad Sci Vol 456, 1985
5. Carafoli E: Intracellular calcium homeostasis. Annu Rev Biochem 56:395, 1987

STUDY QUESTIONS

Directions: For each of the following multiple choice questions, choose the most appropriate answer.

1. Which of the following statements concerning the plasma membrane is NOT true?
   A. The plasma membrane controls the movement of solutes in and out of cells.
   B. The plasma membrane contains more glycolipids than glycoproteins.
   C. The plasma membranes of some cells contain hormone receptors.
   D. The plasma membrane determines cellular boundaries.

2. The most abundant membrane components are
   A. protein and lipid
   B. protein and carbohydrate
   C. lipid and carbohydrate
   D. protein and glycolipid

3. Liposomes are
   A. lipid monolayer vesicles
   B. spherical lipid assemblies that do not contain an intravesicular compartment
   C. the type of structure that begins to form when the CMC of an amphipathic lipid in water has been reached
   D. lipid bilayer vesicles

4. Integral membrane proteins
   A. are loosely associated with the membrane
   B. may be removed from the membrane by altering the ionic strength of the medium
   C. usually contain β-pleated sheet structures embedded in the lipid portion of the membrane
   D. are usually only removed from the membrane by the use of drastic procedures that disrupt the membrane

5. All of the following changes in the fatty acid composition of membrane phospholipids will cause a decrease in membrane fluidity EXCEPT
   A. linoleic acid → oleic acid
   B. palmitic acid → stearic acid
   C. oleic acid → palmitoleic acid
   D. linolenic acid → linoleic acid

6. All of the following statements about secondary active transport systems are true EXCEPT that
   A. secondary active transport systems require energy
   B. secondary active transport systems can transport a solute against its concentration gradient
   C. secondary active transport systems utilize ATP as their immediate source of energy
   D. secondary active transport systems involve the co-transport of a second solute

Directions: Answer the following questions using the key outlined below.
   (A) if 1, 2, and 3 are correct
   (B) if 1 and 3 are correct
   (C) if 2 and 4 are correct
   (D) if only 4 is correct
   (E) if all four are correct

7. The interaction of membrane lipid molecules with water results in
   1. monolayers
   2. micelles
   3. bilayers
   4. dissolution of most of the lipid in water

8. All of the following statements are true EXCEPT that
   1. lipids undergo lateral diffusion.
   2. intracellular proteins may associate with the membrane but usually have no influence on protein diffusion.
   3. membrane proteins usually undergo lateral diffusion.
   4. membrane proteins routinely flip-flop between monolayers.

9. All of the following statements about plasma membrane asymmetry are true EXCEPT that
   1. proteins are usually located on the exterior monolayer.
   2. carbohydrate is usually located on the exterior monolayer.
   3. integral membrane proteins are randomly embedded in either monolayer.
   4. the exterior and cytoplasmic monolayers usually contain different proportions of certain lipids.

10. Facilitated diffusion
   1. requires a transport protein
   2. may occur via an antiport or symport mechanism
   3. does not require energy
   4. usually only involves lipid-soluble solutes

22

CONNECTIVE TISSUE MACROMOLECULES

Proteins/Proteoglycans/Clinical Correlations

The student should be able to:

1. Identify the major macromolecules of connective tissue.
2. Describe the characteristic amino acid pattern of collagen primary structure.
3. Recognize the triple helix and fibril structures of collagen, and describe the posttranslational modifications involved in the formation of mature collagen.
4. Distinguish proteoglycans from glycoproteins, and know the general function of most proteoglycans.

Learning Objectives

This chapter describes the protein and proteoglycan components of connective tissue. Connective tissue functions to hold together and support tissue cells, and consists mainly of an extracellular *matrix* which contains very few cells. The matrix contains fibrous proteins and proteoglycans. The principal proteins, *collagen* and *elastin*, occur in the form of large, insoluble, extensively cross-linked fibers. Collagen has very high tensile strength but little elasticity, whereas elastin is highly elastic. Proteoglycans—high-molecular-weight conjugates of proteins and polysaccharides—help give the connective tissue its viscoelastic properties. The cells of connective tissue synthesize and secrete the components of the connective tissue matrix.

Perspective: Connective Tissue Macromolecules

PROTEIN COMPONENTS

Collagen

Collagen is the most abundant protein of the human body. A mature collagen fibril consists of many collagen monomers aligned in a staggered array and connected by covalent cross-links. The collagen monomer is sometimes called *tropocollagen*. It is a very long molecule, consisting of three helical polypeptide chains that are wound around each other to form a three-stranded superhelix (Fig. 22-1). Each polypeptide chain is about 1000 residues long.

A number of different collagen polypeptide chains are coded for by different genes. The body contains several kinds of collagen, each characterized by the specific trio of polypeptide chains that makes up the tropocollagen triple helix. Specific collagens are characteristic of particular tissues, although the characteristic collagen of a tissue may change during development. Types I and III collagens are most abundant in adult humans.

Collagen polypeptides have a very distinctive amino acid content and primary structure. Nearly one-third of all residues are glycine, and there is also a high proportion of proline and of the modified amino acids hydroxylysine and hydroxyproline. Every third amino acid of collagen polypetide chains is usually glycine, and the primary sequence of collagen is characterized by the triplet Gly-X-Y, where X is usually proline and Y is usually hydroxylysine or hydroxyproline. The helix adopted by the individual polypeptide chains is different from an α-helix. It has three residues per turn. The three-stranded tropocollagen superhelix is very tight, and is sterically permitted only because every third residue of each polypeptide chain—i.e., every residue that faces the interior of the triple helix—is glycine. The other residues of the chain face outward and may be bulky. The hydroxylated residues form hydrogen bonds that help to hold the strands of the triple helix together. A varying proportion of the hydroxylysine residues of collagen are glycosylated to small oligosaccharides.

Collagen Biosynthesis

Figure 22-2 shows the process of collagen biosynthesis. The precursor polypeptide chains, called *preprocollagens*, are synthesized by membrane-bound ribosomes. They have an amino-terminal signal peptide (Chapter 14) that causes them to be secreted into the lumen of the endoplasmic reticulum, where the signal peptide is cleaved off by a signal peptidase. The resulting *procollagen* has an amino-terminal peptide of about 180 residues and a carboxyl-terminal peptide of about 300 residues which are both cleaved off during maturation. The carboxyl-terminal peptide is globular, whereas the amino-terminal peptide contains helical and nonhelical segments. Both peptides contain cysteine residues. The cysteine residues of the carboxyl-terminal segment form interchain disulfide bridges that are important in the association of pro-

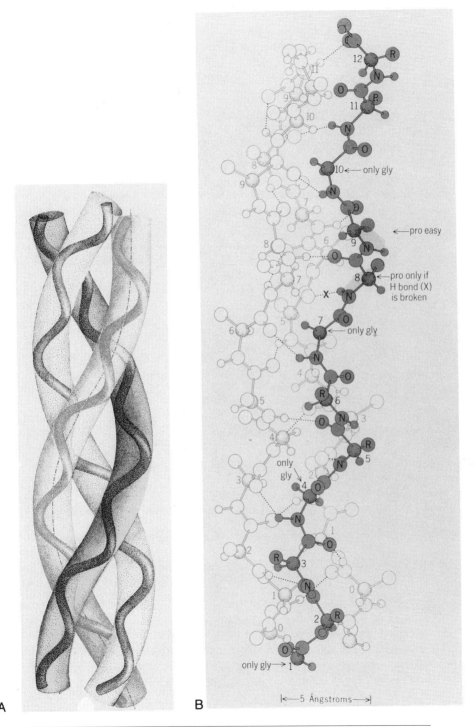

Fig. 22-1. The triple-helix structure of collagen. (**A**) The generalized coiled model. (**B**) Atomic model. (From Dickerson RE, Geis I: The Structure and Actions of Proteins. Benjamin-Cummings, Menlo Park, CA, 1969, with permission.)

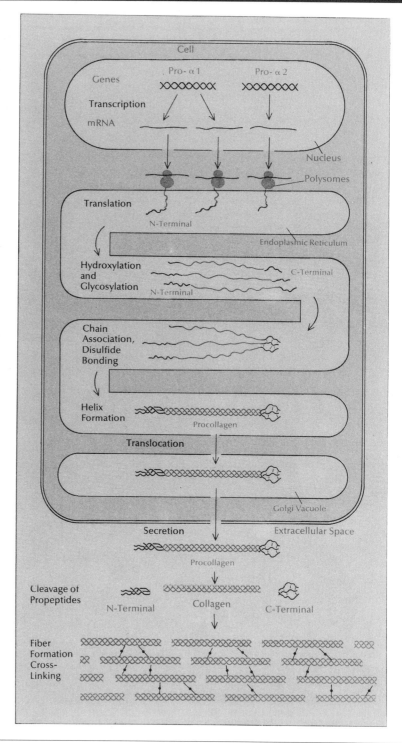

Fig. 22-2. Collagen biosynthesis. (From Prockop DJ, Guzman NA: Collagen diseases and the biosynthesis of collagen. Hosp Pract 12:61, 1977, with permission.)

collagen chains to form the triple helix. The cysteine residues of the amino-terminal segment form intrachain disulfide bridges.

The procollagen chains undergo extensive hydroxylation at proline and lysine residues within the lumen of the endoplasmic reticulum. The hydroxylation reactions are carried out by three enzymes, the products of which are 4-hydroxyproline, 5-hydroxyproline, and 5-hydroxylysine, respectively. Each of these enzymes requires Fe^{2+} and ascorbic acid for activity. In addition, each reaction uses molecular oxygen and α-ketoglutarate as substrates. The 4-hydroxylation of proline occurs as follows:

$$\text{Proline} + O_2 + \alpha\text{-ketoglutarate} \rightarrow \text{Hydroxyproline} + CO_2 + \text{succinate}$$

A portion of the 5-hydroxylysine residues are then glycosylated successively by UDP-galactose and UDP-glucose. The processes of hydroxylation, glycosylation, and polypeptide association all occur in the lumen of the endoplasmic reticulum, but it is not clear to what extent they occur simultaneously.

The hydroxylated and glycosylated procollagen triple helix is next secreted via the golgi apparatus into the extracellular space, where processing continues. First the amino- and carboxyl-terminal peptides are cleaved off to yield collagen monomers. These monomers spontaneously aggregate to form the staggered array typical of collagen fibrils (Fig. 22-3). Note that there are gaps between the ends of the collagen monomers. These gaps are aligned and give the surface of collagen fibrils a distinctive banded appearance.

After the collagen fibril has formed spontaneously, it is stabilized and rendered insoluble by covalent cross-linking (Fig. 22-4). Cross-linking is initiated by the enzyme *lysyl oxidase*, which catalyzes the oxidative deamination of certain lysine and hydroxylysine residues to the corresponding aldehydes (allysine and hydroxyallysine). The resulting aldehydes can react with each other by aldol condensation to form a co-

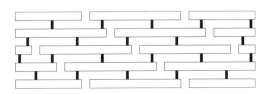

Fig. 22-3. The staggered array of cross-linked collagen monomers in collagen.

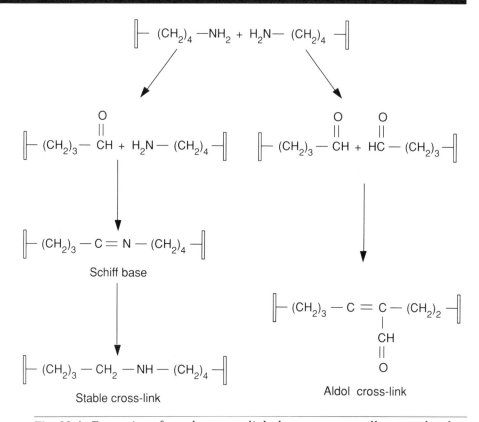

Fig. 22-4. Formation of covalent cross-links between tropocollagen molecules.

valent cross-link. Alternatively, an aldehyde can react with the ε-amino group of a lysine residue to form a Schiff base, which can then undergo rearrangement to yield a stable cross-link. Both of these reactions appear to occur spontaneously. Collagen fibrils subsequently associate to form fibers.

Collagen Degradation

Collagen degradation probably involves several proteolytic enzymes, but it is initiated and controlled by the action of the enzyme *collagenase*. Collagenase cleaves each peptide of the tropocollagen triple helix into two fragments, which are then further hydrolyzed by other proteases. Collagenase is very specific: it cleaves Gly-Leu or Gly-Ile peptide bonds that occur in the sequences Gly-Leu-Ala-Gly or Gly-Ile-Ala-Gly.

Elastin

Elastin fibers make it possible for connective tissues to stretch and rebound. They are especially abundant in blood vessel walls, skin, ligaments, and the lungs. Elastin resembles collagen in some ways. Like collagen, elastin consists of about 30 percent glycine residues. Elastin fibers are built by the association and subsequent cross-linking of a monomer called tropoelastin. Unlike tropocollagen, tropoelastin monomers are single polypeptide chains. Tropoelastin consists of a pliable coiled tertiary structure.

The cross-linking of elastin is initiated by oxidative deamination of lysine to allysine, but the cross-links differ in structure from the cross-links of collagen. In elastin, three allysines usually react with one lysine to form a four-way *desmosine* or *isodesmosine* link (Fig. 22-5). The desmosine and isodesmosine cross-links tie each tropoelastin to several neighbors, producing a three-dimensional elastic structure.

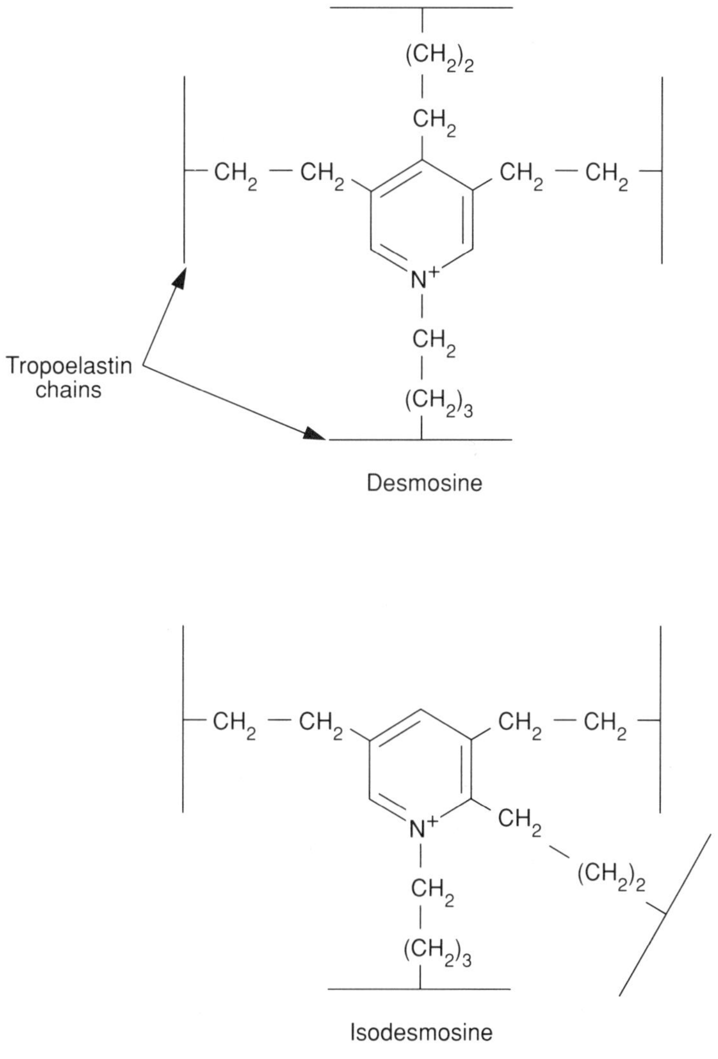

Fig. 22-5. Structures of the desmosine and isodesmosine cross-links found in elastin fibers.

Elastin Biosynthesis

Elastin synthesis involves the formation of *proelastin*, which contains a carboxyl-terminal peptide sequence that is cleaved off to yield tropoelastin. The details of tropoelastin processing are not yet understood.

Elastin Degradation

Elastin is degraded by the enyzme *elastase*. Elastase is secreted by the pancreas as the zymogen proelastase, which is activated by trypsin. Elastase cleaves peptide bonds that involve the carboxyl groups of aliphatic amino acids.

Other Proteins of the Extracellular Matrix

The connective tissue matrix contains a variety of proteins besides collagen and elastin. These proteins are believed to be important in cell adhesion and in the maintenance of tissue structure. The best known member of this group is *fibronectin*. Fibronectin exists in the connective tissue matrix as an insoluble glycoprotein, but it also exists in blood plasma as a soluble glycoprotein. Limited proteolysis fragments fibronectin into three biologically active domains. One of these domains binds certain proteoglycans, one binds collagen, and one binds both the cell surface and proteoglycans. It has been suggested that the interaction of fibronectin and proteoglycans is involved in cell-to-cell adhesion.

PROTEOGLYCANS

As described in Chapter 4, proteoglycans are very large conjugates of glycosaminoglycan polysaccharides with protein. A primary characteristic of glycosaminoglycans is their ability to bind a great deal of water. Proteoglycans are important in determining the viscoelastic properties of the matrix. Because of their ability to bind water and to absorb shock, some proteoglycans function as lubricants in certain areas of the body, for example in the synovial fluid of the joints.

The organization of the protein and polysaccharide components varies among different proteoglycans, and the polysaccharide chains can be linked to the protein component either covalently or noncovalently. As described in Chapter 4, glycosaminoglycans consist of disaccharide repeat units. There are six distinct classes of glycosaminoglycan chains: (1) chondroitin sulfate, (2) hyaluronic acid, (3) heparin, (4) heparan sulfate, (5) dermatan sulfate, and (6) keratan sulfate. Each of these is discussed below.

Chondroitin Sulfate

Disaccharide Repeat Unit

Molecular Weight of Glycosaminoglycan Chains. Up to approximately 60,000.

Proteoglycan Structure. As many as 100 glycosaminoglycan chains may be attached to one protein unit.

Tissue Distribution. Cartilage, bone, skin, blood vessel wall, and cornea.

Special Notes. The chondroitin sulfate chains on a given protein unit display significant heterogeneity, primarily in chain length and the extent of sulfation. Sulfation occurs only on C-4 or C-6 of the galactosamine residues. The glycosaminoglycan chains are covalently linked to the protein via a xyulose-*O*-Ser glycosidic linkage.

Hyaluronic Acid

Disaccharide Repeat Unit

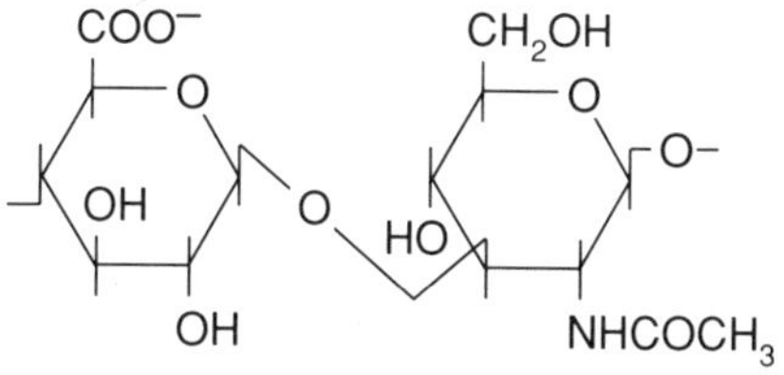

Molecular Weight of Glycosaminoglycan Chains. May range up to several million.

Proteoglycan Structure. The glycosaminoglycan does not appear to bind covalently to its protein component.

Tissue Distribution. Skin, synovial fluid, vitreous humor, and umbilical cord.

Special Notes. Hyaluronic acid glycosaminoglycan chains are not sulfated.

Heparin

Disaccharide Repeat Unit

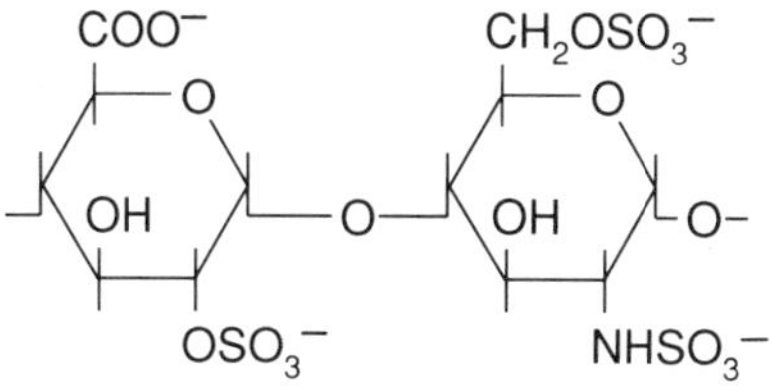

Molecular Weight of Glycosaminoglycan Chains. Up to approximately 70,000.

Proteoglycan Structure. Approximately 15 glycosaminoglycan chains may be covalently linked to the protein unit.

Tissue Distribution. Connective tissue of lung, skin, and inside mast cells of the intestinal mucosa.

Special Notes. Heparin also has anticoagulant activity. It is the only proteoglycan that occurs inside cells.

Heparan Sulfate

Disaccharide Repeat Unit. The repeat units of heparan sulfate are similar to those of heparin except that they have more N-acetyl groups, fewer N-sulfates, and fewer O-sulfates. Heparan sulfate also has additional sulfates on C-3 of glucosamine and C-2 of glucuronic acid.

Molecular Weight of Glycosaminoglycan Chains. Usually 10,000 to 20,000.

Proteoglycan Structure. Only about four or five glycosaminoglycan chains per protein unit.

Tissue Distribution. Blood vessel walls and cell surfaces.

Special Notes. Heparan sulfate appears to be more extracellular in location than heparin.

Dermatan Sulfate

Disaccharide Repeat Unit

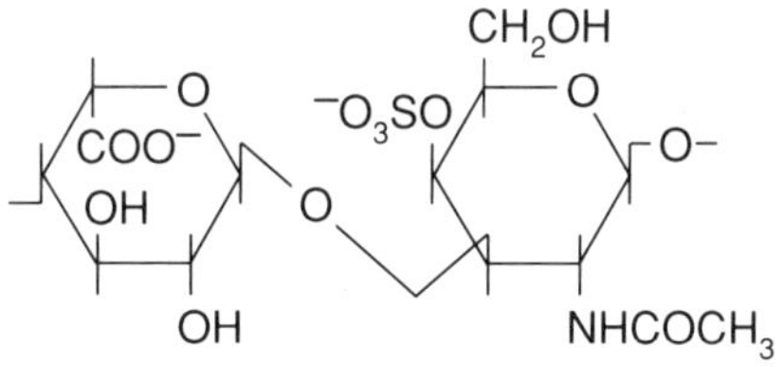

Molecular Weight of Glycosaminoglycan Chains. Approximately 35,000.

Tissue Distribution. Skin, blood vessel walls, heart valves, and tendons.

Special Notes. Dermatan sulfate is similar to chondroitin sulfate except for the predominance of L-iduronic acid. It also contains some sulfates esterified to C-4 of L-iduronic acid.

Keratan Sulfate

Disaccharide Repeat Unit

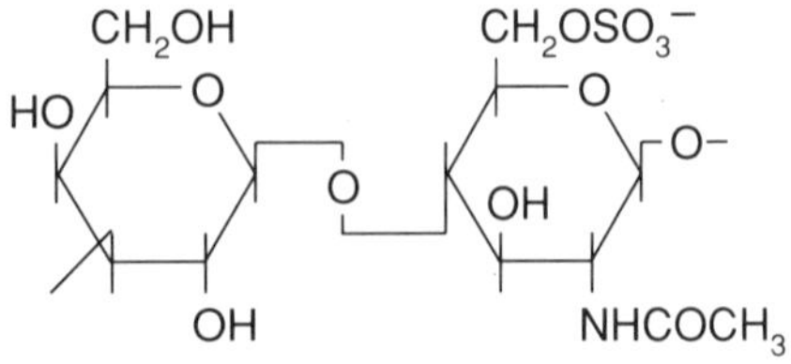

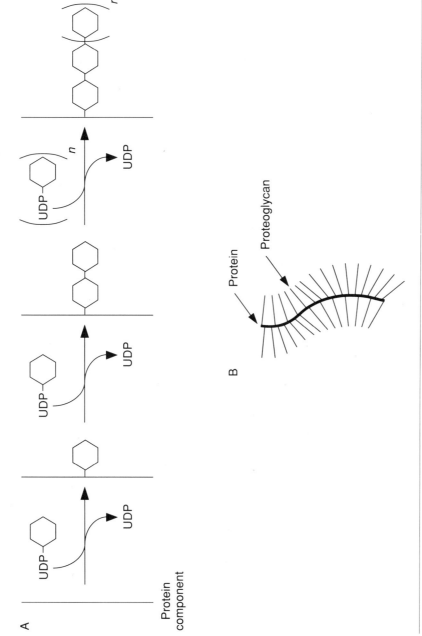

Fig. 22-6. (**A**) A representation of the synthesis of proteoglycans. (**B**) The organization of one type of completed glycoprotein.

Molecular Weight of Glycosaminoglycan Chains. Approximately 20,000.

Tissue Distribution. Cornea and cartilage.

Special Notes. Two types of keratan sulfate have been distinguished. Type I, found in the cornea, is linked to its protein unit by an *N*-acetylgalactosamine. Type II, found in cartilage, is linked to its protein unit by an *O*-glycosidic linkage through either serine or threonine. The monosaccharides mannose, fructose, sialic acid, and galactosamine are also present in keratan sulfate structures. Keratan sulfate chains show a high degree of heterogeneity.

Synthesis of the Proteoglycans

The protein components of proteoglycans are synthesized on ribosomes attached to the endoplasmic reticulum. Glycosylation of the protein components apparently occurs in the lumen of the endoplasmic reticulum/golgi system. The carbohydrate chains are synthesized by a series of specific glycosyltransferases, one for each type of sugar residue. UDP-monosaccharides are used as donors. The carbohydrate chains are synthesized on the protein component: the first enzyme transfers a specific monosaccharide to the correct amino acid residue, then the second monosaccharide is added to the chain, and so forth (Fig. 22-6). Sulfation of the carbohydate residues occurs simultaneously with chain elongation. The sulfating agent is *3'-phosphoadenosine-5'-phosphosulfate* (PAPS) (Fig. 22-7). Iduronic acid residues are created by epimerization of newly added glucuronic acid residues by an epimerase. The completed proteoglycan is secreted from the cell.

CLINICAL CORRELATIONS

Mucopolysaccaridoses

The glycosaminoglycan chains of proteoglycans (mucopolysaccharides) are degraded by specific lysosomal hydrolases and sulfatases that sequentially remove sulfate groups and monosaccharides from the non-

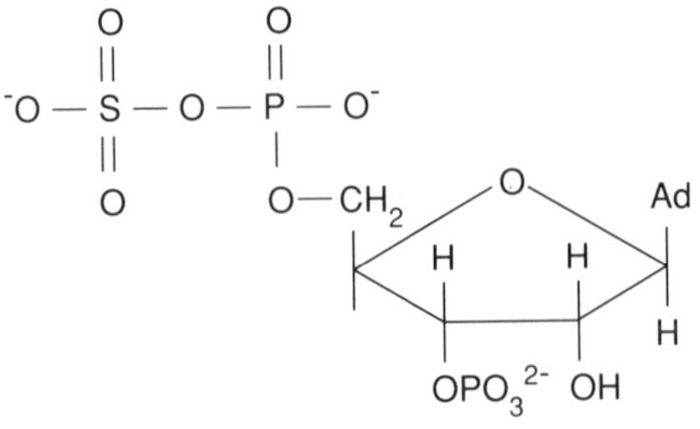

Fig. 22-7. The structure of 3'-phosphoadenosine-5'-phosphosulfate (PAPS).

Table 22-1. The Enzyme Deficiency in Some Ehlers-Danlos Syndromes

Ehlers-Danlos Syndrome	Deficient Enzyme	Clinical Features
Type VI	Lysyl hydroxylase	Hyperextensibility of the skin and joints, poor wound healing
Type VII	Procollagen aminopeptidase	Dislocation of the major joints, skin hyperextensibility
Type V	Lysyl oxidase	Short stature, joint dislocations, cardiac involvement

reducing end of the glycosaminoglycan chain. A genetic deficiency in any of these degradative enzymes will cause the corresponding substrate (a partially degraded glycosaminoglycan) to accumulate in the tissues and eventually appear in the urine. The diseases caused by this type of deficiency are called the *mucopolysaccharidoses*. Clinical observations usually include enlarged liver and spleen, corneal clouding, respiratory infections, gradual loss of neurologic function, and retardation of growth. The symptoms and their severity differ widely both between specific mucopolysaccharidoses and between individuals with the same biochemical abnormality.

Ehlers-Danlos Syndromes

The Ehlers-Danlos syndromes are several diseases caused by the failure to form or stabilize collagen fibrils as the result of a defect in the posttranslational modification of collagen. Table 22-1 lists some of the enzyme deficiencies that cause specific Ehlers-Danlos syndromes. These diseases are most readily characterized by joint hypermobility and by skin that is hyperextensile, fragile, and easily bruised. The precise symptoms and their severity differ for different biochemical defects and also between patients.

SUGGESTED READING

1. Bornstein P, Traube W: The chemistry and biology of collagen. p. 412. In Neurath H, Hill RL (eds): The Proteins. Vol 4. Academic Press, Orlando, FL, 1979
2. Roden L: Structure and Metabolism of Connective Tissue Proteoglycans. p. 267. In Lennarz WJ (ed): The Biochemistry of Glycoproteins and Proteoglycans. Plenum Press, New York, 1980
3. Watts RWE, Gibbs DA: Lysomal Storage Diseases: Biochemical and Clinical Aspects. Taylor Francis, London, 1986

STUDY QUESTIONS

Directions: Answer the following questions using the key outlined below:
- (A) if 1, 2, and 3 are correct
- (B) if 1 and 3 are correct
- (C) if 2 and 4 are correct
- (D) if only 4 is correct
- (E) if all four are correct

1. Collagen polypeptide chains contain high quantities of
   1. glycine
   2. hydroxylysine
   3. proline
   4. hydroxyglycine

2. The hydroxylated residues present in collagen
   1. contribute to the stabilization of the triple helix
   2. form mostly intrachain hydrogen bonds
   3. provide sites for glycosylation
   4. aid in the cleavage of the signal peptide from preprocollagen

3. Substances involved in the synthesis of proteoglycans include
   1. UDP-monosaccharides
   2. glycosyltransferases
   3. PAPS
   4. UDP-SO_4^{2-}

4. Glycosaminoglycan chains
   1. are usually anionic
   2. consist of disaccharide repeat units
   3. always contain glucosamine or galactosamine
   4. are uncharged at neutral pH

Directions: For each of the following multiple choice questions, choose the most appropriate answer.

5. Which substance listed below is NOT a type of glycosaminoglycan chain?
   A. Hyaluronic acid
   B. L-Iduronic acid
   C. Dermatan sulfate
   D. Keratan sulfate

6. Which of the following is not a connective tissue protein?
   A. Collagen
   B. Elastin
   C. Myosin
   D. Fibronectin

7. All of the following statements concerning collagen are true EXCEPT that
   A. procollagen has a triple helix structure.
   B. collagen molecules associate into fibrils.
   C. collagen fibrils are made stable by interchain covalent cross-links
   D. procollagen molecules associate in fibrils to form mature collagen.

23

IMMUNOGLOBULINS

Antibody Production/The Antigen-Antibody Reaction/
Antibody Structure/The Mechanism of Antibody Diversity

The student shoulde be able to:

Learning Objectives

1. Define the terms *antibody, antigen, hapten, antigenic determinant,* and *antibody binding site.*
2. Describe the precipitin reaction.
3. Explain how one antigen may stimulate the synthesis of several different antibody populations.
4. Diagram and describe the organization and domain structure of the IgG antibody molecule.
5. List the different antibody classes.
6. Describe the essential features of the mechanism of antibody diversity.

Perspective: Immunoglobulins

Antibodies (also called *immunoglobulins*) are a class of glycoproteins that constitute a major element of the vertebrate immune system. Antibodies protect the organism against infective agents by recognizing and binding to foreign substances. Bound antibody molecules trigger the destruction of foreign agents by two mechanisms: they stimulate enzyme cascades (the complement pathways) that result in the lysis of foreign cells, and they signal phagocytic cells to ingest and destroy invading substances.

ANTIGENS AND ANTIBODY PRODUCTION

First, it is necessary to define *antigen* and *hapten*:

1. *Antigen.* An antigen is any foreign substance that can elicit in a vertebrate the production of antibody reacting specifically with that substance. The foreign substance may be a protein, polysaccharide, nucleic acid, bacterial cell coat, etc.
2. *Hapten.* Antibodies are not formed in response to most small molecules. However, antibodies specific for the small molecule can be formed if

the small molecule is bound to a carrier macromolecule. Small molecules that are made antigenic in this manner are called haptens.

The Antibody-Antigen Reaction

1. Antibodies bind antigen only at specific sites called *antigenic determinants*.
2. The antigen-antibody reaction is usually noncovalent (although occasionally a reactive group on the antigen may form a covalent bond with a group on the antibody).

The region on an antigen molecule that is recognized and bound by the antibody is the *antigenic determinant*. A hapten bound to a carrier functions as a *haptenic determinant*. The antibody recognizes the antigenic determinant via complementary chemical and/or structural properties. For example, a determinant may consist of a segment of amino acids, or alternatively, of a cluster of amino acids brought together by polypeptide chain folding. The antibody has *antigen binding sites* that are complementary in conformation and hydrophobicity/hydrophilicity to the antigenic determinant. This complementarity is the basis for antigen-antibody binding.

A single antigen particle may stimulate the production of several populations of antibodies, as follows:

1. One antigen may contain several different antigenic determinant sites. (It may also carry multiple copies of a given determinant.)
2. Each different antigenic determinant stimulates the production of a specific antibody population that recognizes and binds that determinant. Note that the binding sites of a given antibody may have limited reactivity with a range of structurally similar determinants; thus, some antibodies show cross-reactivity.

Each antibody has two antigen-binding sites, and can therefore bind to two antigenic determinants. Because most natural antigens have more than one surface determinant site, the antibody-antigen reaction results in the agglutination of the antigen (i.e., in the formation of an antigen-antibody network) (Fig. 23-1). Because these large agglutinations readily precipitate out of solution, the antigen-antibody reaction is also called the *precipitin reaction*.

Antibody Proliferation

How does an antigen introduced into the blood induce the production of large amounts of specific circulating antibody? The body contains lymphocytes called *B cells*. Each B cell carries a unique surface-associated antibody. When this antibody recognizes and binds to an antigen, the B cell is stimulated to proliferate and transform into *plasma cells*. The resulting population of plasma cells synthesizes and secretes circulating antibody with the same antigen specificity as the surface antibody of the parent B cell. Each antigenic determinant stimulates the transformation of a different B cell population and thus elicits the production of a different antibody.

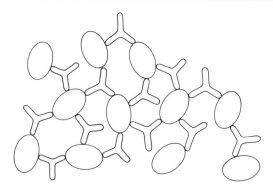

Fig. 23-1. Formation of antigen-antibody networks. Each antibody molecule is able to bind two antigenic determinants.

ANTIBODY STRUCTURE

The five classes of antibody all share a common structure consisting of two identical *heavy chains* (MW 50,000 to 70,000) and two identical *light chains* (MW ~23,000). These chains are joined by interchain disulfide bridges to form the structure shown in Figure 23-2. Each light chain is folded into two domains and each heavy chain into four domains. Each of these domains has a similar tertiary structure and is stabilized by an intrachain disulfide bond.

Analysis of the primary structure of antibodies elicited by different antigens shows that both of the chains are divided into distinct regions of primary structure. The amino-terminal regions of both the light and heavy chains have a highly variable amino acid sequence, whereas the other regions have a primary structure that is quite constant. Each light chain consists of the N-terminal variable region (the V_L region) and one C-terminal constant region (the C_L region). These regions fold to form the globular V_L and C_L domains of the light chain. Each heavy chain consists of the N-terminal variable region (V_H) and three constant regions (C_H1, C_H2, and C_H3). These regions fold to form the V_H, C_H1, C_H2, and C_H3 domains of the heavy chain. The C_H2 and C_H3 domains of one heavy chain associate with the same domains on the other chain to form the stem of the antibody structure. The branches of the structure are formed by the association of the C_H1 and V_H domains of each heavy chain with the C_L and V_L domains, respectively, of a light chain. The two variable domains (V_L and V_H) associate to form a unit that contains the antigen binding site. The constant domains mediate activation of the complement pathways.

Specific antibodies differ in the amino acid sequences of their variable regions. Humans are capable of synthesizing an essentially limitless array of antibodies with different variable regions. Therefore, any foreign substance that enters the body is likely eventually to encounter an antibody that can bind it.

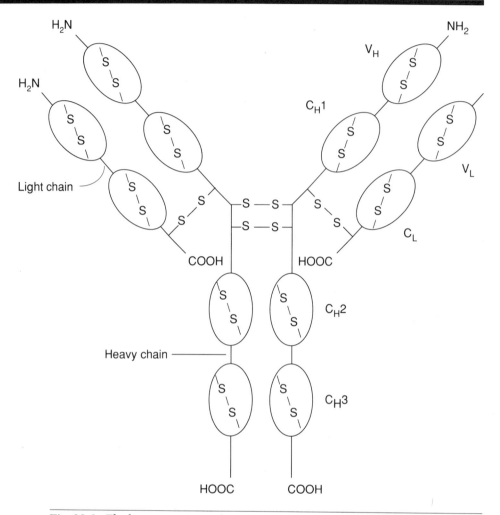

Fig. 23-2. The basic structure of antibody molecules, showing the domain structure of the light and heavy chains.

ANTIBODY CLASSIFICATION

There are five classes of antibodies: IgG, IgA, IgM, IgD, and IgE. Each class has a distinct type of heavy chain (the γ, α, μ, δ, and ε chains, respectively). They all share the same structural motif of two heavy chains and two light chains. However, IgM is a pentamer of five structural units joined by disulfide bridges (Fig. 23-3), and IgA may occur as a dimer or trimer.

Antibody Structure: Key Points

1. The basic structure of antibody molecules consists of two heavy and two light chains joined by disulfide bonds to form a forked structure.
2. Each light chain consists of one constant region (C_L) and one variable

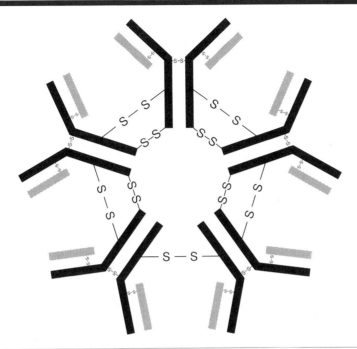

Fig. 23-3. The structure of IgM. (Modified from Zubay G: Biochemistry. 3rd Ed. Macmillan, New York, 1988, with permission.)

region (V_L), which correspond to the C_L and V_L globular domains of the light chain. Each heavy chain consists of three constant regions (C_H1, C_H2, and C_H3) and one variable region (V_H), which correspond respectively to the C_H1, C_H2, C_H3, and V_H globular domains of the heavy chain.
3. The C_H2 and C_H3 domains of the two heavy chains associate to form the stem of the antibody structure. The branches are formed by the association of the V_H and C_H1 domains of the heavy chains with the V_L and C_L domains, respectively, of the light chains.
4. The two variable domains of each arm interact to form the antigen binding site.
5. The five classes of antibody (IgG, IgA, IgM, IgD, and IgE) each have a characteristic heavy chain. They all share a common structural motif, but some are polymers, and they differ in the response they elicit.

MECHANISM OF ANTIBODY DIVERSITY

Vertebrates must be capable of fabricating an essentially endless array of antibody molecules. The diversity of these antibodies is due primarily to diversity in the primary structure of the variable regions. Vertebrates do not have a different gene for every variable-region sequence. Instead, the varible-region genes are divided into several sections which are present in multiple copies that differ in base sequence. These sections recombine and splice during cell maturation to produce a variety of mature genes.

The Light Chain Genes

The variable region of the light chain is coded for by two distinct gene elements, called the V and J elements. The constant region is coded for by a third gene, the C gene (Fig. 23-4A). During cell maturation, the V and J segments combine to form the functional V-J gene for the light chain variable region. Antibody diversity arises because the genome contains up to several hundred V elements and as many as five J ele-

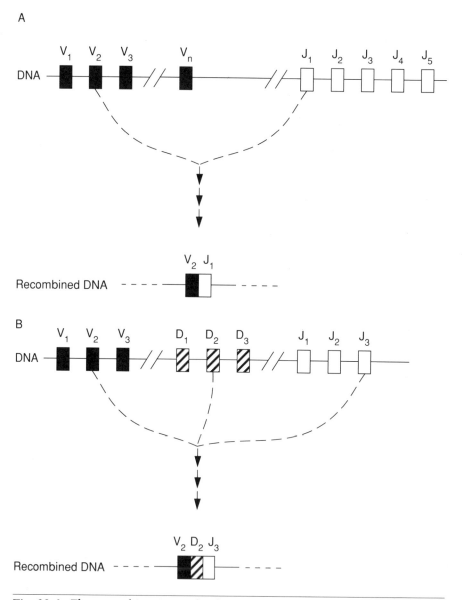

Fig. 23-4. The recombination and splicing events that result in the formation of specific light chain (**A**) and heavy chain (**B**) variable regions genes.

ments. Different combinations of V and J elements give rise to different light chain variable region genes. The completed variable region gene is subsequently joined to the appropriate constant region gene to generate a complete light chain gene.

The Heavy Chain Genes

The heavy chain varible region is coded for by three different gene elements, called the V, D, and J elements (Fig. 23-4B), which combine during cell maturation to form a single functional heavy chain variable region gene. Each of these elements is present in multiple copies of differing base sequence (several hundred V elements, about 20 D elements, and about 4 J elements). Different combinations of V, D, and J elements gives rise to the diversity of heavy chain variable region genes. As with the light chain genes, the completed variable region gene joins the appropriate constant region genes to generate a complete heavy chain gene.

SUGGESTED READING

1. Kabat EA: Structural Concepts in Immunology and Immunochemistry. 2nd Ed. Holt, Reinhart, & Winston, New York, 1976
2. Weissman IL, Hood LE, Wood WB: Essential Concepts in Immunology. Benjamin-Cummings, Menlo Park, CA, 1978
3. Honjo T: Immunoglobulin genes. Annu Rev Immunol 1:499, 1983

STUDY QUESTIONS

Items 1 through 4. Consider the following terms.
  A. Antigen
  B. Hapten
  C. Antibody
  D. Antigenic determinant
  E. Antigen binding site

Match each word or term with the most appropriate definition below.

1. The structural region of a foreign substance that is recognized by antibodies.

2. A small molecule that induces antibody formation only when attached to a larger carrier molecule.

3. A substance that stimulates antibody production.

4. A structural region of antibody molecules.

Directions: For each of the following multiple choice questions, choose the most appropriate answer.

5. All of the following statements concerning the antigen-antibody reaction are true EXCEPT that
  A. the antigen-antibody reaction normally involves noncovalent bonds.
  B. antigenic determinants are recognized by antibody binding sites.
  C. antibodies bind randomly to the surface of antigens, forming large precipitin networks.
  D. antibodies usually bind poorly to antigens other than the one that elicited their synthesis.

6. All of the following statements concerning antigens are true EXCEPT that
  A. there is only one antigenic determinant per antigen.
  B. one antigen may elicit the production of more than one antibody population.
  C. free small molecules are usually very poor antigens.
  D. haptens are good antigens.

Directions: Answer the following questions using the key outlined below:
- **(A)** if 1, 2, and 3 are correct
- **(B)** if 1 and 3 are correct
- **(C)** if 2 and 4 are correct
- **(D)** if only 4 is correct
- **(E)** if all four are correct

7. The fork-like structure of antibody molecules
   1. contains two heavy chains and two light chains.
   2. consists of eight constant domains and four variable domains.
   3. contains carbohydrate moieties.
   4. contains one antigen binding site per molecule.

8. Which of the following statements is (are) true?
   1. Light chain variable regions are coded for by V, D, and J genes.
   2. Genes that code for antibody variable regions are present in multiple copies of slightly different sequences.
   3. Heavy chain variable regions are coded for by V and J genes.
   4. Antibody diversity requires different variable region gene combinations, which ultimately result in different variable regions.

24

BIOCHEMISTRY OF NERVE TISSUE

Structure of the Neuron/Metabolism of Nerve Tissue/
The Action Potential/Synaptic Transmission

Learning Objectives

The student should be able to:

1. Illustrate the structure of the neuron.
2. Know the physiologic roles of different neuronal structures.
3. Describe the action potential.
4. Describe the formation of the myelin sheaths and their effect on the conduction of nerve impulses.
5. Distinguish conduction and synaptic transmission of nerve signals.

Perspective: Biochemistry of Nerve Tissue

Nerve tissue consists of two major cell types: *glial cells* and *neurons*. Glial cells are components of the connective tissue that supports neurons and holds them together. Neurons are involved in responding to stimuli and in transmission of signals from one cell to another. This chapter presents the characteristic biochemical properties of nerve tissue and the mechanisms of neuronal function.

STRUCTURE OF THE NEURON

Figure 24-1 shows the morphology of a typical motor neuron. The component parts may be differently proportioned and arranged in other types of neurons, but they function in much the same way. The anatomy of this typical neuron is as follows.

1. *Cell body.* The cell body of the neuron contains the nucleus and other organelles. Synthesis of neurotransmitters and other biochemical components occurs in the cell body.

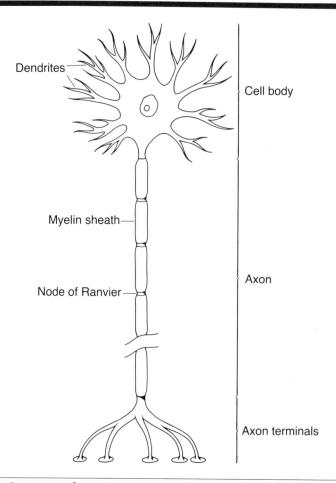

Fig. 24-1. Structure of a motor neuron with a myelinated axon (not all nerve axons are myelinated).

2. *Dendrites.* The cell body bears numerous projections called dendrites that receive signals from other neurons or from sensory cells.
3. *Axon.* Extending from the cell body is a long, tube-like structure, the axon, that conducts nerve impulses away from the cell body. The axons of some neurons are covered with periodic myelin sheaths. Human axons are up to a meter long.
4. *Nerve endings.* The axon ends in a series of branches called axon terminals or nerve endings. At the nerve endings the impulse is transmitted either to another neuron or to an effector cell (such as a muscle fiber or a gland cell). The two cells do not actually contact each other but are separated by a narrow space called the *synaptic cleft*, which is usually about 20 nm wide.

METABOLISM OF NERVE TISSUE

This section describes the characteristic features of the metabolism of nerve tissue.

Carbohydrate

The brain prefers to use glucose as its sole energy source, although during starvation it also metabolizes ketone bodies. Most of the glucose taken up by the brain is oxidized via glycolysis coupled to the TCA cycle. The brain stores only a very small amount of glucose as glycogen, and is almost entirely dependent on the blood for fuel. A drop in blood glucose impairs brain function. An inadequate oxygen supply also impairs brain function, and prolonged oxygen deprivation may cause coma and nerve tissue damage. Nerve cells also oxidize some glucose by the pentose phosphate pathway to generate NADPH, which presumably is used mostly for the synthesis of fatty acids and cholesterol.

Amino Acids and Proteins

The major free amino acids in nerve tissue are aspartate, N-acetylaspartate, glutamine, and γ-aminobutyrate (formed by the α-decarboxylation of glutamate). Glutamate and aspartate are important links between the amino acid pool and the TCA cycle. γ-Aminobutyrate also contributes to the TCA cycle via the γ-aminobutyrate shunt (Fig. 24-2). Other amino acids in nerve cells are converted to neurotransmitters or to substances believed to be neurotransmitters. Glutamate, aspartate, and γ-aminobutyrate have all been implicated as neurotransmitters. Tyrosine is converted to the catecholamines (see Chapters 10 and 15). Tryptophan is the precursor for the neurotransmitter serotonin. His-

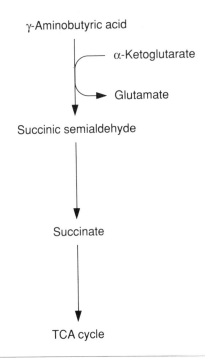

Fig. 24-2. The γ-aminobutyrate shunt.

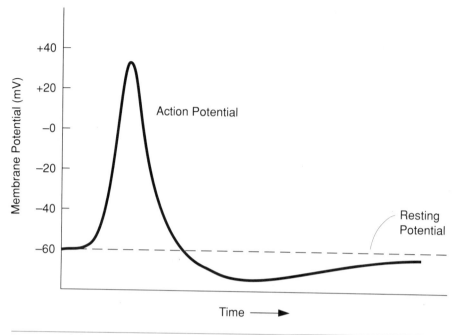

Fig. 24-3. (A) The changes in membrane potential during and after the passage of an action potential. (*Figure continues*.)

tidine is converted to histamine, which displays neurohumoral effects and may also be a transmitter. Like other cells, neurons use amino acids in protein synthesis.

Calmodulin is present in brain cells, and is believed to be involved in the Ca^{2+}-dependent release of certain neurotransmitters from presynaptic vesicles. Dendrites and axons contain prominent cytoskeletal structures called *neurotubules* and *neurofilaments*, which are composed mainly of the proteins tubulin and actin, respectively. Neurotubules are important in the transport of materials within the neuron—for example, the transport of neurotransmitters from the cell body to the synapse.

Four proteins of unknown function occur in strikingly high concentrations in the brain: S-100 protein, an enolase called 14-2-3 protein, myelin basic protein, and myelin proteolipid. S-100 protein and the enolase are acidic proteins. Both the myelin basic protein and the proteolipid are associated with the myelin membranes.

Nucleotides

Brain cells are incapable of the de novo synthesis of pyrimidines because they lack carbamoyl phosphate synthetase II, the initial enzyme of the pathway. The salvage pathway of pyrimidine synthesis is very active. Purine nucleotides are synthesized by both the de novo and the salvage pathways.

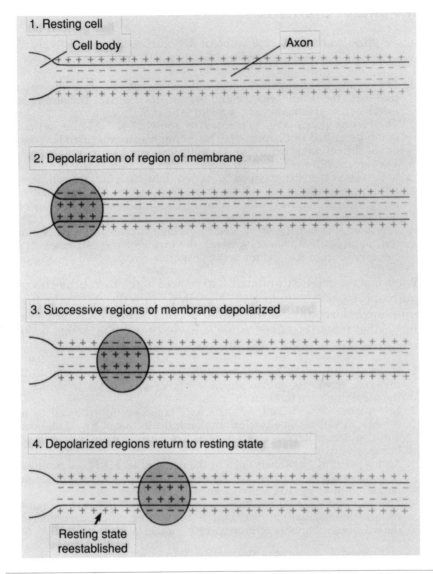

Fig. 24-3. (*Continued*). **(B)** The propagation of the action potential (impulse) down the axon. (Fig. B from Darnell JE, Lodish H, Baltimore D: Molecular Cell Biology. Scientific American Books, New York, 1986, with permission.)

Lipids

The lipids found in nerve tissues are primarily membrane components. They include cholesterol, phosphoglycerides, plasmalogens, and sphingolipids. Sphingolipids are unusually abundant and include sphingomyelin, gangliosides, and cerebrosides. The myelin sheath and various regions of the neuron membrane have distinctive lipid compositions. The lipids of nerve cell membranes turn over very slowly.

THE ACTION POTENTIAL

A nerve impulse consists of a wave of depolarization that travels along the nerve cell membrane from the dendrites to the axons.

1. The resting (unstimulated) nerve cell membrane is electrically polarized, with the inside negative and the outside positive. The *resting potential* across the membrane is about -60 mV.
2. The resting potential is generated by the Na^+-K^+ ATPase antiport pump described in Chapter 21. This pump transports three Na^+ ions out of the cell and two K^+ ions into the cell per ATP hydrolyzed.
3. The cell membrane is impermeable to Na^+ but somewhat permeable to K^+. Thus, the Na^+ pumped out cannot flow back into the cell. Some K^+ flows out of the cell. However, because of the electrical gradient across the membrane (inside negative), the outflow of K^+ stops before it achieves osmotic equilibrium.
4. Therefore, the outside of the cell has a high concentration of Na^+ as well as a positive charge, whereas the inside of the cell has a high concentration of K^+ and an overall negative charge.

When a nerve cell is stimulated, a local area of the membrane becomes slightly more permeable to Na^+. Na^+ flows into the cell, reducing the membrane polarization. If this initial depolarization exceeds a specific value, called the *threshold potential*, a nerve impulse or *action potential* is triggered. The action potential occurs as follows (Fig. 24-3):

1. If the eliciting stimulus decreases the resting potential by about 20 mV—i.e., from -60 mV to about -40 mV—special Na^+ *channels* in the membrane are induced to open.
2. Na^+ rushes into the cell. Because Na^+ is pulled into the cell by an osmotic as well as an electrical gradient, the inflow of Na^+ causes not only depolarization but also a brief *hyperpolarization* of about $+30$ mV (inside positive).
3. At this point, K^+ channels open in the membrane. K^+ flows out down its concentration and electrical gradient, returning the membrane polarization to a negative value of about -75 mV.
4. The Na^+ and K^+ channels then close, and the Na^+-K^+ ATPase pump restores the normal ion distribution and resting potential, rendering the membrane capable of propagating another action potential.

Propagation of the Nerve Impulse

An action potential occurring in one part of the membrane stimulates an action potential in an adjacent region of the membrane. The impulse therefore moves down the axon membrane to the nerve endings. However, an action potential is triggered only if the initial stimulus causes a depolarization that exceeds the threshold potential (about 20 mV).

THE MYELIN SHEATH

Myelin sheaths (Fig. 24-4) are formed when a type of glial cell called the *Schwann cell* wraps an extension of its plasma membrane around the axon to form a tight spiral. The Schwann cell spirals are separated

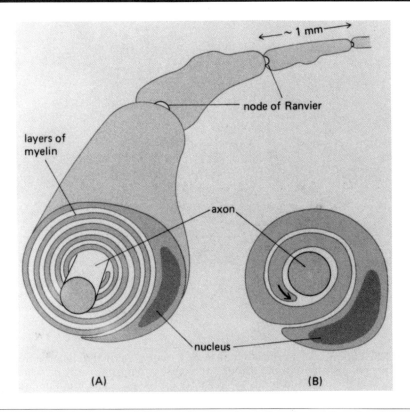

Fig. 24-4. (A) Structure of the myelin sheaths and nodes of Ranvier. **(B)** Formation of a myelin sheath by a Schwann cell. (From Alberts B, Bray D, Lewis J et al: Molecular Biology of the Cell. Garland, New York, 1983, with permission.)

by short unmyelinated axon segments called *nodes of Ranvier*. Because the myelin sheath consists mostly of lipid, it forms an electrical insulator around the axon. The nerve impulse is conducted between the nodes of Ranvier as a fast positive current which is strong enough to trigger an action potential at the next node. The nerve impulse thus hops from node to node, and is slowed only by the time required to form an action potential at each node. Nerve conduction is much faster on myelinated than on unmyelinated fibers.

Perspective: Synaptic Transmission

When the action potential arrives at the end of the axon, it causes *neurotransmitter* molecules to be released into the synaptic cleft (Fig. 24-5). The neurotransmitter molecules diffuse across the synaptic cleft and bind to specific receptor sites on the postsynaptic cell (a neuron or other type of cell), triggering the appropriate physiologic response.

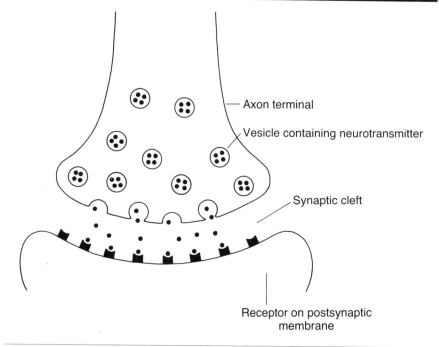

Fig. 24-5. Function of a typical cholinergic synapse. Arrival of a nerve impulse at the presynaptic membrane stimulates release of neurotransmitter, which diffuses across the cleft and binds to receptors on the postsynaptic membrane.

MECHANISM OF SYNAPTIC TRANSMISSION

1. Presynaptic nerve endings are rich in vesicles that contain neurotransmitter and usually also ATP.
2. When the action potential arrives at the nerve ending, the presynaptic membrane becomes more permeable to Ca^{2+}. The influx of Ca^{2+} causes neurotransmitter vesicles to fuse with the presynaptic membrane and to release their contents into the synaptic cleft.
3. The neurotransmitter diffuses across the cleft and binds to specific receptors on the postsynaptic membrane. Binding of the transmitter causes a response in the postsynaptic cell.
4. As transmitter molecules dissociate from the postsynaptic membrane, they are either metabolized in the synaptic cleft, taken back up by the presynaptic membrane, or removed from the synaptic cleft by another mechanism. The effect of the transmitter on the postsynaptic membrane is thus relieved.
5. The synapse is now ready to transmit another nerve impulse.

Synaptic transmission is best understood in motor neurons. In motor neurons the transmitter is acetylcholine and the postsynaptic cell is a muscle fiber. Binding of acetylcholine to the postsynaptic receptors causes Na^+ to diffuse into the postsynaptic cell, triggering a wave of depolarization. This wave of depolarization in turn stimulates muscle contraction (Chapter 19). The enzyme acetylcholinesterase in the synaptic cleft metabolizes acetylcholine by hydrolysis:

$$(CH_3)_2N-CH_2-CH_2-O-\overset{\overset{O}{\|}}{C}-CH_3 \rightarrow$$

Acetylcholine

$$(CH_3)_2N-CH_2-CH_2-OH + CH_3COO^-$$

Choline

This reaction promotes dissociation of acetylcholine from postsynaptic receptors. The synapse and muscle fiber return to their resting state and are ready to transmit the next nerve impulse.

SUGGESTED READING

1. Katz B: Nerve, Muscle and Synapse. McGraw-Hill, New York, 1966
2. Bennett MVL (ed): Synaptic Transmission and Neuronal Interaction. Raven Press, New York, 1974
3. Siegel GJ, Albers RW, Agranoff BW, Katzman R (eds): Basic Neurochemistry. 3rd Ed. Little, Brown, Boston, 1981
4. Kuffler SW, Nicholls JG: From Neuron to Brain. Sinauer, Suderland, MA 1976
5. McIlwain H, Blanchard HJ: Biochemistry and the Nervous System. 4th ed. Churchill Livingstone, Edinburgh, 1973
6. Hille B: Ion channels of excitable membranes. Sinauer, Suderland, MA, 1984

STUDY QUESTIONS

Directions: Answer the following questions using the key outlined below:
- **(A)** if 1, 2, and 3 are correct
- **(B)** if 1 and 3 are correct
- **(C)** if 2 and 4 are correct
- **(D)** if only 4 is correct
- **(E)** if all four are correct

1. All of the following statements are true EXCEPT that:
   1. nerve cells that conduct impulse signals are called axons.
   2. myelin is derived from Schwann cells.
   3. dendrites are structural components of glial cells.
   4. axon terminals are involved in signal transmission.

Items 2 through 5. Consider the following terms
   A. Nodes of Ranvier
   B. Synaptic cleft
   C. Neurotransmitter
   D. Dendrite

Match each term with the most appropriate definition below.

2. Regions of the axon between myelin sheaths

3. The space between the axon terminal (presynaptic membrane) and the target cell (postsynaptic membrane)

4. Substance released into the synaptic cleft from the presynaptic membrane which transfers the signal between cells.

5. Structural region of the neuron which receives signals from other neurons

Directions: For the following multiple choice questions, choose the most appropriate answer.

6. All of the following statements concerning nerve cell metabolism are true EXCEPT that
    A. under normal conditions, glucose is the sole energy source of brain cells.
    B. the Ca^{2+}-dependent release of neurotransmitters from presynaptic vesicles is believed to be independent of calmodulin.
    C. brain cells are incapable of de novo pyrimidine synthesis.
    D. dendrites and axons contain structural elements called neurotubules and neurofilaments which are primarily composed of tubulin and actin, respectively.

7. Which of the following statements about the resting (unstimulated) axon is NOT true?
    A. the axonal membrane is electrically polarized.
    B. Na^+ is more concentrated outside the neuron.
    C. The axonal membrane is permeable to K^+.
    D. The potential difference across the axonal membrane is $+30$ mV.

8. All of the following statements about the stimulated axon are true EXCEPT that
    A. the membrane becomes locally depolarized.
    B. the interior of the axon becomes negatively charged.
    C. the potential across the membrane is lowered.
    D. the membrane becomes more permeable to Na^+.

9. All of the following statements about the action potential are true EXCEPT that
    A. during the action potential, the membrane potential may go from -60 mV to $+30$ mV.
    B. in order for the action potential to occur, the membrane potential must first decrease by about 20 mV.
    C. the occurrence of an action potential in one region of the axon will stimulate an action potential in an adjacent region.
    D. the overshoot of depolarization associated with the action potential is due to the large influx of K^+ ions into the axon.

25

BIOCHEMISTRY OF VIRUSES

Types of Viruses/Mechanisms of Viral Multiplication

The student should be able to:

1. List the classes of viruses and the essential features of their biochemistry in host cells.
2. Distinguish the multiplication cycles of (+) and (−) RNA viruses.
3. Distinguish the properties of SV40 infection in permissive and nonpermissive host cells.
4. Distinguish two mechanisms for the release of viral progeny from infected host cells.
5. Describe the major events of retroviral multiplication.

Learning Objectives

Perspective: Biochemistry of Viruses

A virus consists at the minimum of a nucleic acid genome surrounded by a protective shell that is constructed partly or entirely of protein. Some viruses infect bacteria whereas others infect eukaryotic cells. Viruses are usually specific as to the type of host cell they will infect. A wide variety of animal, human, and plant diseases, some severe, are caused by viruses. The type of effect that a virus has on a particular host cell is determined by several factors, including

1. The ability of the virus to infect the host. If the viral genome cannot enter the host, nothing will happen.
2. The nature of the postinfective events in the host cell.

Viruses are believed to recognize specific receptors on the host cell, and cannot infect cells that lack these receptors. Either the entire virus, a subparticle containing the viral genome, or the naked genome enters the cell. Once the infective particle has entered the host cell, viral biochemistry begins and the impact on the host cell becomes apparent. This chapter reviews the biochemistry of the major types of virus and of their effects on host cell metabolism and integrity.

TYPES OF VIRUS

Viruses are divided into those that contain DNA (DNA viruses) and those that contain RNA (RNA viruses). These two classes are subdivided according to whether the viral nucleic acid is single-stranded or double-stranded. The genome of single-stranded RNA viruses is further classified according to whether the RNA strand in the virus can serve as mRNA. In general, an RNA strand that can serve as mRNA is called (+), whereas its complement is called (−). Therefore, if the RNA strand from a virus can serve as mRNA, it is called (+) RNA, whereas if strands complementary to viral genomic RNA must be synthesized after infection and used as mRNA, the virus contains (−) RNA. The classification of viruses may be summarized as follows:

1. DNA viruses
   a. Double-stranded
   b. Single-stranded
2. RNA viruses
   a. Double-stranded
   b. Single-stranded
      i. (+)RNA
      ii. (−)RNA

MECHANISMS OF VIRAL MULTIPLICATION

After entering a host cell, the virus uses appropriate elements of the host cell's transcription and translation machinery to synthesize new viral particles. In most (though not all) cases, the expression of the host genome is seriously altered or turned off, resulting in disease. Different viruses multiply in their hosts by different mechanisms. The following features are general:

1. The viral genome is expressed, leading to the synthesis of viral proteins.
2. The viral genome is also replicated.
3. New virus particles (virions) are assembled, each including a copy of the viral genome as well as viral structural proteins.
4. The new virions are released from the host.

Some representative examples of viral infection and replication are explored in the following sections.

DNA VIRUSES

Double-Stranded DNA viruses

Two well-studied examples of double-stranded DNA viruses are bacteriophage T4, which infects *E. coli*, and simian virus 40 (SV40), which infects rhesus monkey cells.

Bacteriophage T4

As shown in Figure 25-1, bacteriophage T4 is a relatively complicated virus, consisting of a protein head that contains the double-stranded

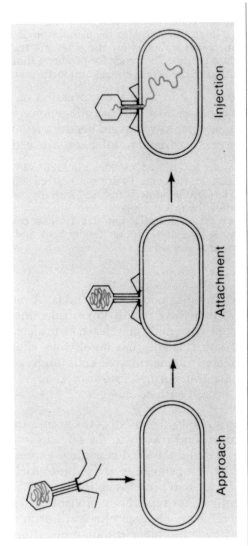

Fig. 25-1. Infection of a host cell by bacteriophage T4. (From Friefelder D: Molecular Biology. 2nd Ed. Jones and Bartlett, Boston, 1987, with permission.)

DNA genome, a tail, and fibers attached to the tail. Infection by T4 occurs as follows.

1. T4 recognizes and binds to *E. coli* by its tail fibers. The hollow tail core then penetrates the cell wall and injects the genome into the host cell.
2. The viral genome is transcribed and translated by the host cell machinery. Different regions of the T4 genome are expressed at different times. One group, the *early genes*, is expressed early and codes for enzymes needed to replicate the viral DNA. A second group, the *late genes*, is expressed later and codes for the proteins necessary to assemble the virion. Some genes, called *quasi-late genes*, are transcribed in both the early and late groups. They code for products that are needed for both the replication of the viral genome and virus assembly.

Within minutes of infection, the expression of the viral early genes begins and the expression of the host cell genome is terminated by viral gene products. Late gene expression begins a few minutes after the initiation of early gene expression. Infection proceeds as follows:

3. New virions are assembled. As shown in Figure 25-2, the head, tail, and tail fibers of bacteriophage T4 are synthesized separately and then assembled. The viral genome is inserted into the completed head before the head joins the tail.
4. After the new virions are all assembled, the host cell is lysed in a process that involves a viral gene product (endolysin), and about 1,000 to 2,000 new virions are released.

SV40

SV40 contains a circular double-stranded DNA molecule. Host cells respond in one of two ways to this virus. Infection of a *permissive* host cell results in cell lysis and the release of projeny virions. Infection of a *nonpermissive* host usually has no obvious effect. However, a small fraction of infected nonpermissive cells undergo *oncogenic transformation*—they are transformed into tumorous cells displaying unregulated growth and cell surface changes.

Infection of a Permissive Host. SV40 is taken into the host cell by a pinocytotic mechanism and travels to the nucleus, where the protein coat of the virus is shed and the viral genome is expressed. As in the case of bactcriophagc T4, the genome is divided into early and late genes. The early genes code for two proteins called large T and small t. The large T protein is required for replication of viral DNA, and both the large T and small t proteins are necessary for transformation in nonpermissive hosts. The early gene products initiate replication of the viral genome, and after an interval the late genes, which code for the three protein components of the viral coat, are expressed. New virions are synthesized as coat proteins become available, and the accumulation of new virus in the host cell eventually results in cell lysis.

Infection of a Nonpermissive Host. The behavior of SV40 in a nonpermissive host is similar to its behavior in a permissive host up through the expression of the large T and small t genes. However, all or part of the

Biochemistry of Viruses 571

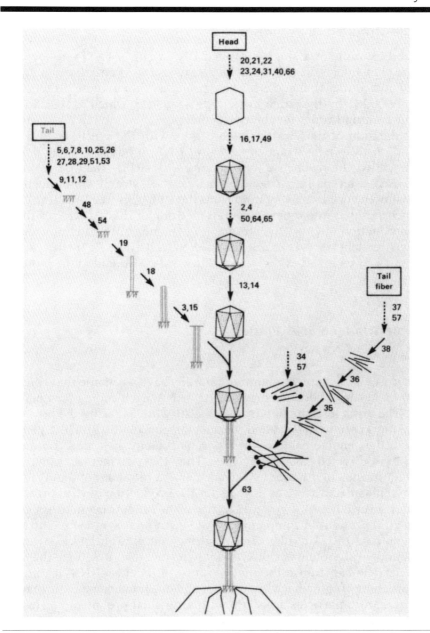

Fig. 25-2. Assembly of bacteriophage T4. The numbers refer to the specific genes required at each step. (From Wood WB, Bishop RJ. In Fox CF, Robinson WS (eds): Virus Research. Academic Press, Orlando, FL, 1973, with permission.)

viral DNA is integrated into the cellular genome instead of being replicated. New virions are not made. The early genes continue to be expressed along with the host genes, and the viral genome is replicated and passed on to daughter cells with the host genome. A small percentage of nonpermissive cells with an integrated viral genome undergo oncogenic transformation.

Single-Stranded DNA Viruses

A good example of a single-stranded DNA virus is φX174, a bacteriophage that infects *E. coli*. It has a single-stranded circular genome of 5,370 nucleotides that codes for nine proteins. Two of these proteins are largely coded by the same nucleotide sequence, which is read in different reading frames for the two proteins.

Replication of the viral genome begins with the synthesis of a ($-$) strand on the single-stranded ($+$) viral DNA, producing a ($\pm$) duplex circle called the *replicative form intermediate* (RFI). The ($-$) strand is then used as a template to synthesize a new ($+$) strand, simultaneously releasing the original ($+$) strand from the RFI. This free ($+$) strand can be incorporated into a new virion. Repetition of this process results in the production of ($+$) DNA strands and the multiplication of progeny virions, which are eventually released by cell lysis. As in the case of phage T4, lysis involves the action of a viral-coded endolysin protein.

RNA VIRUSES

Double-Stranded RNA Viruses

Reovirus

Reoviruses, which infect mammalian cells, have a genome that consists of 10 molecules of double-stranded ($\pm$) RNA, each coding for one protein. The virus loses its outer protein coat while entering the host cell, releasing a core particle called the *subviral particle* (SVP). The SVP consists of the genome enclosed in a porous protein shell, and contains a viral RNA-directed RNA polymerase. This RNA polymerase transcribes the ($-$) strands to produce ($+$) strands which move out of the SVP into the cytoplasm and serve as mRNAs for the production of viral proteins.

After several hours of protein synthesis, the mRNAs begin to associate with viral proteins. A complementary ($-$) strand is synthesized on each to yield a ($\pm$) RNA duplex. The resulting complex of double-stranded RNA and viral protein is called an *immature viral particle*. Like the SVP, this particle can transcribe its ($-$) strand to yield mRNA, leading to the production of further viral protein and duplex genomes. Ultimately, the immature particles associate with viral shell proteins to form mature virions.

Single-Stranded (+) RNA Viruses

Examples of single-stranded ($+$) RNA viruses are the RNA phages Qβ, R17, F2, and MS2, which infect *E. coli*, and the picornaviruses, which infect animal cells.

E. coli RNA Phages

As soon as the viral genome of an *E. coli* RNA phage enters the bacterial cell, it functions as mRNA and directs the synthesis of viral proteins.

These proteins include a replicase that replicates the genome, a coat protein, a maturation protein required for the assembly of new virions, and a protein that lyses the host cell membrane to release the progeny virions. The coding sequences for some of the viral proteins overlap and are translated using different reading frames. The viral (+) RNA strand serves as a template for the synthesis of a complementary (−) strand, which serves as a template for the synthesis of many (+) strands to be used in the assembly of new virions. Several thousand virions emerge from each infected host bacterium.

Poliovirus

Poliovirus, a picornavirus, is the causative agent of polio. The (+) RNA genome of the poliovirus functions as an mRNA in the host cell, directing the synthesis of viral proteins. Immediately after viral proteins begin to be produced, the synthesis of host protein is inhibited by a viral protein that inactivates an initiation factor needed for the synthesis of host protein. The viral genome is replicated as follows. A complementary strand is synthesized on the (+) RNA strand, yielding a double-stranded *replicative form* (RF). As shown in Figure 25-3, the (−) strand then directs the simultaneous synthesis of numerous daughter (+) strands, forming a structure called the *replicative intermediate* (RI). The daughter (+) strands can either serve as mRNA or be packaged into nascent virions. The entire viral genome is translated as a single polypeptide chain called the *precursor polyprotein*, which is cleaved to yield individual proteins.

Single-Stranded (−) RNA Viruses

Viruses that contain single-stranded (−) RNA include members of the rhabdovirus and paramyxovirus groups. One of the better understood viruses of this type is vesicular stomatitis virus (VSV), a rhabdovirus that infects a variety of animals and humans. Humans infected with the virus usually have been exposed to an infected domestic animal.

Vesicular Stomatitis Virus

The VSV virion contains an inner particle, the *nucleocapsid*, which consists of a (−) RNA strand of about 11,000 nucleotides surrounded by a lipid and protein coat. The nucleocapsid is sheathed by a lipid bilayer envelope that contains two kinds of protein. The virion enters a host cell by binding to a coated pit in the plasma membrane, followed by invagination. The envelope is shed in this process, releasing the nucleocapsid into the host cytoplasm. Entry into the cytoplasm activates a viral RNA polymerase that transcribes the (−) genome. Two different populations of (+) RNA are produced:

1. Some (+) strands contain the complete genome and serve as templates for the production of new (−) strands to be incorporated into virions.
2. In other cases the (−) genome is transcribed as five (+) mRNAs, one for each viral protein. Each of these mRNAs contains a poly A tail.

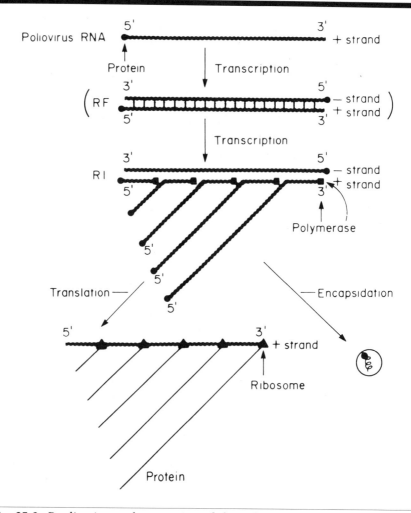

Fig. 25-3. Replication and expression of the poliovirus genome. (From Joklik WK: Virology. 2nd Ed. Appleton-Century-Crofts, Norwalk, CT, 1985, with permission.)

VSV uses host ribosomes and tRNAs to synthesize viral proteins. The two envelope proteins are transported to regions of the host cell membrane where they replace host proteins and form patches of viral protein. The other viral proteins join with a viral (−) genome to form nucleocapsids. As shown in Figure 25-4, newly formed nucleocapsids migrate to the areas of the host cell membrane that contain viral protein, and mature virions are released by a process of budding off.

Retroviruses

Retroviruses are single-stranded (+) RNA viruses. They differ from other single-stranded (+) RNA viruses in that the viral genome is transcribed into a duplex DNA that becomes incorporated into the host

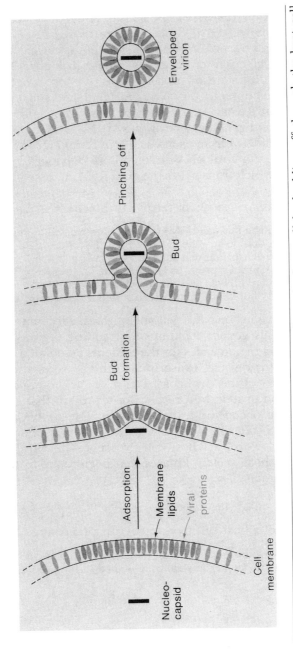

Fig. 25-4. The release of vesicular stomatitis virions from the host cell by budding off through the host cell membrane. (From Friefelder D: Molecular Biology. 2nd Ed. Jones and Bartlett, Boston, 1987, with permission.)

genome. New viral (+) RNA is then synthesized by the host's normal transcription apparatus. All the RNA viruses that have been shown to cause tumors in vertebrate hosts, for example avian sarcoma virus and avian and murine leukemia viruses, are retroviruses. Human immunodeficiency virus (HIV), the causative agent of acquired immune deficiency syndrome (AIDS), is also a retrovirus.

Retrovirus particles are approximately spherical and are surrounded by a lipid-glycoprotein coat. As shown in Figure 25-5, the retrovirus genome is diploid: the virion contains two copies of the (+) RNA strand joined by complementary base pairing over a short region. Low molecular weight RNAs and a variety of tRNAs may also be associated with the retroviral genome. One of these tRNAs functions as a primer in the synthesis of DNA from each viral RNA template. The retrovirus particle also contains a viral RNA-directed DNA polymerase, called *reverse transcriptase*, which transcribes the viral RNA genome into double-stranded DNA.

The peculiar aspects of retrovirus biochemistry are as follows:

1. The single-stranded RNA genome is converted by reverse transcriptase into a double-stranded DNA called the *provirus*.
2. The provirus is integrated into the host genome.
3. The integrated provirus is transcribed into mRNA, which directs viral protein synthesis and the assembly of new viral particles. The cell is usually not killed, but instead produces and secretes new virions.

Figure 25-5 shows the genome of avian sarcoma virus, a typical retrovirus. This genome contains four coding regions: *gag*, *pol*, *env*, and *src*. The *gag* region codes for the proteins associated with the viral core; *pol* codes for reverse transcriptase; and *env* codes for the glycoprotein components of the viral coat. The *src* gene is an oncogene. It codes not for a virion protein but, rather, for a protein that causes the host cell to undergo oncogenic transformation. The *src* oncogene of avian sarcoma virus codes for a protein kinase that phosphorylates tyrosine residues on a variety of host cell proteins. This viral tyrosine kinase is similar to host protein kinases that participate in the transmission of growth signals.

The Retrovirus Infective Cycle

Infection of a host cell by a retrovirus proceeds as follows (Fig. 25-6).

1. The virion binds to a specific receptor on the host cell membrane and enters the cell. The viral coat is shed in the process.

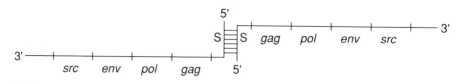

Fig. 25-5. The diploid retroviral genome of avian sarcoma virus, showing the four coding sequences.

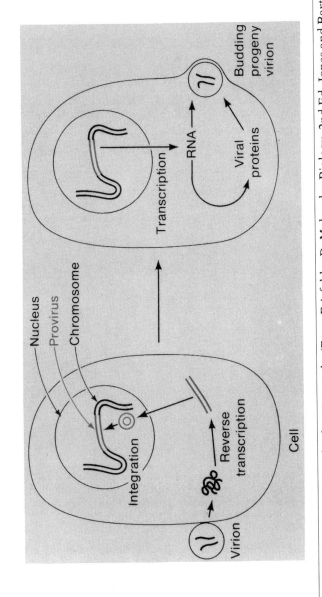

Fig. 25-6. The retroviral multiplication cycle. (From Friefelder D: Molecular Biology. 2nd Ed. Jones and Bartlett, Boston, 1987, with permission.)

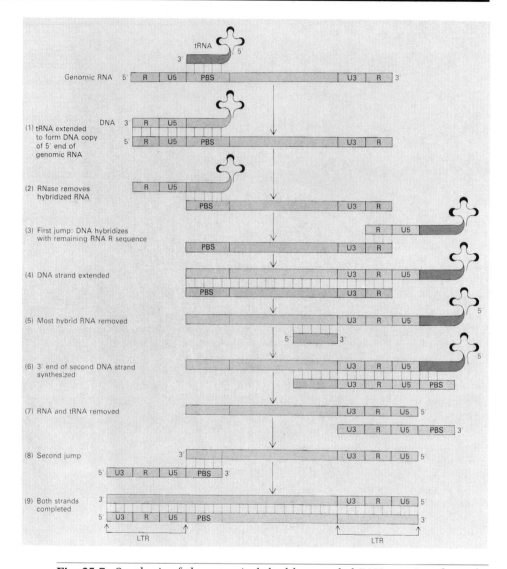

Fig. 25-7. Synthesis of the retroviral double-stranded DNA provirus from the single-stranded RNA genome. (From Darnell JE, Lodish H, Baltimore D: Molecular Cell Biology. Scientific American Books, New York, 1986, with permission.)

2. In the cytoplasm, the (+) RNA genome serves as a template for the synthesis of double-stranded DNA proviruses.
3. The resulting linear double-stranded DNA enters the nucleus, where it becomes circularized. The circular form is incorporated into the host genome via recombination. Multiple copies of the provirus may be inserted into the host genome, and the sites of insertion are apparently random. The integrated provirus has identical segments at each end called *long terminal repeats* (LTRs). Four base pairs of the host genome are also repeated at either end of the LTRs.

4. Once integrated into the host genome, the provirus is transcribed by the host cell's RNA polymerase II. The LTR segments of the provirus contain very strong promoter sequences that enhance transcription of the viral genome. The integrated provirus is also replicated with the host genome.
5. The viral RNA transcript can be used to synthesize mRNAs that subsequently direct the synthesis of viral proteins. Alternatively, the transcript may be used as a genome for incorporation into a new virion. The mRNAs and the viral genomic RNAs both carry 5' caps and 3' poly A tails.
6. The assembly of new virions starts with the association of the RNA genome with core proteins to form the viral core particle. The virus coat is obtained from the plasma membrane during release from the cell via a budding mechanism. Therefore, complete virus particles are found only outside of cells. The host cell usually survives and continues to produce virions.

The synthesis of double-stranded DNA from the (+) RNA genome is quite complicated. It occurs as follows (Fig. 25-7):

1. A host cell tRNA binds by complementary base pairing to a region near the 5' end of the viral genome called the *primer binding site*. This tRNA serves as a primer. Starting at the primer, reverse transcriptase synthesizes a DNA strand until the 5' end of the viral RNA is reached.
2. The RNase activity of reverse transcriptase removes the RNA from the resulting short RNA/DNA hybrid.
3. Because the two ends of the viral genome (called R segments) have identical sequences, the 3' end of the DNA segment is complementary to the 3' end of the genomic RNA. The DNA strand accordingly moves to the 3' end of the RNA strand, where it base pairs with the complementary R sequences of the RNA genome.
4. Reverse transcriptase elongates the DNA strand until the 5' end of the RNA strand is reached.
5. The RNase activity of reverse transcriptase then removes most of the RNA from the RNA/DNA hybrid.
6. The remaining segment of RNA serves as a primer for the synthesis of a second strand of DNA, which extends to the 5' end of the first DNA strand.
7. All the remaining RNA is removed by reverse transcriptase.
8. The two DNA strands undergo rearrangement so that their PBS segments align and base pair.
9. The 3' ends of both strands are extended to yield the complete double-stranded DNA. Note that the double-stranded DNA is longer than the RNA genome because it contains a long terminal repeat segment at each end.

SUGGESTED READING

1. Luria SE, Darnell JR Jr., Baltimore D, Campbell A: General Virology. 3rd Ed. Wiley, New York, 1978
2. Joklik WK (ed): Virology. 2nd Ed. Appleton & Lange, East Norwalk, CT, 1985
3. Klein G (ed): DNA-Virus Oncogenes and Their Action. Raven Press, New York, 1983

4. Marin SJ: The Biochemistry of Viruses. Cambridge University Press, NY, 1978
5. Tooze J, Hynes RD: The Molecular Biology of Tumor Viruses. 2nd Ed. Cold Spring Harbor Laboratory, Cold Spring Harbor, NY, 1980
6. Mathew CK: Bacteriophage Biochemistry. Van Nostrand, New York, 1971
7. Fields BN, Knipe DM (eds): Fundamental Virology. Raven Press, New York, 1986

STUDY QUESTIONS

Directions: Answer the following questions using the key outlined below:
- **(A)** if 1, 2, and 3 are correct
- **(B)** if 1 and 3 are correct
- **(C)** if 2 and 4 are correct
- **(D)** if only 4 is correct
- **(E)** if all four are correct

1. A (+) RNA virus contains
   1. double-stranded RNA
   2. single-stranded RNA
   3. RNA that must be transcribed prior to its use as mRNA
   4. RNA that may be used as mRNA

2. Infection of *E. coli* by phage T4 involves
   1. introduction of only the viral DNA into the bacterial cell
   2. expression of the viral genome with very little effect on the expression of the *E. coli* genome
   3. independent synthesis of T4 head, core, and tail fibers components
   4. release of viral particles by budding

3. Infection of a permissive host by SV40 is associated with
   1. synthesis of large T and small t proteins
   2. assembly of new viral particles
   3. lysis of the host cell
   4. integration of viral DNA into the host cell genome

Directions: For the following multiple choice questions, choose the most appropriate answer.

4. All of the following statements concerning poliovirus are true EXCEPT that
   A. poliovirus inhibits host-cell protein synthesis by inactivating an initiation factor required for host-cell translation.
   B. a single replicative intermediate may be used to direct protein synthesis and subsequently used to generate new genomic (+) RNA.
   C. the viral genome is translated into one precursor polyprotein.
   D. poliovirus is a single-stranded (+) RNA virus.

5. All of the following statements concerning vesicular stomatitis virus (VSV) are true EXCEPT that
   A. VSV enters host cells by invagination.
   B. VSV codes for five proteins.
   C. only the nucleocapsid enters the host cell cytoplasm.
   D. VSV progeny virions are released from the host cell by lysis.

6. All of the following statements concerning retroviruses are true EXCEPT that
   A. the retroviral genome is diploid.
   B. the single-stranded retroviral genome is converted to a double-stranded DNA (provirus).
   C. the double-stranded DNA is inserted into the host-cell genome.
   D. One provirus is inserted per host genome.

Items 7 through 10. Consider the following terms:
   A. Bacteriophage T4
   B. Bacteriophage φX714
   C. Reovirus
   D. R17
   E. Poliovirus

Match each statement below with the appropriate term.

7. A (+) RNA virus that infects *E. coli*

8. A double-stranded RNA virus

9. A single-stranded DNA virus that infects *E. coli*

10. A double-stranded DNA virus

APPENDIX 1

Water, Acids and Bases, and Buffers

WATER

The major component of living organisms is water. Cellular reactions occur in a water (aqueous) environment. Water participates directly in many biochemical reactions, and in others it influences the reactivities and properties of enzymes and proteins.

Properties of Water

The oxygen of water (H_2O) is very electronegative. As a result, the electrons involved in the H–O–H covalent bonds are unevenly shared between the oxygen and the hydrogens. They are more closely associated with oxygen than with hydrogen, so the oxygen of water carries a partial negative charge and the hydrogens carry partial positive charges. Because of these opposite partial charges, there are strong intermolecular interactions between hydrogens and oxygens of different water molecules. These interactions are called hydrogen bonds (Fig. A-1). Hydrogen bonding among water molecules results in the transient formation of large networks of water molecules.

Because of the strong electronegativity of the oxygen atom, water molecules exhibit a slight tendency to dissociate, as follows:

$$H_2O \rightleftharpoons H^+ + OH^- \qquad (1)$$

The dissociated proton does not exist as such in solution but instead interacts with another water molecule to form a protonated species, the hydronium ion:

$$H^+ + H_2O \rightleftharpoons H_3O^+ \text{ (Hydronium ion)} \qquad (2)$$

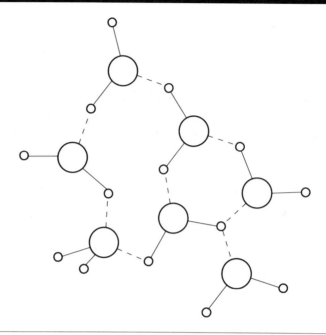

Fig. A-1. A representation of hydrogen bonding (dotted lines) among water molecules.

The extent to which pure water dissociates can be calculated as follows:

$$K = \frac{[H^+][OH^-]}{[H_2O]} \tag{3}$$

where K is the equilibrium constant for the reaction. Since the concentration of water in pure water is constant (55.5 moles/liter) the concentration of water can be incorporated into the equilibrium constant, and equation 3 can be re-written as

$$K_w = K[H_2O] = [H^+][OH^-]$$

or

$$K_w = [H^+][OH^-] \tag{4}$$

where K_w is called the ion product of water. In pure water the values of $[H^+]$ and $[OH^-]$ are the same and are equal to 1×10^{-7} M. Therefore,

$$K_w = 1 \times 10^{-14} \text{ mole}^2/\text{liter}^2$$

The ion product of water remains constant at 1×10^{-14} mole²/liter². Therefore, if $[H^+]$ increases, $[OH^-]$ must decrease to the same extent, and vice versa.

ACIDS AND BASES

For biochemical purposes, acids are best defined as proton donors and bases are best defined as proton acceptors. Consider the following reaction:

$$HA \rightleftharpoons H^+ + A^- \tag{5}$$

The greater the tendency for HA to dissociate to form H^+ and A^-, the stronger is the acid. The following reaction demonstrates the properties of a base:

$$B + H^+ \rightleftharpoons BH^+ \tag{6}$$

The greater the tendency for B to bind H^+, the stronger base it is.

Strengths of Acids and Bases

In equation 5, HA is the acid and A^- is its conjugate base. The tendency of HA to dissociate corresponds to its strength as an acid. The tendency of A^- to bind H^+ corresponds to its strength as a base. Therefore, the weaker HA is as an acid, the stronger A^- is as a base. The tendency of HA to dissociate is given by the equation

$$K = \frac{[H^+][A^-]}{[HA]} \tag{7}$$

K is large for strong acids and small for weak acids. $K = 0$ for compounds which do not dissociate and these substances are not considered acids. K cannot be calculated for very strong acids that dissociate completely. However, for compounds in which there is a measurable reversible dissociation into $[H^+]$ and the conjugate base, K can be determined from the hydrogen ion concentration associated with the compound in solution. In order to make this obvious, equation 7 may be rearranged as follows:

$$[H^+] = K \times \frac{[HA]}{[A^-]} \tag{8}$$

The pH of a solution is defined as the negative of the log of $[H^+]$, and the pK of an acid is defined as the negative of the log of its K. Therefore, if the negative log is taken of both sides of equation 8, the result is

$$-\log[H^+] = -\log K + \log \frac{[A^-]}{[HA]}$$

or

$$pH = pK + \log \frac{[A^-]}{[HA]} \tag{9}$$

Equation 9 is known as the *Henderson-Hasselbalch equation* and may be used to calculate the pH or pK of acid solutions. It may also be used to calculate the extent of dissociation of a given acid at a particular pH.

BUFFERS

Buffers are defined as solutions that resist pH changes. A buffer (also called a *buffering system* or *couple*) usually consists of an acid and its conjugate base. For any buffer system, maximum buffering is seen in the pH range from one pH unit above to one pH unit below the pK of the acid component. Buffering capacity is also increased by increasing the concentration of the buffer system components.

Calculations

The Henderson-Hasselbalch equation can be used to calculate the pH at any point along the titration curve except at the initial and final extremes. It can also be used to calculate pK and the extent of acid dissociation. In order to calculate the pH of solutions of weak acids, it may be assumed that $[H^+] = [A^-]$ and that the [HA] at equilibrium equals [HA] that was added to the solution minus $[H^+]$.

Appendix 2

REDUCTION POTENTIALS

When electrons move from one compound to another, energy is released. The release of energy associated with electron transfer is the basis for the functioning of *electrochemical cells*. To understand how electrochemical cells work, it is important to be familiar with their design. Some factors essential to the design and functioning of electrochemical cells are as follows:

1. An electrical current can be generated and maintained by the flow of ions as well as by the flow of electrons.
2. In the truest sense, an operating electrochemical cell is a closed electrical circuit.
3. An electrochemical cell consists of two solutions of ions (called *half-cells*) that are connected by a *salt bridge* which allows ions to flow between the two solutions. A salt bridge may consist, for example, of a tube filled with agar through which the ions can diffuse.
4. The two half-cells are also connected by external wiring that allows for the direct flow of electrons between the solutions.

Now, let us consider the reaction

$$\text{Pyruvate} + \text{NADH} + \text{H}^+ \rightarrow \text{lactate} + \text{NAD}^+$$

Electrons are transferred from NADH to pyruvate. Lactate/pyruvate and NADH/NAD$^+$ are *oxidation-reduction pairs* or *couples*. For the purpose of electrochemistry, the above reaction is composed of two *half-reactions*, each involving one of the oxidation-reduction couples. These half-reactions are:

1. Pyruvate + 2 e$^-$ → lactate
2. NADH → NAD$^+$ + 2 e$^-$

An electrochemical cell can be used to measure the amount of energy released as a result of the reduction of pyruvate by NADH. Two half-cells (ion solutions) are first prepared. Half-cell 1 will contain 1 M pyruvate and 1 M lactate. Half-cell 2 will contain 1 M NADH and 1 M NAD^+. In half-cell 1, lactate is the reductant, and in half-cell 2, NADH is the reductant. The two half-cells are connected by a salt bridge. An electrode is immersed in each half-cell, and finally the two electrodes are connected by external wiring. When this external circuit is closed, electrons will flow either from the pyruvate/lactate half-cell to the $NADH/NAD^+$ half cell or vice versa. The direction of electron flow depends on the relative tendencies of lactate to reduce NAD^+ and of NADH to reduce pyruvate—i.e., on which pair has the greater *reduction potential*. If lactate has a greater tendency (potential) to reduce NAD^+ than NADH has to reduce pyruvate, electrons will flow from half-cell 1 to half-cell 2, and vice versa.

If a voltmeter is inserted into the external circuit of the electrochemical cell and the circuit is closed, the direction of electron flow will be known and the voltage difference between the two cells can be measured. This voltage difference corresponds to the energy released by the electrochemical reaction (the *electromotive force* or *cell potential*, measured in volts). The cell potential is the difference between the potentials of the two half-cells. The energy released by each half-cell is referred to as the *half-cell potential*. Half-cell potentials can be treated as either oxidation or reduction potentials; the absolute value of the potential is the same in both cases, but the sign is different.

It is convenient to express all half-cell potentials as *electrode potentials*. Cell potentials can be calculated from known electrode potentials as follows:

$$\text{Cell potential} = E_C = E_1 - E_2$$

where E_1 and E_2 are the electrode potentials of half-cells 1 and 2, respectively. The voltage across the external circuit of the electrochemical cell gives the total cell potential, but not the individual half-cell potentials. By convention, a half-cell potential is the potential across an electrochemical cell consisting of the half-cell in question and a standard half-cell which contains H_2 gas at 1 atm and 1 M H^+. This standard half-cell is called the *standard hydrogen electrode (SHE)*, and is arbitrarily assigned a half-cell potential of 0.0 V. Therefore, the half-cell potential for half-cell 1 is given by:

$$\text{Cell potential} = 0.0 - E_1 = \text{half-cell potential}$$

The *reduction potentials* of many different oxidation-reduction couples have been determined by comparing them in this way with the SHE. The more negative the reduction potential, the greater the tendency of that substance to donate its electrons, and the more positive the reduction potential, the greater the tendency of the substance to accept electrons. When electrons are transferred between substances, the direction of spontaneous electron flow is always toward the compound with the more positive reduction potential. In this way, reduction potentials can be used to predict the direction of electron flow in oxidation-reduction reactions.

As stated earlier, electron-transfer reactions are accompanied by the release of free energy. The amount of energy released is directly proportional to the number of electrons transferred and to the difference between the reduction potentials of the two half-reactions involved. The exact relationship is

$$\Delta G = -nF\Delta E$$

where n is the number of electrons transferred, F is the Faraday constant, and ΔE is the difference in E for the two half-reactions:

$$\Delta E = E_{\text{electron acceptor}} - E_{\text{electron donor}}$$

ANSWERS

Chapter 1

1. **B** Amino acids with branched side chain are more effective at increasing polypeptide chain rigidity. Therefore, the correct choice is valine. (p. 8)

2. **C** pK_2 refers to the pK of the α-amino group of alanine, which is a basic group. At that pH, 50% of the alanine α-amino groups will be un-ionized, but all of the α-carboxyl groups will be ionized. Therefore, the species present in solution are $^+NH_3CHCH_3COO^-$ and $NH_2CHCH_3COO^-$. (p. 3)

3. **A** Supersecondary structures are combinations of secondary structures. Of the choices given, only the β-meander, which consists of β-strands and reverse turns, is a combination of secondary structures. (p. 18)

4. **B** Reagents used in detecting the amino terminal residues of proteins react specifically with free amino groups. Ninhydrin is not suitable because it also reacts with secondary amino groups, as evidence by its reaction with proline. (p. 11)

5. **A** Most peptide bonds are in the *trans* configuration, which allows minimum steric interference between neighboring side chains. (p. 4)

6. **B** The side chains of Gly, Ser, Val, and Pro are all uncharged at neutral pH. The side chains of Glu and Asp both have a charge of -1, and Lys has a charge of $+1$. There will also be a charge of -1 on the C-terminal carboxyl group and a charge of $+1$ on the N-terminal free amino group. Therefore, the net charge on the peptide will be -1. (p. 2)

7. **A** Under basic conditions, the lysine side chain is uncharged. The charge on the other side chains will be the same as at neutral pH. The N-terminal free amino group, however, will be uncharged, whereas the C-terminal carboxy group will have a charge of -1. The total charge is therefore -3. (p. 2)

8. **A** Peptide bonds are amide linkages, which are covalent; they are resonance stabilized by amide hydrogens. (p. 4)

9. **C** Multisubunit proteins are held together by noncovalent hydrophobic and hydrophilic interactions. (p. 20)

10. **C** α-Helices are stabilized by the system of hydrogen bonds between each carbonyl oxygen and the amide nitrogen four amino acid residues away. α-Helices, which have 3.6 residues per turn, are more stable that 3^{10} helices, which have 3 residues per turn. Proline residues disrupt the α-helix structure. (p. 15)

Chapter 2

1. **A** Enzymes increase reaction rates by lowing the activation energy required for the conversion of substrate to product. (pp. 28, 29)

2. **B** Coenzymes are usually not tightly bound to the enzyme and freely associate and dissociate from the enzyme. (p. 27)

3. **B** Noncompetitive inhibitors decrease only V_{max}. (p. 41)

4. **C** Some but not all allosteric enzymes show cooperativity. (pp. 44, 46)

5. **C** In the ping-pong mechanism, after the first substrate binds, the first product is formed and dissociates from the enzyme. Then the second substrate binds and the second product is formed and dissociates from the enzyme. Therefore, the ping-pong mechanism is a double displacement mechanism. (p. 42)

6. **C** The turnover number is the moles of product formed per minute per mole of enzyme. (p. 37)

7. **A** In their catalytic mechanisms, enzymes employ one or any combination of the following: general acid-base catalysis, covalent catalysis, substrate strain, and orientation effects. (pp. 29, 30)

Chapter 3

1. **A** By definition, the free energy of a reaction is that portion of the energy change that is able to do work—i.e., that must be supplied by the cell to drive the reaction, or that can be used by the cell to drive other endergonic processes. (p. 54)

2. **C** A negative ΔG indicates that the reaction will proceed spontaneously, with release of energy. Therefore, it is exergonic, and can accomplish useful work. (pp. 54, 55)

3. **D** ATP can be broken down to ADP and P_i or to AMP and PP_i. Alternatively, however, free energy can be released during the transfer of the phosphate group to a low energy acceptor. (p. 56)

4. **A** ΔG ia function of [reactants] and [products]. During the course of the reaction, [reactants] and [products] change continuously. Therefore, ΔG also changes continuously. As equilibrium approaches, [reactants] and [products] change less and less. At equilibrium, the quantities no longer change. Therefore, ΔG goes to zero as the reaction goes to equilibrium. (p. 55)

5. **A** The hydrolysis of ATP to ADP makes available about 8.4 kcal/mol of free energy (using $\Delta G^{\circ\prime}$ as an approximation for ΔG). Therefore, hydrolysis of about five ATP molecules should be able to supply the 40 kcal necessary to drive the reaction. (p. 56)

6. **C** Reactions that have an equilibrium constant greater than 1 have a negative $\Delta G^{\circ\prime}$ and are exergonic. To demonstrate this, try substituting values for K_{eq} that are greater than 1 and less than 1 in the

following equation:

$$\Delta G^{\circ\prime} = -RT \ln K_{eq}$$

(p. 55)

7. **D** In order for the entire pathway to proceed spontaneously, the sum of the individual ΔGs must be negative. Since only the sign of the ΔGs is given and not their values, their sum cannot be calculated. Therefore, it cannot be predicted whether or not the pathway will proceed spontaneously. Three of the individual reactions do have negative ΔGs; these individual reactions can proceed spontaneously. (p. 55)

Chapter 4

1. **A** Furanoses are five-membered rings. Structure A (α-D-fructose) is a ketofuranose because it is a hemiketal. Hemiketals are formed by ketones.
(pp. 65, 66)

2. **C** Structure C consists of a five-membered ring and it is a hemiacetal. Hemiacetals are formed by aldehydes.
(p. 65, 66)

3. **D** Structures B and D are both Haworth projections of D-glucose. When sugars are drawn as Haworth projections, the anomeric carbon is in the β configuration if the hydroxyl points up, and in the α configuration if it points down. (p. 66)

4. **B** The glycosaminoglycan chains of proteoglycans consist of disaccharide repeat units, and they are polyanions. They do not always contain sulfate, and in some cases they are not covalently linked to the protein component. (p. 71)

5. **A** The activated forms of phosphorylase kinase, inhibitor 1, and protein kinase are all involved in the stimulation of glycogen phosphorylase, the regulated enzyme of glycogen degradation. Phosphoprotein phosphatase 1, on the other hand, dephosphorylates (deactivates) glycogen phosphorylase. (p. 78)

6. **C** Binding of epinephrine to α-adrenergic receptors results in the formation of inositol triphosphate from phosphatidyl inositol. Inositol triphosphate stimulates the release of Ca^{2+}, presumably from the endoplasmic reticulum. Ca^{2+} is an allosteric activator of phosphorylase kinase. (p. 78)

7. **C** Glucose derived from the metabolism of muscle glycogen is utilized within muscle to meet the energy needs of the muscle during contraction. (p. 73)

8. **D** A is false; glycogen contains α(1 → 4) and α(1 → 6) linkages but no β(1 → 6) linkages. B is false; glycogen is a glucose homopolymer. C is false because epinephrine functions to raise blood glucose level in response to stress or anticipated stress, not in response to a fall in blood glucose. D is true: the enzyme protein kinase phosphorylates phosphorylase kinase, inhibitor 1, and glycogen synthase. Phosphorylase kinase and inhibitor 1 are activated by phosphorylation, whereas glycogen synthase is inactivated. Phosphorylase kinase and inhibitor 1 turn on glycogen phosphorylase, and thus promote glycogen degradation, whereas inhibitor 1 helps keep glycogen synthase in the phosphorylated, inactivated state, thus inhibiting glycogen synthesis. (pp. 78–81)

9. **B** B is false; glucagon and insulin function in a coordinated manner, but they have opposing effects on glycogen

metabolism. Glycogen stimulates glycogen breakdown, whereas insulin (among other effects) stimulates glycogen synthesis and enhances muscle and adipose tissue uptake of glucose. (p. 80)

10. **B** Muscle cells have only β-adrenergic receptors, whereas liver has both α- and β-adrenergic receptors. Therefore, hormonal stimulation of glycogen breakdown can occur in liver via both the cAMP and the inositol triphosphate pathways, whereas in muscle it occurs only via cAMP. (p. 78)

Chapter 5

1. **A** Humans can digest only polysaccharides joined by α(1 → 4) links or α(1 → 4) links with α(1 → 6) branch points. Thus, humans can digest glycogen and the starches amylose and amylopectin, but not celluloses. (p. 88)

2. **D** The structure of amylose is very similar to that of glycogen. Both are glucose homopolymers consisting of α(1 → 4) chains with α(1 → 6) branch points. Amylose is less densely branched than glycogen. Maltose and maltotriose are oligosaccharides, and cellulose is an unbranched glucose homopolymer with β(1 → 4) links. (p. 88)

3. **C** Saliva contains an α-amylase, but food does not remain in the mouth long enough for significant starch digestion to occur. (Salivary α-amylase is inactivated in the stomach.) Disaccharidases are located on the luminal surface of intestinal mucosal cells. (p. 88)

4. **C** Fructose 2,6-bisphosphate is the most important effector of phosphofructokinase, which is the major regulated enzyme of glycolysis; thus it may be considered the most important effector of glycolysis. Citrate and AMP are also effectors of phosphofructokinase. Glucose 6-phosphate is an inhibitor of hexokinase, but not of liver glucokinase. (p. 103)

5. **D** Gluconeogenesis is distinct from glycolysis at the three steps that are irreversible in glycolysis. These steps are catalyzed by hexokinase/glucokinase, phosphofructokinase, and pyruvate kinase in glycolysis. In gluconeogenesis, pyruvate carboxykinase and phosphoenolpyruvate carboxykinase circumvent the pyruvate kinase reaction of glycolysis. Glucose 6-phosphatase and fructose 1,6-bisphosphatase are also distinct gluconeogenetic enzymes. (pp. 107, 109)

6. **B** The function of the glycerol phosphate shuttle is to convert NADH in the cytosol to $FADH_2$ equivalents in the mitochondrial matrix. The process is mediated by a reduction/oxidation cycle involving cytosolic glycerol 3-phosphate and cytosolic dihydroxyacetone phosphate. (p. 99)

7. **A** Lactate is produced by glycolysis when oxygen is limiting (i.e., when the rate of aerobic respiration in the mitochondria is limited by the supply of oxygen). When the oxygen supply is adequate, pyruvate is produced. (The exception is erythrocytes, which lack mitochondria and therefore perform only lactate fermentation.) (p. 90)

8. **B** Lactate fermentation of glucose uses two ATPs and produces four, giving a net gain of two. (p. 98)

9. **C** Since the concentration of free glucose in liver cells is in equilibrium with the concentration in the blood, the high K_M of glucokinase (about twice normal plasma glucose levels) reveals that the enzyme is relatively inactive when plasma glucose levels are in the normal

range, and that significant liver glycogen synthesis (i.e., formation of glucose 6-phosphate) does not occur until blood glucose rises above normal. (pp. 93, 94)

10. **D** AMP and fructose 2,6-bisphosphate participate in the coordinate regulation of glycolysis and gluconeogenesis. They both act on phosphofructokinase and fructose 1,6-bisphosphatase, activating the former and inhibiting the latter. (pp. 101, 102, 111)

11. **A** Most dietary glucose is immediately taken up by the liver from the portal circulation, and stored as glycogen. This glycogen is gradually degraded to maintain blood glucose; glucose 6-phosphatase is an essential enzyme of this pathway. Citrate is an allosteric inhibitor of phosphofructokinse but not of pyruvate kinase. (p. 90)

Chapter 6

1. **C** Except for succinate dehydrogenase (which is located on the inner mitochondrial membrane), all the TCA cycle enzymes are located in the mitochondrial matrix; thus, the TCA cycle occurs in the matrix. Oxidative phosphorylation is catalyzed by the enzyme ATP synthetase, an integral protein of the inner mitochondrial membrane. Only the inner membrane is folded into cristae.
(pp. 122, 123)

2. **B** Succinate, malate, and oxaloacetate are all TCA cycle intermediates; ketocitrate is not. (p. 126)

3. **A** Each turn of the cycle produces 3 NADH, 1 FADH$_2$, and 1 GTP. (p. 124)

4. **D** The reaction catalyzed by malate dehydrogenase is the only reaction in the TCA cycle with a strongly positive $\Delta G^{\circ\prime}$ (about 7.1 kcal/mol). (p. 128)

5. **B** NAD$^+$ and FAD both normally carry two electrons. Coenzyme Q can carry either one electron (semiquinone form) or two electrons (ubiquinol). Cytochrome c is not capable of carrying more than one electron. (pp. 131, 134)

6. **B** The TCA cycle occurs in the mitochondria, which have both an inner and an outer membrane. The inner membrane of mitochondria is highly impermeable to polar solutes and ions, but the outer membrane is permeable to most solutes. The pentose phosphate pathway, as discussed in Chapter 5, occurs in the cytosol. (p. 122)

7. **B** Oxidative phosphorylation cannot occur without the proton gradient set up by electron transport. The rate of electron transport is also normally tightly coupled to the rate of oxidative phosphorylation. Therefore, inhibitors of ATP synthetase, which block oxidative phosphorylation, will also inhibit electron transport. Uncouplers are substances that allow protons to re-enter the matrix without flowing through ATP synthetase. In the presence of an uncoupler, electron transport and oxygen consumption will proceed at a very high rate, but oxidative phosphorylation will be depressed.
(pp. 143, 144)

Chapter 7

1. **B** Because lipids are highly reduced substances, more energy can be stored per unit weight. Since lipid is often stored for extended periods of time, it is beneficial that they are relatively nonreactive substances. (p. 150)

2. **A** Triacylglycerols are neutral lipids, the major form of stored lipid in the body, and they contain fatty acids esterified to glycerol. (p. 153)

3. **D** High carbohydrate diets are believed to stimulate the synthesis of the fatty acid synthetase complex. (p. 179)

4. **D** VLDLs transport endogenously synthesized triacylglycerol from the liver to adipose tissue. (p. 171)

5. **A** Lipoprotein lipase hydrolyzes fatty acids from positions 1 and 3 of triacylglycerols. (p. 171)

6. **D** Two NADPHs are required to add one acetyl CoA to a growing fatty acid chain. (p. 160)

7. **D** Due to the operation of the citrate shuttle, both acetyl CoA and NADPH are ultimately generated in the cytosol. (p. 163)

8. **B** Fatty acids are transported from adipose tissue to the liver bound to albumin. (p. 171)

9. **B** Only 1NADH and 1 $FADH_2$ are produced during one complete cycle of the β-oxidation of fatty acids. (p. 174)

10. **C** The availability of acetyl CoA largely regulates the observed level of fatty acid β-oxidation. The extent of acetyl CoA utilization by the TCA cycle significantly determines the availability of acetyl CoA. (pp. 177, 178)

Chapter 8

1. **B** Triacylglycerol synthesis requires glycerol 3-phosphate or dihydroxyacetone phosphate. Glycerol 3-phosphate can be synthesized from free glycerol and ATP via glycerol kinase. Adipose tissue does not contain glycerol kinase, so it cannot use free glycerol as a precursor to triacyglycerols. (p. 188)

2. **A** Phosphocholine is synthesized from ATP and choline by choline kinase. The resulting phosphocholine is then used to form CDP-choline, which ultimately reacts with 1,2-diacylglycerols to form phosphatidylcholine. (pp. 191, 192)

3. **D** Triacylglycerol synthesis in the liver and adipose tissue involves both phosphatidate and 1,2-diacylglycerol as intermediates. (pp. 191–194)

4. **C** Phospholipase C is specific for hydrolyzing the glycerol–phosphate bond of phospholipids. In the case of phosphatidylcholine, phosphocholine and 1,2-diacylglycerol would be produced. (p. 200)

5. **A** Globosides are ceramide oligosaccharides. The other major class of glycosphingolipids, the cerebrosides, are ceramide monosaccharides. (p. 205)

6. **A** Cholesterol synthesized in the liver is packaged into VLDLs. After removal of triacylglycerols, VLDLs become LDLs, which ultimately deliver cholesterol to the tissues. Dietary cholesterol arising from the intestine is packaged into chylomicrons. HDLs are believed to transport cholesterol from tissues back to the liver. (pp. 211, 213)

7. **D** The initial products of the lipoxygenase reaction are HPETES. HPETES are then converted to HETES. (pp. 220, 221)

8. **A** Free cholesterol feedback-inhibits HMG CoA reductase and represses its synthesis. The enzyme is also under the hormonal influence of both glucagon and insulin. (p. 211)

9. **B** Bile salts are the terminal products of cholesterol metabolism. They are effective emulsifying agents that are important in the solubilization and there-

fore the absorption of dietary lipids and fat-soluble vitamins. (pp. 214, 215)

10. **B** Arachidonic and dihomo-γ-linoleic acids are immediate fatty acid precursors of prostaglandins. Both these fatty acids may be synthesized endogenously from dietary linoleic acids. (pp. 216, 217)

11. **A** Cholesterol biosynthesis occurs in the cytosol. Acetyl CoA is the only starting material for cholesterol synthesis. Acetyl CoA is initially used to form isopentyl intermediates which subsequently condense to form the larger cholesterol intermediates and ultimately cholesterol. (pp. 210, 211)

Chapter 9

1. **B** The synthesis of glutamine is the major fate of free ammonia in mammals. Glutamine is subsequently used in the synthesis of nitrogenous compounds in the body. The kidney also takes up glutamine and hydrolyzes it to yield free ammonia, which is used to maintain the pH of urine. Ammonia in excess of the body's needs is synthesized into urea and eliminated in the urine. (p. 235)

2. **A** The muscle and brain are major sites of uptake and metabolism of the branched chain amino acids (valine, leucine, and isoleucine). (pp. 231, 232)

3. **C** Transaminases catalyze the transfer of amino groups among amino acids. They are important in the production of free ammonium in combination with the glutamate dehydrogenase reaction. (pp. 232, 233)

4. **D** Ornithine decarboxylase is not involved in the synthesis of urea. Instead, it catalyzes the conversion of ornithine to putrescine in the polyamine synthesis pathway (Chapter 10). (p. 236)

5. **B** Aspartate, cysteine, and histidine are each degraded only to intermediates of carbohydrate metabolism. (p. 239)

6. **B** Aspartate and its amide derivative asparagine are both converted to oxaloacetate. Histidine, arginine, and proline, on the other hand, may be converted to α-ketoglutarate. (p. 239)

7. **A** Alanine and serine both are capable of producing pyruvate, but are nutritionally nonessential. Isoleucine is nutritionally essential, but does not produce pyruvate on degradation. Threonine can yield pyruvate on degradation and is nutritionally essential. (pp. 239, 252)

8. **C** Genetic deficiency of any of the urea cycle enzymes usually results in elevated levels of ammonium in the blood and urine, because the conversion of ammonium to urea is at least partially blocked. (p. 255)

9. **E** All of the proteolytic enzymes listed are active in the degradation of dietary protein. (pp. 228, 229)

10. **A** The major sources of endogenous one-carbon units are serine, glycine, and histidine. (pp. 250, 252)

Chapter 10

1. **C** The syntheses of serotonin and dopamine require both decarboxylation and hydroxylation of the amino acid precursor. γ-Aminobutyric acid and histamine, on the other hand, are formed by amino acid decarboxylation only. (p. 262)

2. **A** Catecholamine degradation can proceed by several pathways, which involve conjugation, methylation, oxidation, or any combination of these. (p. 264)

3. **C** Heme exerts feedback repression on δ-aminolevulinic acid synthetase, and also represses its synthesis. The enzyme is induced by allylisopropylacetamide. Succinyl CoA and iron do not influence the enzyme. (p. 277)

4. **C** Bilirubin generated in extrahepatic tissues is transported to the liver complexed to albumin. Heme oxygenase catalyzes the reaction that opens the heme ring. This reaction releases ferrous iron and generates carbon monoxide and biliverdin. (pp. 278, 280)

5. **D** Of the choices given, only porphobilinogen is synthesized in the cytosol. The other substances are formed in the mitochondria. (pp. 277, 278)

6. **D** Creatine formation involves a transamidination reaction followed by *N*-methylation. The amidine group of arginine is transferred to glycine to form guanidinoacetic acid (the transamidination). The α-nitrogen of guanidinoacetic acid is then methylated to yield creatine. (p. 269)

7. **B** Monoamine oxidase does not participate in polyamine degradation. The enzymes of polyamine degradation are diamine oxidase and polyamine oxidase. Ornithine is decarboxylated to yield putrescine. *S*-adenosylmethionine is decarboxylated and subsequently used in the synthesis of spermidine and spermine. (pp. 271, 272)

8. **D** Protoporphyrin is the last intermediate of heme synthesis prior to heme formation. Therefore, it is synthesized after porphobilinogen, which is formed immediately after synthesis of δ-aminolevulinic acid. (pp. 278, 279)

Chapter 11

1. **B** Thymine is a pyrimidine base, adenosine is a purine nucleotide, and dTMP is a pyrimidine nucleotide. (pp. 287–289)

2. **A** Only glutamine, aspartate, and CO_2 are required in the syntheses of both purine and pyrimidine nucleotides. (p. 290)

3. **D** IMP is initially formed. It is then used to synthesize AMP and GMP. (p. 290)

4. **B** PRPP amidotransferase is the rate-limiting enzyme in the synthesis of the purine nucleotides. It is subject to feedback inhibition by IMP, GMP, and AMP. (p. 298)

5. **A** dTMP is actually a deoxyribonucleotide. It is synthesized from dUMP by thymidylate synthetase. (p. 305)

6. **C** Phosphoribosyltransferase enzymes catalyze purine and pyrimidine base salvage reactions. In salvage reactions, the base is converted to the nucleotide monophosphate in the presence of PRPP and the phosphoribosyltransferase enzyme. PP_i is also produced during the reaction. (pp. 309–310)

7. **D** TMP contains the pyrimidine base thymine, and pyrimidine bases are ultimately degraded to β-aminoisobutyrate. (pp. 308, 309)

8. **E** CDP, UDP, GDP, and ADP are substrates for nucleotide diphosphate reductase. This enzyme will not reduce any other diphosphates. (p. 302)

9. **A** Adenosine is a component in the structure of NAD, NADP, FAD^+, and

coenzyme A. Adenine nucleotides are involved in the synthesis of these coenzymes. (p. 312)

10. **D** Biochemical abnormalities associated with gout cause an increased synthesis of purine nucleotides, which ultimately results in the increased production and deposition of uric acid in the joints. (p. 316)

Chapter 12

1. **E** The term *nuclease* refers to the class of enzymes that cleave the phosphodiester bonds of nucleic acids. Some nucleases are very general and some are very specific in their substrate requirements. (p. 322)

2. **B** G-C base pairs are more difficult to disrupt because they are held together by three hydrogen bonds, as opposed to two in A-T base pairs. DNA with a high G + C content is very stable and more difficult to denature. (p. 326)

3. **A** Mammalian cells contain right-handed and left-handed DNA double helices. There are three forms of right-handed helix: A-DNA, B-DNA, and C-DNA. (p. 326)

4. **C** Palindromes are recognized by restriction endonucleases and by DNA methylases. After methylation, however, they are no longer substrates for restriction endonucleases. (p. 335)

5. **A** RNA molecules from mammalian cells are single-stranded polynucleotides with variable amounts of secondary structure. Some RNAs are folded into tertiary conformations. Both the secondary and the tertiary structure of tRNAs are known in detail. (pp. 338, 340)

6. **E** The annealing of nucleic acid strands from different sources, or the annealing of RNA and DNA strands, is called hybridization. Hybridization involves chain alignment, which involves nucleation, which in turn involves hydrogen bonding. (p. 330)

7. **D** Superhelices are present in both circular and linear DNA molecules, and may be positive or negative. Superhelices alleviate the strain introduced into DNA molecules by over- or underwinding. (pp. 330, 331)

8. **A** The histone core of nucleosomes consists of two copies each of H2A, H2B, H3, and H4. (p. 332)

9. **D** The greater the G+C content, the more stable is the DNA double helix and the more difficult it is to denature (i.e., the higher the temperature required for denaturation). (p. 326)

10. **C** DNA sequences with a higher degree of repetitiveness renature faster, giving smaller $C_0 t_{\frac{1}{2}}$ values. The reason is that chain alignment is easier the more repetitive the DNA. (pp. 335–336, 337)

Chapter 13

1. **B** Both strands of parental DNA are replicated simultaneously at the replication fork. Eukaryotic chromosomes have multiple replication initiation sites. Prokaryotic chromosomes have only one such site. Helicases unwind supercoiled DNA and cause the complimentary strands to separate, making single stranded templates available for DNA polymerase. DNA replication results in the formation of parent strand-daughter strand hybrids, which is characteristic of semiconservative replication. (pp. 346, 347)

2. **B** The primasome is a complex of several proteins, one of which is the enzyme primase. The function of the primasome is to synthesize RNA primers that are required to initiate replication during the formation of the replication fork and at the beginning of each Okazaki fragment. (p. 347)

3. **E** DNA polymerases I and III, primase, and Okazaki fragments are all associated with the replication fork. DNA polymerase I removes the RNA primers from the Okazaki fragments and replaces them with complementary deoxyribonucleotides. DNA polymerase III catalyzes the elongation of the leading strand and of the Okazaki fragments. Primase is the enzyme component of the primasome responsible for synthesizing the RNA primers. Okazaki fragments are the segments of daughter DNA attached to RNA primers on the lagging strand. (pp. 346, 347)

4. **B** General recombination does require long stretches of homologous sequences between recombining strands and it also involves the formation of Holliday structures. (p. 360)

5. **D** Site-specific recombination requires only short segments of homologous sequences between recombining DNAs. Recombination occurs within these homologous sequences. (p. 360)

6. **A** Xeroderma pigmentosum is due to defective ability to repair pyrimidine dimers. It is associated with unusual sensitivity to sunlight, and causes an increased incidence of skin cancer. Patients should protect themselves from exposure to sunlight. (p. 368)

7. **D** Primase is not involved in the repair of thymine dimers. Rather, it is involved in the synthesis of RNA primers for DNA replication. (p. 354)

8. **C** Transitions result in the replacement of one purine-pyrimidine base pair by another one. (p. 352)

9. **B** When an additional base is inserted or deleted, the reading frame is altered, resulting in a frameshift mutation. (p. 352)

10. **A** In a transversion, a purine-pyrimidine base pair is replaced by a pyrimidine-purine base pair. Therefore, mutation A is the correct choice. (p. 352)

11. **C** In a transition, one purine-pyrimidine base pair is replaced by another purine-pyrimidine base pair. Therefore, mutation C is the correct choice. (p. 352)

12. **C** DNA that is to be cloned must first be inserted into an appropriate vector. DNA inserted into a cloning vector is called passenger DNA. The synthesis of double-stranded cDNA is one method of generating DNA for the purpose of cloning. (p. 363)

13. **D** Restriction endonucleases are very important in the construction of vector-passenger DNA adducts. (p. 365)

14. **A** The initial strand of cDNA is synthesized by reverse transcriptase, which uses mRNA as a template to synthesize a complementary DNA strand. (p. 364)

15. **B** A genomic library consists of fragmented genomic DNA that has been inserted into cloning vectors. Therefore, it is a form of recombinant DNA. (p. 366)

Chapter 14

1. **B** RNA polymerase II synthesizes mRNA in eukaryotic cells. rRNA and tRNA are synthesized by RNA polymerases I and III, respectively. (RNA polymerase III also synthesizes 5S rRNA.) (p. 373)

2. **A** Most prokaryotic promoters are characterized by two highly conserved consensus sequences centered at the -35 and -10 positions. The sequence at -10 is the Pribnow box. (p. 374)

3. **D** Only release factor proteins are required to terminate translation once a termination codon has been encountered. Factor ρ is involved in the termination of transcription, not translation. (pp. 375, 394)

4. **C** Polyribosomes are structures consisting of multiple ribosomes bound to a single mRNA. They form because each ribosome moves away from the mRNA ribosome binding site after translation has commenced, rendering the binding site accessible to another ribosome. (p. 398)

5. **C** TATA and CAAT boxes are characteristic of promoters recognized by RNA polymerase II. (p. 375)

6. **B** Both prokaryotic and eukaryotic RNA polymerases recognize promoter and terminator sequences. Eukaryotic RNA polymerases do not require ρ factor, but it sometimes required by prokaryotic RNA polymerases. (p. 374, 375)

7. **B** RNA polymerases do require single-stranded DNA as template, and they synthesize 5′ to 3′ phosphodiester bonds between ribonucleotides. (p. 373)

8. **A** RNA processing in the conversion of the primary transcript into mature RNAs. It may include the removal of flanking sequences, intergene spacers, and intragene introns. Exons consist of coding sequence and are not excised. (p. 380)

9. **A** The nuclear genetic code is degenerate, universal, has start and stop punctuation codons, and allows for wobble in codon-anticodon base pairing. (pp. 390, 391)

10. **D** Erythromycin, streptomycin, and rifampicin all are antibiotics that specifically interfere with bacterial gene expression. α-Amanitin, on the other hand, is a toadstool toxin that interferes with eukaryotic RNA polymerases II and III. (pp. 378, 399, 400)

11. **D** Poly A tails are attached to most eukaryotic mRNAs during posttranscriptional processing. (p. 384)

12. **A** Repressor proteins regulate gene expression by binding to operator regions. The binding of repressors is regulated by either corepressors or inducers. (p. 387)

13. **D** The operator is a segment of DNA that is involved in the control of gene expression in prokaryotes; it binds repressors which prevent transcription. (p. 387)

Chapter 15

1. **D** The peptide hormones and catecholamines bind to surface receptors, while the lipid-soluble hormones bind to intracellular receptors. Norepinephrine is a catecholamine, whereas the other three are all lipid soluble. (p. 404)

2. **A** cAMP activates cAMP-dependent protein kinases, which phosphorylate specific proteins. The resulting phosphorylated proteins usually express modified activity. (p. 406)

3. **B** The mineralocorticoid aldosterone functions to regulate the Na^+/K^+ ratio. The other steroid hormones do not regulate water balance and would not be secreted. (pp. 411, 414)

4. **A** The Leydig cells of the testes are the primary sites of androgen synthesis, and the follicle cells of the ovary are the pri-

mary site of estrogen synthesis. However, small amounts of androgens are synthesized in the ovary and small amounts of estrogens are synthesized in the testes. (p. 411)

5. **A** ACTH stimulates the synthesis of cortisol, and CRF stimulates the secretion of ACTH. Therefore, when blood levels of cortisol decrease, CRF will be secreted from the hypothalamus and will stimulate the pituitary to secrete ACTH. (p. 414)

6. **D** Parathyroid hormone increases the absorption of Ca^{2+} by the kidney. (p. 417)

7. **D** Low blood levels of T_3 and T_4 stimulate the secretion of TRF from the hypothalamus. TRF then stimulates TSH from the pituitary. TSH finally stimulates the synthesis and secretion of T_3 and T_4 by the thyroid gland. High blood levels of T_3 and T_4 exert feedback inhibition over the secretion of TSH and TRF. (Somatostatin also inhibits the secretion of TSH.) (p. 421)

8. **C** The secretion of hypothalamic hormones is controlled by both neuronal and hormonal stimuli. (p. 429)

9. **D** Oxytocin is secreted by the posterior pituitary. (p. 434)

10. **C** The cholesterol desmolase complex is the rate-limiting enzyme of steroid hormone biosynthesis. It catalyzes the reactions that convert cholesterol to pregnenolone. (p. 412)

11. **A** Steroid hormone inactivation occurs mostly in the liver where the active species may undergo oxidation and/or reduction prior to conjugation with sulfate or glucuronate and excretion in the urine. (pp. 415, 416)

12. **C** Thyroxine and triiodothyronine stimulate the basal metabolic rate and their receptor proteins are located in the nucleus of target cells. (p. 420)

13. **D** Insulin lowers serum glucose, increases the synthesis of triacylglycerols, and increases protein synthesis. Glucagon stimulates glycogen degradation, gluconeogenesis, and protein and lipid degradation. (pp. 423, 424)

14. **B** Norepinephrine and epinephrine oppose the action of insulin and support the action of glucagon. The catecholamines are synthesized and stored in granules in the adrenal medulla until secretion. They are secreted in emergency circumstances and bind to both α and β-adrenegic receptors. (pp. 427, 428)

15. **E** FSH and LH directly participate in stimulating the synthesis of the steroids. However, the secretion of FSH and LH is stimulated by FSHRF and LHRF, respectively. (pp. 415, 433)

Chapter 16

1. **B** Albumin participates in the transport of a variety of substances in the plasma and in the regulation of plasma osmotic pressure. (p. 440)

2. **D** Tissue enzymes are usually released into the plasma from of tissues damaged by injury or disease. The presence and levels of such plasma enzymes are often useful diagnostic tools. (p. 441)

3. **C** Both dilute [H^+] and 2,3-bisphosphoglycerate can bind to hemoglobin, and tend to cause oxygen to dissociate. The displacement of O_2 by [H^+] is part of the Bohr effect. (p. 444)

4. **B** Hemoglobin reacts with CO_2 to form carbamino groups. It also promotes

the dissociation of H_2CO_3 to H^+ and HCO_3^- by binding the released H^+, thus allowing HCO_3^- to be formed and carried to the lungs without significantly changing the plasma pH. (pp. 445, 446)

5. **D** The major protein fraction of blood plasma is albumin. (p. 440)

6. **A** Hemoglobin is normally present in red cells and not in plasma. (p. 441)

7. **D** β-Carboxylation of glutamate does not occur. Each of the other reactions is important to the formation of blood clots. (p. 454)

8. **A** The intrinsic pathway is triggered when Factor XII makes contact with a surface. (p. 453)

9. **B** The extrinsic pathway clots blood very rapidly (within 10 to 12 seconds). (p. 452)

10. **D** Fibrinolysis is the proteolytic removal of blood clots, which occurs after the injured vessel is sufficiently healed. (p. 454)

Chapter 17

1. **C** Blood plasma and the interstitial fluid are the major subcompartments of the extracellular fluid. Cerebrospinal fluid, lymph, synovial fluid, and aqueous humor are all volumetrically minor. (p. 459)

2. **A** Lowered extracellular pH means that the $[HCO_3^-]/[H_2CO_3]$ ratio is lowered. This change can be compensated by lowering $[H_2CO_3]$. That is accomplished by increasing the rate and depth of breathing, which lowers the alveolar PCO_2 and therefore increases the rate at which CO_2 moves from the blood into the alveoli. More CO_2 is therefore exhaled in the breath. (pp. 461, 462)

3. **E** The kidneys compensate for changes in pH by three different mechanisms, depending on the severity of the acid challenge. A rise in plasma pH is compensated by incomplete resorption of HCO_3^- from the tubules. Low pH is corrected by excretion of titratable acid (low to moderate acid challenge) and by excretion of ammonium salts (severe acid challenge). The kidneys maintain ECF pH in the normal range by resorbing HCO_3^-. (pp. 462, 463, 465)

4. **A** Na^+ is the major extracellular cation. (p. 460)

5. **A** When the Na^+ content of the extracellular fluid increases, the volume (water content) must also increase in order to maintain the proper Na^+ concentration. (p. 460)

6. **D** The ratio of $[HCO_3^-]/[H_2CO_3]$ in extracellular fluid under normal conditions is 20:1. (p. 461)

7. **C** The renal excretion of ammonium salts is the most effective method of eliminating large quantities of acid via the urine. (p. 465)

8. **A** Aldosterone regulates extracellular fluid volume by stimulating Na^+ reabsorption. (p. 460)

9. **B** Vasopressin regulates extracellular fluid osmotic pressure by stimulating water reabsorption. (p. 461)

Chapter 18

1. **D** Fiber consists of the polysaccharides from fruit and vegetable sources that are indigestible or only poorly digestible. Therefore, they cannot be effectively used for energy. (p. 472)

2. **B** The concentration of protein inside plant cells is less than that inside animal cells. (p. 471)

3. **A** Menkes' syndrome is due to copper deficiency resulting from inadequate copper absorption. (p. 486)

4. **A** Thiamine is converted to thiamine pyrophosphate, the coenzyme for α-keto acid decarboxylation reactions. (p. 473)

5. **B** Riboflavin is converted to FAD, which is a participant in many oxidation-reduction reactions. (p. 474)

6. **D** Folic acid is the dietary precursor of tetrahydrofolate, which is the primary cellular carrier of one-carbon units and participates in one-carbon transfer reactions. (p. 476)

7. **C** Pyridoxine is converted to pyridoxal phosphate and to pyridoxamine phosphate. These coenzymes are interconvertible and participate in transamination, amino acid oxidation, and amino acid decarboxylation reactions. (p. 475)

8. **E** Several functions have been attributed to vitamin A, including participation in secretory glycoprotein synthesis, regulation of gene expression, vision, prevention of keratinization, and maintenance of a healthy epithelium. (p. 479)

9. **D** The function of vitamin E is not yet known with certainty, but it is believed to act as an antioxidant that contributes to the prevention of membrane lipid oxidation. (pp. 481, 482)

10. **C** Adult men and nonpregnant, non-lactating women normally do not require iron supplementation. The groups that are more likely to need supplemental iron are children and pregnant and lactating women. (p. 484)

Chapter 19

1. **E** Striated muscle sarcomeres consist of thin and thick filaments. Thin filaments contain actin, tropomyosin, and troponin. Thick filaments consist of myosin bundles. (p. 494)

2. **A** Thin filaments consist of F-actin, tropomyosin, and troponin. (p. 494)

3. **B** Muscle cells contain ATP, phosphocreatine, and creatine. (p. 501)

4. **B** Under conditions of extreme physical exertion, the AMP kinase reaction (2 ADP → ATP + AMP) and the AMP deaminase reaction (AMP + H_2O → IMP + NH_3) combine to replenish depleted ATP levels. Since the AMP kinase reaction is reversible, removal of AMP by AMP deaminase prevents the reversal of the AMP kinase reaction. Under less extreme conditions, AMP kinase may act alone to replenish ATP. (p. 502)

5. **D** There is no S line in striated muscle. (p. 494)

6. **B** Actomyosin is only formed during the contraction cycle when tropomyosin changes position, allowing actin and myosin to interact. (pp. 500, 501)

7. **B** Both myosin globular heads interact with one actin filament. (p. 501)

Chapter 20

1. **C** Both the cornea and the lens are transparent structures. (pp. 505, 506)

2. **D** The retina contains both cone and rod cells. Both are photoreceptor cells. Rod cells are responsible for vision in dim light, and cone cells are responsible for vision in bright light and for color vision. (p. 506)

3. **A** The cornea is exposed to the environment. (p. 505)

4. **A** The transparency of the cornea is due to the fine structure, uniform size, and uniform distribution of collagen fibrils. (p. 505)

5. **D** Rod cells are responsible for vision in dim light. Cone cells are responsible for vision in bright light and for color vision. (p. 506)

6. **C** During the normal vision mechanism in rod cells, rhodopsin dissociates into opsin and all-*trans*-retinal. (pp. 506, 508)

Chapter 21

1. **B** The plasma membrane contains more glycoproteins than glycolipid. (p. 515)

2. **A** The major components of biologic membranes are protein and lipid. (p. 514)

3. **D** Liposomes are bilayer vesicles that contain an intravesicular space. (p. 516)

4. **D** Integral membrane proteins usually have helical structures embedded in the membrane, and can only be removed from the membrane by very drastic treatments. (pp. 514, 519)

5. **C** In order for a change in fatty acid composition to increase membrane fluidity, the change must be from a longer fatty acid to a shorter one or from a less unsaturated fatty acid to a more unsaturated one. The only change consistent with this is oleic acid → palmitoleic acid. (p. 517)

6. **C** Secondary active transport systems utilize the energy released during the cotransport of the second solute down its concentration gradient to drive the transport of the first solute against its concentration gradient. (p. 521)

7. **A** Monolayers, micelles, and bilayers may result from the interaction of lipids with water. Only very few individual lipid molecules will be dissolved. (pp. 515, 516)

8. **B** Membrane proteins do not flip-flop between monolayers. The movement of membrane proteins is more likely to be influenced by the association of intracellular proteins with the membrane. (p. 519)

9. **B** Individual proteins may be associated with either monolayer, but they are only associated with the membrane in certain orientations. (p. 520)

10. **A** Facilitated diffusion does involve a transport protein, but it does not require the expenditure of energy, and it is not limited to lipid-soluble solutes. Facilitated diffusion may also involve antiport and symport mechanisms. (p. 521)

Chapter 22

1. **A** Collagen polypeptide chains contain large numbers of glycine, hydroxylysine, and proline residues. The approximate formula for collagen polypeptide chains is (gly-X-Y)$_n$, where X and Y are often proline and hydroxyproline or hydroxylysine, respectively. (p. 530)

2. **B** Hydroxylated residues present in collagen contribute to the stabilization of the triple helix by forming interchain hydrogen bonds. They also provide sites for glycosylation. (p. 530)

3. **A** Monosaccharides are transferred one at a time from their UDP derivatives to the growing glycosaminoglycan chain by specific glycosyltransferases. PAPS is used as a sulfating agent in the syn-

thesis of sulfated glycosaminoglycan chains. (p. 540)

4. **A** Glycosaminoglycan chains are usually anionic. They consist of disaccharide repeat units that contain glucosamine or galactosamine. (p. 536)

5. **B** L-Iduronic acid is a monosaccharide carboxylic acid. Hyaluronic acid, dermatan sulfate, and keratan sulfate are different types of glycosaminoglycan chains. (p. 536)

6. **C** Myosin is a muscle protein that participates in muscle contraction. (p. 494)

7. **D** Procollagen is first cleaved to yield tropocollagen, and then tropocollagen associates into fibrils. (pp. 530, 532, 533)

Chapter 23

1. **D** Antigens contain regions called antigenic determinants which are recognized by antibodies. (p. 546)

2. **B** Haptens are usually small molecules, and only stimulate antibody production when attached to a large carrier molecule. (p. 545)

3. **A** Antibody production is stimulated by antigens. (p. 546)

4. **E** Antigen binding sites are structural regions of antibody molecules that recognize and bind antigenic determinants. (p. 546)

5. **C** Antibodies bind antigens only at the antigenic determinant sites. (p. 546)

6. **A** Some antigens contain several antigenic determinants. (p. 546)

7. **A** The fork-like structure of antibody molecules contains two heavy and two light chains which are folded into eight constant domains (one per light chain, three per heavy chain) and four variable domains (one per chain). The heavy chains carry carbohydrate moieties. Each antibody molecule contains two antigen binding sites. (pp. 547, 548)

8. **C** Light chain variable regions are coded for by V and J genes. Heavy chain variable regions are coded for by V, D, and J genes. (pp. 550, 551)

Chapter 24

1. **B** Cells that conduct nerve impulse signals are called neurons. Dendrites are components of neurons. (pp. 555, 556)

2. **A** The axonal regions between myelin sheaths are called nodes of Ranvier. (pp. 556, 561)

3. **B** The space between the axon terminals and the target cell membrane is called the synaptic cleft. (pp. 556, 561, 562)

4. **C** Substances that are released into the synaptic cleft from the presynaptic membrane and that transmit the signal to the target cells are called neurotransmitters. (p. 561)

5. **D** Dendrites receive signals from other neurons or sensory cells. (p. 556)

6. **B** Calmodulin is believed to be involved in the Ca^{2+}-dependent release of neurotransmitters from the presynaptic vesicles. (p. 558)

7. **D** The potential difference across the resting axon membrane is about -60 mV. (p. 560)

8. **B** When the axon becomes stimulated, Na^+ ions enter the axon, causing its interior to become positively charged. (p. 560)

9. **D** The overshoot of depolarization is due to the large influx of Na$^+$ ions into the axon. (p. 560)

Chapter 25

1. **C** (+) and (−) RNA viruses contain single-stranded genomes. The genome of a (+) RNA virus, as opposed to a (−) RNA virus, corresponds to mRNA which directs viral protein synthesis. (p. 568)

2. **B** Infection of *E. coli* by bacteriophage T4 causes the expression of host cell genes to be terminated. Progeny T4 viruses are released by host cell lysis. (p. 570)

3. **A** Integration of viral DNA into the host genome occurs during the infection of a nonpermissive host. (p. 570)

4. **B** The (+) RNA strand associated with the replicative intermediate may be used either to direct protein synthesis or to generate new genomic DNA, but not both. (p. 573)

5. **D** VSV progeny are released from host cells by a budding mechanism. (p. 574)

6. **D** More than one provirus may be inserted per host cell genome. (p. 578)

7. **D** R17 is a (+) RNA phage that infects *E. coli*. (p. 572)

8. **C** Reovirus is a double-stranded RNA virus. (p. 572)

9. **B** Bacteriophage φX174 is a single-stranded DNA virus that infects *E. coli*. (p. 572)

10. **A** Bacteriophage T4 is a double-stranded DNA virus. (p. 568)

INDEX

Page numbers followed by f indicate figures; those followed by t indicate tables

A

A band, 494, 497f
Abetalipoproteinemia, 181
A block, 375
Acceptor stem, tRNA, 340
Acetic acid, from amino acids, 239
Acetoacetate
 amino acid degradation, 244, 245f
 tyrosine degradation, 246f, 247
Acetoacetate-succinyl CoA CoA transferase, 179
Acetoacetyl CoA, amino acid degradation, 248, 248f–249f
Acetone breath, 183
Acetylcholine, 496, 562–563
Acetyl CoA, 123, 124f
 amino acid degradation, 244, 245f
 from amino acids, 239
 binding to fatty acid synthetase complex, 160–161
 glycolysis, 91
 ketone body synthesis, 179, 180f
 oxidation, 121, 122f
 sources for palmitate synthesis, 163, 164f, 165
Acetyl CoA carboxylase
 diet effects, 179
 fatty acid synthesis, 178
 inhibition by palmitoyl CoA, 178
N-Acetylglucosamine, structure, 67
N-Acetylneuraminic acid, see Sialic acid
Acetyl transacetylase, 161
Acid
 definition, 585
 excretion, extracellular pH, 463, 464f, 465
 pK, definition, 585
 strength, 585
Acid-base catalysis, 30, 31f
Acid-base properties, of amino acids, 2–3, 2f–3f
Acid mucopolysaccharides, 63
Aconitase, TCA cycle reactions, 125
ACTH, 414, 431, 432f
Actin, 492, 500f
Action potential, 560
Activators, 387–388
Actomyosin, 498
N-Acylation, of amino acids, 268
Acyl carrier protein, 161
Acyl CoA dehydrogenase, 173
Acyl CoA synthetase, 157
Adduct formation, fatty acid biosynthesis, 161–162
Adenine
 in double helix, 323, 325f
 poly A tail, 384
 structure, 288f
Adenine phosphoribosyltransferase, 309
Adenohypophysis, 429
Adenosine, structure, 288f
Adenosine diphosphate, see ADP
Adenosine 5'-monophosphate, see AMP
Adenosine triphosphate, see ATP
S-Adenosylhomocysteine, 242, 243f
Adenylate, 290
Adenylate cyclase, activation, 405–406, 406f
Adenylosuccinate synthetase, inhibition, 297
Adipocyte
 glycolysis role, 91
 lipid mobilization and transport, 170–171
 structure, 150, 150f
 triacylglycerol synthesis, 188

609

Adipose tissue, *see* Adipocytes
A-DNA, 326, 327f
ADP, 56–57, 56
  energy status of cell, 103
  phosphorylation, 90
α-Adrenergic receptor, 78, 428, 428t
β-Adrenergic receptor, 76, 78, 427, 428t
Adrenocorticotropic hormone, 414, 431, 432f
Alanine, 5, 8, 232
  degradation, 240
  structure, 5f
  synthesis, 252
  tryptophan degradation, 247, 248f
β-Alanine, formation, 308f, 309
Albinism, 257
Albumin, 440
Aldofuranose, 66
Aldolase, glycolysis pathway, 95
Aldopyranose, 66
Aldose, 63
Aldosterone
  Na$^+$ renal excretion and, 460–461
  synthesis regulation, 414–415
Alkaptonuria, 257
Allopurinol, gout treatment, 316–317
Allosteric effectors
  gluconeogenesis, 112f
  TCA cycle control, 129, 130f
Allosteric regulation, enzymes, 44, 45f
Allosteric site, 44
Allosterism, cooperativity and, 44, 46
All-*trans*-retinal, 508, 508f
α-helix, of protein, 15, 16f
  bacteriorhodopsin, 519–520, 519f
  coiled coil, 18, 19f
α-Amanitin, RNA polymerases, 378
Aminoacetone pathway, 244
Amino acids, 2–10, 227–258; *see also* Glycogenic amino acids; Ketogenic amino acids; individual amino acids
  absorption, 230, 230f
  acid-base properties, 2–3, 2f–3f
  N-acylation, 268
  aliphatic, 5, 5f–6f, 8
  aliphatic hydroxyl, 6f–7f, 9
  aromatic, 6f, 8
  branched chain, enzyme deficiencies, 257–258
  circulation and uptake, 231–232, 231f
  classification, degradation products, 239
  collagen polypeptides, 530
  decarboxylation and hydroxylation, 261–264, 263f, 265f–267f, 268
    GABA, 261–262
    histamine, 262–263, 263f
    serotonin, 264, 267f, 268
    tyrosine derivatives, 263f, 264, 265f–266f
  definition, 2
  digestion of dietary protein, 228–229, 229f
  Edman degradation procedure, 13, 14f, 15
  endogenous protein digestion, 230–231
  essential, 250, 252, 252t
    dietary deficiency, 470
  fluorometric detection, 11–12
  found in proteins, 5f–7f
  gluconeogenesis from, 111
  ionized, 7f, 9
  isoelectric point, 3
  metabolism, 227, 228f
    alanine, 240
    alkaptonuria, 257
    arginine, 241
    asparagine, 241
    aspartate, 241
    branched chain, 244, 245f
    cystathionuria, 257
    cysteine, 240, 240f
    glutamate, 242
    glutamine, 242
    glycine, 240
    histidine, 241, 242f
    isoleucine, 244, 245f
    leucine, 244, 245f
    lysine, 248, 249f, 250
    methionine, 242, 243f, 244
    ornithine, 241, 241f
    phenylalanine, 246–247
    phenylketonuria, 256, 256f
    proline, 241, 241f
    propionyl CoA, 244, 246
    serine, 240
    sulfite oxidase deficiency, 257
    threonine, 244
    tissue distribution, 231–232, 231f
    tyrosine, 246f, 247
    tryptophan, 247, 248f
    tyrosinase deficiency, 256–257
    valine, 244, 245f
  modified, 9
  nerve tissue, 557–558, 558f
  ninhydrin detection, 11–12, 13f
  nonessential, 250, 252, 252t
    synthesis, 252–254, 253f–254f
  as nutrients, 470–471
  one-carbon chemistry, 250, 251f
  peptide bond, 4, 4f
  peptide chain, 4–5
  pK of ionizable group, 3
  sequence
    determination, 11–13, 13f–14f, 15
    in proteins, 1
  side chain, 2
  side chain polarity, 10
  structure, 2, 2f
  titration curve, 3, 3f
  transamidination, creatine synthesis, 269, 269f
  zwitterionic form, 2, 2f

D-Amino acid oxidase, 233–234
L-Amino acid oxidase, 233
Aminoacyl tRNAs, synthesis, 392
γ-Aminobutyrate shunt, 557, 557f
γ-Aminobutyric acid, 261–262
2-Amino-3-carboxymuconic semialdehyde, 247
β-Aminoisobutyrate, formation, 308f, 309
δ-Aminolevulinic acid, 275, 277
δ-Aminolevulinic acid dehydrase, 277
α-Aminolevulinic acid synthase, synthesis, 282
δ-Aminolevulinic acid synthetase, 274–275, 277
Aminopeptidases, 229
Aminotransferases, 232
Ammonia
   amino acid degradation, 232
   metabolism, 232–235, 235f
   toxicity, 256
Ammonium salts, renal excretion, 464f, 465
AMP, 56
   degradation, 307, 307f
   energy status of cell, 103
   structure, 288f
   synthesis, 296, 297f
AMP deaminase, as energy source for contraction, 502
Amphipathic lipids, 149, 515
   as surface active agents, 150–151
AMP kinase, as energy source for muscle contraction, 502
α-Amylase, 88
Amylopectin, 88
   degradation, 88, 89f
Amylose, 88
   degradation, 88, 89f
Anabolic processes, 53
Anaplerotic reactions, TCA cycle, 129–131
Androgens, 409
   synthesis site, 411
Anemia
   copper deficiency, 484
   folic acid deficiency, 476
   iron deficiency, 484
   vitamin B_{12} deficiency, 477
Angiotensin I and II, 414–415
Angiotensinogen, 414
Annealing, DNA, 328
Anomer, α and β, 64, 64f
Anomeric carbon, 64, 64f
   reducing end, 67
Antibiotics, that interfere with gene expression, 399–400
Antibody, 545
   antigen binding sites, 546
   classification, 548, 549f
   binding with antigen, 546, 547f
   light chain, 547, 548f
   mechanism of diversity, 549–551, 550f
      heavy chain genes, 550, 551
      light chain genes, 550–551, 550f
   proliferation, 546
   structure, 547–549, 548f
Anticoagulants, 455
Anticodon, tRNA, 388, 391
Anticodon loop, tRNA, 340
Antigen, definition, 545
   reaction with antibody, 546, 547f
Antigen binding sites, 546
Antigenic determinant, 546
α_2-Antiplasmin, 455
Antiport systems, 522, 523f
Apolipoprotein B deficiency, 181
Apolipoproteins, 157
Apurinic site, 358f, 359
Apyrimidinic site, 358f, 359
Arachidonic acid, 216–217, 217f
Arginase, 236, 241
   deficiency, 254–255
Arginine, 9, 236
   degradation, 241
   structure, 7f
   synthesis, 252
Arginine-glycine amidotransferase, 269
Argininosuccinate lyase, 236
   deficiency, 254
Argininosuccinate synthetase, 236
   deficiency, 254
Argininosuccinic aciduria, 254
Aromatase, 412
Aromatic amino acids, 6f, 8
Arthritis
   alkaptonuria, 257
   gouty, 316
Ascorbic acid, structure, 479f
Asparaginase, 241, 252
Asparagine, 9
   degradation, 241
   N-glycosidic bond of glycoproteins, 68–69
   hydrolysis, 234
   structure, 7f
   synthesis, 252
Asparagine synthetase, 252
Aspartate, 9
   degradation, 241
   malate-aspartate shuttle, 99
   structure, 7f
   synthesis, 252
Aspartate carbamoyltransferase, 300
Atherosclerosis, dietary recommendations, 471–472
ATP
   energy status of cell, 103
   gluconeogenesis, reactions requiring, 106
   hydrolysis, 56, 58
   net yield, from palmitate oxidation, 174
   regeneration, 57
   structure, 57f

ATP (*Continued*)
  synthesis, 56, 122, 139
    glycolysis, 90, 93
    oxidative phosphorylation, 141
ATPase, activity of myosin, 493
ATP synthetase complex, 139, 141
  oxidative phosphorylation, 141, 142f
Avidin, 476
Axon, 556, 556f

B

Bacteriophages, 363
Bacteriophage T4, 568, 569f, 570, 571f
  assembly, 570, 571f
Bacteriophage ɸX174, 572
Bacteriorhodopsin, helical segments inserted into bilayer, 519–520, 519f
Bases, 585
Bathorhodopsin, 508
B cells, 546
B-DNA, 326, 327f
Benzoate therapy, urea cycle enzyme deficiencies, 255
Beriberi, 473
$\beta\alpha\beta$ unit, 18, 19f
β-barrel, 16
β-meander, 18, 20f
β-pleated sheet, 15–16, 17f
β-turns, 16, 18, 18f
Biantennary structure, 71
Bicarbonate
  blood pH and, 446–447
  resorption, 463, 464f
Bile acids, 214–215, 214f–215f
Bile salts, 155
Bilirubin, from heme, 278, 279f
Bilirubin diglucuronide, synthesis, 278, 280, 280f
Biliverdin, reduction, 278, 279f
Biliverdin reductase, 278
Bioenergetics, 53–58
  biochemical thermodynamics, 54
  cellular energetics, 56
  definition, 54
  high energy phosphate compounds, 56–58, 57f
Biologic membranes, 513–514; *see also* Plasma membranes
  association of cholesterol with lipid bilayers, 517f, 518
  asymmetry, 520
  Ca^{2+} pump, 524
  chemical and physical properties, 514
  composition, 514–515, 515f
  depolarization, 496, 498
  dynamics, 516–520, 517f–519f
  fluidity, 517–518
  fluid mosaic model, 516, 517f
  interaction of integral proteins with lipid bilayer, 519–520, 519f
  interaction of membrane lipids with water, 515–516, 516f
  ionophores, 524–525, 524f–525f
  ion transport systems, 522, 524
  Na^+-K^+ ATPase antiport pump, 522, 524
  protein movement in bilayer, 519
  transport, 520–522, 521f–523f, 522t
    active, 521, 522f–523f, 522t
    antiport systems, 522, 523f
    facilitated diffusion, 521, 522t
    mediated, 521, 521f
    nonmediated, 520–521, 521f
    symport systems, 522, 523f
Biologic value, proteins, 470–471
Biotin, 475–476, 476f
1,3-Bisphosphoglycerate, 96
2,3-Bisphosphoglycerate, 444
Blood
  composition, 439
  glucose levels, 73
  pH and plasma bicarbonate, 446–447
Blood clotting, 447–450, 447f, 450f–454f, 452–454
  activation of fibrinogen to fibrin, 449
  activation of prothrombin to thrombin, 449
  extrinsic pathway, 448, 448f, 452, 452f, 454
  fibrin, 448
    cross-linking, 450, 451f
  fibrinogen, 448
  FSF activation, 449–450
  hemophilia, 455
  inhibition, 455
  intrinsic pathway, 448, 448f, 452–454, 453f
  vitamin K role, 454, 454f, 483
Boat conformation, 66, 66f
Bohr effect, 444
Brain, glycolysis role, 91
Branching enzyme, 76, 77f
Briggs-Haldane equation, assumptions, 33

C

CAAT box, 375
Calcitonin, 416–418, 417t
Calcium, 418, 420f, 483
Calcium ion
  as hormone intracellular messenger, 416
  as second messenger, 407, 408f–409f
Calcium-regulating hormones, 416–418, 417t, 419f–420f
  1,25-dihydroxycholecalciferol, 417–418, 417t, 419f
  parathyroid hormone, 417, 417f
Calmodulin, 407–409, 408*f*
  activation of cellular proteins, 407, 409f
  Ca^{2+} binding sites, 407, 408t
  nerve tissue, 558

cAMP
    mediation of glycogen degradation, 78
    as second messenger, 405–407, 406f
Ca^{2+} pump, 496, 524
Carbamino groups, formation, 445–446
N-Carbamoylaspartate, 300
Carbamoyl phosphate, 236, 299–300
Carbamoyl phosphate synthetase I, 236
    deficiency, 254
Carbamoyl phosphate synthetase II, 300, 302
Carbohydrates, *see also* Monosaccharides;
        Oligosaccharides; Polysaccharides
    in biologic membranes, 515
        asymmetry, 520
    classification, 61
    definition, 61
    dietary, 87, 90, 472
    digestion, absorption, and cellular uptake, 88, 89f, 90
    Fisher projections, 62f, 63
    metabolism, insulin effects, 424
    nerve tissue, 557
    nondigestible, 88
    pyruvate metabolism, 97–98
    structure, 71
Carbon dioxide, transport in blood, 443–444
Carbonic acid
    dissociation, 446–447
    extracellular fluid, 461
Carbonic anhydrase, 444
γ-Carboxylation, glutamate residues, 454, 454f, 482
Carboxypeptidases A and B, 229
Carnitine acyltransferase I
    fatty acid transport, 172
    fatty acid degradation, 178–179
Carnitine acyltransferase II, fatty acid transport, 172–173
Carnitine fatty acyl carrier system, 172–173, 172f
β-Carotene, conversion to retinoic acid, 479, 480f
Cartilage, keratan sulfate, 540
Catabolic processes, 53
Catalase, 233
Cataracts, 509–510
Catecholamines, 404, 427–429, 427f, 428t
    biologic effects, 428–429, 428t
    mechanisms of action, 427–428, 428t
    metabolism and excretion, 264, 266f, 429
    methylation, 264
    receptors, 76, 78
    relative potencies, 427, 428f
    storage and secretion, 429
    synthesis, 264, 265f
Catechol-O-methyltransferase, 264
C-DNA, 326, 327f
cDNA, 363, 364f
CDP-choline:ceramide cholinephosphotransferase, 202

CDP-diacylglycerols, synthesis, 193, 194f, 195, 196f
Cellobiose, structure, 68f
Cellular energetics, 56
Centromere, 335
Ceramidase, 207
Ceramide, 201, 209
    degradation, 207
    sphingolipid degradation to, 207, 208f
    synthesis, 202, 203f
Ceramide monosaccharides, 204f, 205
Ceramide oligosaccharides, 205–206
Cerebrosides, 204f, 205
Ceruloplasmin, 440
    deficiency, 485
cGMP, as second messenger, 407
Chair conformation, 66, 66f
C_H1, C_H2, and C_H3 domains, 547
Chemical mutagenesis, 353, 354f
Chenodeoxycholic acid, structure, 214f, 215
Chloride, dietary, 483
Chloride shift, 446
Cholecalciferol, structure, 481f
Cholecystokinin, 155
Cholesterol, 209–211, 209f–210f, 212f, 213
    association with lipid bilayers, 517f, 518
    biologic roles, 209
    conversion to pregnenolone, 412
    dietary, 211
    excretion, 214–215
    in liver, 213
    metabolism, 211, 214
    plasma free, 223
    structure, 209
    synthesis, 210–211, 210f
    transport and cellular uptake, 211, 213
Cholesterol desmolase, 412
Cholic acid, structure, 214f, 215
Choline kinase, 192
Choline phosphotransferase, 192
Cholinergic synapse, function, 561, 562f
Chondroitin sulfate, 536–537
Chromatin, 332, 333f–334f, 380
Chromatosome, 332
Chromium, dietary, 486
Chylomicrons, 171
    cholesterol transport, 211
    lipid transport in, 157–158, 158f
    remnants, 211
    triacylglycerol removal, 211
Chymotrypsinogens, 228
Chymotrypsins, 229
Citrate, 103, 178
Citrate cleaving enzyme, 165
Citrate lyase, 165
Citrate shuttle, 163, 164f, 165
Citrate synthase, TCA cycle reactions, 125
Citric acid cycle, *see* TCA cycle

Citrulline, 236
Citrullinemia, 254
  carrying desired passenger DNA, 366, 367f, 368
Cloning, passenger DNA, 366, 367f
Cloning vector, 363
  clone detection, 365–366
Clotting factors, 448. See also Blood clotting
Coagulation, see Blood clotting
Cobalamin, structure, 477, 478f
Codon
  initiation, 390–391
  mRNA, 388, 390
  punctuation, 390–391
  reading, 390
  termination, 390–391
Coenzyme A, 123
  panthothenate as precursor, 475
  structure, 124f
  synthesis, 312, 314f–315f
Coenzyme Q, 131, 133, 133f
Coenzymes, 27–28
Cofactors, 27–28
Colipase, 155
Collagen, 529, 530, 531f–534f, 533–534; see also Tropocollagen
  biosynthesis, 530, 532f–534f, 533f
  degradation, 534
  polypeptide chains, 530
  posttranslational modification, defect, 541
  triple-helix structure, 530, 531f
Collagenase, 534
Competitive inhibition, 38–39, 39f
Competitive inhibitors, definition, 38
Complementary DNA, 363, 364f
Condensing enzyme, 161, 163
Cone cells, retinal, 506, 509
Congenital erythropoietic porphyria, 281
Connective tissue, 529
  chondroitin sulfate, 536–537
  collagen, 530, 531f–534f, 533–534
  dermatan sulfate, 538
  elastin, 534–536, 535f
  fibronectin, 536
  heparan sulfate, 538
  heparin, 537–538
  hyaluronic acid, 537
  keratan sulfate, 538, 540
  proteoglycans, 529, 536–538, 539f–540f, 540
Consensus sequence, 374–375
Constitutive proteins, 386–387
Converting factor, 414
Cooperativity, 44, 46
Coordinate control, 82
  gluconeogenesis, 111
Copper, dietary, 484–486
Coproporphyrinogen III, 277, 278f
Coproporphyrinogen oxidase, 276
Corepressor, 387

Cori's disease, 82
Cornea, 505, 540
Corticosteroids, 409, 411, 411t, 416
Corticotropin releasing factor, 414
Cortisol, synthesis regulation, 414
Coupled reactions, high energy phosphate compounds, 58
Covalent catalysis, enzymes, 30, 32f
Covalent modification, enzymes, 46
Creatine, synthesis, 269, 269f
Creatine kinase, 268
Creatine phosphate, see Phosphocreatine
Creatinine, synthesis, 269, 270f
Cristae, 123
Critical micelle concentration, 516
Cross-linking
  elastin, 535
  fibrin, 450, 451f
  tropocollagen, 533–534, 533f–534f
Crystallins, 506
CTP, 57
CTP synthetase, 302
Cyanide poisoning, 145
Cyclic adenosine monophosphate, see cAMP
Cyclization, arachidonic acid, 216
γ-Cystathionase, 257
Cystathione-β-synthase, 257
Cystathionine-γ-lyase, 253
Cystathionuria, 257
Cysteine, 8–9
  degradation, 240, 240f
  γ-glutamyl, 230
  residue, 9
    carboxyl-terminal segment, 530, 533
  structure, 6f
  synthesis, 252, 253
Cytidine nucleotides, synthesis, 302
Cytidylate, 290
Cytochrome aa_3, cyanide poisoning and, 145
Cytochrome oxidase, carbohydrates, 137, 138f
Cytochrome reductase, electron transport chain, 135, 137
Cytochromes, electron transport reactions, 134
Cytosine, in double helix, 323, 325f
Cytosine triphosphate, 57
Cytosol, urea cycle, 237f
Cytosolic malate dehydrogenase, 99
Cytosolic receptors, 408–409, 410f

D

Deamination, 232–234
Debranching enzyme, 74, 75f
Degeneracy, 391
Dehydroascorbic acid, structure, 479f
7-Dehydrocholesterol, 481
Dehydrogenase, 124

Deiodination, thyroid hormones, 423
Demineralization, bone
    copper deficiency, 485
    vitamin D deficiency, 481
Denaturation, DNA, 326–328, 328f
Dendrites, 556, 556f
Deoxycholic acid, structure, 215f
Deoxyhemoglobin, 441, 444
Deoxyribonucleic acid, *see* DNA
Deoxyribonucleoside 5'-monophosphates, 290
Deoxyribose, 290
Deoxythymidine monophosphate, synthesis, 304–306, 305f–306f
Deoxythymidine triphosphate, 305–306, 306f
Dermatan sulfate, 538
Desmosine, structure, 535, 535f
Diabetes, cataracts and, 509–510
Diamine oxidase, 270
Diarrhea, chronic, vitamin C and, 478
Dihomo-γ-linoleic acid, structure, 217f
Dihydrobiopterin reductase, 246
Dihydroorotate, 300
Dihydrotestosterone, 412
Dihydroxyacetone, 63
Dihydroxyacetone phosphate, 95, 188, 195
1,25-Dihydroxycholecalciferol
    metabolism and excretion, 418
    physiologic effects, 417t, 418
    synthesis, 417–418, 419f
Dinitrophenol, 144
Dinucleotides, 289
Dipeptidases, 229
Direct deamination, ammonia, 234
Disaccharidase, 88
Disaccharides, 62–63, 67–68, 68f
    dietary, 87
    digestion, 88
    in glycosaminoglycans, 71
    glycosidic bond, 67
    nonreducing end, 67
    reducing end, 67
    repeat unit, 536–538
    structure, 68f
Disulfide bridge, 9
DNA
    cellular, fragmentation, 362–363
    chromosomal, 332, 333f
    complementary, construction, 363, 364f
    denaturation, 326–328, 328f
    excision repair, 357f–358f, 359
    frameshift mutations, 355, 356f
    hybridization, 330
    linker, 332
    mitochondrial, 399
    mutagenesis, 352–355, 354f–356f
    nucleotide sequences, 321, 333, 335, 336f, 338
        highly repetitive, 335
        moderatively repetitive, 335, 336t
        palindromes, 335, 336f
        renaturation, 335, 336f, 338
        single-copy, 335
    operator, 387
    point mutations, 352, 353f
    purine nucleotides, 322
    pyrimidine nucleotides, 322
    renaturation, 328, 329f, 330, 335, 336f, 338
    repair, 357, 357f–358f, 359
        photoactivation repair, 357, 359
    replication, 345–346
        bidirectional, 346, 348f
        DNA winding, 347
        elongation, 347
        E. coli, 346–347, 346f, 348f–350f, 349–351
        eukaryotes, 351–352, 351f
        initiation, 347
        lagging strand, 347, 350–351, 350f
        leading strand, 347, 349–350
        Okazaki fragment joining, 347
        priming, 346–347
        replication fork, 349–351, 349f–350f
        termination, 347, 348f
        unzipping of helix, 346, 346f
    replication errors, 355, 356f
    reversion, 359
    satellite, 335
    structure, 322–324, 323f–325f, 326–328, 327f–329f, 330
        double helix, 323–324, 325f, 326, 327f
    thymine dimer formation, 354–355, 355f
    topology, 330–333, 331f, 333f–334f
        bubble formation, 332
        nucleosomes, 332, 333f
        superhelices, 330–332, 331f
        topoisomerases, 332
    transcriptional units, 378
    unwinding, 347
DNA ligase, 347, 359
DNA methylases, 335
DNA polymerase I, 346, 359
DNA polymerase II, 346, 349
DNA polymerase III, 346, 350–351
DNA promoter, 378–379, 379f
DNA recombination, 359–362, 361f–362f
    general, 359–360, 361f
    site specific, 360, 362f
    transposons, 360–361
DNA viruses, 568, 569f, 570–572, 571f
    bacteriophage T4, 568, 569f, 570, 571f
    double-stranded, 568, 569f, 570–571, 571f
    early genes, 570–571
    late genes, 570
    quasi-late genes, 570
    replicative form intermediate, 572
    single-stranded, 572
Dopamine, synthesis, 264, 265f
Double displacement mechanism, 42

Double helix, DNA, 323–324, 325f, 326, 327f
  antiparallel nature, 324, 325f
  base-pairing, 323, 325f
  left-handed, 326, 327f
  major and minor grooves, 326
  right-handed, 326, 327f
  sugar-phosphate backbone, 324, 326
  unzipping, 346, 346f

E

Edman degradation procedure, 13, 14f, 15
Effector, allosteric, 44, 45f
EF-G, 394
EF-Tu, 394
Ehlers-Danlos syndromes, 541, 541t
Eicospentaenoic acid, structure, 219f
Elastase, 229, 536
Elastin, 529, 534–536, 535f
Electron transfer reactions, 131
Electron transport
  coupled with oxidative phosphorylation, 143–145, 143f
  reactions, 134, 135f
Electron transport chain, 121, 134–135, 136f, 137, 137t, 138f, 139
  complexes I–IV, 135, 137, 138f
  cyanide effects, 145
  cytochrome reductase, 135, 137
  inhibitors, 137, 137t
  NADH-Q reductase, 135
Elongation complex, DNA transcription, 379, 379f
Elongation factors, protein synthesis, 394, 397
Endergonic reactions, 54
Endonucleases, 322, 324f
  restriction, see Restriction endonucleases
  UvrABC, 359
Endoplasmic reticulum, 399, 496
Energy
  gluconeogenesis, 111
  metabolism, see Bioenergetics
  net yield, complete oxidation of glucose, 14–142, 142t
  sources, 159
  storage
    lipids, 149–150, 151f
    triacylglycerols, 159
  yield from glycolysis, 98–99, 100f, 101
Energy of activation, 28
  enzyme effects, 28–30, 28f
Enolase, glycolysis pathway, 97
β-Enoyl-ACP reductase, 162
Enoyl CoA hydratase, 173
env region, 576
Enzyme cascade, 46f, 47
  blood clotting, 447–450
  glycogenolysis, 78, 79f

Enzymes, 25–47
  acid-base catalysis, 30, 31f
  activity, 37
  apoenzyme, 28
  catalysis, 26
  classes, 26t
  coenzymes, 27–28
  cofactors, 27–28
  committed, 44
  covalent catalysis, 30, 32f
  definition, 25
  energy of activation effects, 28–29, 28f
  holoenzyme, 28
  inactivation, 38
  inhibition, 37–42, 39f–41f
    competitive, 38–39, 39f
    noncompetitive, 40–42, 41f
    reversible, 38
    uncompetitive, 39–40, 40f
  kinetics, 30, 33–37, 36f–37f
    determination of K_M and V_{max}, 36, 36f–37f
    enzyme activity, 37
    Michaelis constant, 34–35
    Michaelis-Menten equation, 35
    multisubstrate reactions, 42–43
    rate constants, 33
  orientation and proximity effects, 29–30
  phosphorylation, 406
  physiologic regulation, 43–44, 45f–46f, 46–47
    allosteric regulation, 44, 45f
    cascades, 46f, 47
    cooperativity, 44, 46
    covalent modification, 46
    feedback inhibition, 44
  plasma, as diagnostic tools, 441
  reaction rate, 43
  specific activity, 37
  specificity, 26–27, 27f
  substrate strain, 29
  turnover number, 37
Enzyme-substrate complex, 33
Enzyme units, 37
Epimers, definition, 63
Epinephrine
  structure, 427f
  synthesis, 264, 265f
Equilibrium constant, 55, 584
Ergocalciferol, structure, 481f
Erythromycin, interference with gene expression, 400
Erythrose 4-phosphate, 115
Escherichia coli, 345
  DNA replication, 346–347, 346f, 348f–350f, 349–351
    bidirectional mode, 346
  general recombination, 360, 361f
  lac operon, 388, 389f
  RNA phages, 572–573

RNA polymerase, 377–378
    site-specific recombination, 360, 362f
Esterases, 155
Estradiol, 412
Estrogens, 409
  metabolism and excretion, 416
  synthesis site, 411
Ethanolamine, structure, 190f
Ethanolamine plasmalogen, synthesis, 195, 197f
Eukaryotes
  DNA replication, 351–352, 351f
  gene expression, transcriptional, 388
  mRNA processing, 384, 385f–386f, 386
  promoters, 374–375, 376f, 377
  RNA polymerases, 378
  rRNA processing, 382, 382f
  transcription, 380
  translation, 394–397, 396f
    initiation factors, 395
  tRNA processing, 383
Eukaryotic cells, 322
Excision repair, DNA, 357f–358f, 359
Exergonic reactions, 54
Exons, mRNA, 384
Exonucleases, 322, 324f
Extracellular fluid, 459
  electrolyte composition, 459, 460t
  osmolarity regulation, 461
  pH regulation, 461–463, 464f, 465
    bicarbonate resorption, 463, 464f
    renal compensation, 462–463, 464f, 465
    renal excretion of ammonium salts, 464f, 465
    respiratory compensation, 462
    titratable acid excretion, 463, 464f, 465
  sodium regulation, 483
  volume regulation, 460–461
Extracellular matrix, 529
  proteins, 563; *see also* Collagen; Elastin

F

F-actin, 492, 494, 495f
Factor V, 454
Factor V_a, 449
Factor VII_a, 452
Factor VIII, 454
  deficiency, 455
Factor IX_a, 453
Factor X_a, 449, 452–454
Factor XI_a, 453
Factor XII_a, 453
Factor XIII, activation, 449–450
FAD^+, 131, 132f–133f, 133
$FADH_2$, 56, 133, 312, 313f
Fast-twitch fibers, 492
Fat cells, *see* Adipocytes

Fatty acids, 151, 152f, 153, 153t
  activation, 172
  biosynthesis, 159–163, 160f, 164f, 165
    acetyl CoA binding to fatty acid synthetase complex, 160–161
    adduct formation, 161–162
    adduct transfer to condensing enzyme, 163
    alkyl chain formation, 162
    malonyl CoA donor binding, 161
    malonyl CoA formation, 160
    palmitate, 159, 160f
  chain length, membrane fluidity and, 517
  containing *cis* double bonds, β-oxidation, 174–175, 175f
  degradation, regulation, 178–179
  desaturation, 167, 169, 169f
  α,ω-dicarboxylic, 176–177
  double bonds between β and γ carbons, β-oxidation, 175, 176f
  elongation, 165, 166f, 167, 168f
    microsomal, 165, 167, 168f
    mitochondrial, 165, 166f
  endogenously synthesized, fate, 170
  essential, 167, 471
    deficiency, 472
  exchange reactions, 195, 198f–199f
  hydroxylation, 169
  medium- and short-chain, 158
  methylated, 169
  monoenoic, 151, 167
  nomenclature, 151, 153t
  oxidation, 170, 171–177, 172f, 175f–177f
    activation, 172
    carnitine fatty acyl carrier system, 172–173, 172f
    containing odd number of carbons, 175–176, 177f
    entry into mitochondria, 172–173, 172f
    hydration, 173
    net ATP yield, 174
    thiolysis, 174
    unsaturated fatty acids, 174–175, 175f–176f
  α-oxidation, 176
  ω-oxidation, 176–177
  polyenoic, 151
    formation, 167, 169, 169f
  polyunsaturated, 167, 472
  saturated, 151, 471
  structure, 151, 152f
  synthesis
    acetyl CoA and NADPH sources, 163, 164f, 165
    regulation, 177–178
    short- and medium-chain, 169
  unsaturated, 151
    classes, 169, 169f
    degree of, membrane fluidity and, 517
    oxidation, 174–175, 175f–176f

Fatty acid synthase complex, 160, 163
  binding of acetyl CoA, 160–161
  diet effects, 179
Fatty acyl CoA cholesterol acyltransferase, 213
Feedback inhibition, enzymes, 44
Fermentation, lactate, 91
Ferritin, 484
Ferrochelatase, 277
  deficiency, 281
Fiber, dietary, 472–473
Fibrin, 448
  cross-linking, 450, 451f
  fibrinogen activation to, 449
  hard clot, 448
    proteolysis, 454
  soft clot, 449
Fibrinogen, 448–449
Fibrinolysis, 454–455
Fibrin stabilizing factor, activation, 449–450
Fibronectin, 536
Fisher projections, 62f, 63
Flavin adenine dinucleotide, 56, 133, 312, 313f
Flavin mononucleotide, 131, 132f–133f, 133
Fluidity, biologic membranes, 517–518
Fluid mosaic model, 516, 517f
Fluoride, dietary, 487
FMN, 131, 132f–133f, 133
$FMNH_2$, 133
Folic acid, 476–477
  structure, 47, 49f, 476f
Folicin, see Folic acid
Follicle stimulating hormone, 415, 433
Follicle stimulating hormone releasing factor, 415
Formate, tryptophan degradation, 247, 248f
N-Formylmethionine, 391
Frameshift mutations, 352, 353, 355, 356f
Free energy change, 54
  absolute value, 55–56
  for biologic reactions, 55
  equilibrium constant, 55
  minimum, oxidative phosphorylation, 141
  negative, 55
  positive, 55
  standard, 55, 131
Fructokinase, 105
Fructose
  entry into glycolysis, 105, 105f
  gluconeogenesis contribution, 111
  intolerance, 117
D-Fructose, structure, 62f, 65f
Fructose 1,6-bisphosphate, 2, 95, 103
  deficiency, 116
  in gluconeogenesis, 109
Fructose 2,6-bisphosphate, glycolysis regulation, 103
Fructose 6-phosphate, 94–95
Fructose 1-phosphate aldolase, 105
  deficiency, 117
Fructose 6-phosphate-1-kinase, 94

FSF, activation, 449–450
FSH, 415, 433
FSHRF, 415
Fumarase, TCA cycle reactions, 128
Fumarate, 239, 246f, 248
Furan, 65f, 66
Furanose, 66

G

GABA, 261–262
G-actin, 492
gag region, 576
Galactocerebroside, synthesis, 204f, 205
Galactokinase, 104
Galactose
  entry into glycolysis, 104–105, 104f
  gluconeogenesis contribution, 111
D-Galactose, structure, 62f
Galactose 1-phosphate, cataracts and, 509
Gangliosides, 201, 205
  biologic function, 202
  oligosaccharide sequences found in, 206–207, 207t
  synthesis, 206–207, 206f, 207t
Gene expression, see also Transcription
  antibiotics that interfere with, 399–400
  mitochondria, 399
  transcriptional control
    activators, 387–388
    E. coli lac operon, 388, 389f
    eukaryotes, 388
    prokaryotes, 386–388, 389f
    repressors, 387
  translational control, 399
Genetic code, 390–391, 390t, 392t
  degeneracy, 391
  differences between nuclear and mitochondrial, 392t
  punctuation codons, 390–391
  universality, 391, 392t
  wobble concept, 391
Genomic library, 366
Globosides, 205–206
Globulins, 440–441
Glomerular filtration rate, ECF Na^+ content and, 460
Glucagon, 423–424, 423t, 426
  actions, 426
  biosynthesis and secretion, 426
  catabolic effects, 424
  fatty acid synthesis, 178
  gluconeogenesis regulation, 112
  glycogen metabolism effect, 82
  mechanism of action, 426
Glucocerebroside, synthesis, 204f, 205
Glucocorticoids, synthesis, 414, 414f

Glucogenic amino acids, *see* Glycogenic amino acids
Glucokinase, 74, 93–94, 101
Gluconeogenesis, 106, 107f, 108–113, 110f, 112f
 allosteric effectors, 112f
 from amino acids, 111
 contribution of galactose, fructose, and mannose, 111
 coordinate regulation, 111
 energy considerations, 111
 fructose 1,6-bisphosphatase, 109
  deficiency, 116
 glucose 6-phosphatase, 110
 from lactate, 108–110, 108f
 pathway, 107f
 phosphoenolpyruvate carboxykinase, 108f, 109
 in pyruvate, 110–111, 110f
 pyruvate carboxylase, 109
 reactions
  requiring ATP, 106
  requiring GTP, 108
  requiring NADH, 108
 regulation, 111–112, 112f
Glucose
 blood levels, 73
 chair conformation, 66f
 complete oxidation, net energy yield, 141–142, 142t
 α and β configurations, 64f, 65
 phosphorylation, 93
D-Glucose, structure, 62f
α-D-Glucose, structure, 64f–65f
Glucose permease transport systems, insulin effect, 424–425, 425t
Glucose 6-phosphatase, 74, 94
 deficiency, 316
 in gluconeogenesis, 110
Glucose 6-phosphate
 conversion to fructose 6-phosphate, 94
 hydrolysis, 74
 oxidation, 113
 regulation, 101
Glucose 6-phosphate dehydrogenase, 113
Glucose phosphate isomerase, glycolysis pathway, 94
Glucose 1-phosphate uridylyl transferase, 74
Glucose residues, in glycogen, 73
D-Glucuronic acid, structure, 67
Glutamate, 9
 amino acid disaccharides, 241, 241f–242f
 degradation, 242
 malate-aspartate shuttle, 99
 oxidative deamination, 233
 residues, γ-carboxylation, 482
  vitamin K-dependent, 454, 454f
 structure, 7f
 synthesis, 235, 252–253
Glutamate α-decarboxylase, 262

Glutamate dehydrogenase, 130, 233, 252
Glutamic acid, decarboxylation, 261–262
Glutaminase, 242, 253
Glutamine, 9, 232
 degradation, 242
 hydrolysis, 234
 structure, 7f
 synthesis, 235, 253
Glutamine synthetase, 235, 253
γ-Glutamyl cycle, 230, 230f
ψ-Glutamylcysteinylglycine, 270, 272, 273f, 274
Glutathione, 270, 272, 273f, 274
Glutathione-dependent peroxidase, prostaglandin synthesis, 217
Glutathione S-transferase, 274
Glyceraldehyde, 63, 105, 105f
Glyceraldehyde 3-phosphate, 95–96, 115
Glyceraldehyde 3-phosphate dehydrogenase, glycolysis pathway, 96
Glycerol kinase, 188
Glycerol 3-phosphate, triacylglycerol synthesis, 188
Glycerol 3-phosphate dehydrogenase, 99
Glycerol phosphate shuttle, 99, 100f
Glycine, 5, 5f, 232
 degradation, 240
 source of one-carbon units, 250
 synthesis, 253
Glycine synthase, 253
Glycocholic acid, structure, 214f, 215
Glycogen, 63, 72f, 73; *see also* Glycogenolysis
 digestion, 88, 89f
 in liver, 73
 metabolism, 80, 82
 in muscles, 73
 structure, 72f, 73
 synthesis, 73–74, 76, 77f
  coordinate control, 82
  regulation, 80, 81f
 type III storage disease, 82
Glycogenic amino acids, 238–239, 239t
Glycogenolysis, 73–74, 75f
 coordinate control, 82
 enzyme cascade, 78, 79f
 hormonal regulation
  in liver, 76, 78, 79f
  in muscle, 78, 80
 inhibition, 78, 80
Glycogen phosphorylase, 73–74, 75f, 76
Glycogen synthase, 76, 77f
Glycogen synthase D, 80
Glycogen synthase I, 80
Glycolipids, 71
Glycolysis, 90–91, 92f, 93–97
 energy yield, 98–99, 100f, 101
 fructose entry, 105, 105f
 galactose entry, 104–105, 104f
 mannose entry, 106

Glycolysis (*Continued*)
  NADH shuttle mechanisms, 98–99, 100f, 101
  pathway, 92f, 93–97, 107f
    aldolase, 95
    energy investment portion, 92f, 93
    energy-trapping reaction, 96
    enolase, 97
    glucokinase, 93–94
    glucose phosphate isomerase, 94
    glyceraldehyde 3-phosphate dehydrogenase, 96
    hexokinase, 93–94
    lactate dehydrogenase, 97
    payoff portion, 92f, 93
    phosphofructokinase, 94–95
    phosphoglycerate kinase, 96
    phosphoglyceromutase, 96
    pyruvate kinase, 97
    rate-limiting step, 95
    triose phosphate isomerase, 95–96
  regulation, 101, 102f, 103
  respiratory control, 144
Glycoproteins, 68–69, 69f–70f, 71
  microheterogeneity, 71
  N- and *O*-glycosidic linkages, 68–69, 69f
Glycosaminoglycans, 71
  chains, molecular, 537–538
Glycosidic bond, 67
  N- or *O*-, 68–69, 69f–70f
    linking, DNA excision, 359
α-Glycosidic linkage, glycolysis, 205–206
Glycosphingolipids, 71, 201
  biologic function, 202
  neutral and acidic, 201
  synthesis, 204f–206f, 205–207, 207t, 208f, 209
    ceramide degradation, 207
    cerebrosides, 204f, 205
    gangliosides, 206–207, 206f, 207t
    globosides, 206
    sphingolipid degradation, 207, 208f
    sphingosine degradation, 207, 209
    sulfatide, 205–206, 205f
  types, 205
N-Glycosylases, 359
Glycosylation, procollagen, 533
Glycosyltransferases, 205
GMP
  degradation, 307f, 309
  synthesis, 296, 297f
Goiter, 486–487
Gout, 316–317
G protein, 405
Gramicidin A, structure, 525f
Growth failure, folic acid deficiency, 476
Growth hormone, 432, 433f
GTP, 57, 108
Guanine, in double helix, 323, 325f
Guanine-nucleotide-binding regulatory protein, 405

Guanosine triphosphate, 57, 108
Guanylate, 290

H

Hairpin loops, 16, 18, 18f
Hapten, definition, 545–546
Haptenic determinant, 546
Haworth projections, 65f, 66
HDLs, as cholesterol scavengers, 213
Heat denaturation, DNA, 326
Heavy chain
  antibody, 547, 548f
  genes, antibodies, 550f, 551
  myosin, 493
Helicase II, 347
Helicases, 347
3_{10} Helix, 15
Heme
  biosynthesis, 274–277, 278f
    interruptions in pathway and porphyrias, 281, 282f
    intracellular compartmentation, 277, 278f
    iron, incorporation, 277
    porphobilinogen synthesis, 274–275
    regulation, 277
    tetrapyrrole ring formation and modification, 275–276
  degradation, 278, 279f, 280–281
  elimination, 278, 279f–280f, 280–281
Heme-heme interaction, 442
Heme oxygenase, 278
Heme pocket, 441
Heme synthetase, 277
Hemiacetal, formation, 64, 64f
Hemiketal, formation, 64, 64f
Hemoglobin, 441
  α and β subunits, 441
  function, 442–447, 442f–443f, 445f
    Bohr effect, 444
    carbamino group formation, 445–446
    carbonic acid dissociation, 446–447
    chloride shift, 446
    cooperativity, 442–443, 442f–443f
    CO_2 transport, 445–446
    isohydric shift, 446
    oxygen and CO_2 transport and delivery, 443–445, 445f
    plasma bicarbonate and blood pH, 446–447
  Hill equation, 442
  loading and unloading of oxygen, 441
  oxygen-binding curve, 442, 442f
    H^+ and BPG effect, 444, 445f
  R and T state, 443, 443f
  sickle cell, 21–22
Hemopexin, 441
Hemophilia, 455

Heparan sulfate, 538
Heparin, 455, 537–538
HETEs, 220–221, 221f, 223
    synthesis, 221–222, 221f
Hexokinase, 74, 93–94, 101
Hexose monophosphate shunt, see Pentose phosphate pathway
HGPRTase deficiency, 316
High density lipoproteins, as cholesterol scavengers, 213
High-mannose structure, 69
Hill coefficient, 442
Hill equation, 442
Hippurate, formation, 255–256
Histamine, 262–263
Histidine, 9
    degradation, 234, 241, 242f
    source of one-carbon units, 250
    structure, 7f
Histidine-ammonia lyase, 234
Histidine decarboxylase, 262
Histones, 332
HMG-CoA reductase, synthesis, 211
Hogness box, 375
Holliday structure, 360, 361f
Homocysteine methyltransferase, 244
Homogentisate dioxygenases, 246f, 247
Homogentisate oxidase, deficiency, 257
Hormone action. See also individual hormones
    lipid-soluble hormones, 408–409, 410f
    second messengers, 404–408, 406f, 408f–409f
        calcium ion, 407, 408f–409f
        cAMP, 405–407, 406f
        cGMP, 407
Hormone-receptor complex, 404
Hormone system, 404
HPETE, 220
Human immunodeficiency virus, 576
Hyaluronic acid, 537
Hybridization, DNA, 330
Hydrogen bonds, 583, 584f
    between base-paired nucleotides, 323–324, 325f
Hydrogen ion, binding, 444
Hydrolases, 26t
Hydrolytic deamination, ammonia, 234
Hydronium ion, formation, 583
Hydroperoxyeicosatetraenoic acid, 220
β-Hydroxyacyl CoA dehydrogenase, 173
Hydroxyeicosatetraenoic acid, see HETEs
5-Hydroxyindoleacetic acid, 267f, 268
β-Hydroxyl-ACP dehydratase, 162
β-Hydroxyl acyl CoA dehydrogenase, fatty acid degradation, 178–179
Hydroxylation, 533
o-Hydroxyphenylacetate, 256, 256f
p-Hydroxyphenylpyruvate, 246f, 247
17-α-Hydroxypregnenolone, 412
Hydroxytryptophol, 267f, 268

Hyperarginemia, 254–255
Hyperchromic effect, 327, 328f
Hyperthyroidism, 434–435
Hyperuricemia, 316
Hypothalamic hormones, 429, 430t, 431
Hypothalamic inhibiting factors, 429, 430t
Hypothalamic releasing factors, 429, 430t
Hypoxanthine-guanine phosphoribosyltransferase, 309

I

I band, 494, 497f
IDLs, 211, 213
IgA, 548
IgM, 21f, 548, 549f
Imidazoleacetic acid, 263
Immunoglobulins, see also Antibody
    IgA, 548
    IgM, 21f, 548, 549f
IMP, 290
IMP dehydrogenase, inhibition, 297
Indoleacetic acid, 263
Induced-fit model, 27, 27f
Inducer, 387
Inducible proteins, 387
Infection, 570–571
Inhibiting factors, hypothalamic, 429, 430t
Initial-rate assumption, 33
Initiation, transcription, 378–379
Initiation codon, 390–391
Initiation complexes, 392, 394, 397
Initiation factors, 392, 395
Inosine monophosphate, 290
Insulin, 424–426, 425f–426f
    actions, 80, 423–424, 423t
    biosynthesis, 425, 426f
    fatty acid synthesis, 178
    gluconeogenesis regulation, 112
    glucose permease transport system effect, 424–425, 425f
    inhibition of glycogenolysis, 80
    mechanisms of action, 424–425, 425f
    secretion, 425–426
    stimulation of glycogen synthesis, 80
Intermediate density lipoproteins, 211, 213
Intracellular fluid, 459
    electrolyte composition, 460t
Intrinsic factor, vitamin B_{12} deficiency and, 477
Introns, 335, 386
Iodine, dietary, 486–487
Ion-carrier ionophores, 524, 524f
Ion-channel ionophores, 525
Ionized amino acids, 7f, 9
Ionophores, 524–525, 524f–525f
Ion transport systems, 522, 524
Iron, dietary, 484, 485f

Iron-sulfur centers, non-heme, 134, 135f
Irradiation, mutagenesis, 353–355, 355f
Isocitrate dehydrogenase, 125–126, 178
Isodesmosine, structure, 535, 535f
Isoelectric point, amino acids, 3
Isohydric shift, 446
Isoleucine, 8
   degradation, 244, 245f
   enzyme deficiencies and, 258
   structure, 6f
Isologous interactions, 20
Isomaltose, 67, 68f
Isomerases, 26t
Isozymes, 47

K

Kallikrein, 453
Keratan sulfate, 538, 540
Keratosis, xeroderma pigmentosum, 368
α-Keto acid, 233
α-Keto acid dehydrogenase, deficiency, 258
Ketoacidosis, 472
β-Ketoacyl-ACP reductase, 162
β-Ketoacyl-ACP synthetase, 161–162
α-Ketobutyrate, from threonine, 244
3-Ketodihydrosphingosine, 202
3-Ketodihydrosphingosine synthetase, 202
Ketofuranose, 66
Ketogenic amino acids, 238–239, 239t
Ketoglutarate, from amino acids, 239
α-Ketoglutarate, 233, 242
α-Ketoglutarate dehydrogenase, TCA cycle reactions, 127
Ketone bodies, 179, 180f, 472
Ketonemia, 183
Ketonuria, 183
Ketose, 63
Ketosis, 183
Kidneys, extracellular fluid
   regulation, 462–463, 464f, 465
   volume control, 460
Krebs cycle, see TCA cycle

L

lac operon, E. coli, 388, 389f
Lactate
   fermentation, 91
   generation, 97
   gluconeogenesis from, 108–110, 108f
Lactate dehydrogenase, glycolysis pathway, 97
Lactose, 67, 68f
Lagging strand, DNA, 347, 350–351, 350f
LCAT, 213
   deficiency, 223
LDLs, see Low density lipoproteins

Leader sequence, RNA, 380, 382
Leading strand, 347, 349–350
Lecithin:cholesterol acyltransferase, 213
   deficiency, 223
Lens, 506, 509
Leucine, 8
   degradation, 244, 245f
   enzyme deficiencies and, 258
   structure, 6f
Leukemia, treatment with methotrexate, 47, 48f–49f, 49
Leukopenia, folic acid deficiency, 476
Leukotrienes, 220–221, 222f
LH, 415, 433
LHRF, 415
Ligases, 26t
Light chain
   antibody, 547, 548f
   genes, antibodies, 550–551, 550f
   myosin, 493, 493f
α-Limit dextrins, 88
Lineweaver-Burk equation, 36, 37f
   competitive inhibition, 38–39, 39f
   noncompetitive inhibition, 41–42, 41f
   uncompetitive inhibition, 40, 40f
Linoleic acid, 167, 169
Linolenic acid, 167, 169
Lipases, 155, 170–171
Lipids, see also Fatty acids
   abetalipoproteinemia, 181
   absorption, 156f, 157–159
   amphipathic, 515
   in biologic membranes, 514
      asymmetry, 520
      association with cholesterol, 517f, 518
      bilayer structure, 516
      functions, 150
      interaction with integral proteins, 519–520, 519f
      interaction with water, 515–516, 516f
      movement in, 518, 518f
      protein movement, 519
   complex, 187
   digestion, 155, 156f, 157–159
   excess dietary, 471
   functions, 149–151, 150f
      biologic responses, 151
      energy storage, 149–150, 150f
      membrane structure, 150
      surface active agents, 150–151
   ketosis, 183
   metabolism, 154, 154f, 424
   mobilization, 170–171
   monolayer, 515, 516f
   mucosal cell membrane, 157
   nerve tissue, 559
   as nutrients, 471–472
   Refsum's disease, 181, 182f, 183
   transport in chylomicrons, 157–158, 158f

Lipoprotein lipases, 171
Lipoproteins, 440; see also Chylomicrons
  classes, 171
  high density, 213
  intermediate density, 211, 213
  low density, 181, 211, 213
  structure, 157, 158f
  triacylglycerol mobilization and transport, 171
  very low density, 171, 211
Liposome, 516, 516f
Lipoxygenases, 220
Lithocholic acids, structure, 215f
Liver
  glucokinase role, 93–94
  glycogen in, 73
  glycolysis role, 91
  hormonal regulation of glycogen degradation, 76, 78, 79f
  triacylglycerol synthesis, 188, 189f
Lock-and-key model, 27, 27f
Long terminal repeats, 578–579
Low density lipoproteins, 181, 211, 213
LTRs, 578–579
Lungs
  extracellular fluid regulation, 462
  hemoglobin, function, 447
Luteinizing hormone, 415, 433
Luteinizing hormone releasing factor, 415
Lyases, 26t
Lymphocytes, B cells, 546
Lysine, 9
  degradation, 248, 249f, 250
  structure, 7f
Lysosomal hydrolases, mucopolysaccharidoses, 540–541
Lysosomal sulfatases, mucopolysaccharidoses, 540–541
Lysosomes, cholesterol in, 213
Lysyl oxidase, 533

M

Magnesium, dietary, 484
Malate-aspartate shuttle, 99, 100f
Malate dehydrogenase, 128, 165
Malic enzyme, 130, 165
Malonyl CoA, 160–161
Malonyl transacetylase, 161
Maltose, 67, 68f
Mannose, 106, 111
D-Mannose, structure, 62f
Maple syrup urine disease, 258
Melanin, synthesis, 256, 263f, 264
Melting temperature, 327–328
Mercapturic acid, synthesis, 273f, 274
Messenger RNA, see mRNA
Metal ions, electron transport reactions, 134

Methionine, 8
  degradation, 242, 243f, 244
  structure, 6f
Methionine synthetase, 244
Methotrexate
  leukemia treatment, 47, 48f–49f, 49
  structure, 47, 49f
Methylation, catecholamines, 264
N^5,N^{10}-Methylenetetrahydrofolate, 240–241, 242f, 253
5'7-Methylguanylate cap, 338f, 384
N^1-Methylimidazoleacetic acid, 263
Methylmalonyl CoA mutase, 176
Methylmalonyl CoA racemase, 175–176
Micelles, 515–516, 516f
Michaelis constant, 34–35
Michaelis-Menten equation, 35
  assumptions, 33
  competitive inhibition, 38
  noncompetitive inhibition, 41
  plot, 36, 36f
  uncompetitive inhibition, 40
Microsomal fatty acid elongation, 165, 167, 168f
Mineralocorticoids, 411
Minerals, 469–470
  calcium, 483
  chloride, 483
  magnesium, 484
  phosphate, 483
  potassium, 483
  sodium, 483
  trace, see Trace minerals
Minor groove, DNA, 326
Mitochondrion, 122–123, 123f; see also Electron transport chain; TCA cycle
  electron transfer reactions, 131
  entry of fatty acyl CoAs into, 172–173, 172f
  fatty acid elongation, 165, 166f
  gene expression, 399
  inner membrane, 122–123, 123f
  intermembrane space, 122, 123f
  matrix, 123, 123f
  outer membrane, 122, 123f
  pyruvate dehydrogenase role, 123–124, 124f
  structure, 122, 123f
  urea cycle, 237f
Mitosol, 122
Mixed micelles, 157
M line, 494, 497f
Molybdenum, dietary, 486
Monoamine oxidase, 264
Monocistronic mRNA, 384
Mononucleotides, 289
Monosaccharides, 62, 62f, 63–67, 64f–67f
  absorption, 88
  aldehyde, 63
  definition, 61
  derived, 62, 66–67, 67f

Monosaccharides (*Continued*)
   in disaccharides, 67
   in glycoproteins, 69
   ketone, 63
   nomenclature, 63
   structure, 62f, 63–66, 64f–66f
Motor neurons, synaptic transmission, 562
mRNA, 321
   codon, 388, 390
   exons, 384
   introns, 386
   5'7-methylguanylate cap, 338f, 384
   poly A tail, 384
   polygenic, 384
   primary transcripts, 384, 385f
   processing
      eukaryotes, 384, 385f–386f, 386
      prokaryotes, 384
   structure, 338f–339f, 340
   transcription, RNA polymerase II, 375, 376f
   viral, 579
Mucopolysaccharidoses, 540–541
Mucosal cell membrane, lipids, 157
Muscle
   cardiac, 491
   fast-twitch fibers, 492
   fibrils, 496
   glycogen in, 73
   glycolysis role, 90–91
   hormonal regulation of glycogen degradation, 78, 80
   red fibers, 492
   slow-twitch fibers, 492
   smooth, 491–492
   striated, 491–492, 494
   structure, 496, 498f–499f
   T tubules, 496
   white fibers, 492
Muscle contraction, 491
   actin, 492
   AMP deaminase, 502
   AMP kinase, 502
   energy sources, 501–502
   interactions of actin and myosin, 500f
   mechanism, 496, 498–499, 500f, 501–502
   membrane depolarization, 496, 498
   myofilaments, 494, 495f–496f
   myosin, 493, 493f
   phosphocreatine, 501
   proteins, 492–493, 493f
   sarcomere, 494, 497f
   tropomyosin, 493
   troponin, 493
Mutagenesis
   chemical, 353, 354f
   DNA, 352–355, 354f–356f
   irradiation, 353–355, 355f
Mutarotation, 65

Mutations, 352–353, 359
Myelin basic protein, 558
Myelin proteolipid, 558
Myofibrils, 496, 498f
Myofilaments, 494, 495f–496f
   thick filaments, 494, 496f
Myosin, 493, 493f
   globular heads, 496, 498–499
   interactions during muscle contraction, 500f
   thick filaments, 494, 496f

N

Na^+ channels, 560
Na^+ concentration, extracellular fluid, volume regulation, 460
NAD^+, 131, 132f–133f, 133
   oxidized and reduced forms, 132f
   regeneration, 91
   synthesis, 310f–311f, 312
NADH, 56, 133
   cytosolic, 99
   functional roles in cells, 113
   gluconeogenesis, reaction requiring, 108
   glycerol phosphate shuttle, 99, 100f
   malate-aspartate shuttle, 99, 100f
   mitochondrial, 99
   oxidation, 141
   shuttle mechanism, 98–99, 100f, 101
      gluconeogenesis, 110–111, 110f
      glycolysis energy yield and, 101
NADH-Q reductase, electron transport chain, 135
$NADP^+$, 131, 132f–133f, 133
   oxidized and reduced forms, 132f
   synthesis, 312
NADPH, 56
   functional roles in cells, 113
   in pentose phosphate pathway, 115
   sources for palmitate synthesis, 163, 164f, 165
Na^+-K^+ ATPase antiport pump, 522, 524, 560
Nerve endings, 556, 556f
Nerve tissue, 555
   action potential, 558f–559f, 560
   amino acids, 557–558, 558f
   γ-aminobutyrate shunt, 557, 557f
   calmodulin, 558
   carbohydrate, 557
   cell body, 555, 556f
   glial cell, 555
   lipids, 559
   metabolism, 556–559, 557f–559f
   neuron, 555–556, 556f
   nucleotides, 558
   myelin sheath, 560–561, 561f
   neurofilaments, 558
   neurotubules, 558
   nodes of Ranvier, 561, 561f

Schwann cell, 560–561, 561f
structure, 555–556, 556f
TCA cycle, 557
Neurofilaments, 558
Neurohypophysis, 429
Neuromuscular junction, 496
Neuron, 555–556, 556f
Neuropathy, vitamin B_{12} deficiency, 477
Neurophysins, 434
Neurotransmitter molecules, 561–562
Neurotubules, 558
Niacin, 474
Nicotinamide adenine dinucleotide, see NADH
Nicotinamide adenine dinucleotide phosphate, see NADPH
Nicotinamide mononucleotide, 312
Nicotinate mononucleotide, 312
Nicotinic acid, 474
Night blindness, vitamin A deficiency, 479
Nitrite, cyanide poisoning treatment, 145
Nodes of Ranvier, 561, 561f
Noncompetitive inhibition, 41–42, 41f
Noncompetitive inhibitors, definition, 38
Non-heme iron-sulfur centers, 134, 135f
Nonoxidative deamination, ammonia, 234
Nonpermissive host, 570
Nonreducing end, disaccharides, 67
Norepinephrine, 264, 265f, 427f
N protein, 405
N-terminal residue, 4, 11, 12f
Nuclear receptors, 408–409
Nucleases, 322, 324f
Nucleation, 328
Nucleic acids, 321–322
purine and pyrimidine bases, 289t
structure, 340–341
Nucleocapsid, 573–574
Nucleofilament, 333, 334f
Nucleoside diphosphate kinase, 57, 127
Nucleoside diphosphate reductase, 304, 304t
Nucleoside phosphorylases, 310
Nucleosides, 289
Nucleosomes, DNA, 332, 333f
Nucleotide coenzymes, synthesis, 310f–311f, 312, 313f–315f
Nucleotide diphosphate kinase, 302
Nucleotide diphosphate reductase, 302–303, 304f
Nucleotide kinases, 311
Nucleotides, 558; see also DNA, nucleotide sequences; Purine nucleotides; Pyrimidine nucleotides; specific nucleotides
base-pairing properties and nitrous acid, 353, 354f
degradation, 306, 310–311
de novo synthesis, 290
function, 287

metabolism, 290
cancer chemotherapy and, 312, 316
regulation, 311
nomenclature, 289
structure, 288f, 288–290, 288f–289f, 289t
synthesis, 290
Nus A, 379
Nutrients, see also Minerals; Trace minerals; Vitamins
carbohydrate, 472
definition, 469
energy-yielding, 470–472
essential, 469
fiber, 472–473
lipids, 471–472
nonessential, 469
protein, 470–471
types, 469
Nutrition, 469–470

O

Obesity, 487
Ochronosis, 257
Okazaki fragment, 347, 349, 351
Oleic acid, 167
Oligopeptide
absorption, 230
definition, 4
degradation, 229
non-mRNA-dependent synthesis, 270, 272, 273f, 274
Oligosaccharidase, 88
Oligosaccharides, 63, 68–69, 69f–70f, 71; see also Glycoproteins
complex structure, 69
core structure, 69, 70f
definition, 61
dietary, 87
digestion, 88
glycolipids, 71
high-mannose structure, 69
sequences found in gangliosides, 206–207, 206t
OMP, 290
Oncogenic transformation, 570
One-carbon chemistry, 250, 251f
Open complex, DNA promoter, 379, 379f
Operator, DNA, 387
Operon, 384
activators, 387
lac, 388, 389f
Opsin, 508, 508f
Origin of replication
E. coli DNA, 347
eukaryotic DNA, 351, 351f
Ornithine, 236, 241, 241f
Ornithine decarboxylase, 270

Ornithine transcarbamoylase, 236
  deficiency, 254
Orotate, 300–301
Orotidine 5'-monophosphate, 301
Orotidine monophosphate, 290
Osmolarity, extracellular fluid, 461
Osmoreceptors, 461
Overwinding, superhelices, 332
Oxaloacetate
  from amino acids, 239
  in gluconeogenesis, 108f, 109
  malate-aspartate shuttle, 99
Oxidation, see also Mitochondrial oxidations
  energy released during, 121
  fatty acids, see Fatty acids, oxidation
α-Oxidation, fatty acids, 176
β-Oxidation, fatty acids, 173–174
ω-Oxidation, fatty acids, 176–177
Oxidative deamination, ammonia, 232–233
Oxidative phosphorylation, 122, 139, 140f, 141–142, 142f, 142t
  ATP synthetase complex, 141, 142f
  chemiosmotic model, 139, 140f, 141
  coupled with electron transport, 143–145, 143f
  energy transformations, 139
  inhibitors, 144–145
  minimum free energy change, 141
  net energy yield, complete oxidation of glucose, 141–142, 142t
  uncouplers, 144
Oxidoreductases, 26t
Oxygen transport. See hemoglobin
Oxygen binding, by hemoglobin, 442–443, 442f–443f
Oxyhemoglobin, 441–442
Oxytocin, 434

P

Palindrome, 335, 336f
  terminator, 375
Palmitate
  carbon atom sources, 160f
  oxidation, net ATP yield, 174
  synthesis, 159, 160f
    acetyl CoA and NADPH sources, 163, 164f, 165
Palmitoleate, formation, 167
Palmitoyl CoA, acetyl CoA carboxylase inhibition, 178
Palmitoleic acid, 167
Pancreatic hormones, 423–427, 423t, 425f–426f
Pancreatic lipase, 155
Panthothenate, 475
  coenzyme A synthesis, 312, 314f
Panthothenic acid, 475
Papain, peptide bond hydrolysis, 30, 32f

Parathyroid hormone, 416–417, 417f
Passenger DNA, 363
  cloning, 366, 367f
  detection of clones carrying, 366, 367f, 368
  insertional inactivation, 366
Pellagra, 474
Pentose phosphate pathway, 113, 114f–116f, 115–116
  cellular roles, 115
  nonoxidative phase, 113, 115, 116f
  oxidative phase, 113, 115f
  tissue utilization, 116
Pepsin, 228
Pepsinogen, 228
Peptide
  bond
    formation, 4, 4f
    hydrolysis, 30, 32f
    resonance stabilization, 4
  signal, 398
Peptide hormones, 404
Peptidyl transferase, 394, 397
Permissive host, 570
Pernicious anemia, 477
Peroxidase, glutathione-dependent, prostaglandin synthesis, 217
PGA, 216
PGE, 216
PGF, 216
PGH$_2$
  conversion to other prostaglandins, 217, 218f
  synthesis of thromboxanes from, 217, 219, 220f
pH
  definition, 585
  enzyme reaction rate effects, 43
  extracellular, see Extracellular fluid
  optimum, 443
Phenylacetate, 256, 256f
Phenylalanine, 8
  degradation, 246–247
  structure, 6f
Phenylalanine hydroxylase, 246
Phenylketonuria, 256, 256f
Phenyllactate, 256, 256f
Phenylthiocarbamoyl peptide, 13, 14f
Phenylthiohydantoin amino acids, 13, 14f
Phosphate
  dietary, 483
  exchange between ATP/ADP and phosphocreatine, 268
Phosphate compounds, high energy, 56–58, 57f
Phosphatidic acid, 188
Phosphatidylcholine, 190, 190f–192f, 192
Phosphatidylethanolamine, synthesis, 192, 193f–194f
Phosphatidylethanolamine transferase, 193
Phosphatidylinositol, synthesis, 195, 196f
Phosphatidylserine, synthesis, 193

3′-Phosphoadenosine–5′-phosphosulfate, structure, 540f
Phosphocholine cytidylyl transferase, 192
Phosphocreatine. See Creatine phosphate
Phosphodiesterases, 407
Phosphoenolpyruvate, 97
Phosphoenolpyruvate carboxykinase, 108f, 109
Phosphofructokinase, glycolysis
   regulation, 101, 103
   pathway, 94–95
Phosphofructokinase 2, 103
Phosphoglucomutase, 74
6-Phosphogluconate dehydrogenase, 113
Phosphogluconate pathway, see Pentose phosphate pathway
6-Phosphogluconolactone, 113
2-Phosphoglycerate, 96–97
3-Phosphoglycerate, 96, 253, 254f
Phosphoglycerate kinase, glycolysis pathway, 96
Phosphoglycerides, 190–193, 190f–194f, 195, 196f–200f, 200
   degradation, 195, 200, 200f
   exchange, 195, 199f
   fatty acid exchange reactions, 195, 198f–199f
   in intestinal lumen, 155
   structure, 190f
   synthesis, 191–193, 191f–194f, 195, 196f
      CDP-diacylglycerols, 193, 194f, 195, 196f
      phosphatidylcholine, 191f, 192, 192f
      phosphatidylethanolamine, 192, 193f–194f
      phosphatidylserine, 192
      plasmalogens, 195, 197f
Phosphoglycerols, see Phosphoglycerides
Phosphoglyceromutase, glycolysis pathway, 96
Phospholipase A_2, 155, 157, 217
Phospholipases, 195, 200, 200f
Phospholipids, see Phosphoglycerides
Phosphomannose isomerase, 106
Phosphoprotein phosphatases, 78, 406–407
5-Phosphoribosyl-1-amine, 291, 293
5′-Phosphoribosyl-5-aminoimidazole, 294
5′-Phosphoribosyl-5-aminoimidazole-4-carboxamide, 295–296
5′-Phosphoribosyl-5-aminoimidazole-4-carboxylate, 294–295
5′-Phosphoribosyl-5-aminoimidazole-4-N-succinocarboxamide, 295
5′-Phosphoribosyl-5-formamidoimidazole-4-carboxamide, 296
5′-Phosphoribosylformylglycinamide, 293–294, 294
5′-Phosphoribosylglycinamide, 293
5-Phosphoribosyl pyrophosphate, synthesis, 291
5-Phosphoribosyl pyrophosphate amidotransferase, 293
5-Phosphoribosyl pyrophosphate synthetase, 291
Phosphorylase kinase a, 78
Phosphorylation
   enzymes, 406

oxidative, see Oxidative phosphorylation
   substrate-level, 90, 96
Phosphorylation sites, 141
Photolyase, 357, 359
Photoreactivation repair, thymine dimers, 357, 359
Phytanic acid hydroxylase deficiency, 181, 182f, 183
Phytol, 181
Ping-pong mechanism, 42
pI, 3
Pituitary hormones, 429, 430t, 431–434, 432f, 433t
   anterior, 430t, 431–434, 432f, 433t
      ACT, 431, 432f
      FSH, 433
      growth hormone, 432, 433f
      LH, 433
      prolactin, 433–434
      thyrotropin, 431–432
   posterior, 434
pK
   amino acid ionizable group, 3
   definition, 585
PKU, 256, 256f
Plasma, 439
   albumin, 440
   B cells, 546
   enzymes as diagnostic tools, 441
   globulins, 440–441
   pH
      bicarbonate and, 446–447
      renal compensation, 462–463
   proteins, 440–441
   solutes, 439
Plasmalogens, synthesis, 195, 197f
Plasma membrane
   functions, 513–514
   structure, lipids, 150
Plasmid, 363
Plasmin, 454–455
Plasminogen, 454
Platelet plug, 447
Poliovirus, 573, 574f
pol region, 576
Polyamine oxidase, 270
Polyamines, 269–270, 271f–272f
   degradation, 270, 272f
   physiologic roles, 269–270
Poly A polymerase, 384
Poly A tail, 384
Polycistronic mRNA, 384
Polygenic mRNA, 384
Polypeptides
   chains
      amino acids in, 4
      collagen, 530
      degradation, 229
Polyribosomes, 397

Polysaccharides, 63, 71, 72f, 73
  definition, 61
  dietary, 87
Polysomes, 397
P/O ratio, 141
Porphobilinogen, synthesis, 274–275
Porphyrias, 281–283, 282f
  congenital erythropoietic, 281
  intermittent acute, 282–283
  protoporphyria, 281
  toxic acquired, 281–282
Potassium, dietary, 483
Precipitin reaction, 546
Pregnenolone, 411–412
Preinitiation complex, 396
Prekallikrein, 453
Prelumirhodopsin, 508
Preprocollagens, 530
Preproinsulin, synthesis, 425, 426f
Pribnow box, 374
Primary structure, 11–13, 13f–14f, 15
Primary transcription, 378
Primase, 347
Primasome, 347
Primer binding site, 579
Probenecid, gout treatment, 316
Procarboxypeptidases A and B, 228
Procollagen, 530, 533
Proelastase, 228
Proelastin, 536
Progesterone, 411, 416
Progestins, 409
Proinsulin, 425, 426f
Prokaryotes
  gene expression, transcriptional control, 386–388, 389f
  mRNA processing, 384
  promoters, 374, 377
  RNA polymerases, 377–378
  rRNA, processing, 380, 381f
  transcription, 378–380, 379f
  translation, 392, 393f, 394f, 395f
  tRNA processing, 383, 383f
Prolactin, 433–434
Prolactin inhibiting factor, 430t, 431
Prolidase, 229
Proline, 8, 232
  hydroxylation, 533
  residues, α-helix and, 15
  structure, 6f
  synthesis, 253, 253f
Proline oxidase, 241
Promoters, transcription, 374–375, 376f, 377
Propionyl CoA
  amino acid degradation, 244, 245f
  metabolism, 175, 177f, 244, 246
Propionyl CoA carboxylase, 175
Prostaglandin A, 216

Prostaglandin E, 216
Prostaglandin F, 216
Prostaglandins, 215–217, 217f–219f, 219, 221
  classes, 216, 221
  α-configuration, 216
  degradation, 219
  regulation, 219
  structure and nomenclature, 216
  synthesis, rate-limiting step, 221
Prostaglandin synthetase complex, 217
Prosthetic groups, 27–28
Protease inhibitors, 440
Protein, 10–21
  activation by calmodulin, 407, 409f
  from animal tissues, 471
  in biologic membranes, 514–515, 515f
    asymmetry, 520
  biologic value, 470–471
  classification, 1
  connective tissue components, see Connective tissue
  constitutive, 386–387
  digestion, 228–229, 229f
  endogenous, digestion, 230–231
  extrinsic, 515, 515f
  inducible, 387
  integral, 514–515, 515f
  integral interaction with lipid bilayer, 519–520, 519f
  intrinsic, 514–515, 515f
  metabolism, insulin effects, 424
  movement in bilayer, 519
  nerve tissue, 557–558, 557f
  N-terminal residue determination, 11, 12f
  as nutrient, 470–471
  peripheral, 515, 515f
  plant, 471
  posttranslational modification, 397
  primary structure, 11–13, 13f–14f, 15
    Edman degradation procedure, 13, 14f, 15
  quaternary structure, 11, 20–21
  repressible, 387
  scaffold, 333
  secondary structure, 10, 15–16, 16f–17f, 18, 18f
    α-helix, 15, 16f
    β-sheet, 15–16, 17f
    3_{10} helix, 15
    reverse turns, 16, 18, 18f
  structure, 10–12
    domains, 20, 21f
  subunit aggregation, 20
  supersecondary structure, 18, 19f–20f
  synthesis. See also Translation
    for export, 398–399, 398f
  tertiary structure, 10, 18, 20, 21f
Protein kinase, 78, 406
Proteoglycans, 63, 71
  connective tissue, 529, 536–538, 539f–540f, 540
  synthesis, 539f–540f, 540

Prothrombin, activation to thrombin, 449, 450f
Protoporphyria, 281
Protoporphyrin IX, 276–277
Protoporphyrinogen oxidase, 276
Provirus, 576
   synthesis, 578f, 579
   transcription and RNA polymerase II, 579
PRPP amidotransferase, inhibition, 297
PRPP synthetase, genetic variants, 316
Pseudogenes, 335
PTC amino acids, 13, 14f
Punctuation codons, 390–391
Purine base, 288, 289f, 289t
   5'-phosphoribosyl derivative, 289
   ribosyl and deoxyribosyl derivatives, 289
   RNA, 338
Purine nucleotide/aspartate cycle, 235, 235f
Purine nucleotides
   degradation, 290, 306–307, 307f, 309
   de novo synthesis, 291, 292f, 293–299, 297f–299f
      AMP and GMP formation, 296, 297f
      interconversion, 298–299, 299f
      regulation, 296–297, 298f
   DNA, 322
   gout, 316
   salvage, 309
   synthesis, 558
Purine ring, atom sources, 289f
Putrescine, synthesis, 270, 271f
Pyran, structure, 65f, 66
Pyranose, 66
Pyridoxal phosphate, 232
Pyridoxine, 475
Pyrimidine, synthesis, salvage pathway, 558
Pyrimidine base, 288–289, 338
Pyrimidine dimers, repair deficiency, xeroderma pigmentosum, 368
Pyrimidine nucleotides
   degradation, 308f, 309
   de novo synthesis, 299–302, 303f
      cytidine nucleotides, 302
      localization, 301
      regulation, 302, 303f
   DNA, 322
   interconversion, 305–306, 306f
   salvage, 309–310
Pyrimidine phosphoribosyltransferase, 309
Pyrimidine ring, atom sources, 289f
Pyrophosphatase, 76
Δ^1-Pyrroline-5-carboxylic acid, 253, 253f
Pyruvate, 97
   from amino acids, 239
   cysteine degradation, 240, 240f
   gluconeogenesis from, 110–111, 110f
   metabolism, 97–98
Pyruvate carboxylase, 109, 130, 163
Pyruvate dehydrogenase, 123–124, 124f, 141, 142t, 163
   regulation, 123, 130f
   TCA cycle and, 124, 126f
Pyruvate kinase, glycolysis
   regulation, 103
   pathway, 97

Q

Quasi-late genes, 570
Quaternary structure, proteins, 20–21

R

Reactions
   free energy charge, see Free energy charge
   intermediate, 28
Recombinant DNA, 359, 362–363, 364f–367f, 365–366, 368
   cloning, 363, 365, 365f–366f
   gene preparation, 362–363, 364f
      cDNA construction, 363, 364f
      cellular DNA fragmentation, 362–363
      direct chemical synthesis, 363
Recombination, DNA, see DNA recombination
Red cells
   carbonic acid in, 446
   glycolysis role, 91
Redox reactions, 131
Reducing end, disaccharides, 67
Reduction potential, standard, 131
Refsum's disease, 181, 182f, 183
Regulatory gene, 388
Releasing factors, 391, 429, 430t
Renaturation
   DNA, 328, 329f, 330, 335, 336f, 338
   rate, 329f, 330
      repetitive DNA, 335, 336f, 338
Renin, 414
Renin-angiotensin system, 414
Reovirus, 572
Replication errors, DNA, 355, 356f
Replication fork, 346, 346f, 349–351, 349f–350f
   diagram, 349f
   lagging strand, 350–351, 350f
   leading strand, 349–350
Replicative form, 573, 574f
   intermediate, 572, 573, 574f
Rep protein, 347
Repressible proteins, 387
Repressors, 387
Respiratory control ratio, 144
Resting potential, 560
Restriction endonucleases, 335
   cloning and, 363, 365, 365f

Restriction endonucleases (*Continued*)
  DNA fragmentation, 363
  DNA recombination, 359
Retina, 506
11-*cis*-Retinal
  regeneration and maintenance, 508–509, 509f
  structure, 506, 507f
Retinoic acid, 479, 480f
Retinol, 480
Retinol-binding protein, 440
Retinol phosphate, 479
Retroviruses, 573, 576, 576f–578f, 578–579
  biochemistry, 576
  diploid genome, 576, 576f
  infective cycle, 576, 577f–578f, 578–579
  long terminal repeats, 578–579
  reverse transcriptase, 576, 579
  synthesis, 578f, 579
Reverse transcriptase, 363
  retrovirus, 576, 579
Reverse triiodothyronine, 423
Reversion, DNA, 359
Rhodanese, cyanide poisoning treatment, 145
Rhodopsin
  light-induced dissociation, 508, 508f
  regeneration, 508–509, 509f
Riboflavin, 312, 313f, 474
Ribonucleic acid, *see* RNA
Ribonucleoside 5'-monophosphates, 290
Ribonucleotides, 302
Ribose, 290
D-Ribose, structure, 62f
β-D-Ribose, structure, 65f
Ribose 5-phosphate, 291
Ribosomal RNA, *see* rRNA
Ribosomes, viral protein synthesis, 574
Ribulose 5-phosphate, 113, 115
Rifampin, interference with gene expression, 400
RNA
  classes, 321–322
  cleavage, acid-base catalysis, 30, 31f
  messenger, *see* mRNA
  monocistronic, 384
  nascent, 379
  ribosomal, *see* rRNA
  splice site sequence, 386, 386f
  structure, 337f–339f, 338, 340
  synthesis, *see* Transcription
  transfer, *see* tRNA
  viral genomic, 579
RNA-DNA hybrid, 379
RNA polymerase I, rRNA synthesis, 374
RNA polymerase II
  mRNA synthesis, 375, 376f
  provirus transcription, 579
RNA polymerase III, 375
RNA polymerases, 347
  α-amanitin, 378
  binding, 387
  closed complex, 378–379
  eukaryotic, properties, 378
  immature viral particle, 572
  open complex, 379, 379f
  prokaryotic, properties, 377–378
  RNA synthesis, 373–374
  RNA viruses, 572
  σ subunit, 377–378
  transcriptional factors, 378
RNA primer, 347
RNA processing, 380, 381f–383f, 382–384, 385f–386f, 386
  mRNA, 384, 385f–386f, 386
  rRNA, 380, 381f–382f, 382
  splicing, immature tRNAs and rRNAs, 383
  tRNA, 383, 383f
RNase D, 383
RNase P, 383
RNA viruses, 572–574, 574f–578f, 578–579
  double-stranded, 572
  E. coli RNA phage, 572–573
  poliovirus, 573, 574f
  precursor polyprotein, 573, 574f
  reovirus, 572
  replicative form, 573, 574f
  replicative intermediate, 573, 574f
  retroviruses, 573, 576, 576f–578f, 578–579
  single-stranded (+), 572–573, 574f
  single-stranded (−), 573–574, 575f
  subviral particle, 572
  vesicular stomatitis virus, 573–574, 575f
Rod cells, retinal, 506
  hyperpolarization of membrane, 508
  light detection by, 506, 507f–509f, 508–509
  structure, 506, 507f
rRNA, 322
  immature splicing, 383
  processing
    bacterial cells, 380, 381f
    eukaryotic cells, 382
  structure, 340
  transcription, RNA polymerase I and II, 374–375

S

Saccharopine pathway, 249f, 250
Sarcolemma, 496
Sarcomere, 494, 497f
  contraction, 498–499
Sarcoplasmic reticulum, 496
Satellite DNA, 335
Schiff base, 534, 534f
Schwann cell spirals, 560–561, 561f
Scurvy, 477–478
Secondary structure, *see* Protein, secondary structure

Second messengers, 404–408, 406f, 408f–409f
  calcium ion, 407, 408f–409f
  cAMP, 405–407, 406f
  cGMP, 407
Sedoheptulose 7-phosphate, 115
Selenium, dietary, 486
Serine, 9
  deamination, 234
  degradation, 240
  source of one-carbon units, 250
  structure, 6f, 190f
  synthesis, 253, 254f
Serine-threonine dehydratase, 234, 240, 244
Serine transhydroxymethylase, 250, 253
Serotonin
  degradation, 264, 267f, 268
  synthesis, 264, 267f
Serum monoacylglycerol hydrolase, 171
Sex hormones, 409, 411
  synthesis regulation, 415, 415f
Shuttle mechanism, NADH, 98–99, 100f, 101
  gluconeogenesis, 110–111, 110f
  glycolysis energy yield and, 101
Sialic acid, structure, 201f
Sickle cell anemia, 21–22
Side chain, amino acids, 2, 10
Signal peptidase, 398, 530
Signal peptide, 398, 530
Single-copy DNA, 335
Single-stranded binding proteins, 347
Skin cancer, xeroderma pigmentosum, 368
Slippage errors, 355
Slow-twitch fibers, 492
Smell, decreased acuity, zinc deficiency, 486
Sodium, dietary, 483
Somatomedins, 432
Somatostatin, 423, 426–427, 430t, 431
  functions, 423–424, 423t
Somatotropin, 432, 433f
Specific activity, enzymes, 37
Spermidine, synthesis, 270, 271f
Spermine, synthesis, 270, 271f
Sphingolipids, 71, 200–202, 201f, 203f–204f, 205–207, 205f–206f, 207t, 208f, 209, 559
  biologic function, 202
  concentration, 207
  degradation to ceramide, 207, 208f
  glycosphingolipid synthesis, 204f–206f, 205–207, 207t, 208f, 209
  structure, 200–201, 201f
  synthesis, 202, 203f
Sphingomyelin, 201–202, 209
Sphingosine, 200, 201f
  degradation, 207, 209
  synthesis, 202, 203f
S-100 protein, 558
Squalene, 210
  conversion to cholesterol, 211, 212f
  synthesis, 210–211, 210f

src oncogene, 576
Starch, 88, 89f
Start codons, 390–391
Stearoyl CoA desaturase, 167
Steroid hormones, 404, 409, 411–412, 411t, 413f–415f, 414–416
  action, 409, 410f
  biosynthesis, 411–412
    branching pathway, 412, 413f
    regulation, 412, 414–415, 414f–415f
  main site of synthesis, 411t
  metabolism and excretion, 415–416
  transcription stimulation, 388
  transport in blood, 412
Stop codons, 390–391
Stroma, 505
Structural domains, protein, 20, 21f
Substrate binding, lock-and-key and induced-fit models, 27, 27f
Substrate strain, energy of activation effects, 29
Subviral particle, 572
Succinate dehydrogenase, 127–128
Succinate thiokinase, 127
Succinyl CoA synthetase, 127
Sucrose, structure, 68f
Sugar, *see* Disaccharides; Monosaccharides
Sugar-phosphate backbones, DNA, 324, 326
Sulfatide, 201
  synthesis, 205–206, 205f
Sulfite oxidase deficiency, 257
Sulfur-containing amino acids, 6f, 8–9
Supercoils, positive and negative, 331–332, 331f
Superhelix, 330–332, 331f
  collagen, 530, 531f
Supersecondary structure, proteins, 18, 19f–20f
Surface active agents, amphipathic lipids, 150–151
SV40, 570
SVP, 572
Symport systems, 522, 523f
Synaptic cleft, 556, 561, 562f
Synaptic transmission, 561–563
Synthesis rate, enzyme-substrate complex, 34

T

T_3, *see* Triiodothyronine
T_4, *see* Thyroxine
Taste, decreased acuity, zinc deficiency, 486
TATA box, 375
TCA cycle, 121, 124, 124–131, 126f, 130f
  equilibrium, 128–129
  fatty acid synthesis and degradation, 177–179
  link with urea cycle, 238f
  nerve tissue, 557
  pyruvate dehydrogenase and, 124, 126f

TCA cycle (*Continued*)
  reactions, 125–128, 126f
    aconitase, 125
    citrate synthase, 125
    fumarase, 128
    isocitrate dehydrogenase, 125–126
    α-ketoglutarate dehydrogenase, 127
    malate dehydrogenase, 128
    succinate dehydrogenase, 127–128
    succinate thiokinase, 127
  regulation, 129–131, 130f
  respiratory control, 144
Temperature
  DNA melting, 327–328
  enzyme reaction rate effects, 43
  membrane fluidity and, 517
Termination codons, 390–391
Terminators, transcription, 375, 377, 377f
Tertiary structure, proteins, 10
Testosterone
  conversion to estradiol, 412
  metabolism and excretion, 416
Tetracyclines, interference with gene expression, 400
Tetrahydrobiopterin, regeneration, 246–247
Tetrahydrofolate, one-carbon units carried by, 250, 251f
Tetrapyrrole ring, 275–276
Thiamine, 473–474, 473f
Thiamine pyrophosphate, 473f
Thioesterase, 163
Thiolase, 174, 178–179
Thiolysis, fatty acids, 174
Thioredoxin, 303
Thioredoxin reductase, 303, 304f
Threonine, 9
  deamination, 234
  degradation, 244
  structure, 7f
Threshold potential, 560
Thrombin, prothrombin activation to, 449, 450f
Thromboxane, 215–217, 219, 220f, 221
  regulation, 219
Thromboxane A_2, 217, 219
Thromboxane B_2, 217, 219
Thrombus, 447
Thyroid hormones, 404
THS, *see* Thyrotropin
Thymidine monophosphate, 47, 48f
Thymine
  degradation, 290
  in double helix, 323, 325f
Thymine dimers
  formation, 354–355, 355f
  photoreactivation repair, 357, 359
Thyroglobulin, 421
Thyroid gland, hormones, 418, 420–421, 420f–422f, 423

Thyroid hormones, action, 409
Thyroid peroxidase, 421
Thyroid releasing factor, 421, 423
Thyroid stimulating hormone, *see* Thyrotropin
Thyroid stimulating immunoglobulin, 434
Thyrotoxicosis, 434–435
Thyrotropin, 423
  physiologic actions, 431
  subunits, 433
  synthesis and secretion, 431–432
Thyroxine
  actions, 420–421
  metabolism and excretion, 422f, 423
  structure, 420f
  synthesis regulation, 421, 421f, 423
Tocopherols, 481–482, 482f
Topoisomerases, 332, 347
Toxic acquired porphyrias, 281–282
TψC loop, 340
Trace minerals, 470
  chromium, 486
  copper, 484–486
  fluoride, 487
  iodine, 486–487
  iron, 484, 485f
  molybdenum, 486
  selenium, 486
  zinc, 486
Trailer sequence, RNA, 380, 382
Transacylase, 163
Transaldolase, 115
Transamidination, phosphocreatine, 268
Transamination, 232–233
Transcription, 373–375, 376f–377f, 377–380, 379f, 381f–383f, 382–384, 385f–386f, 386; *see also* RNA polymerases
  chain elongation, 379, 379f
  closed complex, 378–379
  eukaryotes, 380
  eukaryotic promoters, 374–375, 376f, 377
  initiation, 378–379
  nus A, 379
  open complex, 379, 379f
  primary transcription, 378
  prokaryotes, 378–380, 379f
  prokaryotic promoters, 374, 377
  RNA processing, *see* RNA processing
  termination, 380
  terminators, 375, 377, 377f
  transcriptional factors, 380
  transcriptional unit, 378
Transcriptional factors, 378, 380, 388
Transcriptional unit, 378
Transferases, 26t
Transferrin, 441, 484
Transfer RNA, *see* tRNA
Transition state intermediate, 28
Transition temperature, 328

Transketolase, 115
Translation, 388
  eukaryotes, 394–397, 396f
    elongation, 397
    initiation, 394–397, 396f
    termination, 397
  gene expression, control, 399
  polyribosomes, 397
  posttranslational modification of proteins, 397
  prokaryotes, 392, 393f, 394f, 395f
    elongation, 394, 395f
    initiation, 392, 393f, 394
    termination, 394
  tRNA mediation of, 388
Translocase, 394, 397
Transport
Transposons, 360–361
Transverse tubules, 496
Triacylglycerols, 150, 153, 153f, 200
  in chylomicrons, hydrolysis, 172
  degradation, 190
  digestion, 155
  energy storage, 159
  lipoprotein, 171
  removal from chylomicrons, 211
  synthesis, 188, 189f, 190
    adipose tissue, 188
    inhibition, 171
    intestinal mucosa, 188, 190
    in liver, 188, 189f
Triantennary structure, 71
Tricarboxylic acid cycle, see TCA cycle
Triiodothyronine
  actions, 420–421
  biosynthesis, 421
  metabolism and excretion, 422f, 423
  reverse, 423
  structure, 420f
  synthesis regulation, 421, 421f, 423
Trinucleotides, 289
Triose phosphate isomerase, glycolysis pathway, 95–96
tRNA, 321–322
  aminoacyl, synthesis, 392
  anticodon, 388, 391
  cloverleaf secondary structure, 337f, 338, 340
  immature, splicing, 383
  mitochondrial, 399
  processing, 383
  structure, 337f, 338, 340
  transcription, RNA polymerase III, 375
  translation mediation, 388
  viral protein synthesis, 574
Tropocollagen, 530
  cross-linked, 533–534, 533f–534f
Tropoelastin, 534
Tropomyosin, 493
  in resting myofibril, 496
  thin filaments, 494, 495f

Troponin, 493
  C subunit, 498
  thin filaments, 494, 495f
Trypsin, 229
Trypsinogen, 228–229
Tryptophan, 8
  degradation, 247, 248f
  structure, 6f
TSI, 434
T tubules, 496
Turnover number, enzymes, 37
Type I and II topoisomerases, 332
Type III glycogen storage disease, 82
Tyrosinase, 264
  deficiency, 256–257
Tyrosine, 8
  degradation, 246f, 247
  structure, 6f
  synthesis, 252, 254
Tyrosine hydroxylase, 264
Tyrosine kinase, viral, 576

U

Ubiquinol, 133
Ubiquinone, 131, 133, 133f
UDP-galactose, 104–105, 104f
UDP-glucose, 74, 76
Ultraviolet irradiation
  mutagenesis, 354–355, 355f
  xeroderma pigmentosum, 368
UMP, 288f, 301
UMP kinase, 302
Uncompetitive inhibition, 40, 40f
Uncompetitive inhibitors, definition, 38
Uncouplers, 144–145
Underwinding, superhelices, 332
Uracil, 288f, 290
Urea, synthesis, 235
Urea cycle, 236, 237f–238f, 238
  enzyme deficiencies, 254–256
  link with citric acid cycle, 238f
  regulation, 238
Uric acid, gout, 316
Uricosuric drugs, 316–317
Uridine, structure, 288f
Uridine diphosphate-glucose, 74, 76
Uridine 5'-monophosphate, 288f, 301
Uridine triphosphate, 57, 74
Uridylate, 290
Urobilinogens, 280
Uroporphyrinogen, overproduction, 281
Uroporphyrinogen decarboxylase, 276
Uroporphyrinogen III, formation, 275
Uroporphyrinogen III cosynthase, 276
Uroporphyrinogen I synthase, 276
Uroporphyrinogen I synthetase, deficiency, 283

UTP, 57, 74
UvrABC endonuclease, 359

V

Valine, 8
  degradation, 244, 245f
  enzyme deficiencies and, 258
  structure, 5f
Valinomycin, 524, 525f
Variable loop, tRNA, 340
Vasopressin, 434, 461
Vector-passenger DNA adducts, formation, 363, 365, 365f
Very low density lipoproteins, 171, 211
Vesicular stomatitis virus, 573–574, 575f
V_H domain, 547
Virus, 567–568, 579; *see also* DNA viruses; RNA viruses
Vision
  light detection
    by cone cells, 509
    by rod cells, 506, 507f–509f, 508–509
  mechanism, 506
Vitamin A, 479–480
Vitamin B_1, 473–474, 473f
Vitamin B_2, 312, 313f, 474
Vitamin B_3, 474
Vitamin B_5, 475
Vitamin B_6, 475
  deficiency, 257
Vitamin B_9, 476–477
Vitamin B_{12}, 477, 478f
Vitamin C
  biologic function, 477
  deficiency, 477–478
  distribution, requirement, sources, 478–479
  structure, 479f
Vitamin D, 480–481; *see also* 1,25-Dihydroxycholecalciferol
  hydroxylation, 418
Vitamin E, 481–482, 482f
Vitamin H, 475–476, 476f
Vitamin K
  biologic functions, 482
  blood clotting role, 454, 454f
  deficiency, 483
  distribution, requirement, sources, 483
  structure, 482f

Vitamins, 469
  fat-soluble, 469
    vitamin A, 479–480, 480f
    vitamin D, *see* Vitamin D
    vitamin E, 481–482, 482f
    vitamin K, 482–483, 482f
  water-soluble, 469
    biotin, 475–476, 476f
    folic acid, 476–477, 476f
    niacin, 474
    panthothenic acid, 475
    pyridoxine, 475
    riboflavin, 312, 313f, 474
    thiamine, 473–474, 473f
    vitamin B_{12}, 477, 478f
    vitamin C, 477–479, 479f
V-J gene, 550–551, 550f
VLDLs, 171, 211
V_L region, 547
VSV virion, 573–574, 575f

W

Water
  equilibrium constant, 584
  excretion regulation, 461
  intake regulation, 461
  interaction with membrane lipids, 515–516, 516f
  ion product, 584
  properties, 583–584, 584f
White fibers, 492
Wilson's disease, 485–486
Wobble concept, 391
Wobble position, 391

X

Xeroderma pigmentosum, 368–369
Xerophthalmia, vitamin A deficiency, 479

Z

Z-DNA, 326
Zinc, 486
Z line, 494, 497f
Zwitterion, 2, 2f